2011 IEEE International Conference on IC Design & Technology

(ICICDT 2011)

Kaohsiung, Taiwan
2-4 May 2011

IEEE Catalog Number: CFP11412-PRT
ISBN: 978-1-4244-9019-6

Copyright © 2011 by the Institute of Electrical and Electronic Engineers, Inc
All Rights Reserved

Copyright and Reprint Permissions: Abstracting is permitted with credit to the source. Libraries are permitted to photocopy beyond the limit of U.S. copyright law for private use of patrons those articles in this volume that carry a code at the bottom of the first page, provided the per-copy fee indicated in the code is paid through Copyright Clearance Center, 222 Rosewood Drive, Danvers, MA 01923.

For other copying, reprint or republication permission, write to IEEE Copyrights Manager, IEEE Service Center, 445 Hoes Lane, Piscataway, NJ 08854. All rights reserved.

***This publication is a representation of what appears in the IEEE Digital Libraries. Some format issues inherent in the e-media version may also appear in this print version.**

IEEE Catalog Number: CFP11412-PRT
ISBN 13:   978-1-4244-9019-6

Additional Copies of This Publication Are Available From:

Curran Associates, Inc
57 Morehouse Lane
Red Hook, NY 12571 USA
Phone: (845) 758-0400
Fax: (845) 758-2633
E-mail: curran@proceedings.com
Web: www.proceedings.com

Session A: I/O Circuits and ESD Protection

Transient-to-Digital Converter to Detect Electrical Fast Transient (EFT) Disturbance for System Protection Design 001

Cheng-Cheng Yen[1], Wan-Yen Lin[1], Ming-Dou Ker[1,2], Ching-Ling Tsai[3], Shih-Fan Chen[3], Tung-Yang Chen[3]
[1]Institute of Electronics, National Chiao-Tung University, Hsinchu, Taiwan
[2]Department of Electronics Engineering, I-Shou University, Kaohsiung, Taiwan
[3]Himax Technologies Inc.,Taiwan

ESD RF Protections in Advanced CMOS Technologies and its Parasitic Capacitance Evaluation 005

Ph. Galy[1], J. Jimenez[1], P. Meuris[2], W. Schoenmaker[2], O. Dupuis[2]
[1]STMicroelectronics, Crolles, France
[2]MAGWEL NV, Leuven, Belgium

Design of Low-Leakage Power-Rail ESD Clamp Circuit with MOM Capacitor and STSCR in a 65-nm CMOS Process 009

Po-Yen Chiu[1], Ming-Dou Ker[1,2]
[1]Institute of Electronics, National Chiao-Tung University, Hsinchu, Taiwan
[2]Dept. of Electronic Engineering, I-Shou University, Kaohsiung, Taiwan

Session B: Advanced Transistor/Material

Invited Paper: Low power UTBOX and Back Plane (BP) FDSOI technology for 32nm node and below 013

C. Fenouillet-Beranger[1,2], P. Perreau[1,2], L. Tosti[2], O. Thomas[2], J-P. Noel[2], T. Benoist[1], O. Weber[2], F. Andrieu[2], A. Bajolet[1], S. Haendler[1], M. Cassé[2], X. Garros[2], K.K. Bourdelle[3], F. Boedt[3], O. Faynot[2], F. Boeuf[1]
[1]STMicroelectronics, Crolles, France
[2]CEA/LETI-Minatec, Grenoble, France
[3]SOITEC, Parc Technologique des Fontaines, Bernin, France

Electrical Characteristic Fluctuation of 16 nm MOSFETs Induced by Random Dopants and Interface Traps 017

Yung-Yueh Chiu[1], Fu-Hai Li[1], Hui-Wen Cheng[1], Yiming Li[2]
[1]Institute of Communications Engineering, National Chiao-Tung University, Hsinchu, Taiwan.
[2]Department of Electrical Engineering and Institute of Communications Engineering, National Chiao-Tung University,Hsinchu, Taiwan

Invited Paper: Excellent Silicon Thickness Uniformity on Ultra-Thin SOI for controlling Vt variation of FDSOI 021

W. Schwarzenbach, X. Cauchy, F. Boedt, O. Bonnin, E. Butaud, C. Girard, B.-Y. Nguyen, C. Mazure, C. Maleville
SOITEC, Parc Technologique des Fontaines – Bernin – Crolles – France

Variability Analysis of UTB SOI Subthreshold SRAM Considering Line-Edge Roughness, Work Function Variation and Temperature Sensitivity 024

Vita Pi-Ho Hu, Ming-Long Fan, Pin Su, Ching-Te Chuang
Department of Electronics Engineering & Institute of Electronics, National Chiao Tung University, Hsinchu, Taiwan

Invited Paper: 3D Integrable Nanowire FET Sensor with Intrinsic Sensitivity Boost 028

Chi On Chui, Jorge Kina, Kyeong-Sik Shin
Department of Electrical Engineering, University of California, Los Angeles, CA, USA

On The Magnitude of Random Telegraph Noise in Ultra-Scaled MOSFETs 032

K.P. Cheung, J.P. Campbell
Semiconductor Electronics Division, NIST, Gaithersburg, MD, USA

Session C: DFM/DFT/DFR/DFY

Invited Paper: Timing Error Prevention using Elastic Clocking 036
Kwanyeob Chae, Chang-Ho Lee, Saibal Mukhopadhyay
Georgia Institute of Technology, Atlanta, GA, USA

Time and Workload Dependent Device Variability in Circuit Simulations 040
D. Rodopoulos, S. Mahato, V. Valduga de Almeida Camargo, B. Kaczer, F. Catthoor, S. Cosemans, G. Groeseneken, A. Papanicolaou, D. Soudris
IMEC, Leuven, Belgium

Invited Paper: An On-Chip Waveform Capturer for Diagnosing Off-Chip Power Delivery 045
Kumpei Yoshikawa[1], Takushi Hashida[1], Makoto Nagata[1,2]
[1]Graduate School of System Informatics, Kobe University, Nada-ku, Kobe, Japan
[2]CREST, JST, Japan

Interconnect Test for Core-based Designs with Known Circuit Characteristics and Test Patterns 049
Tung-Hua Yeh[1], Sying-Jyan Wang[1], Katherine Shu-Min Li[2]
[1]Department of Computer Science and Engineering, National Chung-Hsing University Taichung, Taiwan
[2]Department of Computer Science and Engineering, National Sun-Yat Sen University Kaohsiung, Taiwan

Invited Paper: Architectural-Level Error-Tolerant Techniques for Low Supply Voltage Cache Operation 053
Shih-Lien Lu, Alaa Alameldeen, Keith Bowman, Zeshan Chishti, Chris Wilkerson, Wei Wu
Intel Labs, Hillsboro Oregon, USA

Session D: 3D Integration

Invited Paper: Special Considerations for 3DIC Circuit Design and Modeling 058
Sally Liu, Yung-Chow Peng, Fu-Lung Hsueh
Taiwan Semiconductor Manufacturing Company, Hsinchu Science Park, Hsinchu, Taiwan, ROC

A Single TSV-rail 3D Quasi Delay Insensitive Asynchronous Signaling 061
M. Belleville, E. Beigne, A. Valentian
CEA, LETI, MINATEC Campus, Grenoble, France

Invited Paper: Smart Stacking™ Technology: an Industrial Solution for 3D Layer Stacking 065
C. Lagahe Blanchard[1], I. Radu[1], M. Sadaka[2], K. Landry[1]
[1]SOITEC, Parc Technologique des Fontaines, Bernin, Crolles Cedex, France
[2]SOITEC USA Inc., Austin, Texas, USA

TSV Number Minimization Using Alternative Paths 068
Chun-Hua Cheng, Chih-Hsien Kuo, Shih-Hsu Huanga
Department of Electronic Engineering, Chung Yuan Christian University, Chung Li, Taiwan, R.O.C.

Invited Paper: Through Silicon Via Technology using Tungsten Metallization 072
G. Parès[1], N. Bresson[2], S. Minoret[1], V. Lapras[1], P. Brianceau[1], J.F. Lugand[1], R. Anciant[1], N. Sillon[1]
[1]CEA/LETI – Minatec, France
[2]SOITEC France

Session E: CAD

Statistical Delay Calculation with Multiple Input Simultaneous Switching 077
Qin Tang, Amir Zjajo, Michel Berkelaar and Nick van der Meijs
Circuits and Systems Group, Delft University of Technology, The Netherlands

Balanced Truncation of a Stable Non-Minimal Deep-Submicron CMOS Interconnect 081
Amir Zjajo, Qin Tang, Michel Berkelaar, Nick van der Meijs
Circuits and Systems Group, Delft University of Technology, The Netherlands

Enabling TLM-2.0 Interface on QEMU and SystemC-based VirtualPlatform 085
Tse-Chen Yeh, Zin-Yuan Lin, Ming-Chao Chiang
Department of Computer Science and Engineering, National Sun Yat-sen University, Kaohsiung, Taiwan

A Fast Custom Network Topology Generation with Floorplanning for NoC-based Systems 089
Katherine Shu-Min Li, Shu-Yu Chen, Liang-Bi Chen, Ruei-Ting Gu
Department of Computer Science and Engineering, National Sun Yat-Sen University, Kaohsiung, Taiwan

Session F: Advanced Memory Device

Invited Paper: Evolution of Embedded Flash Memory Technology for MCU 093
Hideto Hidaka
Renesas Electronics Corp. Itami, Hyogo, Japan

Impacts of Intrinsic Device Variations on the Stability of FinFET Subthreshold SRAMs 097
Yin-Nien Chen, Chien-Yu Hsieh, Ming-Long Fan, Vita Pi-Ho Hu, Pin Su and Ching-Te Chuang
Department of Electronics Engineering, National Chiao Tung University, Hsinchu, Taiwan

Invited Paper: Low-Cost Embedded Flash Memory Technology 101
Wein-Town Sun[1], Cheng-Jye Liu[1], Chun-Yuan Lo[1], Yun-Jen Ting[1], Ying-Je Chen[1], Tai-Yi Wu[1], Eng-Huat Toh[2], Xiao-Hong Yuan[2], Ko-Li Low[2], Qiu Han[2], Young-Seon You[2], Ying-Keung Leung[2], Swee-Tuck Woo[2]
[1]eMemory Technology Inc., Hsinchu County, Taiwan
[2]GLOBALFOUNDRIES Singapore Pte. Ltd., Singapore

Session G: Reliability/Plasma Induced Damage

Invited Paper: Crystallization Technique of Epitaxial HfO2 Thin Films on Si Substrates and their Potential for Advanced High-k Gate Stack Technology 106
Shinji Migita, Hiroyuki Ota
Nanodevice Innovation Research Center, National Institute of Advanced Industrial Science and Technology, Tsukuba, Ibaraki, Japan

A New Prediction Model for Effects of Plasma-Induced Damage on Parameter Variations in Advanced LSIs 110
Koji Eriguchi, Yoshinori Takao, Kouichi Ono
Kyoto University, Yoshida-Honmachi, Sakyo-ku, Kyoto, Japan

Invited Paper: Impact of La on the Bias-Temperature Instability of the HfSiO High-κ N-MOSFET 114
D. S. Ang[1], G. A. Du[2]
[1]School of Electrical and Electronic Engineering, Nanyang Technological University, Singapore
[2]GLOBALFOUNDRIES Singapore Pte. Ltd., Singapore

Separation of NBTI Component from Channel Hot Carrier Degradation in pMOSFETs Focusing on Recovery Phenomenon 118
Y. Mitani[1], S. Fukatsu[2], D. Hagishima[3], K. Matsuzawa[1]
[1]Advanced LSI Technology Laboratory, Corporate R&D Center, Toshiba Corporation, Yokohama, Japan
[2]Device Process Development Center, Toshiba Corporation, Japan
[3]Toshiba Semiconductor Company, Japan

Session H: High Power/High Voltage

On the Impact of the Edge Profile of Interconnects on the Occurrence of Passivation Cracks of Plastic-Encapsulated Electronic Power Devices 122
Jan Ackaert[1], Daniel Vanderstraeten[1], Bart Vandevelde[2]
[1]Corporate R&D, ON Semiconductors, Oudenaarde, Belgium
[2]Imec, Leuven, Belgium

Domestic Indirect Feedback Compensation of Multiple-Stage Amplifiers for Multiple-Voltage Level-Converting Amplification 126
Shang-Hsien Yang, Chua-Chin Wang
Department of Electrical Engineering, National Sun Yat-Sen University, Kaohsiung, Taiwan

Session I: Low Power

Microwatt Low-noise Variable-Gain Amplifier 130
Chun-Yi Li, Yu-Bin Lin, Robert Rieger
Electrical Engineering Department, National Sun Yat-Sen University, Kaohsiung, Taiwan

Invited Paper: SRAM Bitcell Design for Low Voltage Operation in Deep Submicron Technologies 134
Young Hwi Yang[1], Jisu Kim[1], Hyunkook Park[1], Joseph Wang[2], Geoffrey Yeap[2], Seong-Ook Jung[1]
[1]School of Electrical and Electronic Engineering, Yonsei University, Seoul, Korea
[2]Qualcomm Inc., San Diego, CA, USA.

An Ultra-Low Power K-Band Low-Noise Amplifier Co-Designed With ESD Protection in 40-nm CMOS 138
Ming-Hsien Tsai[1,2], Shawn S. H. Hsu[1], Fu-Lung Hsueh[2], Chewn-Pu Jou[2], Tzu-Jin Yeh[2], Ming-Hsiang Song[2], Jen-Chou Tseng[2]
[1]Dept. of Electrical Engineering and Institute of Electronics Engineering, National Tsing Hua University, Hsinchu, Taiwan
[2]Design Technology Division, Taiwan Semiconductor Manufacturing Company, Hsinchu, Taiwan

Invited Paper: Low Power Embedded Memory Design – Process to System Level Considerations 142
Esin Terzioglu, Sei Seung Yoon, ChangHo Jung, Ritu Chaba, Venu Boynapalli, Mohamed Abu-Rahma, Joseph Wang, Sam Yang, Giri Nallapati, Aaron Thean, Chidi Chidambaram, Michael Han, Geoffrey Yeap, Mehdi Sani
Qualcomm Inc., San Diego, CA, USA

65nm PD-SOI Glitch-Free Retention Flip-Flop for MTCMOS Power Switch applications 147
J. Le-Coz[1], P. Flatresse[1], S. Clerc[1], M.Belleville[2], A. Valentian[2]
[1]STMicroelectronics, Crolles, France
[2]CEA LETI, MINATEC campus, Grenoble, France

Session J: RF & Analog, Mixed signal

An Ultra-Low Energy Capacitive DAC Array switching Scheme for SAR ADC in Biomedical Applications 151
Chao Yuan, Yvonne Y. H. Lam
School of Electrical and Electronics Engineering, VIRTUS, IC Design Centre of Excellence, Nanyang Technological University, Singapore

Slew-Rate Controlled Output Stages for Switching DC-DC Converters 155
Jia-Ming Liu, Yi-Cheng Huang, Yu-Chun Ying, Tai-Haur Kuo
Department of Electrical Engineering, National Cheng Kung University, Tainan City, Taiwan

Temperature Dependence of Device Mismatch and Harmonic Distortion in Nanoscale Uniaxial-Strained PMOSFETs 159
Jack J.-Y. Kuo, William P.-N. Chen, Pin Su
Department of Electronics Engineering & Institute of Electronics, National Chiao Tung University, Hsinchu, Taiwan

A 8-bit 50-Msamples/s Switched-Current Pipelined ADC with Residue Generator and Interlaced Stage 163
Guo-Ming Sung, Ying-Tzu Lai
Department of Electrical Engineering, National Taipei University of Technology, Taipei, Taiwan

Continuously Auto-Tuned and Self-Ranged Dual-Path PLL Design with Hybrid AFC 167
Min Wang, Bo Zhou, Woogeun Rhee, Zhihua Wang
Institute of Microelectronics, Tsinghua University, Beijing, China

Session K: SoC/MPSoC/SIP

An Integrated HDTV Predictive Pixel Compensator for H.264/AVC Decoder 171
Ting-Chi Tong, Yun-Nan Chang
Department of Computer Science and Engineering, National Sun Yat-sen University, Kaohsiung, Taiwan

Invited Paper: IBM zEnterprise™ Energy Efficient 5.2Ghz Processor Chip 175
H. Wen[1], J. Warnock[2], Y. Chan[3], G. Mayer[4], B. Truong[1], T. Strach[4], T. Slegel[3], S. Carey[3], G. Salem[3], F. Malgioglio[3], D. Malone[3], D. Plass[3], B. Curran[3], Y.-H. Chan[3], M. Mayo[3], W. Huott[3], P. Mak[3]
[1]IBM Systems and Technology Group, Austin, TX, USA
[2]IBM Systems and Technology Group, Yorktown Heights, NY, USA
[3]IBM Systems and Technology Group, Poughkeepsie, NY, USA
[4]IBM Systems and Technology Group, Boeblingen, Germany

Ultra-Low Power FIR Filter using STSC-CVL Logic 180
Sajib Roy, Md. Murad Kabir Nipun, J Jacob Wikner
Division of Electronic Systems, Linkoping University, Sweden

Design of a Low-Cost Floating-Point Programmable Vertex Processor for Mobile Graphics Applications Based on Hybrid Number System 184
Shen-Fu Hsiao, Chan-Feng Chiu, Chia-Sheng Wen
Department of Computer Science and Engineering, National Sun Yat-Sen University, Kaohsiung, Taiwan

Session L: I/O Circuits and ESD Protection

A Low Jitter Active Body-Biasing Control-based Output Buffer in 65nm PD-SOI 188
Dimitri Soussan[1,2], Sylvain Majcherczak[1], Alexandre Valentian[2], Marc Belleville[2]
[1]STMicroelectronics Crolles, Crolles, France
[2]CEA LETI, Minatec campus, Grenoble, France

Adaptable Stimulus Driver for Epileptic Seizure Suppression 192
Ming-Dou Ker[1,2], Wei-Ling Chen[1], Chun-Yu Lin[1]
[1]Institute of Electronics, National Chiao-Tung University, Hsinchu, Taiwan
[2]Department of Electronic Engineering, I-Shou University, Kaohsiung, Taiwan

Gate-Driven 3.3V ESD Clamp Using 1.8V Transistors 196
Guang-Cheng Wang, Chia-Hui Chen, Wen-Hsin Huang, Kuo-Ji Chen, Ming-Hsiang Song, Ta-Pen Guo
Taiwan Semiconductor Manufacturing Corp., Hsinchu, Taiwan

Beta-Matrix ESD Network: throughout End of Placement Rules? 199
J. Bourgeat, P. Galy, B. Jacquier
STMicroelectronics , Crolles , France

Invited Paper: Design of on-chip Transient Voltage Suppressor in a Silicon-based Transceiver IC to meet IEC System-Level ESD Specification 203

Ryan Hsin-Chin Jiang, Tang-Kuei Tseng, Chi-Hao Chen, Che-Hao Chuang
Amazing Microelectronic Corp., HsinChiu, Taiwan R.O.C.

Session M: Soft Error Rate

Invited Paper: Soft Error Modeling, Simulation, and Testing at Advanced Technology Nodes 207

B. L. Bhuva, W. T. Holman, L. W. Massengill
Vanderbilt University, Nashville, TN, USA

Layout Optimization to Maximize Tolerance in SEILA: Soft Error Immune Latch 210

Taiki Uemura, Tsunehisa Sakoda, Hideya Matsuyama
Fujitsu Semiconductor Ltd., Boulder, Akiruno, Tokyo, Japan

Comparative Analysis of Flip-Flop designs for Soft Errors at Advanced Technology Nodes 214

B. L. Bhuva[1], K. Lilja[2], J. Holts[3], S.-J. Wen[3], R. Wong[3], S. Jagannathan[1], T. D. Loveless[1], M. McCurdy[1], Z. J. Diggins[1]
[1]Vanderbilt University, Nashville, TN, USA
[2]Robust Chip, Inc., Pleasanton, CA, USA
[3]Cisco Systems, Inc. San Jose, CA, USA.

Session N: Emerging Technologies

Invited Paper: Silicon Quantum Well Light-Emitting Diode 218

S. Saito
Institute for Photonics-Electronics Convergence System Technology (PECST), Photonics Electronics Technology Research Association (PETRA), and Central Research Laboratory, Hitachi, Ltd., Kokubunji, Tokyo, Japan

A Frequency-Shift Readout System for FPW Allergy Biosensor 222

Chia-Hao Hsu, Yain-Reu Lin, Yue-Da Tsai, Yun-Chi Chen, Chua-Chin Wang, Department of Electrical Engineering, National Sun Yat-Sen University, Kaohsiung, Taiwan

Invited Paper: Scaled Nanoelectromechanical (NEM) Hybrid Devices 226

Hiroshi Mizuta[1,2], Mario A. Garcia-Ramirez[2], Zakaria Moktadir[2], Yoshishige Tsuchiya[2], Shunichiro Sawai[3], Jun Ogi[3], Shunri Oda[3]
[1]School of Materials Science, Japan Advanced Institute of Science and Technology (JAIST), Ishikawa, Japan
[2]NANO Group, Electronics and Computer Science, Faculty of Physical and Applied Sciences, University of Southampton, Highfield, Southampton, U.K.
[3] Quantum Nano Electronics Research, Center, Tokyo Institute of Technology, Ookayama, Meguro-ku, Tokyo, Japan

Evaluation of DC and AC Performance of Junctionless MOSFETs in the Presence of Variability 230

Xin Qian[1], Yinglin Yang[1], Zhiwei Zhu[1], Shi-Li Zhang[1,2], Dongping Wu[1]
[1]State Key Laboratory of ASIC and System, Fudan University, Shanghai, People's Republic of China
[2]Solid-State Electronics, The Ångström Laboratory, Uppsala University, Uppsala, Sweden

Welcome

The International Conference on IC Design and Technology is the global forum for interaction and collaboration of IC design and technology for "accelerating product time-to-market". Close collaboration of the multi-discipline technical fields - design/device/process - accelerates the implementation of new designs and new technologies into manufacturing.

IC industry trends toward specializing system design and manufacturing outsourcing - such as fabless design house, wafer foundry, design automation tool/software house, and semiconductor processing tool supplier - created the needs for individuals with multi-discipline technical skills for collaborations. Furthermore, advanced IC technology no longer can offer the same level of control over many parameters that have direct adverse impact on circuit behavior. New IC designs also push the limit of technology, and in some cases require specific fine-tuning of certain process modules in manufacturing. Thus the traditionally separated communities of design and technology - design/device/process - are increasingly intertwined. Issues that require close interaction and collaboration for trade-off and optimization by all design/device/process fields are addressed in this conference. They are:

- Design/device/process optimizations and trade-off for leakage current, power consumption, & noise issues in mixed-signals, large scale IC devices, or design re-use.
- Incorporation of new materials (i.e. dual gate, multi-material active layers, etc.) in IC cell library and design of advanced transistor structures (i.e. Double Gate FDSOI, FinFET, etc.).
- Implementation of IC design and manufacturing process of new device structures (i.e. PDSOI, FDSOI, MRAM, etc.).
- Reduction of process & plasma induced damage or reduction of device/process parameter fluctuation through the optimization of circuit design & layout, device structure, manufacturing process, and semiconductor processing tool.

As IC design & process technology continue to advance for increased performance, lower power, and accelerated time-to-market, the engineering activities, traditionally separated along the boundary of design and process technology, will have difficulties in meeting the shrinking window of product optimization tasks. The International Conference on IC Design & Technology provides a forum for engineers, researchers, scientists, professors and students to cross this boundary through interactions of design and process technology on product development & manufacturing. The unique workshop style of the conference provides an opportunity to technologists and product designers to exchange breakthrough ideas and collaborate effectively. Two days of technical presentations and workshops will be preceded by a one-day tutorial program of value to both the expert and the beginner.

The venue of 2011 ICICDT will be **No.202, Ming-Sheng 2nd Road, Kaohsiung City, Taiwan.**

ICICDT **2011**

Committees

Organizing Committee

General Chair:
Marc Belleville, CEA-LETI

Conference Chair:
Chua-Chin Wang, National Sun Yat-Sen University

Executive Committee Chair:
Thuy Dao, Freescale Semiconductor

Secretary:
Dina H. Triyoso, Global Foundries

Local Arrangement Chairs:
Po-Ming Lee, Southern Taiwan University
Robert Rieger, National Sun Yat-Sen University
Katherine Shu-Min Li, National Sun Yat-Sen University

Publication Chair:
Thomas Ea, ISEP

Publicity Chair:
Terrence Hook, IBM

Tutorial Chair:
Geoffrey Yeap, Qualcomm

Keynote & Invited Papers Chair:
Ming-Dou Ker, National Chiao Tung University/ I-Shou University

Award Chair:
David Pan, University of Texas

Treasurer:
Soon-Jyh Chang, National Cheng Kung University

ICICDT **2011**

Committee

Technical Program Sub-Committee Chairs

Advanced Transistors/Materials:
Dong-Won Kim, Samsung Electronics

Advanced Memory Devices:
Susumu Shuto, Toshiba Corporation
Hideto Hidaka, Renesas

CAD:
Mehrdad Manesh, Qualcomm

DFM/DFT/DFR/DFY:
Keith Bowman, Intel

Emerging Technologies:
Simon Deleonibus, CEA-LETI
Hiroshi Mizuta, University of Southampton

High Power/High Voltage:
Jan Ackaert, ON Semiconductor

3D Integration:
Bich-Yen Nguyen, SOITEC

I/O Circuits and ESD Protection:
Ming-Dou Ker, National Chiao Tung University/ I-Shou University
Dimitri Linten, IMEC

Low Power:
Philippe Royannez
Michael Han

Reliability/Plasma-Induced Damage:
Yuichiro Mitani, Toshiba Corporation
Koji Eriguchi, Kyoto University

RF & Analog, Mixed Signal:
Didier Belot, ST Microelectronics
Andrea Mazzanti, Università di Pavia

SoC/MPSoC/SIP:
Dac Pham, Freescale Semiconductor
Masaya Sumita, Panasonic-Matsushita

Soft Error Rate:
Eishi Ibe, Hitachi

Committee

Technical Program Committee

Jan Ackaert, ON Semiconductor

Laurent Alacoque, CEA

Simone Alba, Numonyx

Ingo Aller, IBM

Himanshu Arora, Marvell Semiconductor

Marc Belleville, CEA LETI

Didier Belot, ST Microelectronics

Keith A Bowman, INTEL

Chen Hung Chang, TSMC

Ty Chen, Amazing Microelectronic Corporation

Shui-Ming Cheng, TSMC

Kin P. Cheung, National Institute of Standard & Technology

Minsik Cho, IBM

Thuy Dao, Freescale Semiconductor

Emeric Defoucauld, CEA

Simon Deleonibus, CEA LETI

Shurong Dong, Zhejiang University

David Duarte, Intel

Koji Eriguchi, Kyoto University

Thomas Ernst, CEA

Véronique Ferlet, ESA

Phillipe Galy, STMicroelectronics

Mark Hall, Freescale

Frederic Hameau, CEA

Michael Han, Qualcomm

Hideto Hidaka, Renesas

Toshiro Hiramoto, University of Tokyo

Terrence Hook, IBM

Eishi Ibe, Hitachi

Atsuki Inoue, Fujitsu

Ben Kaczer, Imec

Rouwaida Kanj, IBM

Ming-Dou Ker, National Chiao-Tung University

Dong-Won Kim, Samsung

Yoshinori Kumura, Toshiba

Didier Lattard, CEA

Committee

Jean-Luc Leray, CEA

Dimitri Linten, IMEC

Prashant Majhi, SEMATECH

Mehrdad Manesh, Qualcomm

Andrea Mazzanti, Università di Modena e Reggio E.

Yuuichiro Mitani, Toshiba Corporation

Hiroshi Mizuta, University of Southampton

Dominique Morche, CEA

Bich-Yen Nguyen, SOITEC-USA

Mototsugu. Okushima, Renesas

David Pan, The University of Texas at Austin

Juergen Phlle, IBM

Ruchir Puri, IBM

Marina Reyboz, CEA

Bruno Robisson, CEA

Elyse Rosenbaum, University of Illinois at Urbana-Champaign

Anne-sophie Royet, CEA

Olivier Rozeau, CEA

Ashoka Sathanur, Imec

Andrea Scarpa, NXP Semiconductors

Rich Shen, eMemory

Susumu Shuto, Toshiba Corporation

Alecandre Siligaris, CEA

Charles Slayman, Opsalacarte

Masaya Sumita, Panasonic-Matsushita

Takayanagi Toshinari, Apple

Olivier Thomas, CEA

David Trémouilles, LAAS

Dina Triyoso, Global Foundries

Jean-Chou Tseng, TSMC

Taiki Uemura, Fujitsu Semiconductor

Alexandre Valentian, CEA

Vladislav Vashchenko, Maxim

Chua-Chin Wang, National Sun Yat-Sen University

Piet Wessels, NXP

Geoffrey Yeap, Qualcomm

ICICDT **2011**

Tutorials

Tutorial 1 - Dr. Yuanjin ZHENG – Nanyang Technological University

Integrated RF Transceiver Circuit and System for Communication, Sensing and Biomedical Applications

In this tutorial, firstly, various RF transceiver architecture and key circuits are reviewed. Secondly, ultra-Wideband (UWB) impulse transceiver circuits and systems for different applications such as WPAN, WSN and WBAN are presented. Thirdly, case studies of RF transceivers for other applications like biomedical wearable sensor and implantable devices are introduced.

Yuanjin Zheng received the B.Eng. and M.Eng. degrees from Xi'an Jiaotong University, China, in 1993 and 1996, respectively, and the Ph.D. from Nanyang Technological University, Singapore, in 2001. From July 1996 to April 1998, he worked in the National Key Lab of Optical Communication Technology, University of Electronic Science and Technology of China. In 2001, he joined the Institute of Microelectronics, A*STAR, as a Senior Research Engineer, and was then promoted to a Principle Investigator. In IME, he has led and developed various projects like CMOS RF transceivers, baseband SoC for WLAN, WCDMA, ultrawideband, and low power medical radio, etc. In July 2009, he joined Nanyang Technological University as an Assistant Professor. His research interest is on GHz RFIC and SoC design, UWB system and circuits, Bio-IC system and circuits, adaptive signal and image processing algorithm and ASIC. He has published more than 70 international journal and conference papers, 11 patents filed and one book chapter (Springer). He has successfully led and contributed numerous public funded research and industry projects.

Tutorials

Tutorial 2 - Dr. Chun – Ming HUANG --- CIC

MorPACK: A Multi-Die Heterogeneous Integration Platform for System Prototyping

In this talk, we will present a multi-die multi-substrate system-in-package platform for heterogeneous system integration and prototyping, namely, MorPACK (Morphing PACKage). Logically, MorPACK provides an OS-ready system design platform which includes: embedded processor(s), SDRAM, NOR Flash, peripherals, system connection fabrics, and hardware IP connector(s). Users can follow the MorPACK design guideline to design their hardware IP(s) and easily integrate the hardware IP(s) into the MorPACK design platform. Physically, the MorPACK system platform and user designed hardware IP(s) are partitioned and implemented as individual dies via same or different processes, all the fabricated dies will be packaged and assembled as a multi-substrate structure to form a single system module. We will show the MorPACK physical structure, platform architecture, connection scheme, hardware IP design flows, system integration flow, measurement results, and finally, some thermal analysis results.

Chun-Ming Huang received the B.S. degree in mathematical science from National Chengchi University, Taipei, Taiwan, R.O.C., in 1990, and the M.S. and Ph.D. degree, both in computer science, from the National Tsing-Hua University, Hsin-Chu, Taiwan, R.O.C., in 1992 and 2005, respectively. Since 1993, he has been with the National Chip Implementation Center (CIC), where he is currently a researcher and department manager in the Design Service Division (DSD). His research interests include VLSI design and testing, platform-based SOC design methodologies, system integration technologies, and multimedia communication. Dr. Huang has served as a reviewer for the Design Automation Conference (DAC), the IEEE ISCAS, the IEEE Transactions on Computers, the ACM Transactions on Design Automation of Electronic Systems (TODAES), the ACM Transactions on Embedded Computing Systems (TECS), and the Journal of Marine Science and Technology (JMST). Dr. Huang is a member of Phi Tau Phi Scholastic Honor Society.

Tutorials

Tutorial 3 - Dr. David WANG – ChipMOS Technologies, INC.

3D TSV Technology Introduction - A Packaging Perspective

Driven by improved performance and mobile computing needs, new innovations of IC design are being explored continuously. As 3D IC using through-silicon-via (TSV) technology provides advantages in smaller form factor, lower power consumption and high level of integration, it could offer a less capital extensive alternative towards "More than Moore". By using vertical via interconnections, one may achieve wide bandwidth interconnections while significantly reduces the power consumption and package dimensions, allowing further heterogeneous integration. Thus smaller, smarter and greener devices can be more readily incorporated in variety of advanced platforms, such as: smart phone, tablet PC, MEMS, CMOS image sensor, servers etc.

In this session, subjects to be covered include a brief introduction of electronic packaging technologies evolution, future market trends and requirements, design and process requirements for interposer and 3D TSV thin-die stack used in related applications.

Dr. David W. Wang joined ChipMOS in 2007 as Vice President, Research and Strategy Development Center, ChipMOS TECHNOLOGIES, INC. one of the leading back-end assembly and test service providers for advanced memories, LCD driver semiconductors, and mixed-signal products. He is a member of SEMI Taiwan Packaging and Test Committee.

Prior to ChipMOS, he was Vice President of Fibera, Inc. a Silicon Valley startup. He served as Senior Director at Lam Research where his responsibilities included 300mm new product introduction, system automation, field escalation and management of regional teams. David also worked for IBM's Microelectronics Division for 13 years at facilities located in Endicott and East Fishkill, New York where he as Senior Engineering Manager led advanced packaging materials/process development and marketing organizations.

He was Chairman of the Board and President of Chinese American Semiconductor Professional Association (CASPA) from 2003 to 2004 and presently is a member of International Advisors. David received his Ph.D. / M.S. from the University of Michigan and B.S. from Fu Jen University. He holds 45 U.S. patents.

Tutorials

Tutorial 4 - Geoffrey YEAP – Qualcomm Inc.

Advanced CMOS Process/ Device and Circuit Co-Design for Mobile Wireless Devices

VP of Technology of VLSI Technology, Qualcomm Inc. He is a member of SIA International Technology Roadmap for Semiconductors (ITRS) Process Integration & Device Structures (PIDS) Technical Working Group working on addressing semiconductor device scaling trends for logic and memory processes, and identify potential solutions for difficult challenges. He has Ph.D., Electrical & Computer Engineering from The University of Texas at Austin, Austin, Texas. Dr. Yeap held over twenty US patents. He published over sixty journal and conference publications in the field of advanced semiconductor device design, simulation and modeling.   He is a member of the following organizations:

•Member, 2009 and 2010 IEDM CMOS Devices & Technology sub-committee.

•Member, 2011 VLSI Technology Symposium committee

•Member, 2002 and 2003 IEDM Integrated Circuits & Manufacturing sub-committee.

•Senior Member, IEEE (Electron Devices, Solid-State Circuits, and Circuits & Systems societies).

• Distinguished Member of Technical Staff, Motorola Inc. Engineering Technical Ladder. Motorola DigitalDNA Laboratories' On-the Spot Award for outstanding performance. Motorola Silver Quill program for promoting innovation and engineering excellence..

•Dean's Honor List, Eta Kappa Nu (National Electrical Engineering Honorary), Tau Beta Pi (National Engineering Honorary).

•Conference scholarship award by Graduate Engineering Council at The Univ. of Texas at Austin.

Transient-to-Digital Converter to Detect Electrical Fast Transient (EFT) Disturbance for System Protection Design

Cheng-Cheng Yen, Wan-Yen Lin, Ming-Dou Ker, Ching-Ling Tsai, Shih-Fan Chen, and Tung-Yang Chen

Abstract—**A new on-chip 4-bit transient-to-digital converter for electrical fast transient (EFT) protection design has been proposed. The converter is designed to detect EFT-induced transient disturbances and transfer different EFT voltages into digital codes under EFT tests. The experimental results in silicon chip have confirmed the successful digital output codes.**

Index Terms —**electromagnetic compatibility (EMC), electrical fast transient (EFT) test, transient detection circuit, transient-to-digital converter.**

I. INTRODUCTION

Recently, the electrical fast transient (EFT) tests have become an important reliability issue in the microelectronic systems equipped with CMOS integrated circuits (ICs) [1]-[5]. It has been investigated that, the EFT-induced electrical transients can cause transient-induced latchup (TLU) failure in inevitable parasitic silicon controlled rectifier (SCR) in CMOS ICs. It has been reported that, for a super twisted nematic (STN) liquid crystal display (LCD) driver circuit, the electrical transient disturbance can couple into power and ground pins and cause upset states of panel. Furthermore, during transient disturbance conditions, some of EFT-induced or ESD overshooting/undershooting transients can cause locked states or malfunction on the CMOS ICs inside the microelectronics products [6]-[13].

This reliability issue results not only from the integration of more electrical functions into single chip but also from the strict requirements of reliability test standards, such as the EFT test. Typically, if display panel products are required to achieve the immunity of "level 4" in the IEC 61000-4-4 standard [14], the equipment under test (EUT) should sustain the EFT voltage level of ±2 kV under EFT tests. During EFT tests, the power lines of the CMOS ICs in the microelectronic products no longer maintained normal voltage levels, but an exponential voltage pulse with the amplitude of several tens volts occurred. Such EFT-induced transients are quite large and can randomly couple into power, or signal pins of microelectronic circuits.

C.-C. Yen, W.-Y. Lin, and M.-D. Ker are with the Institute of Electronics, National Chiao-Tung University, Hsinchu, Taiwan. M.-D. Ker was also with Department of Electronics Engineering, I-Shou University, Kaohsiung, Taiwan.

C.-L. Tsai, S.-F. Chen, and T.-Y. Chen are with the Himax Technologies Inc., Taiwan.

According to the IEC 61000-4-4 standard, for the simplified circuit diagram of the EFT generator, only the output impedance resistor (50 Ω) and the dc blocking capacitor (10 nF) are fixed. The repetitive EFT test is a test with bursts consisting of a number of fast pulses. With the repetition frequency of 5 kHz and 100 kHz, the burst repeats every 300 ms and the application time is not less than 1 minute. With a 50 Ω as output loading, due to the impedance matching, the measured pulse peak is half of the input EFT pulse voltage. The waveform of a single pulse has a rise time of about ~5ns and the pulse duration of ~50 ns. The EFT levels for testing power supply ports and for testing I/O, data, and control ports of the equipment are listed in Table I.

In order to solve such EFT issues, the traditional solution is to add some board-level noise filters into the microelectronic products to decouple, bypass, or absorb the electrical transients under EFT tests. However, with more functions integrated into a system-on-a-chip (SOC), such additional discrete noise-bypassing components may not be integrated into a single chip due to the limitation of chip area and substantially increase in the fabrication cost of microelectronic products. Therefore, to meet high EFT specifications for microelectronic products, the chip-level solutions without additional discrete components on the printed circuit board (PCB) are highly desired by the IC industry.

In this paper, a new on-chip 4-bit transient-to-digital converter is proposed to detect the fast electrical transients and convert different EFT voltages into digital codes under EFT tests. The test chip fabricated in a 0.13-μm CMOS process has verified the circuit performance and output digital codes.

TABLE I
VOLTAGE LEVELS AND REPETITION RATES OF EFT TESTS

Level	On Power and PE (Protective Earth) Ports		On I/O (Input/Output) Signal, Data, and Control Ports	
	Voltage Peak (kV)	Repetition Rate (kHz)	Voltage Peak (kV)	Repetition Rate (kHz)
1	0.5	5 or 100	0.25	5 or 100
2	1	5 or 100	0.5	5 or 100
3	2	5 or 100	1	5 or 100
4	4	5 or 100	2	5 or 100
X	Specified by Customer	Specified by Customer	Specified by Customer	Specified by Customer

II. TRANSIENT DETECTION CIRCUIT

Fig. 1 shows the on-chip CR-based transient detection circuit The CR-based circuit structure is designed to detect EFT-induced transient disturbance. The NMOS (M_{nr}) is used to set the initial output voltage levels at nodes V_{OUT} and V_A as 1.8 V with the V_{DD} of 1.8 V. During the normal operating condition, the node V_G is biased at 0 V by connecting to V_{SS} line through poly resistor. The two-inverter latch is designed to memorize the logic state before and after EFT events. The MOSFET capacitor is designed by PMOS device and connected between V_{DD} line and node V_G to sense transient disturbance coupled on 1.8-V power line. Under the EFT tests with an overshooting transient voltage, the node V_G will be coupled with positive voltage by MOSFET capacitor coupling. Then, the NMOS device (M_{n1}) will be turned on by the overshooting EFT voltage to pull down the voltage level at the node V_A from 1.8 V to 0 V. Therefore, the logic level stored at the node V_B can be further pulled up from logic "0" to logic "1" to memorize the EFT events. With the buffer inverters, the final output voltage of the proposed detection circuit is changed from logic "1" to logic "0" to detect the occurrence of EFT-induced transient disturbance.

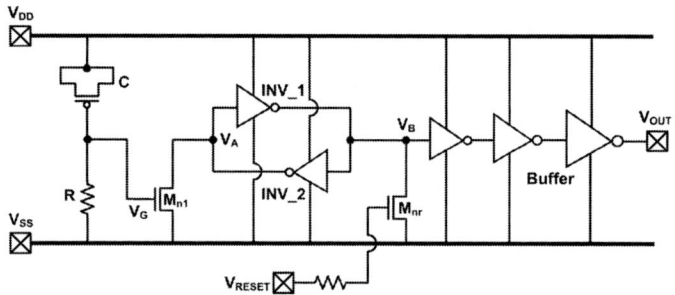

Fig. 1. The on-chip CR-based transient detection circuit.

III. EXPERIMENTAL RESULTS

In order to simulate the EFT-induced transient disturbance on CMOS ICs inside the microelectronic products, the attenuation network with -40 dB degradation is used in this work. The amplitude of EFT-induced transients can be adjusted through the attenuation network. The measurement setup for EFT test combined with attenuation network is shown in Fig. 2. EFT generator is connected to the device under test (DUT) through the attenuation network with V_{DD} of 1.8 V.

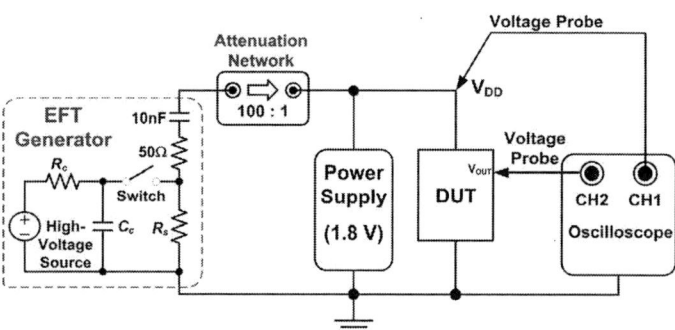

Fig. 2. Measurement setup for EFT test combined with attenuation network.

The V_{DD} and V_{OUT} transient responses of the CR-based transient detection circuit are monitored by the digital oscilloscope. Before each EFT test, the initial output voltage (V_{OUT}) is reset to 1.8 V. After each EFT test, the output voltage (V_{OUT}) level is monitored to check the final voltage level and to verify the detection function.

The measured V_{DD} and V_{OUT} waveforms of the CR-based detection circuit under EFT test with EFT voltage of +200 V (-350 V) are shown in Fig. 3(a) (Fig. 3(b)). As shown in Fig. 3(a) (Fig. 3(b)), under the EFT test with positive (negative) EFT voltage, V_{DD} begins to increase (decrease) rapidly from 1.8 V. Meanwhile, V_{OUT} begins to greatly increase (decrease) with positive (negative) exponential voltage pulse coupled on V_{DD} power line. Finally, after the EFT-induced transient disturbance, the output voltage (V_{OUT}) of the CR-based detection circuit is changed from 1.8 V to 0 V. Therefore, the detection circuit can successfully memorize the occurrence of the EFT event with positive (negative) EFT voltage.

From the EFT test results, the CR-based on-chip transient detection circuit can successfully memorize the occurrence of EFT-induced transient disturbance events.

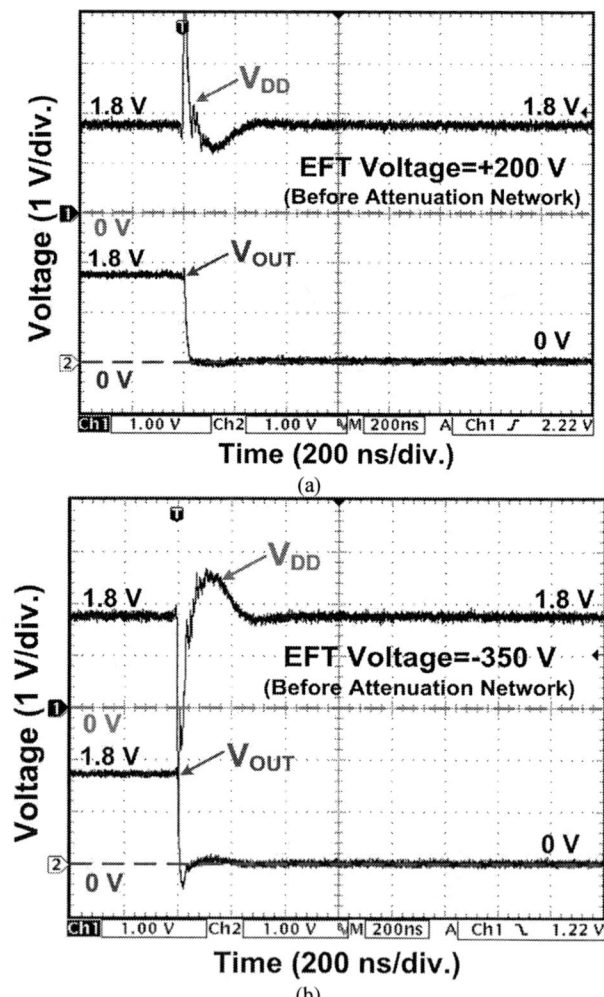

Fig. 3. Measured V_{DD} and V_{OUT} waveforms on the new proposed on-chip transient detection circuit under EFT tests with (a) +200-V, and (b) -350-V, EFT voltages combined with attenuation network.

978-1-4244-9019-6/11 $26.00 © 2011 IEEE

IV. APPLICATION IN DISPLAY PANEL

A. Circuit Structure

Fig. 4 shows the proposed on-chip transient-to-digital converter. For the CR-based transient detection circuit shown in Fig. 1, the EFT energy coupled to node V_G can be further adjusted by different resistive voltage dividers. Therefore, the minimum EFT voltage to cause transition at the output (V_{OUT}) of the proposed CR-based transient detection circuit can be designed for each stage in the transient-to-digital converter.

It has been also investigated that noise filter networks can reduce the susceptibility of CMOS ICs against transient disturbance by decoupling or bypassing EFT-induced noise energy. The noise filter network can suppress the transient peak voltages on power lines, which has influence on EFT voltages to cause transition at the output (V_{OUT}) of the transient detection circuit. In the previous design, 10-pF on-chip decoupling capacitor is used in noise filter network [10]. In this work, a 3-pF on-chip capacitor with a current mirror is adopted to replace the large 10-pF decoupling capacitor between power lines to avoid the gate leakage issue in nanoscale CMOS processes. With different device ratios in the current mirror, different EFT levels on V_{DD} line will reach to each transient detection circuit. Under the EFT zapping conditions, the four transient detection circuits will have different output voltage responses. Therefore, by combining with different filter networks, the proposed transient-to-digital converter can be designed to detect different EFT voltage levels and transfer output voltages into digital codes under EFT tests.

B. Measurement Results

The EFT generator combined with attenuation network shown in Fig. 2 is used to evaluate the digital codes of the proposed on-chip transient-to-digital converter under EFT tests. The V_{OUT1}, V_{OUT2}, V_{OUT3}, and V_{OUT4} transient responses of the proposed on-chip transient-to-digital converter are monitored by the digital oscilloscope. Before EFT test zapping, the initial output voltages of the proposed on-chip transient-to-digital converter are all reset to 1.8 V. After each EFT zapping, the output voltage levels are measured to check the final voltage levels and to verify the transferred digital codes.

The measured V_{OUT1}, V_{OUT2}, V_{OUT3}, and V_{OUT4} waveforms of the proposed transient-to-digital converter under EFT test with EFT voltage of +400 V are shown in Fig. 5. During the EFT-induced disturbance, V_{OUT1}, V_{OUT2}, V_{OUT3}, and V_{OUT4} are disturbed simultaneously. Finally, V_{OUT1} will be changed from 1.8 V to 0 V, while V_{OUT2}, V_{OUT3}, and V_{OUT4} are still kept at 1.8 V. Therefore, under EFT test with EFT voltage of +400 V zapping, the detection output voltages can be transferred into a digital code of "1110."

Similarly, under EFT test with EFT voltages of +500 V, +700 V, and +2000 V, the transferred digital codes of proposed converter are "1100," "1000," and "0000," respectively. Therefore, under EFT tests with positive EFT voltages, the proposed on-chip transient-to-digital converter can successfully transfer different EFT voltage levels into digital codes.

The measured V_{OUT1}, V_{OUT2}, V_{OUT3}, and V_{OUT4} voltage waveforms of the proposed transient-to-digital converter under EFT test with EFT voltage of -700 V are shown in Fig. 6. During the high-energy fast transient of EFT stress, all transient detection circuits can detect the EFT-induced transient disturbance coupled on V_{DD} line. Finally, when V_{DD} returns into the normal operation voltage level of 1.8V, V_{OUT1}, V_{OUT2}, V_{OUT3}, and V_{OUT4} have been pulled down from 1.8 V to 0 V. The four outputs of CR-based transient detection circuit transit from logic "1" to logic "0."

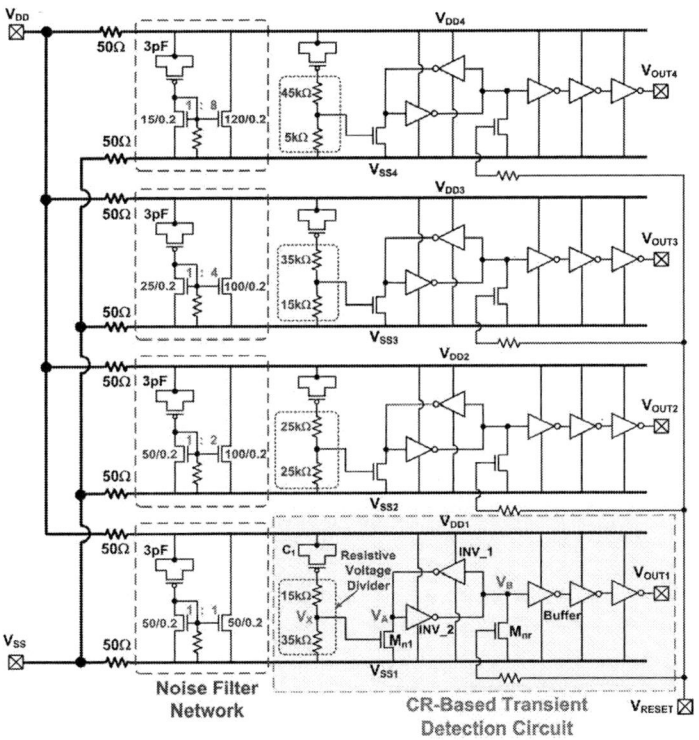

Fig. 4. The proposed 4-bit transient-to-digital converter realized with four CR-based transient detection circuits and four different noise filter networks.

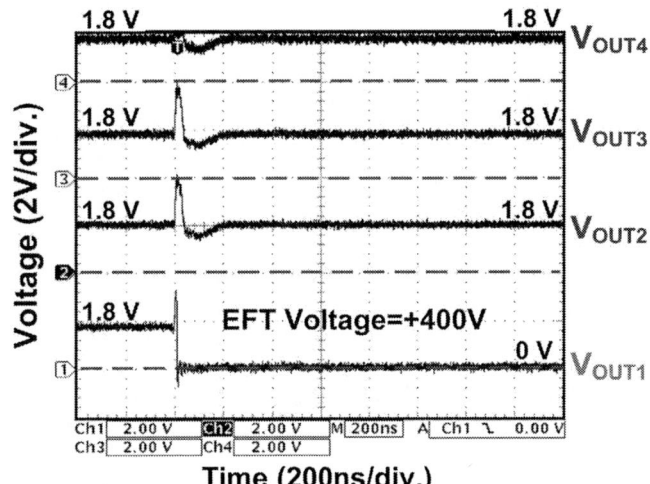

Fig. 5. Measured V_{OUT1}, V_{OUT2}, V_{OUT3}, and V_{OUT4} transient voltage waveforms under EFT tests with EFT voltages of +400V.

From the measurement results under EFT test with EFT voltage of -700 V, the output responds of the proposed 4-bit transient-to-digital converter can be transferred into a digital code of "0000."

Similarly, under EFT test with EFT voltages of -410 V, -450 V, and -500 V, the transferred digital codes of proposed converter are "1110," "1100," and "1000," respectively. Therefore, under EFT tests with negative EFT voltages, the proposed on-chip transient-to-digital converter can successfully transfer different EFT voltage levels into digital codes.

Table II depicts the EFT voltage to digital code characteristic of the proposed 4-bit transient-to-digital converter. Under EFT tests, larger EFT voltage levels can response to higher digital codes. The digital code goes from "1110" to "0000" as the magnitude of EFT voltage increases from +0.4 kV to +2.0 kV and from -0.41 kV to -0.7 kV.

The digital codes from transient-to-digital converter can be temporarily stored as system recovery index for different recovery procedure designs. This hardware/firmware co-design can effectively improve the robustness of the microelectronic products against EFT-induced transient disturbance.

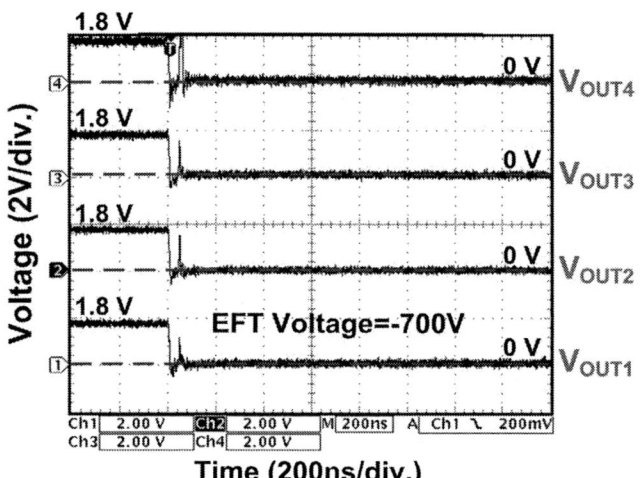

Fig. 6. Measured V_{OUT1}, V_{OUT2}, V_{OUT3}, and V_{OUT4} transient voltage waveforms under EFT tests with EFT voltages of -700V.

TABLE II
EFT VOLTAGE TO DIGITAL CODE CHARACTERISTIC

Digital Codes	Positive EFT Voltage (kV)	Negative EFT Voltage (kV)
1111	< +0.4	> -0.41
1110	+0.4 ~ +0.5	-0.41 ~ -0.45
1100	+0.5 ~ +0.7	-0.45 ~ -0.5
1000	+0.7 ~ +2.0	-0.5 ~ -0.7
0000	> +2.0	< -0.7

V. CONCLUSIONS

A novel transient-to-digital converter composed of four CR-based transient detection circuits and four different noise filter networks has been successfully designed and verified in a 0.13-μm CMOS process with 1.8-V devices. In this filter design, the on-chip capacitor with a current mirror can replace the large decoupling capacitor and avoid the gate leakage issue. The output digital codes can correspond to different EFT voltages under EFT tests. These output digital codes can be used as the firmware index for microelectronic products to execute different system recovery procedures and to solve the EFT-induced transient disturbance events in microelectronic systems equipped with CMOS ICs.

ACKNOWLEDGMENT

This work was partially supported by National Science Council, Taiwan, under Contract NSC 99-2811-E-009-051; and by Himax Technologies Inc., Taiwan.

REFERENCES

[1] A. Wallash and V. Kraz, "Measurement, simulation and reduction of EOS damage by electrical fast transients on AC power," in *Proc. EOS/ESD Symp.*, 2010, pp. 59–64.

[2] J. Koo, L. Han, S. Herrin, R. Moseley, R. Carlton, D. Beetmer, and D. Pommerenke, "A nonlinear microcontroller power distribution network model for the characterization of immunity to electrical fast transients," *IEEE Trans. Electromagn. Compat.*, vol. 51, no. 3, pp. 611–619, Aug. 2009.

[3] M.-D. Ker and S.-F. Hsu, *Transient-Induced Latchup in CMOS Integrated Circuits*, John Wiley & Sons, 2009.

[4] C.-C. Yen, M.-D. Ker, and T.-Y. Chen, "Transient-induced latchup in complementary-metal-oxide-semiconductor integrated circuits under electrical fast transient test," *IEEE Trans. Device Mater. Reliab.*, vol. 9, no. 2, pp. 255–264, Jun. 2009.

[5] G. Cerri, R. Leo, and V. Primiani, "Electrical fast-transient: conducted and radiated disturbance determination by a complete source modeling," *IEEE Trans. Electromagn. Compat.*, vol. 43, no. 1, pp. 37–44, Feb. 2001.

[6] W. Huang, J. Dunnihoo, and D. Pommerenke, "Effects of TVS integration on system level ESD robustness," in *Proc. EOS/ESD Symp.*, 2010, pp. 145–149.

[7] L. Lou, C. Duvvury, A. Jahanzeb, and J. Park, "SPICE simulation methodology for system level ESD design," in *Proc. EOS/ESD Symp.*, 2010, pp. 65–73.

[8] G. Notermans, D. Maksimovic, G. Vermont, M. Maasakkers, F. Pusa, and T. Smedes, "On-chip system level protection of FM antenna pin," in *Proc. EOS/ESD Symp.*, 2010, pp. 83–90.

[9] N. Monnereau, F. Caignet, and D. Tremouilles, "Building-up of system level ESD modeling: impact of a decoupling capacitance on ESD propagation," in *Proc. EOS/ESD Symp.*, 2010, pp. 127–136.

[10] M.-D. Ker and C.-C. Yen, "Transient-to-digital converter for system-level ESD protection in CMOS integrated circuits," *IEEE Trans. Electromagn. Compat.*, vol. 51, no. 3, pp. 620–630, Aug. 2009.

[11] K. Muhonen, P. Erie, N. Peachey, and A. Testin, "Human metal model (HMM) testing, challenges to using ESD guns," in *Proc. EOS/ESD Symp.*, 2009, pp. 387–395.

[12] E. Grund, K. Muhonen, P. Erie, and N. Peachey, "Delivering IEC 61000-4-2 current pulses through transmission lines at 100 and 330 ohm system impedances," in *Proc. EOS/ESD Symp.*, 2008, pp. 132–141.

[13] E. Grund, K. Muhonen, P. Erie, and N. Peachey, "Delivering IEC 61000-4-2 current pulses through transmission lines at 100 and 330 ohm system impedances," in *Proc. EOS/ESD Symp.*, 2008, pp. 132–141.

[14] IEC 61000-4-4 Standard, "*EMC – Part 4-4: Testing and Measurement Techniques – Electrostatic Fast Transient/Burst Immunity Test*," 2004.

ESD RF protections in advanced CMOS technologies and its parasitic capacitance evaluation

Ph. Galy[1], J. Jimenez[1], P. Meuris[2], W. Schoenmaker[2], O. Dupuis[2]

[1] STMicroelectronics, 850 rue Jean Monnet, 38920 Crolles, France
[2] MAGWEL NV, Martelarenplain 13 B-3000 Leuven, Belgium

Abstract— **Electrostatic Discharge (ESD) protection for advanced CMOS technologies is a challenge due to down-scaling which introduces a reduction of the intrinsic robustness. Moreover, another challenge is the RF ESD protection in analogue IO pad. Thus, when you merge both topics the challenges are major. This paper shows a methodology, tools and silicon measurements of ESD RF parasitic capacitance in C65nm & C45nm to reach 10Ghz & 20Ghz bandwidth for 1kV & 2kV HBM.**

Index Terms— **ESD, RF, Diode, SCR, dual SCR Maxwell equations, S parameters**

I. INTRODUCTION

The Electrostatic Discharge (ESD) protection is a big challenge and especially in advanced CMOS technologies with standard oxide gate or High K metal for C32nm. The ESD design window is reduced due to the down-scaling of CMOS process and leads to a degradation of intrinsic ESD robustness of N and PMOS transistors. In addition, the ESD RF protection in a wide and sharp band is another challenge. Obviously, the ESD protection plug on RF signal leads to an introduction of parasitic capacitance. This parasitic capacitance is critical in term of signal band-with and is a contributor of distortion and signal degradation. That is why it is important to develop a methodology and tool to study different solutions and to give an evaluation of the ESD RF parasitic capacitance and its impacts on signal integrity. The characterization is done trough S parameters with 110GHz equipment. Moreover, the ESD robustness is also evaluated through 3D TCAD simulations (the equations set is: Poisson, current continuity and drift/diffusion carrier transport coupled with the Maxwell equations) and silicon integration. Characterization and qualifications are performed through 100ns TLP stresses and standard ESD/LU stress.

II. ESD RF STRATEGY PROTECTION OVERVIEW

The following paragraph gives a short overview of the ESD protection strategy dedicated to the RF domain.

A. Remote protection with STI diodes

An ESD protection strategy uses power devices to dissipate the energy surge. The main basic power device is the ESD diode used in a remote or distributed network. In this first example, the ESD STI diode is used due to the low parasitic capacitance compare to a gated diode. Figure 1 depicts the basic approach of ESD remote network protection for RF IO pad.

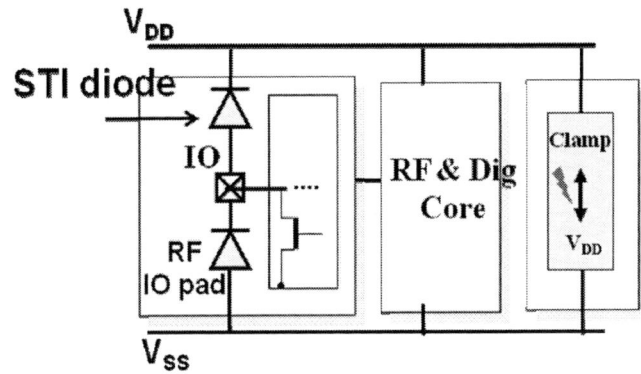

Fig. 1. Basic schematic of RF IO pad with STI diodes

In this example the parasitic capacitances of the diodes are in parallel.

B. Local protection with SCR & STI diode

Another approach to reduce the impact of the ESD protection is to develop a local ESD protection in the IO. For this, a second power device is used like a Silicon Controller Rectifier (SCR). Nevertheless, to address both ESD stress polarities it is necessary to use again the STI diode for reverse surge and the SCR for forward stress. Figure 2 shows a typical example of design in this way of local protection with the associated trigger circuit (TC) of the SCR [1].

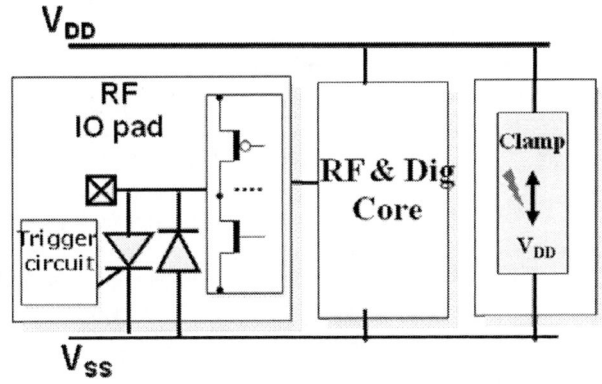

Fig. 2. Basic schematic of RF IO pad with SCR + STI diode

978-1-4244-9019-6/11 $26.00 © 2011 IEEE

III. SET OF EQUATIONS & TOOL

This section is focused on the set of equations to address the field of ESD RF study and is implemented in the MAGWEL tools. This new generation of tools was born within the partnership between the companies STmicroelectronics & MAGWEL

A. Set of equations

The full protection has two main contributors, the first one is the copper metal connection (BEOL) and the second one is the active silicon power device (FEOL). The purpose here, is to couple the Maxwell equations with silicon equations to provide a full evaluation of parasitic capacitance of the RF ESD protection [3]. First of all, the tensor form of the Maxwell equations are:

$$\partial_\mu \mathbf{F}^{\mu\nu} = -4\pi \mathbf{J}^\nu$$

where: $\mathbf{F}^{\mu\nu} = \partial^\mu \mathbf{A}^\nu - \partial^\nu \mathbf{A}^\mu$, $\mathbf{A}^\nu = (\phi, \mathbf{A})$, $\mathbf{J}^\nu = (\rho, \mathbf{J})$
We impose the Lorenz gauge condition:

$$\partial_\nu \mathbf{A}^\nu = 0$$

Thus, the current densities for electrons and holes in the classical drift-diffusion model with the inclusion of the Lorentz force are classically:

$$\mathbf{J}_n = q \, \mu_n \, n \, (\mathbf{E} + \mathbf{v}_n \times \mathbf{B}) + k \, T \, \mu_n \, \nabla n$$

$$\mathbf{J}_p = q \, \mu_p \, p \, (\mathbf{E} + \mathbf{v}_p \times \mathbf{B}) - k \, T \, \mu_p \, \nabla p$$

Where $\mathbf{E} = -\nabla \phi - \partial_t \mathbf{A}$ and $\mathbf{B} = \nabla \times \mathbf{A}$
The currents and velocities are related according to $\mathbf{J}=qn\mathbf{v}$. The previous equations are completed with the current-continuity equations of the drift-diffusion model, i.e.

$$\nabla \cdot \mathbf{J} = -\partial_t \rho + U$$

Where: $\rho \, (\phi, \phi_n, \phi_p)$ is the charge density and $U=R\text{-}G$ is the recombination/generation rate in silicon. Anyway, inside metallic region we use the following expression for the current density: $\mathbf{J} = \sigma\mathbf{E} + \mu_H \mathbf{J} \times \mathbf{B}$

It should be noted that the implicit definition of J must be made explicit by expanding in $|\mu_H B|$ and next applied. Further details can be found in [2]. For contacts we apply Dirichlet boundary conditions and Neumann boundary conditions elsewhere at the boundary of the simulation domain. Moreover, simulations are done with isothermal condition (room temperature).

B. Tool with this set of equations

The new tool developed by MAGWEL allows now to calculate all equations in coupled method in static or quasi-static state. New algorithms and methods are employed in this software (for more details se [4]). The first step is to describe the full ESD RF protection with the Back End (BE) contribution and all electrical parameters (resistivity, dielectric constant, Hall resistance, moblities ...). The second step is the description of the silicon active with all doping/gradient/STI regions defined as Front End (FE). All parameters of process are calibrated.
Thus, the next section is focused on two ESD RF protections with remote and local clamp.

IV. ESD RF DIODES PROTECTION IN REMOTE CLAMP

The first example of ESD RF protection is based on remote clamp using STI diodes as introduced in paragraph II. In this approach, the main purpose is to reduce the parasitic capacitance of STI diode by the spreading of metal connection. This is to reduce the capacitance coupling of metal line. This demonstrator is done in C65nm for an ESD target about 1 kV Human Body Model (HBM).

A. Numerical results in coupled approach

The purpose of this sub section is to evaluate the parasitic capacitance by the numerical solver for the set of coupled equations as described in paragraph III. The left part of figure 3 depicts both ESD layout diodes in the remote clamp protection strategy. The right part shows the 3D extraction of BEOL + FEOL of the full structure.

Fig. 3. Basic Layout of STI diodes with their guard rings and 3D view extraction of BEOL+FEOL in Magwel tool

It's possible to extract interconnects capacitances of the structure which lead to 50fF parasitic capacitance extracted from S parameters up to 35GHz (see figure 4).

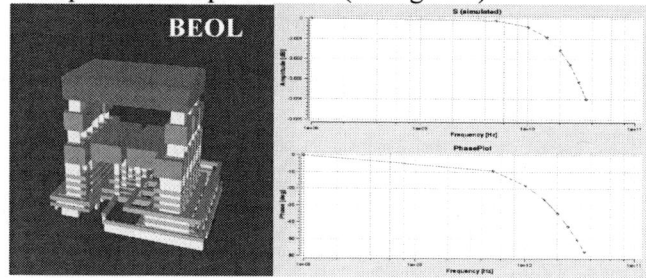

Fig. 4. S parameters& extraction of BEOL contribution (50fF)

In the same view, the FEOL parasitic capacitance extracted from S parameters up to 35GHz is 70fF (See figure 5).

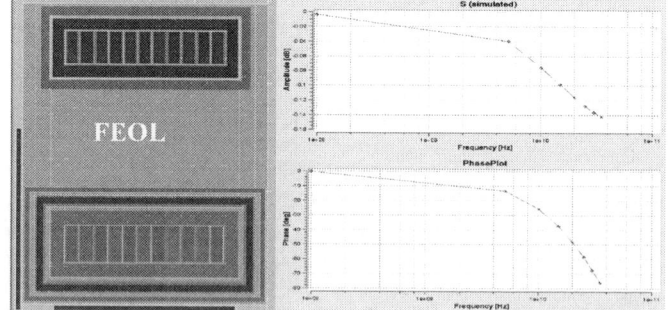

Fig. 5. S parameters & extraction of FEOL contribution (70fF)

978-1-4244-9019-6/11 $26.00 © 2011 IEEE

Now, it is clear that a basic addition of decoupled contributions could not be done and the coupled equations lead to real value. Thus FEOL+ BEOL parasitic capacitance extracted from S parameters up to 35GHz is 190fF (Fig 6.) and not the basic sum of the previous results (120 fF).

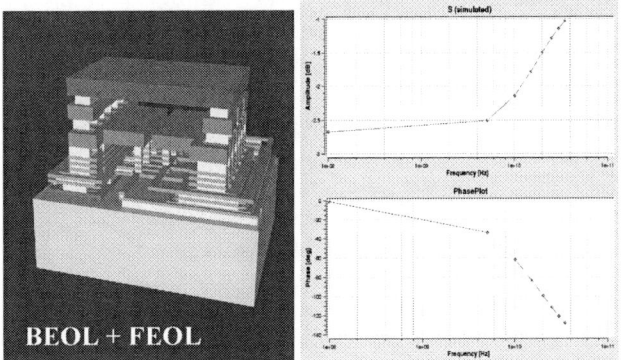

Fig. 5. S parameters & extraction of FEOL+BEOL (190fF)

The final extraction shows that the impact of the coupled contribution is not negligible and in order to achieve an accurate value of an ESD RF protection parasitic capacitance all equations need to be taken into account.

B. Experimental results

Capacitance extraction using 110 GHz network analyzer gives a capacitance for the complete diode network protection of 175 fF. Thus numerical and measurement are close and a fine tune could be performed. Concerning ESD response, it is now possible to predict the behavior of this kind of structure thanks to calibrated 3D TCAD simulation. Measurement of ESD response of diode network coupled to a power clamp is shown in figure 6. The figure of merit (e.g. V_{HBM}/C_{ESD}) is 2/175=0.011.

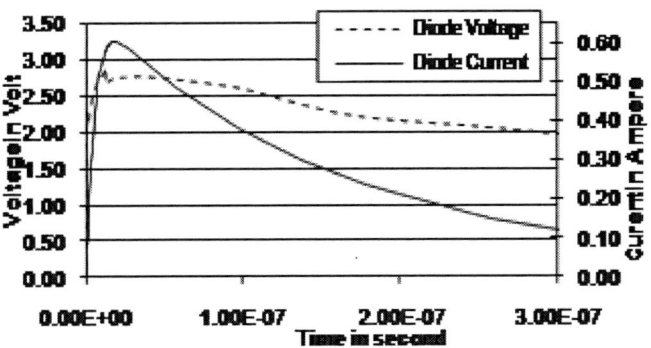

Fig. 6. I/V curves of RF diodes for 1KV HBM stress.

We see excellent correlations between both RF/ESD simulations and measurements using respectively MAGWEL and TCAD Sentaurus. These results help us with the development and fine-tuning of new innovative device topologies dedicated to RF/ESD protections devices with minimum silicon spins.

V. ESD RF (SCR + DIODE) PROTECTION IN LOCAL CLAMP

The second example of ESD RF protection is based on local clamp using STI diodes and a SCR as presented in paragraph II. The reduction of the parasitic capacitance is obtained thanks to a highly compact device to reduce the metal/active area. This area is equivalent to the electrode/plate of the parasitic capacitance. This new demonstrator is done in C45nm for an improved ESD target about 2 kV Human Body Model (HBM).

A. Numerical results in coupled approach

Figure 7 shows the typical Layout of SCR + STI diode and the 3D extraction in Magwel tool.

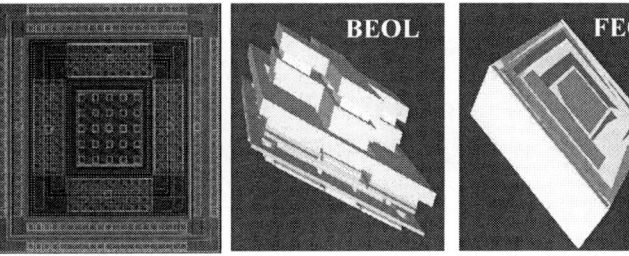

Fig. 7. Topology FEOL & BEOL 3D views

The numerical value of the parasitic capacitance of the compact SCR + STI diode leads to: 80 fF (FEOL+BEOL). Thanks to these numerical values, it is possible to compare the two solutions in advance of silicon test chip. Moreover, the second solution is more efficient that the previous one.

Another interesting point of the tool is that it is possible to extract all physical parameters. As an example, Figure 8 reports the magnetic field **B** and the electric field **E** in the structure with 10 GHz excitation signal. These extractions are done at silicon level. The **B** field is localized in the center whereas the **E** field is spread on the perimeter of the structure.

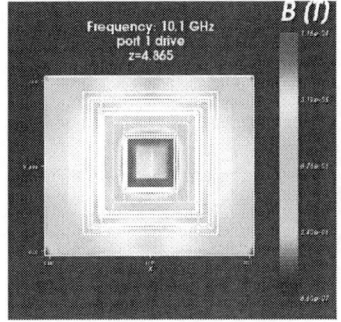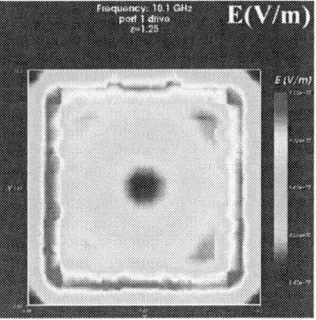

Fig. 8. Magnetic & Electric fiels **B,E** extractions in SCR+STI diode solution at silicon level @ 10 Ghz AC signal stimulus

978-1-4244-9019-6/11 $26.00 © 2011 IEEE

B. Experimental measurements

Capacitance extraction using 110 GHz network analyzer gives a capacitance for the SCR+ diode network protection of 85fF. Figure 9 shows C40nm test chip topology including RF de-embedding structures for capacitance extraction. Thus this structure can address 20 Ghz fast signal toggling.

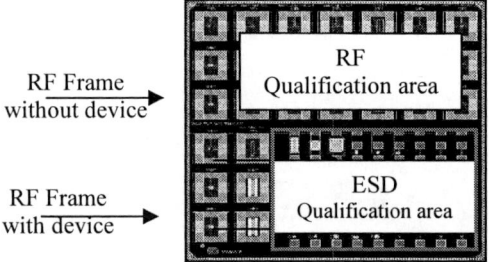

Fig. 9. Layout view of test chip topology and set of devices

ESD performances have been measured using two points HBM stimuli. And device characterizations are performed with TLP 100ns systems. Figure 9 shows 2KV HBM response, the breakdown limit of the protection. Moreover, no Latch Up events are observed in injection modes or power sequences. The figure of merit (V_{HBM}/C_{ESD}) is 2/85=0.023.

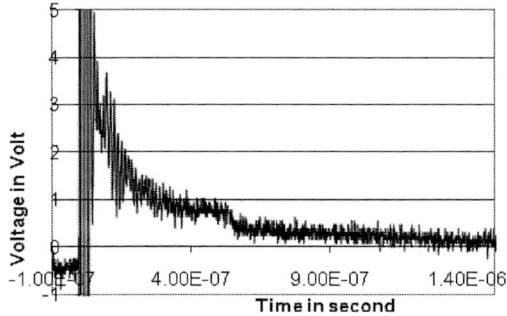

Fig. 9. 2KV positive HBM response (time response)

Moreover, figure 10 shows the TLP response of the SCR with STI diode. Maximum current before breakdown is near 2A which is in line with 2 kV HBM robustness (1.33A + Margin). The Vt1 snapback point is around 4V for a positive stress by R trig and 2V with its trigger in IO. In reverse a classical IV diode response is obtained and not presented here. The ESD robustness margin is 30% which could be reduce to decrease the parasitic capacitance value. This value is to take into account of a future shrink and keep always a safety margin.

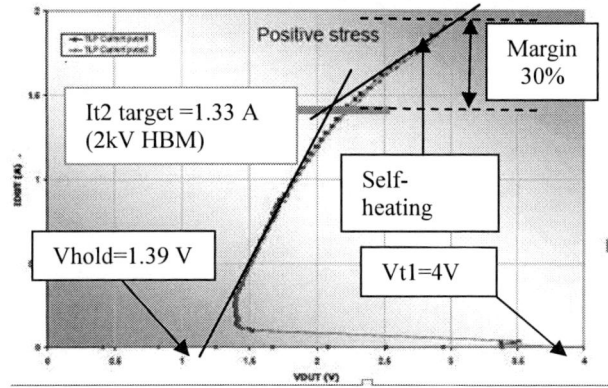

Fig. 10. Positive 100ns/10ns rise time response with R trig.

VI. CONCLUSION

This paper presents the main results of parasitic capacitance evaluation on two RF ESD solutions. The two solutions in C65 nm and C45nm CMOS technologies are with remote clamp using STI diodes and the second one is with (SCR + STI diode) in compact topology. The impact in term of parasitic capacitance is evaluated and compared thanks to the coupled equations (Maxwell + Drift-Diffusion Model) in the Magwel devEM tool. A condensed summary of the equations is given. It is shown that the compact solution (SCR + STI diode) leads to a decrease of 2x of the parasitic capacitance compare to the STI diodes solution. In the same time the ESD HBM robustness is increased by 2x for the SCR + STI diode compare to STI diodes. The measurements of both solutions show close a agreement with the numerical results. Moreover the ESD qualification shows that all ESD targets have been reached. In another point, the total size of SCR + STI diode is (20µm x 20µm) and (33µm x 40µm) for diodes solutions. In a next step we want to do an accurate calibration and to have the possibility to inject an external magnetic field as a perturbation. Anyway, the idea to compact the ESD devices to reduce the parasitic capacitance seems to be a good candidate for RF ESD protection. Thus, the ultra compact ESD device protection could be studied with high accuracy using the coupled equations (Maxwell+SC equations) in the MAGWEL devEM tool.

ACKNOWLEDGEMENT

Authors would like to thank Dündar Dumlugöl and Dave Johnson for setting up the collaboration. This work was partly funded by the MEDEA+ project COSIP.

REFERENCES

[1] Ph. Galy et al; "Development and qualification of ESD protection demonstrators using SCR structure compatible with advanced CMOS technologies". IEW 2010.

[2] W. Schoenmaker, P. Meuris, J. Jimenez, Ph. Galy, "On the Inclusion of Lorentz force Effects in TCAD Simulations". proceeding of the 40th European Solid State Device Research Conference, Sept. 14-16, 2010, Sevilla, Spain, ESSDERC 2010

[3] Sze ; "Physics of Semiconductor Devices". John Wiley and sons 2nd edition 1981; pp 190-234

[4] Wim Schoenmaker, Peter Meuris, Alexander Steinmair and Walter Pflanzl "Evaluation of Electromagnetic Coupling Between Microelectronic Structures Using Computational Electrodynamics", Scientific Computing in Electrical Engineering SCEE 2008. Eds. Janne Roos and Luis Costa, Springer Verlag, Berlin Heidelberg 2010

Design of Low-Leakage Power-Rail ESD Clamp Circuit with MOM Capacitor and STSCR in a 65-nm CMOS Process

Po-Yen Chiu[1] and Ming-Dou Ker[1, 2]

[1] Institute of Electronics, National Chiao-Tung University, Hsinchu, Taiwan.
[2] Dept. of Electronic Engineering, I-Shou University, Kaohsiung, Taiwan.

Abstract — A power-rail electrostatic discharge (ESD) clamp circuit designed with low-leakage consideration has been proposed and verified in a 65-nm low-voltage CMOS process. By using the metal-oxide-metal (MOM) capacitor in the ESD-detection circuit, the power-rail ESD clamp circuit realized with only thin-oxide (1-V) devices has very low stand-by leakage current, as compared to the traditional design. The experimental results in the silicon chip showed that the stand-by leakage current is only 358 nA at room temperature (25 °C) under the power-supply voltage of 1 V, whereas the traditional design realized with the NMOS capacitor is as high as 828 μA under the same bias condition.

I. INTRODUCTION

In order to protect the internal circuits against electrostatic discharge (ESD) stresses, on-chip ESD protection circuits must be equipped at all I/O and power (V_{DD}/V_{SS}) pads. The typical design of on-chip ESD protection scheme in a CMOS IC is illustrated in Fig. 1. In such ESD protection scheme, the power-rail ESD clamp circuit is very important because it can significantly increase the overall ESD robustness of the IC chips [1]. With the turn-on efficient power-rail ESD clamp circuit between the V_{DD} and V_{SS} power lines, the internal circuits of CMOS ICs can be effectively protected against ESD damages. In the traditional design, the power-rail ESD clamp circuit was often realized with the RC-based ESD-detection circuit and one main ESD clamp device (M_{NESD}), as shown in Fig. 1.

With consideration of area efficiency and fabrication cost, the capacitor (C_1) in the ESD-detection circuit was often realized by the MOS capacitor, because MOS capacitors have the largest capacitance per unit area in the baseline CMOS processes. However, in nanoscale CMOS technology, the gate oxide thickness becomes thinner, which makes the gate-tunneling issue more serious and could not be ignored at all. The gate leakage issue had been studied and the gate-direct-tunneling mechanisms had been modeled for circuit simulation [2], [3]. With the increased gate leakage current in the nanoscale CMOS process, the traditional power-rail ESD clamp circuit with a MOS capacitor in the ESD-detection circuit will cause huge leakage issue in CMOS ICs. Some previous works had addressed this leakage issue with the MOS capacitors in the ESD-detection circuit and provided the modified designs to reduce the overall leakage current [4], [5]. However, the MOS capacitor in ESD-detection circuit always has some leakage current, because there is always a voltage drop across it under the normal circuit operating condition with V_{DD} bias. As a result, additional modified design had been proposed to decrease the leakage current by reducing the voltage drop across the MOS capacitor [6]. But those circuit techniques in the ESD-detection circuit would somehow decrease the trigger ability during the beginning of ESD-stress event. Thus, if the capacitor in the traditional RC-based ESD-detection circuit could be replaced by another capacitor which without the gate leakage issue, the traditional RC-based ESD detection circuit is still the most useful and convenient circuit in the ESD protection design.

In this work, a power-rail ESD clamp circuit with the metal-oxide-metal (MOM) capacitor of ultra low leakage is proposed and verified in a 65-nm CMOS process with the thin-gate oxide devices.

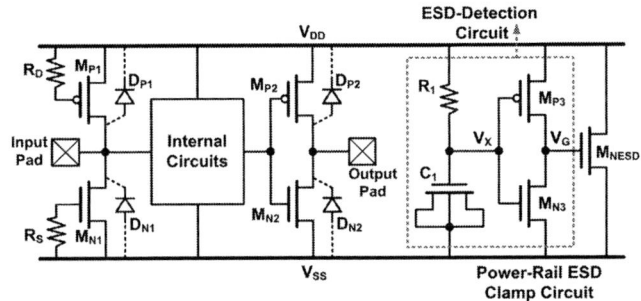

Fig. 1. Typical on-chip ESD protection scheme.

II. GATE-TUNNELING MECHANISMS AND IMPACTS ON TRADITIONAL POWER-RAIL ESD CLAMP CIRCUIT WITH MOS CAPACITOR

Three kinds of gate-tunneling mechanisms (ECB, EVB, and HVB) were reported to explain the gate leakage in CMOS technology [2], [3], as shown in Fig. 2. The ECB is the electron tunneling from the conduction band across the oxide barrier and it basically needs 3.1eV before the electron has the probability to tunnel across the oxide barrier. The EVB is the electron tunneling from the valence band across the oxide barrier and it basically needs 4.2eV before the electron has the probability to tunnel across the oxide barrier. The HVB is the hole tunneling from the valence band across the oxide barrier and basically needs 4.5eV before the hole has the probability to tunnel across the oxide barrier. When the gate-oxide thickness is scaled down, the tunneling carriers increase with a great proportion to cause the gate leakage current. In the nanoscale CMOS processes, the gate-oxide thickness in MOS devices is only a few nanometers and it has been reported to have large leakage current. Due to the gate leakage of the MOS capacitor (C_1) in the ESD-

978-1-4244-9019-6/11 $26.00 © 2011 IEEE

detection circuit, the PMOS (M_{P3}) in the ESD-detection circuit (shown in Fig. 1) cannot be fully turned off, which causes another leakage path through the inverter in the ESD-detection circuit under normal circuit operating conditions. If the ESD clamp device is realized by NMOS (M_{NESD}), the large size ESD clamp NMOS will leak more current because its gate voltage was not fully biased to V_{SS} under normal circuit operating conditions due to the gate leakage on the MOS capacitor (C_1).

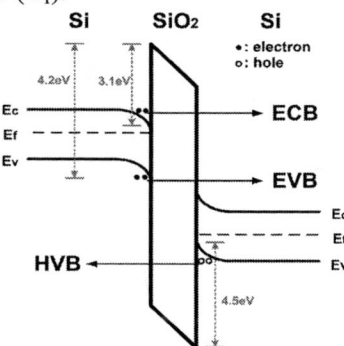

Fig. 2. Gate-tunneling mechanisms in MOS structure.

III. POWER-RAIL ESD CLAMP CIRCUIT WITH MOM CAPACITOR

MOM capacitors have been commonly used in IC design, because the MOM capacitor has higher linearity, higher quality factor (Q), small temperature variation, and almost no leakage current [7]. But in the early generation of CMOS processes, the capacitance density of MOM capacitor is very low, because the lateral and vertical intervals between metal layers are not close enough. As the result, the MOM capacitor would consume more chip area than MOS capacitor to achieve the required capacitance. However, when the dimensions keep shrinking in advanced CMOS processes, the capacitance density of MOM is increased significantly. Thus, to realize the ESD-detection circuit with MOM capacitor will not occupy large chip area. Therefore, the MOM capacitor used in the ESD-detection circuit can solve the leakage issue in the traditional design with MOS capacitor. The proposed low-leakage power-rail ESD clamp circuit with MOM capacitor (C_1) is shown in Fig. 3, which consists of the ESD-detection circuit with MOM capacitor and the substrate-triggered silicon-controlled rectifier (STSCR) as the ESD clamp device. Without the thin gate-oxide structure, SCR has very low leakage current under normal circuit operating conditions. Besides, SCR had been proven to have the highest ESD robustness under the smallest device size [8]. Moreover, SCR can be safely used without latchup danger in advanced CMOS technologies with low supply voltage. The power-rail ESD clamp circuit is designed to provide ESD current path between V_{DD} and V_{SS} during ESD stresses, and to be kept off under normal circuit operating conditions. To meet these requirements, the RC time constant in the ESD-detection circuit is designed to be about 0.1–1 μs to achieve the desired operations.

A. Operation Under Normal Circuit Operating Conditions

Under the normal V_{DD} power-on conditions, the V_{DD} power-on voltage waveform has a rise time in the order of millisecond (ms). With a slow rise time of the normal power-on transition, the voltage level at V_x can follow up the V_{DD} voltage waveform in time to keep M_{P1} off. Simultaneously, the M_{N1} is turned on because its gate terminal is connected to V_x. Therefore, no trigger current is injected into STSCR. As the result, STSCR can be kept off under normal circuit operating conditions. Fig. 4 shows the simulated transient waveforms of the ESD-detection circuit with MOM capacitor under normal power-on transition with a rise time of 0.1 ms. With the power-supply voltage of 1 V, the simulated overall leakage current of the power-rail ESD clamp circuit is only about 307 nA at 25 °C.

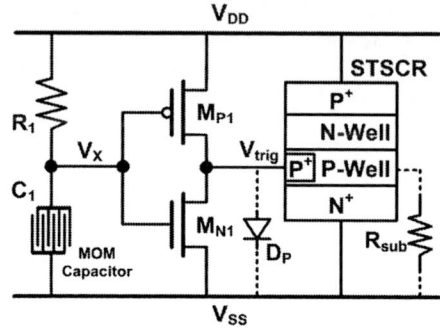

Fig. 3. Low-leakage power-rail ESD clamp circuit with MOM capacitor (C_1).

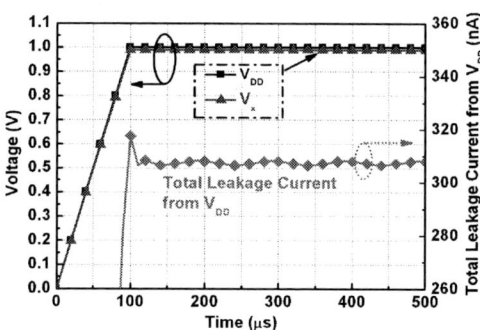

Fig. 4. Simulated transient waveforms of the ESD-detection circuit under normal power-on transition.

B. Operation Under ESD-Stress Conditions

Under the ESD-stress conditions, the ESD voltage has a rise time in the order of nanosecond (ns). The voltage level of V_X in Fig. 3 is increased very slower than the voltage level at V_{DD} power line when the ESD stress is conducted across V_{DD} and V_{SS} power lines. With the relatively lower voltage level kept at V_X due to the RC delay, the PMOS (M_{P1}) in the ESD-detection circuit will be turned on to inject trigger current into the V_{trig} of STSCR. Consequently, the turned-on STSCR provides a low-impedance path to discharge ESD current from V_{DD} to V_{SS}. Although the equivelant circuit model of STSCR device is needed to preciously simulate the quasi-static trigger point and the clamping voltage during high-current conditions, the P-well/N+ diode (D_P) and P-substrate resistor (R_{sub}) can be used to represent the STSCR device before it turned on. Thus, the trigger ability of ESD-detection circuit with D_P and R_{sub} can be simulated. Fig. 5 shows the simulated voltage waveforms and the trigger current of the ESD-detection circuit with MOM capacitor under ESD-like stress condition. When a 5-V voltage pulse with 10-ns rise time and 100-ns pulse width is applied to V_{DD}, which is used to simulate the rising edge of ESD event before device junction breakdown, the ESD-detection circuit can successfully inject the trigger current of ~40 mA to trigger on the STSCR.

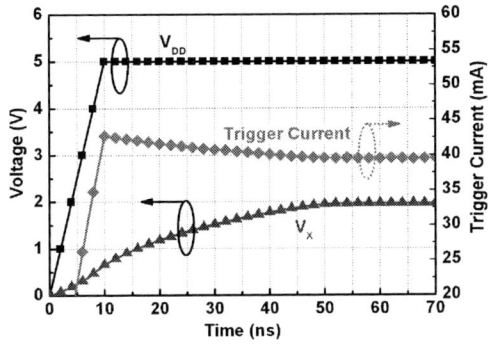

Fig. 5. Simulated transient waveforms of the ESD-detection circuit under ESD-like stress condition.

IV. EXPERIMENTAL RESULTS

The RC-based power-rail ESD clamp circuit with different capacitors have been drawn in the same test chip and fabricated in the same wafer by a 65-nm CMOS process. All devices used in this work are fully-silicided thin-oxide (1-V) devices without using the additional silicide-blocking mask. The device dimensions used in the power-rail ESD clamp circuits are listed in Table I, where these two circuits have the same RC time constant of 200 ns. To achieve 2-pF capacitance, the NMOS capacitor is drawn with 29 μm channel width (W_C) and 28 μm channel length (L_C) to occupy a total layout area of 37 μm x 34 μm. With the same capacitance, the MOM capacitor which realized with 3-layer metals can have similar layout area as compared to that of the MOS capacitor. Thus, each power-rail ESD clamp circuit occupies the same layout area of 45 μm × 75 μm in this work. The chip photograph of the fabricated power-rail ESD clamp circuits and test devices is shown in Fig. 6. Figs. 7(a) and 7(b) show the measured gate leakage currents of the PMOS and NMOS capacitors (W_C = 29 μm and L_C = 28 μm) with gate-oxide thickness of ~20 Å in a 65-nm CMOS low-voltage process, respectively. Under 1-V bias, the gate leakage current of PMOS capacitor (NMOS capacitor) at 25 °C is as high as 21 μA (51 μA). The leakage currents between two power-rail ESD clamp circuits at different temperatures are compared in Fig. 8 with V_{DD} of 1 V. Besides, the leakage currents at 25 °C and 125 °C are also listed in Table II. Comparing with the leakage current of the stand-alone NMOS capacitor, much higher leakage current is observed in

the power-rail ESD clamp circuit with NMOS capacitor (828 μA), which indicates that the leaky MOS capacitor actually causes other leakage path in the ESD-detection circuit. However, the leakage current of the low-leakage design with MOM capacitor has the lowest leakage current of only 358 nA. The leakage current of the design with MOM capacitor is three orders smaller than that with NMOS capacitor from low temperature to high temperature.

TABLE I. DEVICE DIMENSIONS IN POWER-RAIL ESD CLAMP CIRCUITS

Circuit Type	R1	Cap. Layout Area (W/L)	M_{P1} (W/L)	M_{N1} (W/L)	STSCR (W/L)
With NMOS Capacitor	100 kΩ	37 μm / 34 μm	100 μm / 0.15 μm	20 μm / 0.15 μm	40 μm / 7.8 μm
With MOM Capacitor	100 kΩ	39 μm / 36 μm	100 μm / 0.15 μm	20 μm / 0.15 μm	40 μm / 7.8 μm

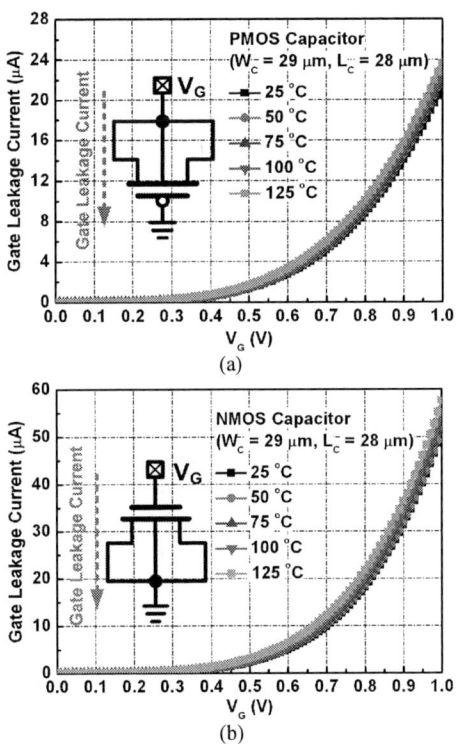

Fig. 6. Chip photograph of the fabricated power-rail ESD clamp circuits and test devices in a 65-nm CMOS process.

Fig. 7. Measured gate leakage current of the 65-nm (a) PMOS and (b) NMOS capacitor at different temperatures.

978-1-4244-9019-6/11 $26.00 © 2011 IEEE

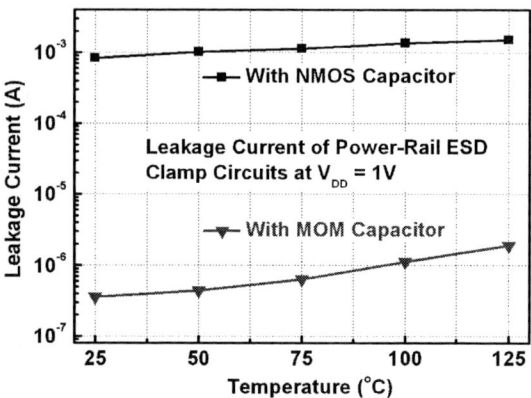

Fig. 8. Measured leakage currents between two fabricated power-rail ESD clamp circuits under different temperatures.

To investigate the turn-on behavior of the ESD clamp device with ESD-detection circuit during ESD stress event, the transmission line pulse (TLP) generator with 100-ns pulse width and 10-ns rise time was used to measure the second breakdown current (I_{t2}) of ESD protection circuits. The TLP-measured I-V characteristics of the STSCR with and without the ESD-detection circuit are shown in Fig. 9. Without any trigger current, the original trigger voltage (V_{t1}) of stand-alone STSCR (width = 40 μm) device is as high as 10.7 V, and the I_{t2} is 2.3 A. However, with the proposed ESD-detection circuit, the V_{t1} of the STSCR device is significantly reduced to 3 V and the I_{t2} is 2.5 A. Therefore, the lower V_{t1} of the power-rail ESD clamp circuit ensures its effective ESD protection capability. In addition, the holding voltage of SCR shown in Fig. 9 is ~2.5 V; therefore this proposed power-rail ESD clamp circuit is free to the latchup issue in the CMOS ICs with V_{DD} of 1V.

The human-body-model (HBM) and machine-model (MM) ESD levels of these two power-rail ESD clamp circuits are evaluated by the ESD simulator (ETS-910). These data are also listed in Table II. The failure criterion is defined as 30 % shift in the leakage current under 1-V V_{DD} bias. The power-rail ESD clamp circuit with SCR width of only 40 μm can achieve ESD robustness of 4 kV in HBM and 350 V in MM, respectively.

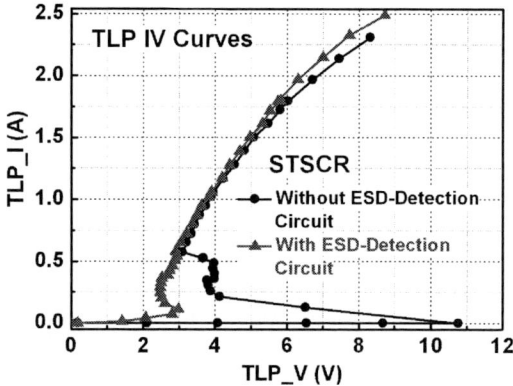

Fig.9. TLP-measured I-V characteristics of the STSCR with and without the proposed ESD-detection circuit.

TABLE II. Measured Leakage Currents and HBM/MM ESD Robustness Between the Fabricated Power-Rail ESD Clamp Circuits

Circuit Type	Leakage Current at V_{DD} = 1V		ESD Robustness	
	25 °C	125 °C	HBM	MM
With NMOS Capacitor	828 μA	1.53 mA	4 kV	350 V
With MOM Capacitor	358 nA	1.91 μA	4 kV	350 V

V. CONCLUSION

A power-rail ESD clamp circuit designed with the consideration of gate-leakage issue has been proposed and successfully verified in a 65-nm low-voltage CMOS process. By using the MOM capacitor in the ESD-detection circuit, the power-rail ESD clamp circuit can achieve low leakage current and keep good ESD protection ability. Besides, comparing with the MOS capacitor, the MOM capacitor does not consume more layout area. The power-rail ESD clamp circuit with MOM capacitor is suitable for on-chip ESD protection design in advanced nanoscale CMOS processes.

ACKNOWLEDGEMENT

The authors would like to thank TSMC University Shuttle Program for providing chip fabrication. This work was supported in part by Ministry of Economic Affairs, Taiwan, R.O.C., under Grant 99-EC-17-A-01-S1-104; and in part by the "Aim for the Top University Plan" of National Chiao-Tung University and Ministry of Education, Taiwan, R.O.C.

REFERENCES

[1] M.-D. Ker, "Whole-chip ESD protection design with efficient VDD-to-VSS ESD clamp circuit for submicron CMOS VLSI," *IEEE Trans. Electron Devices*, vol. 46, no.1, pp. 173-183, Jan. 1999.

[2] K. Cao, W. Lee, W. Liu, X. Jin, P. Su, S. Fung, J. An, B. Yu, and C. Hu, "BSIM4 gate leakage model including source-drain partition," in *IEDM Tech. Dig.*, 2000, pp. 815-818.

[3] W. Lee and C. Hu, "Modeling CMOS tunneling currents through ultrathin gate oxide due to conduction- and valence-band electron tunneling," *IEEE Trans. Electron Devices*, vol. 48, no. 7, pp. 1366-1373, Jul. 2001.

[4] S. Poon and T. Maloney, "New considerations for MOSFET power clamps," in *Proc. EOS/ESD Symp.*, 2002, pp. 1-5.

[5] J. Smith, R. Cline, and G. Boselli, "A low leakage low cost PMOS-based power supply clamp with active feedback for ESD protection in 65-nm CMOS technologies, " in *Proc. EOS/ESD Symp.*, 2005, pp. 298-306.

[6] M.-D. Ker, P.-Y. Chiu, F.-Y. Tsai, and Y.-J. Chang, "On the design of power-rail ESD clamp circuit with consideration of gate leakage current in 65-nm low-voltage CMOS process," in *Proc. IEEE Int. Symp. Circuits Syst.*, 2009, pp. 2281-2284.

[7] H. Samavati, A. Hajimiri, A. Shahani, G. Nasserbakht, and T. Lee, "Fractal capacitors," *IEEE J. Solid-State Circuits*, vol. 33, no. 2, pp. 2035-2041, Dec. 1998.

[8] M.-D. Ker and K.-C. Hsu, "Overview of on-chip electrostatic discharge protection design with SCR-based devices in CMOS integrated circuits," *IEEE Trans. Device and Materials Reliability*, vol. 5, no. 2, pp. 235-249, Jun. 2005.

978-1-4244-9019-6/11 $26.00 © 2011 IEEE

Low power UTBOX and Back Plane (BP) FDSOI technology for 32nm node and below

C. Fenouillet-Beranger, P. Perreau, L. Tosti, O. Thomas, J-P. Noel, T. Benoist, O. Weber, F. Andrieu,
A. Bajolet, S. Haendler, M. Cassé, X. Garros, K.K. Bourdelle, F. Boedt, O. Faynot, F. Boeuf

Abstract— **This paper highlights the interest of FD-SOI with high-k and metal gate as a possible candidate for low power multimedia technology. The possibility of multi-V_T by combining UTBOX with back plane, back biasing, variable TiN thickness and Al_2O_3 in the gate stack is demonstrated. The viability of these approaches is corroborated via mobility and reliability measurements. Dual gate oxide co-integrated devices are reported. The effectiveness of back biasing for short devices is demonstrated through ring oscillators and 0.299µm² SRAM bitcells performance reflecting that the conventional bulk reverse and forward back biasing approaches to manage the circuit static power and the dynamic performances are fully compatible with FDSOI. Finally, thanks to a hybrid FDSOI/bulk co-integration with UTBOX all IP's required in a SOC could be demonstrated for LP applications.**

Index Terms—**MOS devices, Silicon on insulator technology, Thin film devices, High-k gate dielectrics**

I. INTRODUCTION

Thin film devices (FDSOI) are among the most promising candidates for next device generations due to their better immunity to short channel effects (SCE). In addition, the introduction of high-k and metal gate has greatly improved the MOSFETs performance by reducing the CET and gate leakage current. Furthermore, the combination of metal gate with undoped channel greatly improves the variability. Indeed values as low as 1mV.µm [1] have been reported, significantly better than values obtained for typical 45nm LSTP bulk technology 4.7mV.µm [2]. This last point is very important since using devices with a lower variability allows a proper scaling of 6T-SRAM bit-cell layout. However, if midgap metal gate with undoped channel is sufficient to provide a high symmetrical threshold voltage ($V_T \sim 0.45V$) for both NMOS and PMOS devices [3], still one major challenge is to provide Multi-V_T devices and dynamic V_T modulation with an undoped channel in order to satisfy the low power (LP) circuit design requirements [4-6]. To overcome this issue, a simple and novel Multi-V_T strategy has been proposed in [4], combining UTBOX substrate with different back plane (BP) and back biasing (Vb). This method is very attractive thanks to the possibility to re-use the bulk forward (FBB)/reverse (RBB) biasing techniques. However the FBB biasing in order to realize low V_T implies a disruptive circuits design to avoid forward diode biasing in the substrate between the two opposite BP type beneath the BOX [7]. In order to introduce more V_T modulation flexibilities and especially for low V_T (LVT) PMOS, aluminum oxide (Al_2O_3) inserted in TiN gate stack has been proposed for bulk devices [8-9] in a gate first process. In addition to multiple-V_T enablement, analog devices and IP's not fully compatible with thin film devices should be demonstrated in order to integrate a full LP multimedia platform. In this paper we demonstrate experimentally the ability of FDSOI technology for achieving multiple V_T devices. This concept is demonstrated through silicon results across wafers in 45nm technology. In addition I/Os integration is demonstrated. Results are corroborated by reliability, ring oscillator and 0.299µm² 6T SRAM bitcells measurements. Finally, a hybrid FDSOI/bulk co-integration with UTBOX covering all IP's required in a SOC for multimedia LP applications is presented.

II. MULTI-V_T APPROACH

A. Process scheme

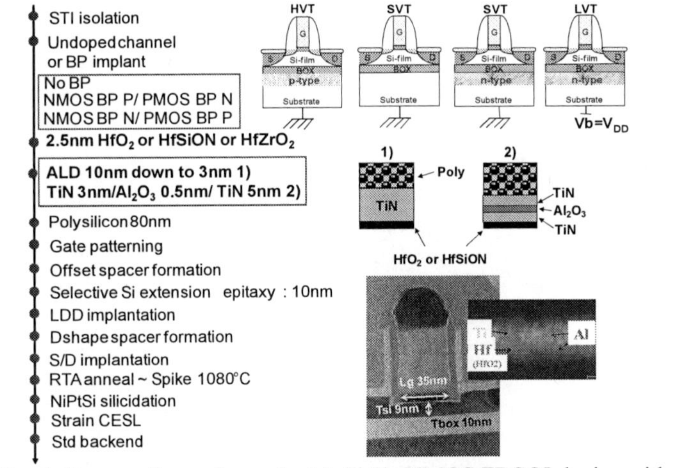

Fig. 1. Process flow scheme for Multi-V_T NMOS FDSOI device with UTBOX. TEM Cross section of a FDSOI device (Lg 35nm).

C. Fenouillet-Beranger & P. Perreau are CEA/LETI-Minatec assignees at STMicroelectronics, Crolles, 38926, France (Phone: 33438923680; e-mail: claire.fenouillet-beranger@st.com). L.Tosti, O. Thomas, J-P. Noel, O. Weber, F. Andrieu, M. Cassé, X. Garros are with CEA/LETI-Minatec, 17 rue des Martyrs, 38054, Grenoble, France. F. Bœuf, A. Bajolet, S. Haendler and T. Benoist are with STMicroelectronics, 850 rue Jean Monnet, 38926, Crolles, France. K.K. Bourdelle & F. Boedt are with SOITEC, Parc Technologique des Fontaines, Bernin, 38926, France.

This work was partially carried out in the frame of the ST/LETI/IBM joint program and partially in the MEDEA DECISIF project founded by the French Ministery of industry, Economy and Finance.

978-1-4244-9019-6/11 $26.00 © 2011 IEEE

The FDSOI devices were processed on 300mm (100) UNIBOND™ SOI wafers with a buried oxide thickness of 10nm, 25nm or 145nm via the process flow scheme presented in Fig.1. The final silicon film thickness under the gate is 8-9nm, the nominal gate length is 35nm with EOT depending on gate stack used (EOT between 1.4-2.9nm). In order to reduce short channel effects as compared to thick BOX devices, UTBOX is combined with a back plane (Fig.2) [10].

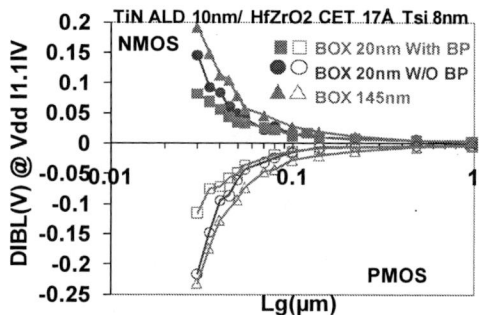

Fig. 2. Comparison of the DIBL for NMOS and PMOS devices for wafers with thin BOX with and without BP and thick BOX

B. Threshold voltage modulation

Two approaches can be used in order to reduce the threshold voltage. The first one consists in the combination of UTBOX with BP and back biasing. The LVT options are based on an n-type BP and p-type BP set to Vdd for NMOS and PMOS respectively. Fig.3 shows that for a BOX thickness of 10nm coupled with BP for LVT, the V_T reduction with a back bias varying between 0V and Vdd is around 208mV/V.

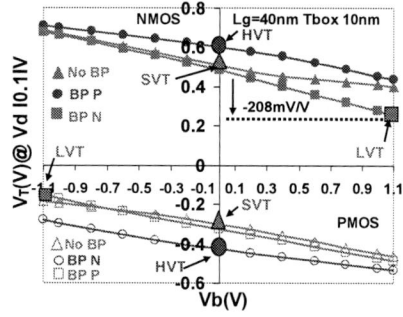

Fig. 3. V_T variation with Vb on NMOS and PMOS devices for Lg 40nm for each BP type [11].

The second approach requires playing with the gate stack materials and thickness. In case of low V_T NMOS, variable TiN thickness to modulate threshold voltage can be used [11-13]. By decreasing the ALD TiN metal thickness from 10nm down to 3nm a V_T reduction of 150mV is obtained on the NMOS transistor (Fig.4).

In addition, similar DIBL and SS are obtained for the different variants [6;11;16]. Furthermore we have demonstrated previously that by reducing the ALD TiN thickness from 10nm down to 3nm the Ion current of the NMOS devices is greatly improved [6]. Regarding the low V_T of the PMOS, the Al_2O_3 material has already been proposed for bulk devices [8-9].

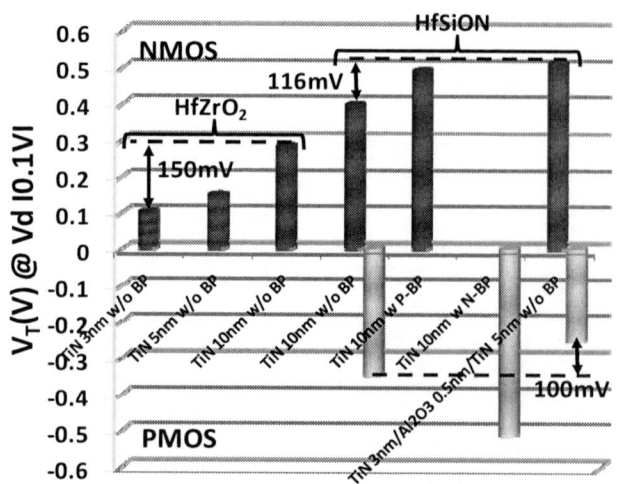

Fig. 4. Linear V_T values versus ALD TiN thickness with or w/o Al_2O_3, with and w/o BP for FDSOI NMOS and PMOS devices (Lg 35nm).

This concept is applied on FDSOI devices where a work-function at only 100mV from the midgap is required as compared to bulk [14]. HfSiON/TiN offers a quasi midgap gate stack. The combination of HfSiON of 2.5nm thickness with a 0.5nm of Al_2O_3 inserted in a sandwich of TiN 3nm and 5nm (see Fig.4) yields a V_T reduction of 100mV and a V_T increase of 116mV for PMOS and NMOS respectively.

Regarding the high V_T (HVT), for NMOS, the UTBOX is combined with a p-type BP set to 0V. By this way the V_T value is increased by around 100mV. The standard V_T (SVT) option does not include BP and the back bias (Vb) is set to 0V. Another alternative for the NMOS SVT is to use an n-type BP with Vb set to 0V (Fig.3). In this case, the well implant to connect the BP should be carefully adapted to avoid forward PN junctions in the substrate under the BOX [15]. For the PMOS device, complementary BP doping type and biasing are applied [4]. Table 1, summarizes the different V_T options proposed.

	HVT (Vb=0)	SVT (Vb=0V)	LVT
NMOS	Midgap gate + BP-P	o Midgap gate w/o BP o Midgap gate + BP-N o Thin metal + BP-P	o Midgap gate with BP-N @ Vb=Vdd o Thin metal + BP-N @ Vb=0V
PMOS	Midgap gate + BP-N	o Midgap gate w/o BP o Midgap gate + BP-P o TiN/Al2O3/TiN + BP-N	o Midgap gate with BP-P @ Vb=Vdd o TiN/Al2O3/TiN + BP-P @ Vb=0V

Table 1. Multi-V_T possibilities

III. MOBILITY AND RELIABILITY

In order to investigate the electron and hole transport behaviour in FDSOI, we plotted the maximum linear transconductance Gm_{max} versus Vg for NMOS and PMOS long devices (Fig.5). Moderate mobility degradation up to 20% on PMOS is seen between devices with and without BP for 10nm of BOX. The introduction of Al_2O_3 in the gate stack reduces the hole mobility by around 14%. A Gm_{max} degradation is observed on NMOS by adding a BP (up to 17% due to indium species). Regarding the devices performance versus TiN thickness, the performance are improved as the TiN is thinner [6] (not seen here).

978-1-4244-9019-6/11 $26.00 © 2011 IEEE

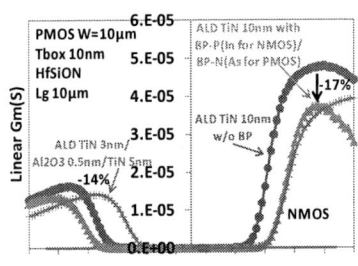
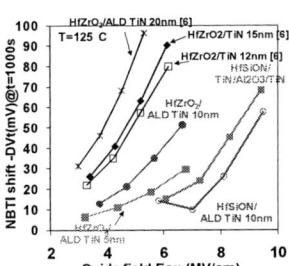

Fig. 5. NMOS and PMOS linear Gm$_{max}$(Vg) curves for devices with BOX 10nm with and w/o BP with and without Al$_2$O$_3$.

Fig. 6. NBTI variation for FDSOI devices with variable metal thickness or with Al$_2$O$_3$ inserted in TiN layers.

NBTI data (Fig.6) show that the thin ALD layers are preferable. However, the intermediate TiN thickness of 10nm combined with HfSiON high-k presents very good NBTI values. In addition, when Al is incorporated inside the TiN gate it allows us to achieve a good trade-off NBTI reliability/device performance [16].

Considering PBTI, it is generally higher for the Al$_2$O$_3$ variants but these values are very good whatever the variants presented in this study [6;16].

IV. I/Os INTEGRATION

In a CMOS platform, FDSOI with high-k/Metal gate stack must be co-integrated with I/Os. In this work, a thick SiON (30Å) is deposited before 25Å of HfO$_2$ (EOT 29Å) at V$_{dd}$=1.8V. Analog gain (g$_m$/g$_d$) for core device and I/O plotted in Figs 7 (EOT 1.9nm) and 8 (EOT 2.9nm) reveals a very good agreement with bulk 45nm devices with poly/SiON (30Å) gate stack [6].

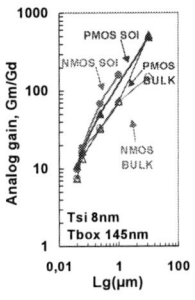
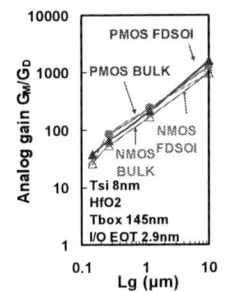

Fig.7. 19Å-24Å/Vdd1.1V nMOS/ pMOS core device analog performances. (FDSOI and Bulk)

Fig.8. 29Å/Vdd1.8V nMOS / pMOS I/O analog performances. (FDSOI and Bulk)

V. BACK BIAS IMPACT ON RO & SRAM

Functional ring oscillators (RO) have been measured with midgap gate with the BP options proposed in table 1. The effectiveness of the FBB and RBB techniques on our multi-V$_T$ strategy has been tested on 73 stages FO1 RO with Lg=40nm, Wn=0.36µm and Wp=0.5µm [11]. In the device analysis, the PMOS reference source voltage is 0V, here, in RO and SRAM cell, the source is pinned to Vdd. Therefore, for Vb between 0V and Vdd, NMOS and PMOS are in forward biasing (as seen on Fig.9). The Vb applied on the chuck is common for NMOS and PMOS. Fig.9 and 10 give the time propagation

delay (Tp) and the static power per gate (Pstat/gate) versus Vb, highlighting their improvement with V$_b$ increase. Thus, the SVT (No BP) and LVT (BP-N NMOS/BP-P PMOS) options boost the Tp by 22% compared to the HVT (BP-N NMOS/BP-P PMOS) option with 200pA of leakage current per stage.

The HVT (BP-P NMOS/BP-N PMOS) option enables to improve Pstat by around 2 decades (Fig. 10).

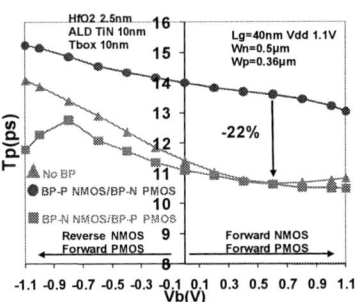

Fig. 9. Tp(Vb) for FO1 RO for each V$_T$ option @ Vdd 1.1V

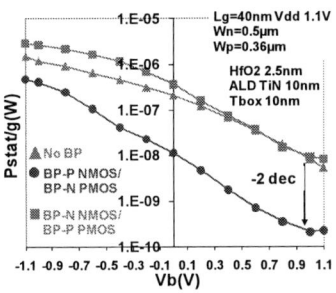

Fig. 10. Pstat/gate(Vb) for FO1 RO for each V$_T$ option @ Vdd 1.1V

The same study is done on the 45nm node 0.299µm^2 6T SRAM bitcells. A TEM picture of the 0.299µm^2 at process end and the layout of the cell are shown on Fig.11.

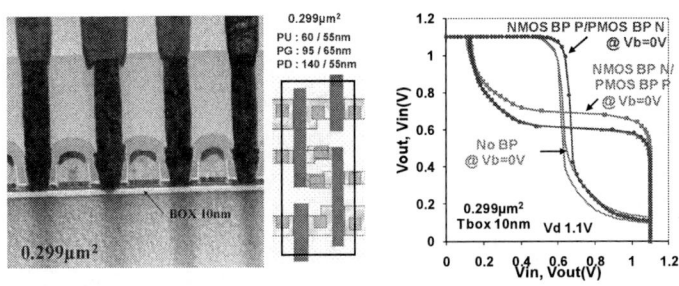

Fig. 11. TEM cross section in a 0.299µm^2 FDSOI high-K/MG UTBOX 6TSRAM cells and 0.299µm^2 SRAM cell butterfly curves for each VT option for Vb=0V

Fig. 11 (right) illustrates the butterfly curves for the three V$_T$ options at Vb=0V. Very good Static-Noise-Margin (SNM) value is obtained (~290mV @ Vb=0V) (Fig.12).

For a fixed SNM the WM is 12% higher for the NMOS LVT option compared to the HVT one. All of these experimental results reflect the effectiveness of the back bias and BP doping type to improve the SRAM bitcell specifications.

978-1-4244-9019-6/11 $26.00 © 2011 IEEE

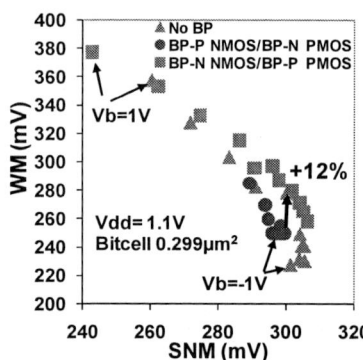

Fig. 12. SNM versus WM with variable Vb for each V_T option @Vdd 1.1V

VI. HYBRID FDSOI/BULK INTEGRATION

In ultra-thin SOI technologies, the thin silicon film and buried oxide degrade the performance of ESD protections. Table 2 shows experimental data of FDSOI gated diodes for various device widths (number of fingers x unitary width). By comparing the breakdown currents It2 measured for bulk and FDSOI devices with the same geometry, we observe that the FDSOI It2 is around four times lower due to the poor thermal dissipation of the BOX and the lower conductivity (1/Ron) of the ultra-thin silicon film [17].

	Bulk without HKMG		FDSOI with HKMG	
width	Ron(Ω)	It2 (mA/um)	Ron(Ω)	It2 (mA/um)
10 x (2.5 um)	1.78	12	16	2.8
10 x (5 um)	1.08	11	7.8	2.5
10 x (10 um)	0.72	11	4.2	2.4
40 x (5 um)	0.32	10	2.7	2.2

Table 2: Robustness (breakdown current It2) and On Resistance (Ron) for FDSOI versus bulk gated diodes with different number of fingers and widths (W= 10x 2,5um; 10x 5um; 10x 10um; 40x 5um) [17].

To encounter this problem, a hybrid co-integration with bulk is proposed.

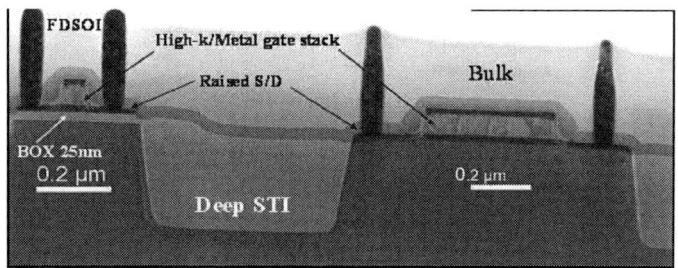

Fig. 13. TEM cross section of the hybrid FDSOI/bulk co-integration in a SRAM cut periphery. The BOX thickness is 25nm [6]

Our approach with ultra-thin BOX is in total agreement with our high and low V_T strategies and consists in integrating I/O's and ESD protections within the substrate region underneath the BOX layer by removing Si film and BOX layer. A small step (<400Å) between FDSOI and bulk is then achieved. The TEM picture (Fig.13) in a SRAM cut periphery shows the FDSOI and Bulk co-integrated devices.

VII. CONCLUSION

This paper reviews the main advantages of FDSOI devices in order to be integrated in a LP multimedia platform. This technology offers multi-V_T availability, good performances in terms of variability, mobility, reliability, dynamic and static functionalities. In addition, the proposed hybrid approach for IP's not fully compatible with ultra-thin film makes this technology one of the best candidates for the future technological nodes.

ACKNOWLEDGMENT

The authors would like to thank, S. Barnola, R. Beneyton, F. Abbate, Y. Campidelli, L. Pinzelli, C. de Buttet, P. Gouraud, C. Borowiak, F. Martin, R. Gassilloud for device fabrication, B.Y. Nguyen and T. Skotnicki for managerial support.

REFERENCES

[1] O. Weber et al, *"High immunity to threshold voltage variability in undoped ultra-thin SOI MOSFETs and its physical understanding"*, IEEE IEDM 2008, pp. 245-248

[2] E. Josse et al, *"Cost effective low power platform for 45nm technology node"*, IEDM 2006, pp. 1-4.

[3] C. Fenouillet-Beranger et al, *"Fully-Depleted SOI Technology using High-K and Single-Metal Gate for32nm Node LSTP Applications featuring 0.179μm² 6T-SRAM bitcell"*, IEEE IEDM 2007, pp. 267-271.

[4] J-P.Noel et al, *"A simple and efficient concept for setting up Multi-V_T devices in thin BOX Fully-Depleted SOI Technology"*, IEEE ESSDERC 2009, pp. 137-140.

[5] Y. Morita et al, *"Smallest Vth Variability Achieved by Intrinsic Silicon on Thin BOX (SOTB) CMOS with Single Metal Gate"*, IEEE VLSI 2008, pp. 166-167.

[6] C. Fenouillet-Beranger et al, *"Hybrid FDSOI/Bulk High-k/Metal gate platform for Low Power (LP) multimedia technology"*, IEEE IEDM 2009, pp. 667-670.

[7] J-P. Noel et al, *"UT2B-FDSOI Device Architecture Dedicated to Low Power Design Techniques"*, IEEE ESSDERC 2010, pp. 210-213.

[8] B.P Linder et al, *"Gate First PFET Poly-Si/TiN/Al2O3 Gate Stacks with Inversion Thicknesses Less than 15Å for High Performance or Low Power CMOS Applications"*, SSDM 2007, pp. 16-17.

[9] T. Morooka et al, *"Vt Variation Controlled Al2O3 Capped HfO2 Gate Dielectrics for Low Vt pMISFETs with High-k/Metal Gate Stacks"*, SSDM 2008, pp. 24-25.

[10] C. Fenouillet-Beranger et al, *"FDSOI devices with thin BOX and ground plane integration for 32nm node and below"*, IEEE ESSDERC 2008, pp. 206-209.

[11] C. Fenouillet-Beranger et al, *"Efficient Multi-VT FDSOI technology with UTBOX for low power circuit design"*, IEEE VLSI 2010, pp. 65-66.

[12] K. Choi et al, *"The effect of Metal Thickness, Overlayer and high-k surface treatment on the Effective Work Function of Metal Electrode"*, IEEE ESSDERC 2005, pp. 101-104.

[13] R. Singanamalla et al, *"On the impact of TiN film variation on the effective work-function of Poly-Si/TiN/SiO2 and Poly-Si/TiN/HfSiON gate stacks"*, IEEE EDL, vol 27, n°5, May 2006, pp. 332-334.

[14] O. Weber et al, *"Work-function engineering in gate first technology for multi-V_T dual gate FDSOI CMOS on UTBOX"*, IEEE IEDM 2010, pp. 341-344.

[15] J-P. Noel et al, *"UT2B-FDSOI device architecture dedicated to low power design"*, IEEE ESSDERC 2010, pp. 210-213.

[16] C. Fenouillet-Beranger et al, *" UTBOX and Ground Plane combined with Al2O3 inserted in TiN gate for V_T modulation in Fully-depleted SOI CMOS transistors"*, accepted to IEEE VLSI-TSA 2011.

[17] T. Benoist et al, *"Improved ESD protection in advanced FDSOI by using hybrid SOI/bulk co-integration"*, EOS/ESD symposium, 2010, pp. 1-6.

978-1-4244-9019-6/11 $26.00 © 2011 IEEE

Electrical Characteristic Fluctuation of 16 nm MOSFETs Induced by Random Dopants and Interface Traps

Hui-Wen Cheng, Student *Member, IEEE*, Yung-Yueh Chiu, Student *Member, IEEE*, and Yiming Li, *Member, IEEE*

Abstract—**In this paper, we estimate the influences of random dopants (RDs) and interface traps (ITs) using experimentally calibrated 3D device simulation on electrical characteristics of high-κ / metal gate CMOS devices. Statistically random devices with 2D ITs between the interface of silicon and HfO₂ film as well as 3D RDs inside the device channel are simulated. Fluctuations of threshold voltage and on-/off-state current for devices with different effective oxide thickness of insulator film are analyzed and discussed. The engineering findings significantly indicate that RDs and ITs govern characteristics, respectively, are statistically correlate to each other and RDs dominate device's variability, compared with the influence of ITs; however, the influence degree varies with IT's number, density and position. The effect of RDs and ITs on device characteristic should be considered together properly. Notably, the position of ITs and RDs results in very different fluctuation in spite of the same number of ITs and RDs.**

Index Terms—**Interface trap, random dopant, fluctuation, threshold voltage fluctuation, high-κ/metal gate, MOSFET, 16 nm gate, 2D interface traps, 3D device simulation, random position, random number, random energy**

I. INTRODUCTION

THE electrical characteristic variability of nanometer scale complementary metal-oxide-semiconductor (CMOS) device increases as feature size reduces, for example, the intrinsic device parameter fluctuations that result from line edge roughness [1,2], the granularity of the polysilicon gate [3-4], random discrete dopants [5-13] effects have substantially affected the device characteristics. With device scaling, various randomness effects resulting from the random nature of manufacturing process have induced significant fluctuations of electrical characteristics in nanoscale MOSFETs. The number of dopants is of the order of tens in the depletion region in a nanoscale MOSFET, whose influence on device characteristic

Manuscript received February 28, 2011. This work was supported in part by National Science Council (NSC), Taiwan under Contract No. NSC-99-2221-E-009-175 and by TSMC, Hsinchu, Taiwan under a 2010-2011 grant.

Hui-Wen Cheng and Yung-Yueh Chiu are with the Institute of Communications Engineering, National Chiao-Tung University, Hsinchu 300 Taiwan. Yiming Li is with the Department of Electrical Engineering and Institute of Communications Engineering, National Chiao-Tung University, Hsinchu 300 Taiwan. (phone: 886-3-5712121 ext. 52974; fax: 886-3-5726639; e-mail: ymli@faculty.nctu.edu.tw).

is large enough to be distinct. Various RDs effects have been recently studied in both experimental and theoretical approaches [5-14]. Fluctuations of characteristics are caused not only by a variation in an average doping density, which is associated with a fluctuation in the number of impurities, but also with a particular random distribution of impurities in the channel region. The randomness of the dopant position and number in device makes the fluctuation of device characteristics have been presented [9]. The RD-induced threshold voltage fluctuation (σV_{th}) up to 40 mV for 20 nm planar CMOS has been experimentally quantified [9]. Therefore, RD fluctuation (RDF) has been recognized as one of the major limitations in device scaling; recently, high-κ / metal gate (HKMG) plays a key technology to reduced intrinsic parameter fluctuation, and leakage current for sub 45 nm generations [9,14]. However, the generation of ITs on the interface of silicon and high-κ introduces a new source of fluctuation for the degradation of device characteristics. Recent studies concerning ITs on planar CMOS technology have been reported [15-24]. The impact of interface states on MOS transistor have been discussed [15], but it only considers 1D channel surface based on 2D device simulation. Effects of ITs and combination of RDs and ITs on characteristic fluctuation for nanoscale CMOS devices with respect to effective oxide thickness (EOT) on 3D device simulation have not been explored yet.

In this work, we investigate the distributions of ITs and RDs induced characteristic fluctuation of 16 nm CMOS devices using an experimentally calibrated 3D device simulation. 2D ITs on the interface of silicon and HfO₂ film and 3D RDs inside the silicon channel are simultaneously considered in the large scale device simulation. New findings of this study on σV_{th} and variability of I_{on}/I_{off} versus the number of ITs and/or RDs are discussed.

II. SIMULATION METHODOLOGY

The validated performance of studied HKMG device, according to ITRS roadmap for low operating power, is experimentally quantified in our recent study [1]. Note that the threshold voltage of 16-nm N-MOSFET is equal to 250 mV.

978-1-4244-9019-6/11 $26.00 © 2011 IEEE

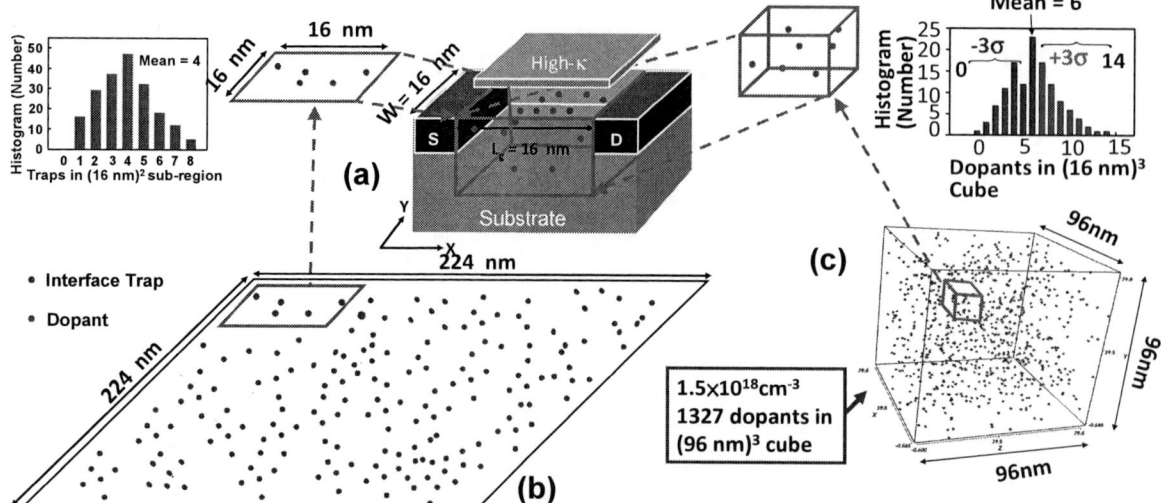

Figure 1. (a) The source of randomness (brown dots are interface traps and blue dots are discrete dopants) and simulation settings for fluctuations of random ITs and RDs. (b) We first generate 753 acceptor-like traps in a large plane, where the trap's concentration in the plane is around 1.5×10^{12} cm^{-2} and the total number of generated traps follow the Poisson distribution. The energy of each trap on the plane is assigned according to distribution of trap's density. Then the entire plane is partitioned into sub-planes (size: 16 nm×16 nm), where the number of traps in all sub-planes may vary from 1 to 8 and the average number is 4. (c) Discrete dopants randomly distributed in (96 nm)^3 cube with the average concentration of 1.5×10^{18} cm^{-3}. There will be 1,327 dopants within the cube and dopants vary from 0 to 14 (the average number is 6) for all 216 sub-cubes. The size of each sub-cube is (16 nm)^3. The total sub-planes and sub-cubes are then mapped into device's 3D channel and 2D surface for ITs' and RDs' position/number-sensitive simulation (a). The settings are similar for P-MOSFET devices.

The devices we tested are the 16 nm planar MOSFETs (width: 16 nm) with amorphous-based TiN/HfO$_2$ gate stacks; an EOT of 0.8 nm and 1.2 nm are shown in Fig. 1(a). For the simulation of IT fluctuation (ITs) over 2D channel surface, based upon device dimension, we first generate 753 acceptor-like traps in a large 2D plane as shown in Fig. 1(b), where the trap's concentration in the large plane is around 1.5×10^{12} cm^{-2} based upon experimental characterization, and the total number of generated traps mainly follows the Poisson distribution. Then, the statistically generated large plane is partitioned into many sub-planes, where the number of traps in the sub-planes may vary from 1 to 8 and the average is 4. The energy of each trap on each sub-plane is assigned according to the distribution of trap density [15-19]. We explore the density of ITs varying from 5×10^{11} to 5×10^{12} cm^{-2}. We repeat this process until all sub-regions are assigned.

Second, RDs in 3D channel region are statistically generated, as shown in Fig. 1(c), which are also incorporated into a 3D device simulation program and performed on our parallel computing system. The detail of RDF simulation technique was reported in our previous works [9,13,25]. Therefore, about 200 statistically generated samples are numerically simulated using the 3D device simulation to estimate the ITs and RDs induced characteristic fluctuation in 16 nm MOSFETs.

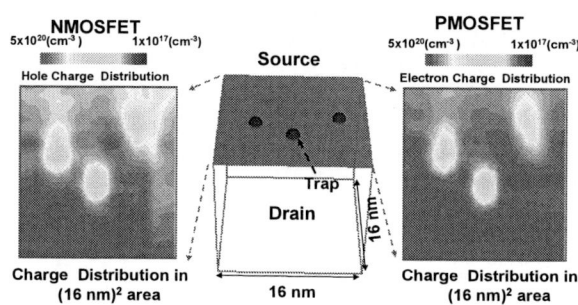

Figure 3. The on-state ($V_G = 0.8$ V and $V_D = 0.8$ V) current density of the channel surface extracted from one of about 200 simulated ITs fluctuated transistors for N- and P-MOSFET devices, where EOT = 0.8 nm. The potential induced by traps between the interface of Si and high-κ. Notably, the barrier is affected by ITF clearly shown.

III. RESULTS AND DISCUSSION

Figures 2 and 3 present the on-state ($V_G = 0.8$ V and $V_D = 0.8$ V) charge distribution and current density of the channel surface extracted from one of about 200 simulated ITs fluctuated transistors of N- and P-MOSFET devices, where they are fluctuated by random ITs with EOT of 0.8 nm. The charge distribution and current density is strongly governed by different position of ITs. Figure 4 shows the potential distribution of the channel surface from one of simulated 16 nm

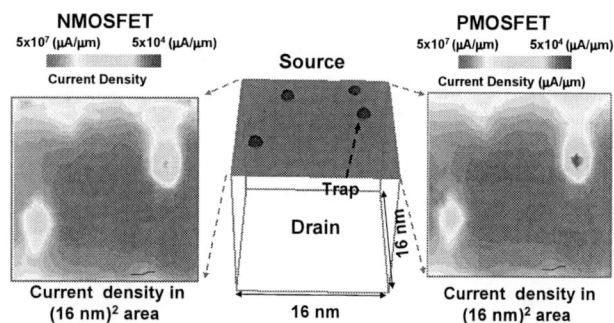

Figure 2. The on-state ($V_G = 0.8$ V and $V_D = 0.8$ V) charge distribution of the channel surface extracted from one of about 200 simulated ITs fluctuated transistors for N- and P-MOSFET devices, where EOT = 0.8 nm. The charge distribution is strongly governed by different position of ITs is clearly shown.

978-1-4244-9019-6/11 $26.00 © 2011 IEEE

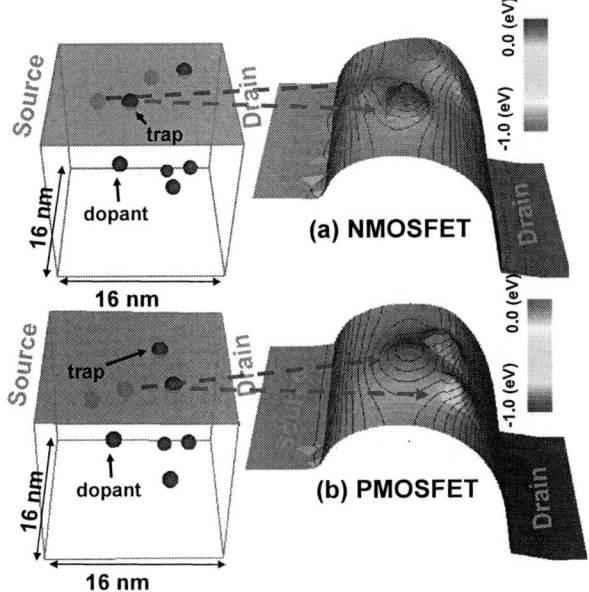

Figure 4. The potential distribution of the channel surface from one of simulated 16 nm transistors, where EOT = 0.8 nm. The device fluctuated by 8 random "RDs+ITs" (i.e., 2 random ITs at Si/High-κ oxide interface, 6 RDs locating inside the silicon channel below the channel surface) simultaneously for N- and P-MOSFET devices. The interaction effect of "RDs+ITs" on the band profile is clearly shown.

transistors, where EOT of 0.8 nm. The device fluctuated by 8 random "RDs+ITs" (i.e., 2 random ITs at Si/High-κ oxide interface, 6 RDs locating inside the silicon channel below the channel surface) simultaneously for N- and P-MOSFETs device. The interaction effect of "RDs+ITs" on the band profile is clearly shown. The totally random ITs, RDs and the combination of ITs and RDs ("RDs+ITs") fluctuated I_{on}-I_{off} characteristics simultaneously for N- and P-MOSFETs with EOT of 0.8 nm and 1.2 nm is shown in Figs. 5((a)and(a')-5(c)and(c')). Figures 5(a), 5(a'), 5(b) and 5(b') show the characteristics of I_{off} versus I_{on} resulting from individual RDF and ITF for N- and P-MOSFETs device with EOT of 0.8 nm and 1.2 nm, respectively. The result clearly shows that I_{on}-I_{off} fluctuation induced by RDs is significantly. The plot of I_{on}-I_{off} characteristics for "RDs+ITs"-induced fluctuations is shown in Figs. 5(c) and 5(c'). The result shows that the device with EOT of 1.2 nm possesses sizeable RDF and ITF due to the weakened metal gate controllability. The large scale statistically simulated threshold voltage as a function of random trap number for the 16-nm N- and P-MOSFET devices are shown in Figs. 6(a) and 6(b). Compared with ITs' number induced σV_{th}, the results imply that RDF influence σV_{th} notably. From the random-dopant-number point of view, the equivalent channel doping concentration increases when the dopant number increases; this substantially alters the threshold voltage and the on/off state currents [9]. Additionally, the position of ITs induces rather different fluctuation in spite of the same number of dopants and traps, as marked in inset of Figs. 6(a) and 6(b). Listed in Table 1 are the RDs-, ITs- and "RDs+ITs"-induced threshold voltage fluctuations of the tested N- and P-MOSFETs device with respect to EOT of 0.8 nm and 1.2 nm

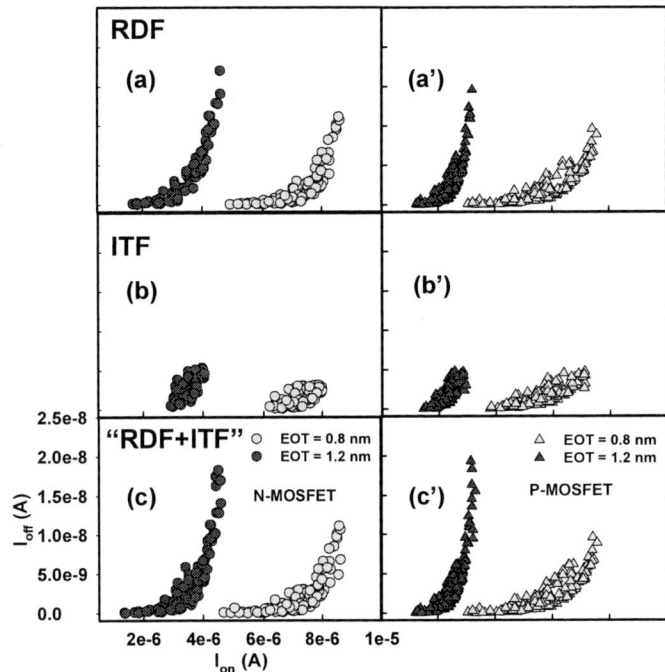

Figure 5. (a),(a'),(b) and (b') The plot of I_{off} versus I_{on} characteristics of about 200 simulated for RDs- and ITs-induced fluctuations individually with EOT of 0.8 nm and 1.2 nm for N- and P-MOSFETs device, respectively. (c) and (c') The combination of RDs- and ITs- ("RDs+ITs") induced I_{on}-I_{off} characteristics fluctuation of about 200 simulated transistors with EOT of 0.8 nm and 1.2 nm for N- and P-MOSFET devices, respectively.

respectively. The device with EOT of 0.8 nm exhibits $\sigma V_{th,RDs}$ = 43 mV, $\sigma V_{th,ITs}$ = 26.3 mV and $\sigma V_{th,RDs + ITs}$ = 45.4 mV. We note $\sigma V_{th,RDs + ITs}$ = 45.4 mV is smaller than the result of statistically independent identical distribution $\sqrt{\sigma^2{}_{V_{th,RDs}} + \sigma^2{}_{V_{th,ITs}}}$ = 50.4 mV due to local charges' interaction between RDs and ITs in N-MOSFETs. Similarly, for device with EOT = 1.2 nm, they are 47.6, 31.4 and 56.2 mV, respectively. Physically, it implies that RDF and ITF should be considered at the same time for simulating HKMG device for N- and P-MOSFETs device.

IV. CONCLUSIONS

In this study, we have explored the effect of RDF and/or ITF on 16 nm HMKG planar CMOS devices. We observed the charge distribution and current density is strongly governed by different position of random ITs and the interaction effect of "RDs+ITs" on the potential distribution and band profile. The σV_{th} induced by ITF is lower than that of RDF owing to a low density of acceptor-like interface traps. The study implies that RDF dominates the characteristic fluctuation when considers both the RDF and ITF together. We have estimated the V_{th} as a function of trap's number for the 16-nm CMOS devices with EOT of 0.8 nm and 1.2 nm; in the meanwhile, the random position ITs induced quite different fluctuation in spite of the same number of traps. Due to randomly localized charges' interaction between RDs and ITs in CMOS devices, RDF and ITF should be considered at the same time. Notably, fluctuations among RDs, ITs and random work function of nanoszied grain of metal gate are currently under examination simultaneously.

978-1-4244-9019-6/11 $26.00 © 2011 IEEE

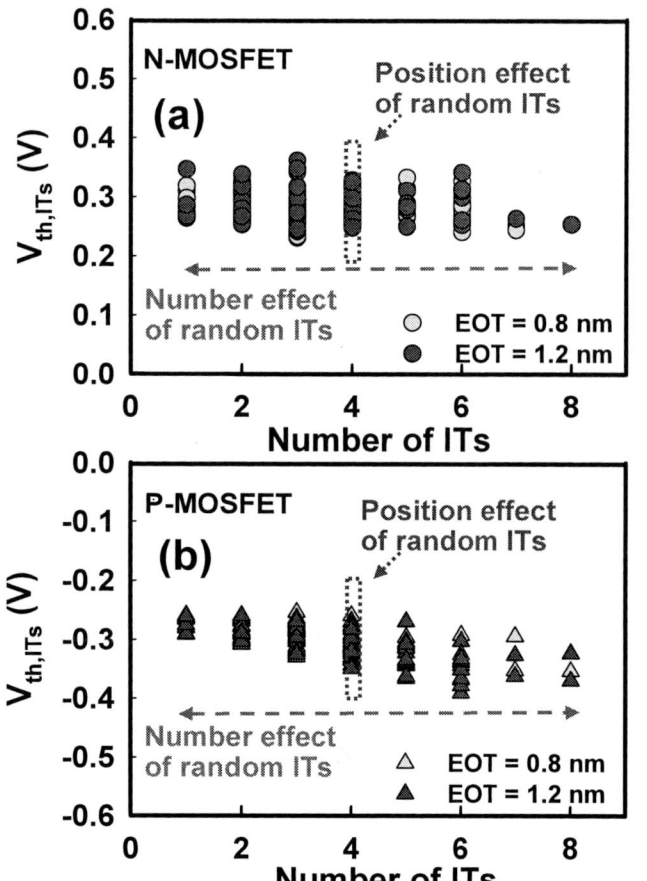

Figure 6. (a) and (b) The large-scale statistically computed V_{th} as a function of trap number with EOT of 0.8 nm and 1.2 nm for N- and P-MOSFET devices, respectively.

TABLE I

Summary of the RDs-, ITs- and RD+ITs-induced threshold voltage fluctuation with EOT of 0.8 nm and 1.2 nm for N- and P-MOSFET devices, respectively.

	$\sigma V_{th, RDs}$		$\sigma V_{th, ITs}$		$\sigma V_{th, RDs+ITs}$	
	EOT = 0.8	EOT = 1.2	EOT = 0.8	EOT = 1.2	EOT = 0.8	EOT = 1.2
N-MOSFET	43 mV	47.6 mV	26.3 mV	31.4 mV	45.4 mV	56.2 mV
P-MOSFET	41 mV	46.5 mV	27.1 mV	32.3 mV	45.1 mV	51.7 mV

REFERENCES

[1] A. Asenov, S. Kaya, and A. R. Brown, "Intrinsic parameter fluctuations in decananometer MOSFETs introduced by gate line edge roughness," *IEEE Trans. on Electron Devices*, vol. 50, no. 5, pp. 1254-1260, 2003.

[2] G. Roy, A. R. Brown, F. Adamu-Lema, S. Roy, and A. Asenov, "Simulation Study of Individual and Combined Sources of Intrinsic Parameter Fluctuations in Conventional Nano-MOSFETs," *IEEE Trans. Electron Device*, vol. 53, no. 12, pp. 3063-3070, Dec. 2006.

[3] H. P. Tuinhout, A. H.Montree, J. Schmitz, and P. A. Stolk, "Effects of gate depletion and boron penetration on matching of deep submicron CMOS transistor," in *IEDM Tech. Dig.*, 1997, pp. 631–634.

[4] A. R. Brown, G. Roy, and A. Asenov, "Poly-Si Gate Related Variability in Decananometre MOSFETs with Conventional Architecture," *IEEE Trans. on Electron Devices*, vol. 54, no. 11, pp. 3056–3063, 2007.

[5] Y. Li and S.-M. Yu, "Comparison of Random-Dopant-Induced Threshold Voltage Fluctuation in Nanoscale Single-, Double-, and Surrounding-Gate Field-Effect Transistors," *Jpn. J. Appl. Phys.*, vol. 45, no. 9A, pp. 6860-6865, Sept. 2006.

[6] Y. Li, and C.-H Hwang, "Discrete-dopant-induced characteristic fluctuations in 16 nm multiple-gate silicon-on-insulator devices", *J. Appl. Phy.*, vol. 102, no. 8, 084509, 2007.

[7] Y. Li, and S.-M. Yu, "A Coupled-Simulation-and-Optimization Approach to Nanodevice Fabrication With Minimization of Electrical Characteristics Fluctuation," *IEEE Trans. Semi. Manufacturing*, vol. 20, no. 4, pp.432-438, Nov. 2007.

[8] N. Sano and M. Tomizawa, "Random dopant model for three-dimensional drift-diffusion simulations in metal-oxide-semiconductor field-effect- transistors," *Appl. Phy. Letter*, vol. 79, 2267, 2007.

[9] Y. Li, S.-M. Yu, J.-R. Hwang and F.-L. Yang, "Discrete Dopant Fluctuated 20nm/15nm-Gate Planar CMOS", *IEEE Trans. Electron Device*, vol. 55, no. 6, pp. 1449-1455, June 2008.

[10] Y. Li, C.-H. Hwang, and H.-M. Huang, "Large-Scale Atomistic Approach to Discrete-Dopant-Induced Characteristic Fluctuations in Silicon Nanowire Transistors," *Physica Status Solidi (a)*, vol. 205, no. 6, pp. 1505-1510, May 2008.

[11] A. Brown and A. Asenov, "Capacitance fluctuations in bulk MOSFETs due to random discrete dopants," *J. Comp. Elect.*, vol.7, no.3, pp.115-118, 2008.

[12] C.L. Alexander, G. Roy, and A. Asenov, "Random Impurity Scattering Induced Variability in Conventional Nano-Scaled MOSFETs: Ab initio Impurity Scattering Monte Carlo Simulation Study," in *Int. Electron Devices Meeting Tech. Dig.*, pp. 1-4, Nov. 2006.

[13] Y. Li, C.-H. Hwang, T.-Y. Li, M.-H. Han, "Process-Variation effect, Metal-Gate Work-Function fluctuation, and random-dopant fluctuation in emerging CMOS technologies," *IEEE Trans. Electron Devices*, vol. 57, no.2, pp. 437-447, 2010.

[14] T. Mizuno, J. Okamura, and A. Toriumi, "Experimental study of threshold voltage fluctuation due to statistical variation of channel dopant number in MOSFETs," *IEEE Trans. Electron Devices*, vol. 41, no. 11, pp. 2216–2221, Nov. 1994.

[15] P. Andricciola, H.P. Tuinhout, B. De Vries, N.A.H. Wils, A.J. Scholten, D.B.M. Klaassen, "Impact of interface states on MOS transistor mismatch," in *IEDM Tech. Dig.*, pp. 711-714, 2009.

[16] A. Appaswamy, P. Chakraborty and J. Cressler, "Influence of Interface Traps on the Temperature Sensitivity of MOSFET Drain-Current Variations," *IEEE Electron Device Lett.*, vol. 31, no. 5, pp. 387-389, May, 2010.

[17] P. Andricciola, H.P. Tuinhout, B. De Vries, N.A.H. Wils, A.J. Scholten, D.B.M. Klaassen, "Impact of interface states on MOS transistor mismatch," in *IEDM Tech. Dig.*, pp. 711-714, 2009.

[18] P.K. Hurley, K. Cherkaoui, S. McDonnell, G. Hughes, A.W. Groenland, "Characterisation and passivation of interface defects in (100)-Si/SiO2/HfO2/TiN gate stacks," *Microelectron. Reliab.*, vol. 47, pp. 1195-1201, 2007.

[19] M. Cassél, K. Tachi1, S. Thiele1and T. Ernst, "Spectroscopic charge pumping in Si nanowire transistors with a high-κ/metal gate," *Appl. Phys. Lett.*, vol. 96, p. 123506, 2010.

[20] M. F. Bukhori, S. Roy and A. Asenov, "Simulation of Statistical Aspects of Charge Trapping and Related Degradation in Bulk MOSFETs in the Presence of Random Discrete Dopants," *IEEE Trans. Electron Device*, vol. 57, no. 6, p. 795-803, 2010.

[21] A.T.M. G. Sarwar, M. R. Siddiqui, R. H. Siddique and Q. D. M. Khosru, "Effects of Interface Traps and Oxide Traps on Gate Capacitance of MOS Devices with Ultrathin (EOT~1 nm) High-κ Stacked Gate Dielectrics," in *IEEE TENCON*, 2009.

[22] Md. Mahbub Satter, A. Haque, "Modeling effects of interface traps on the gate C–V characteristics of MOS devices on alternative high-mobility substrates," *Solid-State Electronics*, vol. 54, pp. 621–627, 2010.

[23] P. Srinivasan, N. A. Chowdhury and D. Misra, "Charge Trapping in Ultrathin Hafnium Silicate/Metal Gate Stacks," *IEEE Electron Device Lett.*, vol. 26, no. 12, pp.913-915, December, 2005.

[24] A.T. Putra, T. Tsunomura, A. Nishida, S. Kamohara, K. Takeuchi and T. Hiramoto "Impact of Fixed Charge at MOSFETs'SiO2/Si Interface on Vth Variation," in *Simulation of Semiconductor Processes and Devices Tech. Dig.*, 2008, pp 241-244.

[25] H.-W. Cheng, F.-H. Li, M.-H. Han, C.-Y. Yiu, C.-H. Yu, K.-F. Lee, and Y. Li, "3D Device Simulation of Work-Function and Interface Trap Fluctuations on High-κ/Metal Gate Devices," in IEDM Tech. Dig., 2010, pp. 379-382.

Excellent Silicon Thickness Uniformity on Ultra-Thin SOI for controlling Vt variation of FDSOI

W. Schwarzenbach, X. Cauchy, F. Boedt, O. Bonnin, E. Butaud, C. Girard, B.-Y. Nguyen, C. Mazure & C. Maleville

SOITEC

Parc Technologique des Fontaines – Bernin – F-38926 Crolles – France
walter.schwarzenbach@soitec.fr

ABSTRACT

Thickness uniformity of the Ultra Thin SOI (UTSOI) substrates is one of the key criteria to control Vt variation of the planar FDSOI devices. We present an evolutionary approach to SmartCut™ technology which already allows achieving a maximum total SOI layer thickness variation of less than ± 10 Å on preproduction volume. Total thickness variation of ± 5 Å is targeted.

SUBSTRATE REQUIREMENT FOR NEXT TECHNOLOGY NODES

For the future 20nm node, standard bulk CMOS technology is facing critical tradeoffs due to increasing random dopant fluctuation, i.e. increasing threshold voltage V_T statistical variability. There is consensus in the IC industry that fully depleted (FD) devices with undoped channel, also known as Ultra Thin Body (UTB) devices [1], are effective solution for eliminating random dopant fluctuation (RDF) in the MOSFET channel, thus significantly reducing threshold voltage V_T variability by over 60% [2].

The foundation of the FD technology is the Ultra Thin SOI (UTSOI) substrate. The starting ultra thin Si thickness (UTSOI) has to be matched to the subsequent FD CMOS processing. Clean, oxidation and etch remove few Si monolayers and it has to be taken into account when specifying the initial UTSOI thickness. The targeted channel Si thickness is typically between 6nm – 7nm for 25nm gate length transistor [2]. The starting UTSOI wafers exhibit SOI layer down to 10 nm and buried oxide layer from 145nm down to 10 nm.

For FD devices the total random V_T variability is the result of gate line edge roughness (LER), workfunction variability and of the channel Si thickness. Since the channel is undoped, there is no significant RDF contribution to V_T variability.

Thus, the thickness uniformity is a key parameter to avoid additional V_T variation of the planar FDSOI device. Typical uniformity requirements include on-wafer uniformity and wafer-to-wafer uniformity. Both of them combined are classified as layer total thickness variation (LTTV) and define the overall manufacturing process window for thickness uniformity. LTTV has to be achieved at the sub-nanometer range for the UTSOI layer for all wafers and all sites in order to meet the FD specifications. UTSOI substrates target high volume production by second half of 2011 to enable the readiness of a 20nm FD CMOS technology platform.

THICKNESS CONTROL REQUIREMENT

Kakhifirooz *et al*, have shown an empirical correspondence between V_T variation on FD-SOI devices and SOI layer thickness variations. [3, 4]

From circuit and device considerations the maximum T_{Si} fluctuation that can be tolerated is < 1nm within-wafer (WiW), total wafer range of the T_{Si} non uniformity, and wafer-to-wafer (WtW), T_{Si} reproducibility. This translates in a SOI T_{Si} thickness maximum wafer-to-wafer variation of ±5 Å.

ADVANCED THICKNESS CONTROL

SOI and BOX (Buried Oxide) layer thickness uniformities are monitored using high resolution UV ellipsometry with KLA-Tencor® F5X tools. Wafer maps are generated with 41 measured points per substrate. The edge exclusion for the ellipsometric monitoring is 3mm to ensure a representative measurement. Fig. 1 illustrates the 41 point distribution.

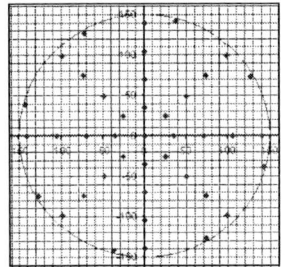

Fig. 1: F5X Thickness measurement mapping

High quality SOI wafers are routinely obtained through the SmartCut™ Process [5], as presented on Fig. 2. FDSOI products are prepared using same process structure as current Partially Depleted Product. The Buried Oxide generation at the "oxidation" step is adjusted versus targeted BOX layer thickness. The Implantation, Splitting & Finishing steps are adjusted to ensure the FDSOI products meeting their tight thickness control requirements.

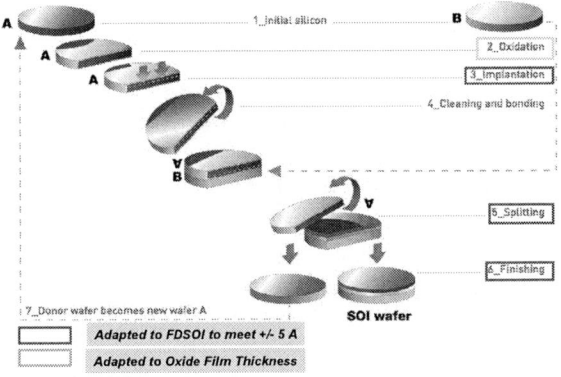

Fig. 2 Schematics of Smart Cut™ process for FDSOI products

978-1-4244-9019-6/11 $26.00 © 2011 IEEE

WAFER-TO-WAFER VARIATION

SOI Thickness control improvement is first obtained by line tool to tool matching. This results in a Thickness WtW variation of +/- 10Å. Fig. 3 shows this SOI thickness distribution. This graph corresponds to the Partially Depleted SOI (PD-SOI) capability, wafers currently in high volume production for microprocessor fabrication at IC makers.

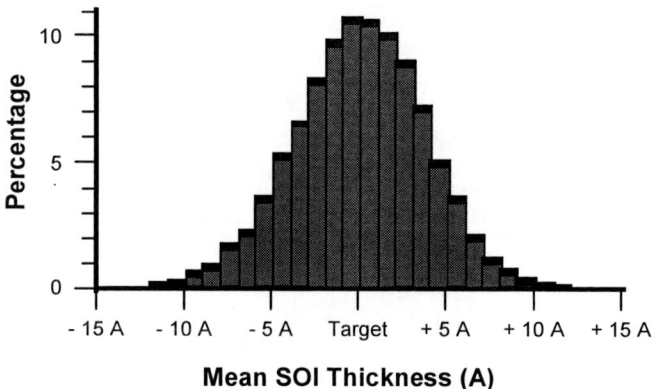

Fig. 3 Raw SOI within-wafer thickness variation achieved by tool-to-tool matching.

However, even through mandatory, this first step is obviously insufficient to reach targeted thickness wafer-to-wafer variation of ± 5Å.

Thus, a specific Tailored Cleaning is implemented. This tailored cleaning module implemented in the UTSOI line does not induce any significant cost & cycle time impact. Processes are fully automated.

Fig. 4 shows UTSOI thickness distribution results. Compared to Fig. 3 the total min-max (7 sigma) Wafer to Wafer (WtW) T_{Si} is now reduced from ± 10 Å to ± 5 Å. Fig. 5 confirms the performance improvement over the UTSOI T_{Si} distribution in pre-production mode quarter after quarter.

Introducing close loop control of chemical thinning, a thickness WtW variation of +/- 2 A is achievable and should be in pre-production mode within the next few months.

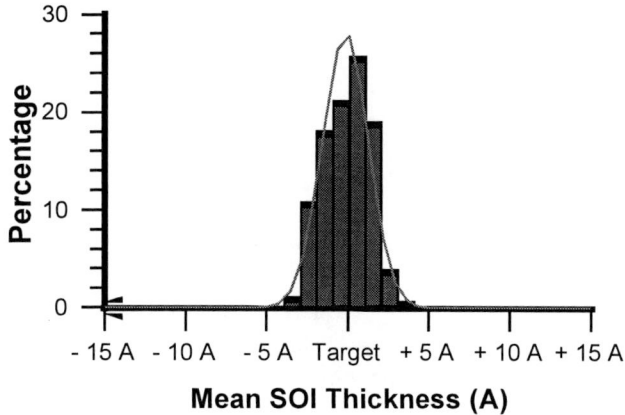

Fig. 4 Improved UTSOI Thickness distribution by tailored cleaning

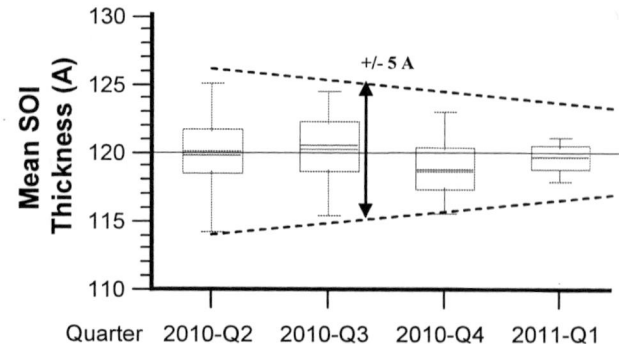

Fig. 5 Month to Month UTSOI thickness variation

WITHIN-WAFER VARIATION

In addition, using 41 pts mapping F5X ellipsometer, On-Wafer uniformity less than 10 Å and down to 7 Å is currently achieved on 12 nm SOI / 25 nm BOX Product. Typical thickness mapping corresponding to these WiW uniformity values are shown in Figure 6. Improvement opportunities are identified to demonstrate 5 Å on-wafer uniformity at maturity phase.

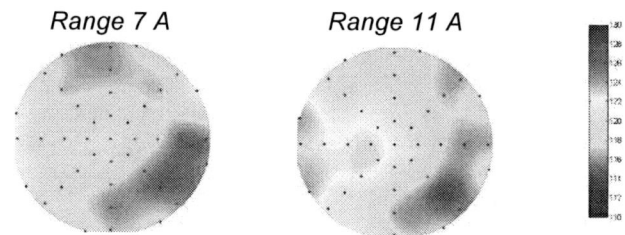

Fig. 6 Typical UTSOI On-Wafer uniformity mapping

TOTAL SOI THICKNESS VARIATION

The total SOI thickness fluctuation for all measured points, all wafers, is obtained by combining wafer-to-wafer and on-wafer contributions. Figure 7 shows performance achieved on pre-production UTSOI volumes. According to above presented results, total SOI thickness fluctuation already reach ± 10 Å.

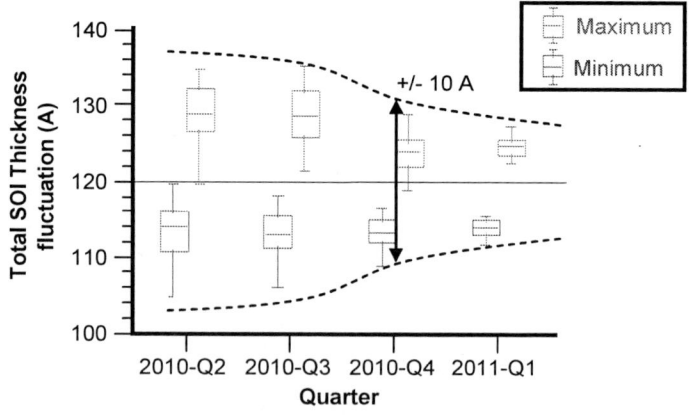

Fig. 6 UTSOI Total SOI Thickness Variation performance

978-1-4244-9019-6/11 $26.00 © 2011 IEEE

CONCLUSION

This work has presented an evolutionary approach to Smart Cut technology to make possible wafer-to-wafer T_{Si} variation maximum of ± 5 Å. Combining wafer-to-wafer and within-wafer uniformities, Total SOI Thickness variation maximum of ± 10 Å is already achieved. This tight controllability of the thickness variation is already demonstrated with pre-production volumes. Additional improvements are targeted within the next few months in order to reach total thickness variation of ± 5 Å. This will secure an industrial mass production process for ultra thin SOI substrate making a FDSOI planar CMOS technology platform a reality.

REFERENCES

[1] C. Shin, M.H. Cho, Y. Tsukamoto, B. Y. Nguyen, C. Mazure, B. Nikolic, T.J. King Liu; "Performance and Area Scaling Benefits of FDSOI Technology for 6T SRAM Cells at the 22nm Node", IEEE TED 57, pp. 1301-1309.

[2] T. Skotnicki; IEDM Short Course "Low Power Logic and Mixed Signal Technology", IEDM 2009.

[3] K. Cheng et al, Extremely Thin SOI (ETSOI) CMOS with Record Low Variability for Low Power System-on-Chip Applications, presented at the IEDM Conference, Baltimore, 2009

[4] A. Khakifirooz *et al, Challenges and Opportunities of Extremely Thin SOI (ETSOI) CMOS Technology for Future Low Power and General Purpose System-on-Chip Applications*, VLSI-TSA Conference, Hsinchu, 2010

[5] C. Maleville and C. Mazure, Solid State Elect., Vol 48-6 (2004) p. 855.

Variability Analysis of UTB SOI Subthreshold SRAM Considering Line-Edge Roughness, Work Function Variation and Temperature Sensitivity

Vita Pi-Ho Hu, *Student Member, IEEE*, Ming-Long Fan, *Student Member, IEEE*,
Pin Su, *Member, IEEE*, and Ching-Te Chuang, *Fellow, IEEE*

Abstract—This paper analyzes stability and variability of Ultra-Thin-Body (UTB) SOI subthreshold SRAMs considering Line-Edge Roughness (LER), Work Function Variation (WFV) and temperature sensitivity. The intrinsic advantages of UTB SOI technology versus bulk CMOS technology with regard to the stability and variability of 6T SRAM cells for subthreshold operation are analyzed. Compared with LER, WFV causes comparable threshold voltage variation and much smaller subthreshold swing fluctuation, hence less impact on the UTB SOI subthreshold SRAMs. Even considering LER, the Lg = 40nm UTB SOI 6T subthreshold SRAM cells still provide sufficient margin (μ RSNM/σ RSNM > 6 at Vdd = 0.3~0.4V). Higher temperature increases the Vread,0 and decrease RSNM because of the degraded subthreshold swing. The RSNM of UTB SOI subthreshold SRAMs show less temperature sensitivity compared with that of bulk subthreshold SRAMs. Due to larger body effect, the back-gating technique is more efficient for the Lg = 40nm and 25nm UTB SOI subthreshold SRAMs compared with the bulk counterparts. By using lower threshold voltage devices with dual band-edge work functions, the Lg = 25nm UTB SOI subthreshold SRAMs show 31.9% reduction in σ RSNM and 55% improvement in μ RSNM/σ RSNM compared with that using single mid-gap work function.

Index Terms—Ultra-Thin-Body SOI, Subthreshold SRAM, Static Noise Margin, Variability

I. INTRODUCTION

The large fraction of chip area often devoted to SRAMs makes low power SRAM design very important for ultra-low power applications, such as portable devices, implanted medical instruments, and wireless sensor networks. Subthreshold operation is an efficient technique to achieve ultra-low power consumption for circuits by lowering the power supply (Vdd) below the threshold voltage [1]. Conventional 6T bulk subthreshold SRAM cells face many challenges with increasing process variations in deep sub-100nm technologies [2-3]. Various 8T and 10T bulk SRAM cells have been proposed to improve the stability, margin and performance in the subthreshold region. Ultra-Thin-Body (UTB) SOI MOSFET with thin buried oxide has emerged as a promising candidate to extend CMOS scaling [4]. Due to its better control of short-channel effects, lower subthreshold swing, and reduced leakage and Random Dopant Fluctuation (RDF) resulting from the use of un-doped (or lightly-doped) thin silicon film, UTB SOI MOSFET is very attractive for subthreshold circuit applications. UTB SOI SRAM cells operating in the subthreshold region show sufficient Static Noise Margin (SNM) [5]. However, the variability of UTB SOI subthreshold SRAMs considering primary

This work was supported in part by the National Science Council of Taiwan under Contract NSC 99-2221-E-009-174, in part by the Ministry of Education in Taiwan under ATU Program, and in part by the Ministry of Economic Affairs in Taiwan under Contract 99-EC-17-A-01-S1-124. The authors are grateful to National Center for High-Performance Computing in Taiwan for computational facilities and software.

The authors are with the Department of Electronics Engineering & Institute of Electronics, National Chiao Tung University, Hsinchu 300, Taiwan (e-mail: vitabee.ee93g@nctu.edu.tw, pinsu@faculty.nctu.edu.tw).

Parameters	UTB SOI	Bulk	UTB SOI	Bulk
Lg [nm]	40	40	25	25
Width [nm]	40	40	25	25
Ti [nm]	1		4.2(HfO$_2$, κ=25)	
Tch [nm]	10	--	5	--
Nch [cm^{-3}]	1E16	3E18	1E17	5E18
TBOX [nm]	10	--	10	--

Table 1. Device parameters of UTB SOI and Bulk MOSFETs with Lg = 40nm and 25nm, respectively.

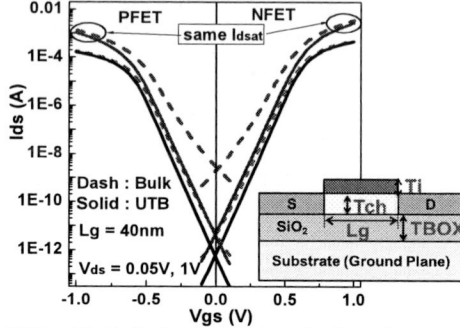

Fig. 1. UTB SOI and bulk devices are designed to have the same Ids,sat at Vgs=Vds=1V. Due to better subthreshold swing, UTB SOI MOSFETs show much smaller Ioff compared with bulk counterparts.

random variations, such as Line-Edge Roughness (LER) and Work Function Variation (WFV), has rarely been examined.

RDF and LER have been recognized as primary causes for random variations in transistor threshold voltage [6]. In addition to RDF and LER, WFV induced by metal grain orientation differences is an emerging problem for state-of-the-art high-k metal-gate MOSFETs [7]. As the channel length scales down, WFV will become much worse and induce comparable or larger threshold voltage variation compared with LER and RDF [8]. However, the impacts of WFV and LER on the variability of subthreshold SRAM cells have rarely been investigated. Our results indicate that the variability of subthreshold SRAM cells is not fully described by considering threshold voltage variations only, and subthreshold swing fluctuations need to be taken into account as well. In this paper, the intrinsic advantages of UTB SOI technology versus bulk CMOS technology with regard to the stability and variability of 6T subthreshold SRAM cells are assessed comprehensively. The temperature sensitivity of the stability/variability of UTB SOI and bulk subthreshold SRAM cells are investigated and compared.

II. DEVICE DESIGN AND TCAD SIMULATION METHODOLOGY

In this paper, two generations of scaled UTB SOI and bulk devices/SRAM cells are examined, and the device parameters are listed in Table 1. The UTB SOI and bulk devices/SRAM cells are analyzed by using 3D TCAD device/mixed-mode simulations [9], and density-gradient model is used to consider the quantum mechanical effects. Fig. 1 shows the Ids-Vgs characteristics for bulk and UTB SOI N/P FETs with gate length (Lg) = 40nm. The bulk and UTB SOI devices are designed to have equal drive current at Vgs=Vds=1.0V for Lg=40nm and 25nm, respectively. A single mid-gap work function is used for the UTB SOI MOSFETs, and the work function values of bulk devices are then selected to adjust the

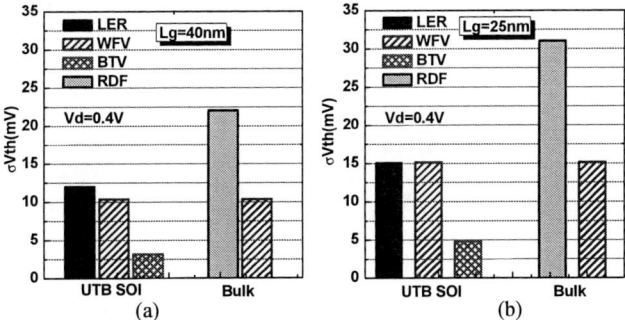

(a)   (b)

Fig. 2. Threshold voltage variations comparisons between UTB SOI and bulk MOSFETs including LER, WFV, BTV and RDF for (a) Lg = 40nm and (b) 25nm. The threshold voltage variation of bulk devices is dominated by RDF for both Lg = 40nm and 25nm. For UTB SOI MOSFEs with Lg = 25nm, LER and WFV show comparable threshold voltage variations.

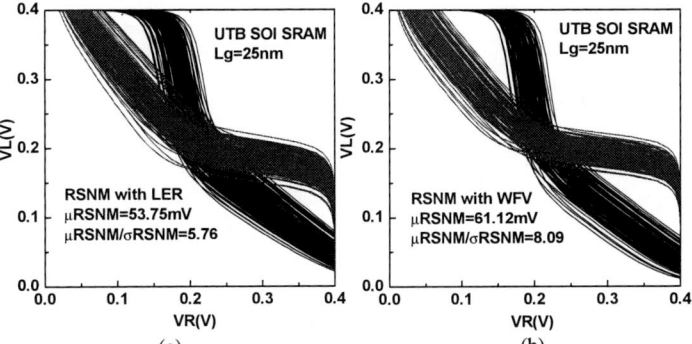

(a)   (b)

Fig. 3. RSNM variations of UTB SOI subthreshold SRAM cells considering (a) LER and (b) WFV.

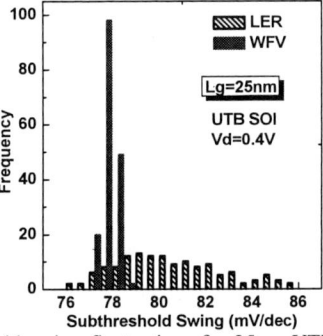

Fig. 4. Subthreshold swing fluctuations for 25nm UTB SOI MOSFETs considering LER and WFV. For 25nm UTB SOI MSOFETs, WFV and LER cause comparable threshold voltage variations as shown in Fig. 2(b); while WFV causes much smaller subthreshold swing fluctuations than LER.

nominal threshold voltage values in order to have the same drive current. As can be seen in Fig. 1, with equal drive current (Idsat) at 1.0V, the off current of the UTB SOI devices are about 2 orders of magnitude lower than their bulk counterparts due to its superior subthreshold swing.

Fig. 2(a) and (b) show the threshold voltage variations (σVth) comparisons for UTB SOI and bulk devices with Lg = 40nm and 25nm, respectively, including LER, WFV, RDF and BTV (Body Thickness Variation). To assess the RDF in bulk devices, atomistic simulations using Monte Carlo approach [10] were carried out. To assess the LER in UTB SOI MOSFETs, the rough line edge patterns are generated using Fourier synthesis approach [11] with correlation length = 20nm and rms amplitude = 1.5 nm, and then atomistic Monte Carlo simulations were performed. To assess the WFV, atomistic simulations considering metal material, possible grain orientations with corresponding probability and work function, metal grain size and device gate size were performed [12]. In this work, TiN is used as the metal gate since this is one of the most

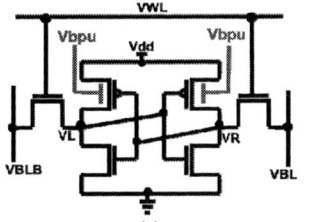   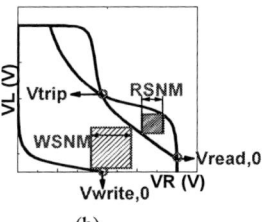

(a)   (b)

Fig. 5(a). The schematic of 6T SRAM cell with back-gating technique, Vbpu is the back-gate bias of pull-up transistors. (b) The definitions of RSNM, WSNM, Vtrip, Vread,0 and Vwrite,0.

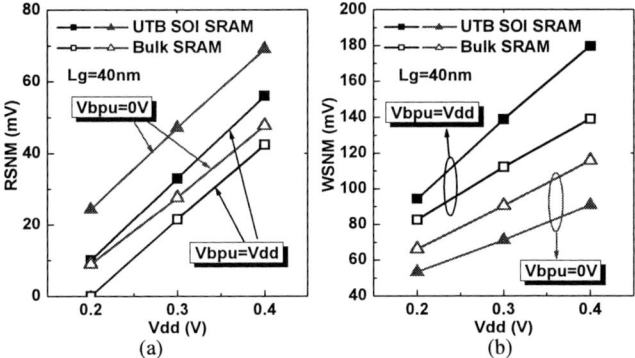

(a)   (b)

Fig. 6(a). RSNM comparisons between UTB SOI and bulk subthreshold SRAMs with Vbpu = 0V and Vdd. (b) WSNM comparisons between UTB SOI and bulk subthreshold SRAMs with Vbpu = 0V and Vdd.

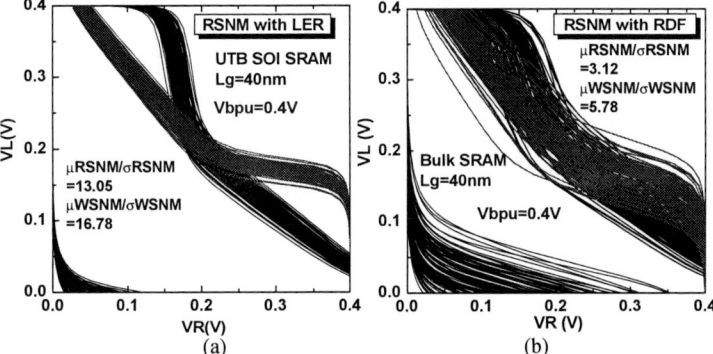

(a)   (b)

Fig. 7. RSNM and WSNM variation for (a) UTB SOI subthreshold SRAMs considering LER and (b) bulk subthreshold SRAMs considering RDF.

widely researched options for the high-k metal-gate stack [13], and the average grain size = 3nm is used [7]. To consider BTV in UTB SOI MOSFETs, the statistical interface roughness patterns are generated between the top and bottom interfaces. The generation is done by employing Fourier synthesis techniques with correlation length = 10nm and rms amplitude = 0.3nm [14]. Atomic-level 3D mixed-mode Monte Carlo simulations with 150 samples are then performed for each case.

III. IMPACTS OF LER AND WFV ON UTB SOI SUBTHRESHOLD SRAM CELLS

As can be seen in Fig. 2, RDF dominates the threshold voltage variations for both Lg = 40nm and 25nm bulk devices. Therefore, RDF dominates the variability of bulk subthreshold SRAM cells. For Lg = 40nm UTB SOI MOSFETs, the threshold voltage variations are dominated by LER (shown in Fig. 2(a)); while for Lg = 25nm UTB SOI MOSFETs, the threshold voltage variations resulting from LER and WFV are comparable (Fig. 2(b)).

Fig. 3(a) and (b) demonstrate the Read Static Noise Margin (RSNM) variations of 25nm UTB SOI subthreshold SRAM cells considering LER and WFV, respectively. μRSNM is defined as the mean of RSNM, and σRSNM is defined as the standard deviation of RSNM. Even though the threshold voltage variations resulting from LER and WFV are comparable for 25nm UTB SOI MOSFETs,

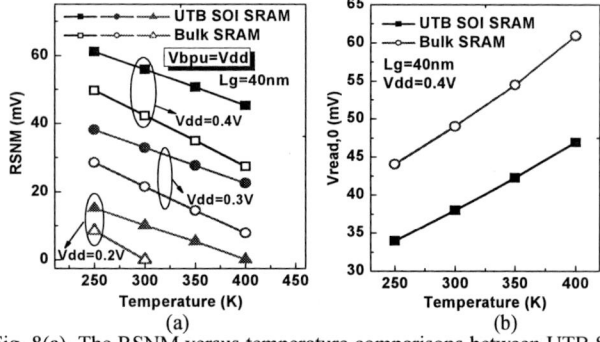

(a)   (b)

Fig. 8(a). The RSNM versus temperature comparisons between UTB SOI and bulk subthreshold SRAMs. (b) Vread,0 versus temperature comparisons between UTB SOI and bulk subthreshold SRAMs.

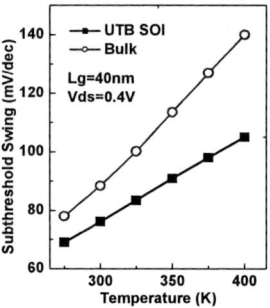

Fig. 9. The subthreshold swing of UTB SOI MOSFETs shows less temperature sensitivity compared with the bulk devices.

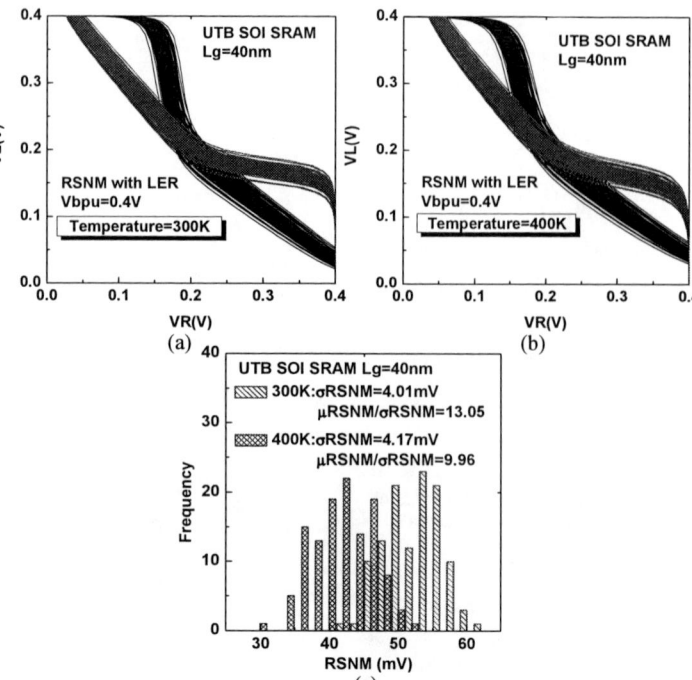

(a)   (b)

(c)

Fig. 10. RSNM variations of UTB SOI subthreshold SRAMs considering LER at (a) temperature = 300°K and (b) temperature = 400°K. (c) RSNM variations comparisons for UTB SOI subthreshold SRAMs with temperature = 300°K and 400°K.

the 25nm UTB SOI subthreshold SRAM cell considering LER shows 12% larger degradation in μRSNM and 28.8% larger degradation in μRSNM/σRSNM compared with the UTB SOI subthreshold SRAM cell considering WFV. This is because that the UTB SOI MOSFETs with LER suffer from larger Short-Channel-Effects (SCE) and hence larger subthreshold swing fluctuations than that with WFV as shown in Fig. 4. The drain current fluctuations in the subthreshold region is not fully described by considering threshold voltage variations only, and subthreshold swing fluctuations need to be taken into account as well[15]. Static Noise Margin is determined by the relative strength (drain current) of N/P FETs in the SRAM cells. Due to its small impact on the subthreshold swing, WFV shows less impact on the variability of UTB SOI subthreshold SRAMs. Therefore, compared with WFV, LER with comparable threshold voltage variation and larger subthreshold swing fluctuations will dominate the variability of UTB SOI subthreshold SRAM cells.

IV. UTB SOI SUBTHRESHOLD SRAM CELLS

In this section, the intrinsic advantages of UTB SOI technology versus bulk CMOS technology with regard to the stability and variability of 6T subthreshold SRAM cells are analyzed.

Fig. 5(a) shows the schematic of 6T SRAM cells with back-gating technique, and Vbpu is the back-gate bias of pull-up transistors. Fig. 5(b) illustrates the static Voltage Transfer Characteristics (VTC) during Read/Write operations. The RSNM is defined as the minimum noise voltage present at each of the cell storage nodes necessary to flip the state of the cell. Vread,0 is the Read disturb voltage determined by the voltage divider effect between pass-gate and pull-down transistors. Vtrip is the voltage needed to flip the cell inverter. Increase in Vread,0 and decrease in Vtrip will degrade the RSNM. Vwrite,0 is determined by the voltage divider effect between pull-up PFET and pass-gate transistors. Lower Vwrite,0 will benefit the Write Static Noise Margin (WSNM).

A. Impact of Back-gating Technique

Fig. 6(a) and (b) show the RSNM and WSNM comparisons between UTB SOI and bulk subthreshold SRAMs with Vbpu = Vdd and 0V at various Vdd. With Vbpu = 0V, UTB SOI subthreshold

SRAM shows larger RSNM than the bulk subthreshold SRAM due to its better subthreshold swing (Fig. 6(a)); while UTB SOI subthreshold SRAM shows smaller WSNM than the bulk one (Fig. 6(b)).

As Vbpu changes from 0V to Vdd, the threshold voltages of pull-up transistors increase. The weaker pull-up transistor lowers Vtrip and degrades RSNM. However, UTB SOI subthreshold SRAM still shows larger RSNM than the bulk counterpart with Vbpu = Vdd as shown in Fig. 6(a). For WSNM, the data holding current from pull-up transistors reduces because of the increased threshold voltage of pull-up transistors which makes the cell easier to flip during Write operation, thus improving the WSNM. In Fig. 6(b), as Vbpu changes from 0V to Vdd, the increase in WSNM of UTB SOI subthreshold SRAM is much larger than that of bulk subthreshold SRAM. This is because the back-gating effect in the UTB SOI MOSFETs is larger than the body effect of deeply scaled bulk MOSFETs. Therefore, as Vbpu changes from 0V to Vdd, the pull-up transistors of UTB SOI SRAM cells become much weaker than pass-gate transistors, thus decreasing Vwrite,0 and resulting in larger WSNM than the bulk counterpart. With Vbpu = Vdd, the UTB SOI subthreshold SRAM shows both larger RSNM and WSNM than the bulk subthreshold SRAM.

Fig. 7(a) and (b) exhibit the RSNM and WSNM variations of Lg = 40nm UTB SOI and bulk subthreshold SRAMs considering their dominated variation sources, respectively. As can be seen, bulk devices with larger threshold voltage variations due to RDF result in significant mismatch of neighboring transistors in SRAM cells, thus severely degrading the cell stability and variability. On the other hand, UTB SOI MOSFETs with better variation immunity can still maintain adequate margin for subthreshold SRAM operation. At Vdd = 0.4V, 40nm UTB SOI subthreshold SRAMs show 4.18X larger in μRSNM/σRSNM and 2.9X larger in μWSNM/σWSNM compared with the bulk counterparts.

B. Impact of Temperature

Fig. 8(a) shows the temperature sensitivity of RSNM comparison between UTB SOI and bulk subthreshold SRAMs at various Vdd. Due to better subthreshold swing, UTB SOI subthreshold SRAM

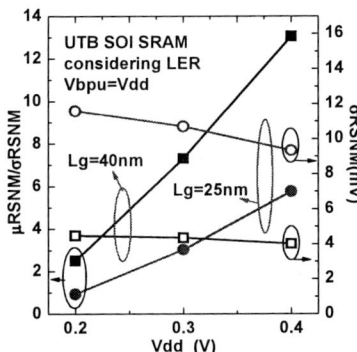

Fig. 11. Impact of Vdd scaling on μRSNM/σRSNM and σRSNM of UTB SOI subthreshold SRAMs with Vbpu=Vdd. As Vdd decreases from 0.4V to 0.2V, the μRSNM/σRSNM degradation is mainly contributed from μRSNM reduction.

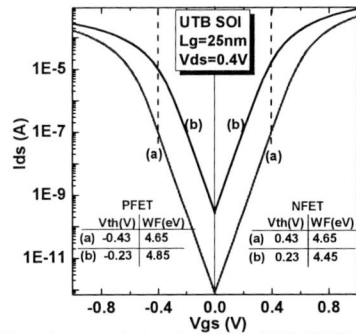

Fig. 12. Ids-Vgs characteristics for UTB SOI MOSFETs with different threshold voltages (Vth)/work functions (WF) at Vds = 0.4V. The |Vth| of NFET and PFET in case (a) and (b) are designed with the same absolute value. (Vgs = Vth @ Ids = 3E-7A.)

shows smaller Vread,0 (Fig. 8(b)) and larger RSNM than the bulk one. As temperature increases, Vread,0 increases and RSNM decreases because of the degraded subthreshold swing. The RSNM of UTB SOI subthreshold SRAM also exhibits less temperature sensitivity than the bulk subthreshold SRAM as shown in Fig. 8(a). This is because the subthreshold swing of UTB SOI MOSFET shows less temperature sensitivity compared with the bulk device shown in Fig. 9. As temperature increases from 300K to 400K, UTB SOI subthreshold SRAM shows 10.68mV reduction in RSNM and bulk subthreshold SRAM shows 14.95mV reduction in RSNM at Vdd = 0.4V.

Fig. 10(a) and (b) show the RSNM variations of UTB SOI subthreshold SRAMs at temperature = 300°K and 400°K, respectively. As temperature increases from 300°K to 400°K, UTB SOI subthreshold SRAM exhibits 23.7% reduction in μRSNM/σRSNM. The σRSNM of UTB SOI subthreshold SRAM only increases slightly as temperature increases from 300°K to 400°K, therefore, the reduction of μRSNM/σRSNM is mainly caused by the decrease in μRSNM shown in Fig. 10(c).

C. Impact of Vdd Scaling

Fig. 11 shows the impact of Vdd scaling on σRSNM and μRSNM/σRSNM for Lg = 40nm and 25nm UTB SOI subthreshold SRAM cells. The μRSNM/σRSNM of the Lg = 40nm UTB SOI subthreshold SRAM cells are larger than 6 at Vdd = 0.3V~0.4V. In other words, as Vdd scales down to 0.3V, the 40nm UTB SOI subthreshold SRAMs still maintain sufficient margin. The 25nm UTB SOI subthreshold SRAMs with larger threshold voltage variations and subthreshold swing fluctuations show μRSNM/σRSNM = 5.76 at Vdd = 0.4V. As Vdd scales down, σRSNM only increases slightly which means the decrease in μRSNM/σRSNM reduction is mainly caused from the decrease in μRSNM for both 40nm and 25nm UTB SOI subthreshold SRAM cells.

D. Impact of Threshold Voltage Design

Fig. 12 shows the Ids-Vgs characteristics of Lg = 25nm UTB SOI MOSFETs at Vds = 0.4V with different threshold voltages. For case (a), the threshold voltages of N/P type UTB SOI MOSFETs with mid-gap work function are +/-0.43V. For case (b), the threshold voltages of N/P UTB SOI MOSFETs with dual band-edge work functions are +/-0.23V. Fig. 13(a) and (b) show the RSNM variations of UTB SOI subthreshold SRAMs considering LER at Vdd = 0.4V with different threshold voltages design. The rough line edge patterns are the same for case (a) and (b). As can be seen, Fig. 13(b) with lower threshold voltage devices (case (b) in Fig. 12) shows smaller RSNM variations compared with Fig. 13(a) with higher threshold voltage devices (case (a) in Fig. 12). This is because the UTB SOI MOSFETs with lower threshold voltage design (case (b)) show

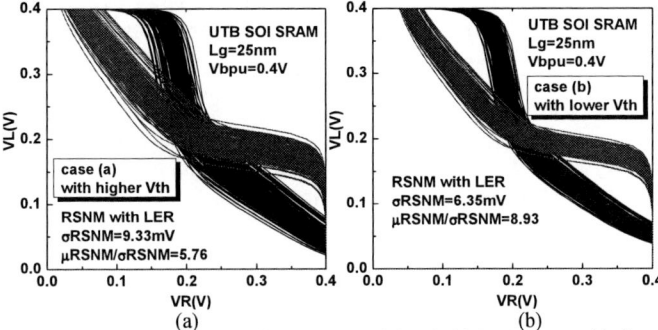

Fig. 13. RSNM variations of UTB SOI subthreshold SRAMs considering LER with (a) higher threshold voltage devices (case (a)) and (b) lower threshold voltage devices (case (b)). The RSNM variability of UTB SOI subthreshold SRAMs can be improved by using dual band-edge work function devices (lower threshold voltages devices).

smaller drain current variation as its operation region is close to the super-threshold region, while the UTB SOI MOSFETs with higher threshold voltage design (case (a)) operating entirely in the subthreshold region show larger drain current variation. By using lower threshold voltage devices, UTB SOI subthreshold SRAMs show 31.9% reduction in σRSNM and 55% improvement in μRSNM/σRSNM. The 25nm UTB SOI subthreshold SRAMs with lower threshold voltage devices show μRSNM/σRSNM = 8.93 at Vdd = 0.4V.

V. CONCLUSION

We have investigated the variability of UTB SOI SRAM cells considering LER, WFV and temperature sensitivity comprehensively. Due to its smaller impact on subthreshold swing fluctuations, WFV shows less impact on the variability of UTB SOI subthreshold SRAMs, and the variability of UTB SOI subthreshold SRAMs is dominated by LER. Compared with the bulk subthreshold SRAMs, the Lg = 40nm UTB SOI subthreshold SRAMs show 4.18X and 2.9X improvement in μRSNM/σRSNM and μWSNM/σWSNM, respectively. The RSNM of UTB SOI subthreshold SRAMs show less temperature sensitivity compared with that of bulk subthreshold SRAMs. The 6T UTB SOI subthreshold SRAMs can provide sufficient margin (μRSNM/σRSNM > 6) even considering LER.

REFERENCES

[1] B. H. Calhoun et al., *JSSC*, vol. 41, no. 7, pp. 1673-1679, 2006.
[2] N. Verma et al., *ISSCC*, 2007 pp. 328
[3] J. P. Kulkarni et al., *JSSC*, vol. 42, no. 10, pp. 2303-2313, Oct. 2007.
[4] Y. Morita et al., *Symp. VLSI Tech. Dig.*, 2008, pp. 166-167.
[5] V. P.-H. Hu et al., *IEEE TED*, vol. 56, no. 9, pp. 2120-2127, 2009.
[6] A. Asenov, *Symp. VLSI Tech. Dig.*, pp. 86-87, 2007.
[7] K. Ohmori et al., *IEDM Tech. Dig.*, 409-412, 2008.
[8] X. Zhang et al., *IEDM Tech. Dig.*, 57-60, 2009.
[9] Sentaurus TCAD, C2009-06 Manual.
[10] D. J. Frank et al., *VLSI Symp.*, pp. 169, 1999.
[11] A. Asenov et al., *IEEE TED*, vol. 50, no. 5, pp. 1254-1260, 2003.
[12] A. R. Brown et al., *IEEE EDL*, vol. 31, no. 11, pp. 1199-1201, 2010.
[13] B. P. Linder et al., *Microelectron. Eng.*, vol. 86, no. 7-9, pp. 1632, 2009.
[14] K. Samsudin et al., *Solid-State Electronics*, vol. 51, pp. 611-616, 2007.
[15] P. Magnone et al., *IEEE TED*, vol. 57, no. 11, pp. 2848-2856, 2010.

3D Integrable Nanowire FET Sensor with Intrinsic Sensitivity Boost

Chi On Chui, *Senior Member, IEEE*, Jorge Kina, and Kyeong-Sik Shin

Abstract—In this paper, we review a recently developed transformative nanowire FET sensor concept and 3D-compatible fabrication technology. Compared to the generic nanowire FET sensors, an intrinsic boost in detection sensitivity is accomplished through the seamless integration of a sensing nanowire with an amplifying nanowire FET. Exclusively enabled by top-down nanofabrication technology, the back-end-of-line compatible sub-450 °C manufacturing processes have been developed. Sensing experimental data have also revealed around 1 order of magnitude sensitivity improvement in solution pH detection. Finally, an ultra-low thermal budget nanowire formation technology has been preliminarily developed for future 3D integration with CMOS.

Index Terms—3D integration, biosensor, field-effect device, nanowire.

I. INTRODUCTION

TANGENTIAL to the semiconductor technology development direction based on dimensional scaling, heterogeneous integrations of hybrid functionality have received an unprecedented attention within the past decade. Envisioned to be included in the so-called System-in-Package (SiP) [1], non-digital components such as sensors, actuators, passives, and others would enhance the CMOS interactions with people and environment.

Leveraging the advance of the heavily invested and mature semiconductor manufacturing technology, integrated biochips hold great promises to be developed into very low-cost and high impact point-of-care (PoC) diagnostic platforms. Challenging requirements for these platforms include high sensitivity and specificity, rapid turnaround time, absence of equipped facilities and trained personnel, low-cost of disposables and fixed equipments, portability, etc. [2]-[3]. As a key component inside these platforms, the biosensor is responsible for capturing the disease biomarker information and translating that into either a chemical or physical output signal [4]. For highly selective detections or captures of these biomarkers, either antigen-antibody specific bindings or complementary nucleic acid hybridizations are widely adopted in common diagnostic platforms. The criteria on the signal transduction device technology include superior sensitivity,

Manuscript received March 25, 2011. This work was supported in part by the Focus Center Research Program-Focus Center on Functional Engineered Nano Architectonics program and in part by the Chinese American Faculty Association Robert T. Poe Faculty Development Award.
C. O. Chui, J. Kina, and K.-S. Shin are with the Department of Electrical Engineering, University of California, Los Angeles, CA 90095 USA (e-mail: chui@ee.ucla.edu).

multiplex and simple operations, response timeliness, compactness, reusability, simple fabrication, and electronic interfaces.

As a contending PoC diagnostic platform component technology, semiconductor nanowire field-effect transistor (FET) sensors [5]-[10] permit visual-label-free and real-time electronic detections of numerous charged biomolecules with high sensitivity and great selectivity. This high detection sensitivity arises from the inherently large surface area-to-volume ratio of the quasi-1D nanowires. The charge-to-current signal transduction occurs at the nanowire FET front-end which thereby minimizes parasitics and thus noise. Moreover, this scheme requires no imaging equipments and the sensors themselves are readily interfaced and integrated with readout electronics.

The sensor detection sensitivity, and especially the limit of detection (LOD) and limit of quantification (LOQ) that take into account the blank and signal noise [11], however, need to be further improved before their PoC diagnostic platform deployment. This criterion is particularly exemplified in detecting samples limited in volume such as those from children and infants [2] or in staging diseases [3]. Furthermore, these sensors are preferred to be 3D-integrated right above the CMOS circuitry to make the most efficient use of silicon real estate. To do so, low thermal budget fabrication processes are mandated and the resultant device integrity should also not be compromised to any extent.

In this paper, we describe a recently developed transformative nanowire FET sensor structure and 3D-compatible fabrication processes [12]-[13]. We summarize here our sensor design and operating mechanism. We also illustrate the low thermal budget fabrication processes. Besides, we demonstrate that the intrinsic amplification built-in to the sensor is able to boost the detection sensitivity by at least 1 order of magnitude over the generic nanowire FET. Finally, we reveal a preliminarily developed, ultra-low thermal budget nanowire formation technology for future 3D integration with CMOS.

II. TRANSFORMATIVE NANOWIRE FET SENSOR STRUCTURE AND AMPLIFICATION MECHANISM

As illustrated in Fig. 1(a), the transformative nanowire FET sensor has a T-shape nanowire structure that seamlessly integrates a sensing nanowire with a nanowire FET, called T-nwFET. This structure exploits the same geometrical advantage in quasi-1D nanowire yet it is able to significantly boost the detection sensitivity through a built-in amplification mechanism with lowest possible parasitics.

The entire sensor is shielded from the ambience except the sensing nanowire surface (shielding not drawn for clarity in illustration). The specific receptors (e.g. antibodies) are first immobilized only onto the sensing nanowire surface. The target analytes (e.g. antigens) present in the sample solution are bound to the receptors with great specificity. The electric field enumerated from the charges on the bound biomolecules is then coupled to modulate the sensing nanowire conductance. The resultant signal change is intimately amplified with minimal parasitics by the built-in nanowire FET to attain very high sensitivity, and then electronically read out.

The intrinsic signal transduction and amplification mechanism in this novel T-nwFET sensor can be understood with the help of the equivalent circuit shown in Fig. 1(b). The node potential at the nanowire intersection is labeled as V_n and each nanowire segment from this node to the respective electrode is represented as an ideal Schottky diode (D_x) in series with the nanowire resistance (R_x). The sensing nanowire current I_B and FET drain current I_D are related to the electrode potentials (e.g. V_B and V_D) [13] by

$$I_B = I_0 \left[\exp\left(\frac{q\left((V_B - V_n) - I_B(R_B + \Delta R_B)\right)}{k_B T} \right) - 1 \right] \quad (1)$$

$$I_D = I_0 \left[\exp\left(\frac{q\left((V_D - V_n) - I_D R_D\right)}{k_B T} \right) - 1 \right] \quad (2)$$

$$I_0 = A A^{**} T^2 \exp\left(-\frac{q\phi_{Bp}}{k_B T}\right) \quad (3)$$

where A, A^{**}, k_B, ϕ_{Bp}, q, R_B, ΔR_B, R_D, and T are respectively the diode area, effective Richardson constant, Boltzmann constant, Schottky barrier height, electronic charge, sensing nanowire resistance, change in R_B due to biomolecular detection, drain nanowire resistance, and absolute temperature.

During the sensing operation, both the drain (V_D) and base (V_B) contacts are fixed and biased positively against the source contact (V_S). The modulation of sensing nanowire conductance due to specific binding of target biomolecules yields a non-zero ΔR_B value and thus changes the I_B according to (2). That in turn varies the potential V_n and modifies the energy band profile between the source and drain contacts. The resultant change in the voltage difference ($V_D - V_n$) causes an exponential change in I_D such that a modest amplification of the detected signal can readily be achieved. Preliminary 3D semiconductor device simulations have been carried out [12] to validate the anticipated operating and amplification mechanism of the proposed novel T-nwFET.

By eliminating V_n from (1) and (2), I_D can be related to I_B, an equation that has to be solved by iteration. The current amplification ratio (dI_D/dI_B) can thus be computed numerically. Since identical Schottky contacts are assumed, the amplification ratio is largely determined by the nanowire

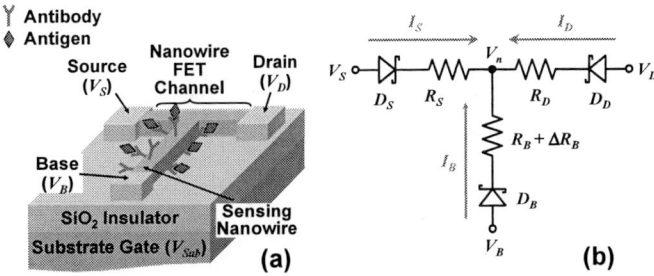

Fig. 1. (a) Device structure and (b) equivalent circuit of the transformative nanowire FET (T-nwFET) sensor integrating a sensing nanowire with a nanowire FET channel for signal amplification [13].

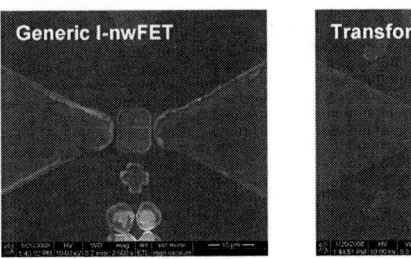

Fig. 2. Fabricated I-nwFET and T-nwFET sensor prototype with a surface passivation layer except above the sensing nanowire surface [12].

resistances R_B and R_D and potentials V_D and V_B, as illustrated in the next section. In other words, the desired amplification amount can be adjusted by the nanowire geometry during structural design and/or voltage biasing during sensing operation.

III. 3D-COMPATIBLE LOW THERMAL BUDGET FABRICATION TECHNOLOGY

The T-nwFET sensor prototypes (Fig. 2) were constructed using sub-450 °C fabrication processes [12]-[13] and standard nanofabrication facility equipments. The starting substrates were (100) oriented silicon-on-insulator (SOI) wafers made by the separation by implanted oxygen (SIMOX) process. The starting SOI layer and buried oxide thickness were respectively 190 nm and 150 nm, and the Si material was p-type with a resistivity of 10-20 Ω-cm. First, the SOI layer was thinned down to 50 nm, and a hydrogen silsesquioxane (HSQ) resist layer was coated on top. This HSQ layer was then defined by electron beam lithography into T-shape nanowire structures with width ranging from 50 nm to 3 ⊐m. Next, the source, drain, and base metal electrodes slightly overlapping the HSQ nanowire structures were formed by photoresist lift-off of a sputtered Pt-on-Ti dual layer. The SOI regions unprotected by HSQ and metals were then removed by reactive ion etching followed by rapid thermal annealing at 450 °C for 10 min to sintering the metal-silicon contacts. With this dopant implantation-free Schottky contact scheme, the fabrication procedures were greatly simplified and the overall thermal budget was also reduced. Finally, all the fabricated sensors were covered with a layer of silicon nitride, except above the sensing nanowire surface (see Fig. 2), to passivate

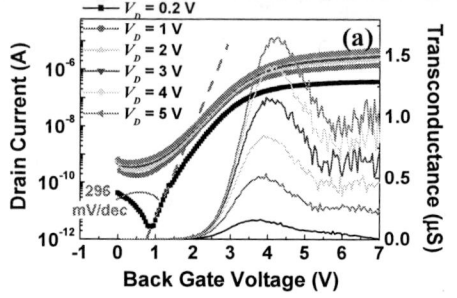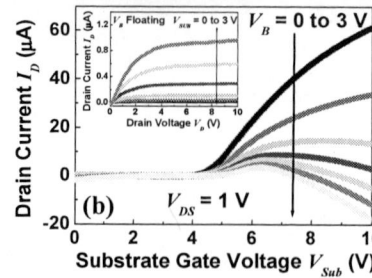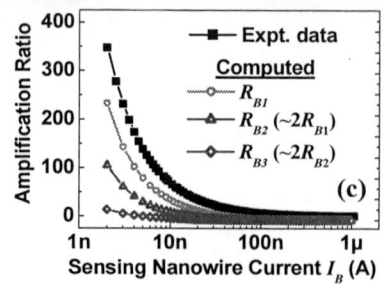

Fig. 3. Baseline electrical characteristics of the T-nwFET sensor prototypes: (a) transfer characteristics with a floating base, (b) transfer characteristics with a connected base, and (c) current amplification [12]-[13].

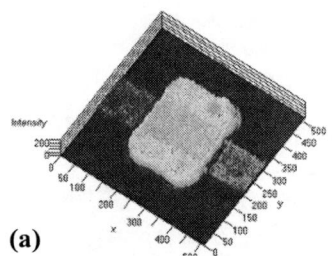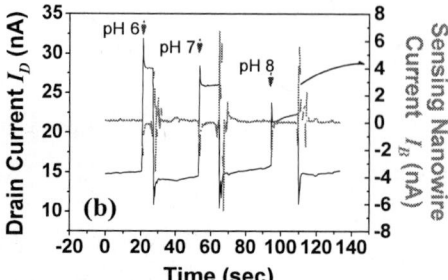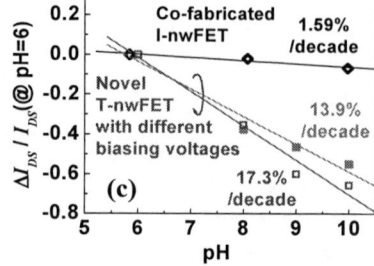

Fig. 4. (a) Confocal micrograph of FITC bindings to a 3 μm wide silicon nanowire after APTES treatment. (b) The real-time pH sensing measurement by the novel T-nwFET prototype. (c) Sensitivity comparisons between I-nwFET and T-nwFET sensors in real-time solution pH detections [12]-[13].

them from any interaction with the analyte solution during the detection measurements. Generic Schottky FET sensors with just an I-shape sensing nanowire channel (denoted as I-nwFET from now on) of the same width and length were co-fabricated for control purposes (Fig. 2(a)).

IV. BASELINE ELECTRICAL CHARACTERIZATIONS

Prior to the solution sensing trails, baseline DC electrical measurements of the T-nwFET sensor prototypes were performed [12]-[13]. As shown in Fig. 3, decent electrical characteristics were extracted from the T-nwFET sensors. When the base electrode (i.e. the sensing nanowire) was floated, the transfer (Fig. 3(a)) and output (inset of Fig. 3(b)) characteristics across the source and drain electrodes with a varying substrate gate voltage V_{Sub} resembled those of a conventional Schottky nwFET [14]. These data assured the baseline integrity of the fabrication process and Schottky contacts. With a non-floating and varying V_B, the T-nwFET transfer characteristics, i.e. I_D-V_{Sub}, were also extracted (Fig. 3(b)). When V_B is small (e.g. <0.5 V), the T-nwFET behaves as a regular nwFET and I_D flows into the channel from the drain. When V_B is increased, V_n will also be raised so as to reduce the voltage difference ($V_D - V_n$). According to (2), I_D would thus decrease to zero and finally becomes negative.

In addition, the anticipated current amplification in the novel T-nwFET was experimentally confirmed (solid square symbols in Fig. 3(c)). The I_D-I_B characteristics were first obtained by injecting a varying amount of I_B from the measurement system followed by taking their derivatives. In fact, the resultant amplification ratio versus I_B behavior could be adequately modeled by computing the dI_D/dI_B ratio (hollow symbols in Fig. 3(c)) as previously described. DC amplification of over 2 orders of magnitude is readily accomplished. More importantly, these amplification ratios can be adjusted either before or after fabrication as explained

in Section II.

V. SOLUTION SENSING CHARACTERIZATIONS

On both transformative T-nwFET and co-fabricated generic I-nwFET sensors, preliminary real-time detections of solution pH were carried out [12]. For pH detections, the sensing nanowire surfaces were first immobilized with 3-aminopro-pyltriethoxysilane (APTES) probes and their stability was confirmed with the fluorescein isothiocyanate (FITC) bindings via confocal microscopy (Fig. 4(a)). Buffer solutions with varying pH value were then sequentially introduced. As shown in Fig. 4(b), the T-nwFET I_D was able to track the corresponding changes in solution pH in practically real time.

The steady-state calibration curves from both devices were also extracted and benchmarked in Fig. 4(c). Being the slope of these calibration curves, the detection sensitivity of different devices or under different biasing conditions can be meaningfully compared. The pre-optimized novel T-nwFET sensor prototypes have exhibited roughly one order of magnitude improvement in solution pH detection sensitivity over the co-fabricated generic I-nwFET sensors. Moreover, the sensitivity of the novel T-nwFET sensor is readily adjusted with different biasing voltages as evident from the two different curves in Fig. 4(c).

VI. ULTRA-LOW THERMAL BUDGET NANOWIRE FORMATION TECHNOLOGY

Towards the future integration of T-nwFET sensors above the metal layers in integrated circuit chips, a back-end-of-line compatible nanowire formation technology is mandated to assure the integrity of the pre-fabricated and underlying circuitry. The most promising approach among others is the wafer-scale stamping transfer of pre-patterned and pre-ordered semiconductor nanostructures and materials [15]-[17].

978-1-4244-9019-6/11 $26.00 © 2011 IEEE

The stamping transfer of lithographically patterned silicon nanowires has been chosen as the technology development driver. As illustrated in 5(a), the silicon nanowires were first formed by electron beam lithography and etching of an SOI layer. A polydimethylsiloxane (PDMS) stamp was then attached onto the nanowires (Fig. 5(b)). After careful removal of the PDMS stamp, the silicon nanowires were released from the original substrate and adhered to the stamp (Fig. 5(c)). It should be noted that nanowires at different sites across the wafer were picked up at the same time (see optical micrographs). Finally, the nanowires adhered to the stamp were successfully transferred onto a foreign substrate upon appropriate contact (Fig. 5(d)). The transfer formation technology of T-shape nanowire is currently being developed and after which the sub-450 °C T-nwFET fabrication processes described in Section III will be integrated together.

VII. CONCLUSION

In this paper, we review the theoretical and experimental development of a highly promising and transformative nanowire FET sensor concept. As a low parasitics amplifying sensor, its structural simplicity permits flexible amplification adjustment in either a pre-fabrication or post-silicon manner. The sub-450 °C fabrication processes are readily deployable for 3D integration above the pre-fabricated CMOS circuitry underneath. Finally, the baseline electrical and solution sensing characterizations discussed have provided convincing evidences that this device might hold the promise to break the current detection sensitivity limitations.

REFERENCES

[1] *International Technology Roadmap for Semiconductors* (http://public.itrs.net)

[2] M. Urdea, L. A. Penny, S. S. Olmsted, M. Y. Giovanni, P. Kaspar, A. Shepherd, P. Wilson, C. A. Dahl, S. Buchsbaum, G. Moeller, and D. C. Hay Burgess, "Requirements for high impact diagnostics in the developing world," *Nature*, vol. S1, pp. 73-79, 2006.

[3] C. D. Chin, V. Linderb, and S. K. Sia, "Lab-on-a-chip devices for global health: Past studies and future opportunities," *Lab Chip*, vol. 7, pp. 41-57, 2007.

[4] D. R. Thévenot, K. Toth, R. A. Durst, G. S. Wilson, "Electrochemical biosensors: recommended definitions and classification," *Biosens. Bioelectron.*, vol. 16, pp. 121-131, 2001.

[5] J.-I. Hahm and C. M. Lieber, "Direct ultrasensitive electrical detection of DNA and DNA sequence variations using nNanowire nanosensors," *Nano Lett.*, vol. 4, pp. 51-54, 2004.

[6] Z. Li, Y. Chen, X. Li, T. I. Kamins, K. Nauka, and R. S. Williams, "Sequence-specific label-free DNA sensors based on silicon nanowires,"*Nano Lett.*, vol. 4, pp. 245-247, 2004.

[7] F. Patolsky, G. Zheng, and C. M. Lieber, :Fabrication of silicon nanowire devices for ultrasensitive, label-free, real-time detection of biological and chemical species," *Nat. Protocols*, vol. 1, pp. 1711-1724, 2006.

[8] Y. L. Bunimovich, Y. S. Shin, W.-S. Yeo, M. Amori, G. Kwong, and J. R. Heath, "Quantitative real-time measurements of DNA hybridization with alkylated nonoxidized silicon nanowires in electrolyte solution," *J. Amer. Chem. Soc.*, vol. 128, pp. 16323–16331, 2006.

[9] E. Stern, J. F. Klemic, D. A. Routenberg, P. N. Wyrembak, D. B. Turner-Evans, A. D. Hamilton, D. A. LaVan, T. M. Fahmy, and M. A. Reed, "Label-free immunodetection with CMOS-compatible semiconducting nanowires," *Nature*, vol. 445, pp. 519-522, 2007.

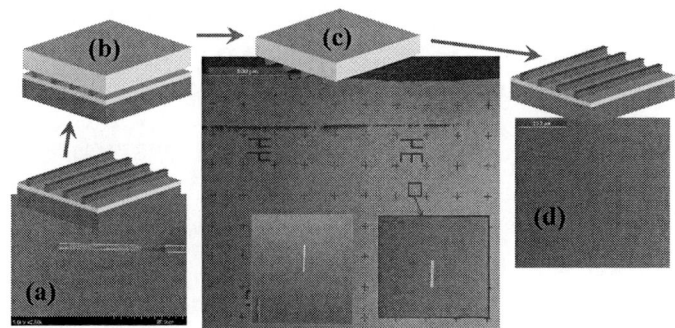

Fig. 5. Process steps in stamping transfer of pre-patterned silicon nanowires. (a) Lithographic patterning of SOI layer to form silicon nanowires. (b) PDMS stamping adhesion of the nanowires. (c) PDMS picking up of the nanowires at multiple sites. (d) Successful transfer of the picked-up nanowires onto a different substrate surface.

[10] G.-J. Zhang, J. H. Chua, R.-E. Chee, A. Agarwal, S. M. Wong, K. D. Buddharaju, and N. Balasubramanian, "Highly sensitive measurements of PNA-DNA hybridization using oxide-etched silicon nanowire biosensors," *Biosens. Bioelectron.*, vol. 23, pp. 1701-1707, 2008.

[11] J. Mocak, A. M. Bond, S. Mitchell, and G. Scollary, "A statistical overview of standard (IUPAC and ACS) and new procedures for determining the limits of detection and quantification: application to voltammetric and stripping techniques," *Pure Appl. Chem.*, vol. 69, pp. 297-328, 1997.

[12] K.-S. Shin, K. Lee, J. Y. Kang, and C. O. Chui, "Novel T-channel nanowire FET with built-in signal amplification for pH sensing," *IEEE Int. Electron Dev. Mtg. (IEDM) Tech. Dig.*, pp. 599-602, 2009.

[13] K.-S. Shin, K. Lee, J.-H. Park, J. Y. Kang, and C. O. Chui, "Schottky contacted nanowire field-effect sensing device with intrinsic amplification," *IEEE Electron Device Lett.*, vol. 31, pp. 1317-1319, 2010.

[14] S.-M. Koo, M. D. Edelstein, Q. Li, C. A. Richter, and E. M. Vogel, "Silicon nanowires as enhancement-mode Schottky barrier field-effect transistors," *Nanotechnol.*, vol. 16, pp. 1482-1485, 2005.

[15] M. A. Meitl, Z.-T. Zhu, V. Kumar, K. J. Lee, X. Feng, Y. Y. Huang, I. Adesida, P. G. Nuzzo, and J. A. Rogers, "Transfer printing by kinetic control of adhesion to an elastomeric stamp," *Nature Mater.*, vol. 5, pp. 33-38, 2006.

[16] J.-H. Ahn, H.-S. Kim, K. J. Lee, S. Jeon, S. J. Kang, Y. Sun, R. G. Nuzzo, and J. A. Rogers, "Heterogeneous three-dimensional electronics by use of printed semiconductor nanomaterials," *Science*, vol. 314, pp. 1754-1757.

[17] M. C. McAlpine, H. Ahmad, D. Wang, and J. R. Heath, "Highly Ordered Nanowire Arrays on Plastic Substrates for Ultrasensitive Flexible Chemical Sensors," *Nature Mater.*, vol. 6, pp. 379-384, 2007.

On The Magnitude of Random Telegraph Noise in Ultra-Scaled MOSFETs

K.P. Cheung, J.P. Campbell

Semiconductor Electronics Division, NIST, Gaithersburg, MD 20899, 301-975-3093   kin.cheung@nist.gov

Abstract

Random telegraph noise (RTN) has been shown to be a more severe scaling issue than the Random Dopant Effect (RDE). However this observation relies heavily on studies which focus only on threshold voltage (V_{TH}) fluctuations. V_{TH} measurements make separation of these two scaling issues (RTN and RDE) difficult. Since future scaled devices may use channels with no or low doping, it is important to examine the impact of RTN without the influence of RDE. In this work, we experimentally verify the "hole in the inversion layer" model of RTN and then use it to examine the magnitude of RTN in ultra-scaled devices without the influence of RDE. This analysis strongly suggests that RTN is a serious issue even in the absence of RDE.

Introduction

The magnitude of RTN in MOSFET drain current is known to become larger as the size of the channel shrinks [1-3]. In recent years, RTN has become a serious concern for memory devices [4-12] due to their aggressively scaled geometries. It is these studies which have led to the conclusion that RTN is a more serious problem than RDE [8]. In these aggressively scaled devices, the measured RTN amplitudes are known to vary drastically [13, 14]. It is this long tail of the statistical distribution of RTN amplitudes which has become such an issue. In many studies, a V_{TH} fluctuation is measured instead of the RTN magnitude. This allows the direct comparison of the impact of RTN with RDE. However, at or below threshold, RDE and RTN are coupled, leading to an anomalously large RTN observation[15]. Since RDE and now RTN are well established scaling issues, it is quite possible future MOSFET's will combat RDE by using channels which are either intrinsic or have very low doping concentrations. Can the removal

of RDE also solve the RTN issues? This is a very important question indeed.

RTN and Screening

A long held belief is that low frequency (1/f) drain current noise is the superposition of many RTN fluctuations. Thus, RTN is often attributed to mobility fluctuation, number fluctuation, or both. However, for very small device dimensions, a third mechanism may be more important [14, 16]. This third mechanism, first proposed by Reimbold [2], suggests that the trapped charge creates a "hole" (or cored-out section) in the inversion layer through Coulombic repulsion (fig. 1). A simple relationship between RTN amplitude and the size of the hole was also given [2].

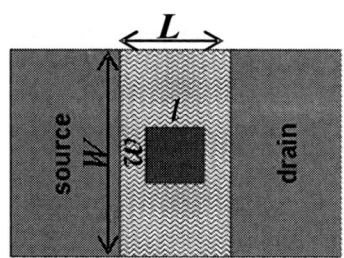

Fig. 1 Defect creates a hole (dark blue square) of the size w x l in the inversion layer.

A more refined equation (eqn. 1) was later introduced by Ohata *et al.* [17] relating the RTN amplitude to the hole size (approximated as a square). In this construct, the screening length determines the size of the hole.

$$\frac{\Delta I_d}{I_d} = \frac{wl(\sigma_0 - \sigma)}{WL(\sigma_0 - \sigma) - w(L-l)(\sigma_0 - \sigma)} \tag{1}$$

where w and l are the width and length of the hole; W and L are width and length of the channel; and σ_0 and σ are the conductivity of the channel and in the cored out region respectively.

Even with the refined equation, Ohata *et al.* found that this "hole in the inversion layer" approach yields (at low gate overdrive) screening lengths which seem unreasonably large [17]. A decade later, anomalously large RTN amplitudes at low gate overdrives have been successfully explained using a channel percolation path concept [15]. A percolation path is the result of random dopant distribution leading to an uneven surface potential in the channel. At low gate overdrives, this percolation path (or uneven channel) has an effective channel width that can be drastically smaller than the drawn width. From equation 1 it is clear that Ohata *et al's* difficulty can be reconciled by using an effective channel width which is much smaller than the as-drawn dimension. In this work we revisit the "hole in the inversion layer" paradigm, and examine RTN in the absence of percolation.

Percolation Path & Screening Length

We utilize nMOSFET's with 1.4 nm SiON gate dielectrics in this study. RTN amplitudes from many highly-scaled devices were investigated. The observed RTN amplitude as a function of gate overdrive is typically non-monotonic (Fig. 2).

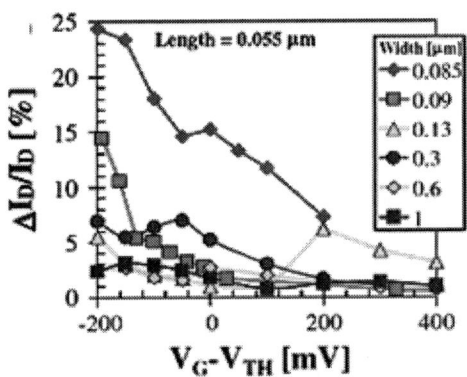

Fig. 2 RTN amplitudes as a function of gate overdrive for a number devices with different width. Typical trends are non-monotonic. Device sizes are as drawn.

This behavior is expected since it is unlikely that a single defect is physically located precisely over the "pinch-off" point in the potential landscape. However, we were able to observe one device with a particularly large RTN amplitude as well as a monotonic dependence on gate overdrive (fig. 3). This device produced very clean 2-level RTN (fig. 4), implying a single defect. This behavior strongly suggests that the defect is positioned entirely

within the active region of the channel, and is located right on top of the percolation path.

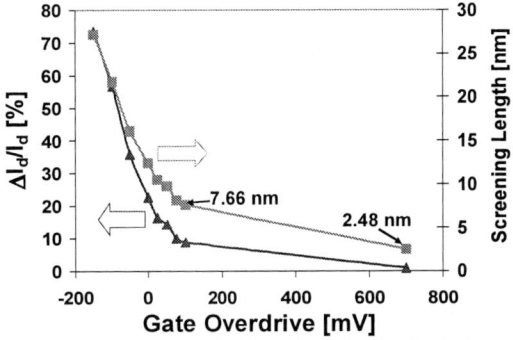

Fig. 3 RTN amplitude (triangles) for this particular device is a smooth monotonic function of gate overdrive. The calculated screening lengths drops rapidly as gate overdrive increase from below threshold to above threshold. The as drawn size of this device is 55 nm x 85 nm. The size used for the calculation is 35 nm x 85 nm.

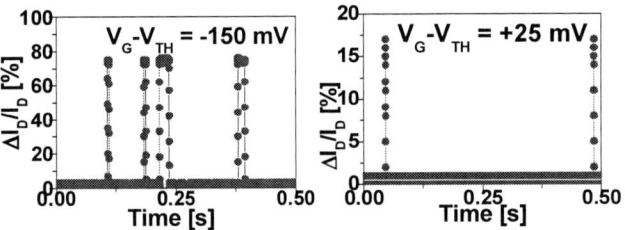

Fig. 4 The RTN signal from the device shown in fig. 3 are very large and very clean. Showing the characteristic of a single defect.

Also shown in fig. 3 are the calculated screening lengths (equation (1), assuming $w = l$, $\sigma = 0$) from the observed RTN amplitudes. At low gate overdrives, the screening length increases rapidly to unphysical lengths. Such large screening lengths led Ohata *et al.* [17] to doubt the "hole in the inversion layer" model. An examination of equation 1 reveals that an ultra-narrow channel width (or a percolation path-induced ultra-narrow effective channel width) can lead to large RTN amplitude, resulting in an artificially large screening length.

As the gate overdrive increases, the percolation path widens and eventually disappears [15]. This suggests that the screening length behavior depicted in fig. 3 is largely due to the widening of the percolation path under the hole (this includes the +150 mV gate overdrive point). To lend support to this assertion, we study the relationship between drain current and channel width (fig. 5). It is clear that a linear relationship is not established until at least +200 mV above threshold. It is also clear that the last point in fig. 3 (+700 mV) is completely free of the percolation effect. Thus the screening length for this point is purely due to the inversion charge density.

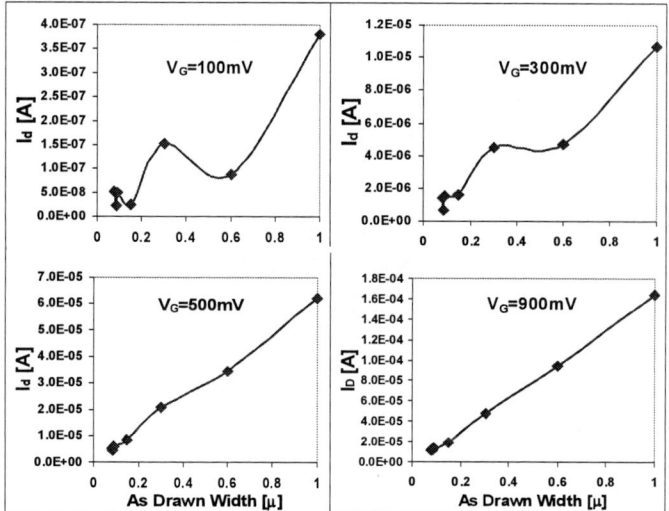

Fig. 5 Drain current as a function of device width for 55 nm (drawn) nMOSFET at various gate overdrives. Deviation from linear relationship is a sign of percolation path which makes the effective channel width smaller than the drawn width, sometimes by a larger factor. The percolation path is expected to disappear at some point above threshold. In this example, some hint of it is still there at V_G=500 mV, which is 200 mV above threshold.

What we have observed and discussed so far is entirely consistent with the simulation done by Asenov *et al.* [15]. They have shown that as gate overdrive increases, the RDE signature of RTN amplitude disappears.

A 2.48 nm screening length at such strong inversion seems large because the Debye length is expected to be ~ 1nm. On the other hand, Debye *et al.*'s Screened Coulomb Potential is a 3-D model [18]. The inversion layer is a 2-D system, and screening length is expected to be larger than the 3-D value [19]. In addition, equation 1 is only an approximation. The actual value may be some what different. The key point is that the calculated screening length from RTN amplitude is no longer outside the reasonable range. Given the crude nature of the equation, this is a strong indication that the "hole in the inversion layer" model is correct and we can now use this data point to anchor an extrapolation to see what lies ahead as we scale the MOSFET further.

Implication to scaling

Here we consider only the conditions in which percolation is absent (its presence will make matters worse). Let $w = l = p/\sqrt{n}$ where n is inversion charge density, and p is proportionality constant, we get

$$\frac{\Delta I_d}{I_d} = \frac{p^2}{n\left(WL - \dfrac{L}{p\sqrt{n}} + p^2 n\right)} \qquad (2)$$

Recall that $n \cong C_{OX,eff}(V_G - V_{TH})$ [20], where $C_{OX,eff}$ is the effective inversion capacitance. If we assume n remains more or less constant with scaling (oxide capacitance increase is offset by gate overdrive decrease), we can use equation 2 to extrapolate to smaller device sizes. Since we can anchor the equation with a known experimental data point (RTN amplitude and device size), the extrapolation should be on solid ground.

In fig. 6, we show the projected RTN amplitude for various technology nodes. Three different channel widths are simulated (W=L, W=3L, and W=5L). If we set the acceptable RTN amplitude to 10%, then the minimum width of $W = L$ can only be used in the 22 nm node. At 16 nm node, it is marginal. Wider channel widths do help. At $W = 5L$, the RTN amplitude is within 10% down to the 8 nm node.

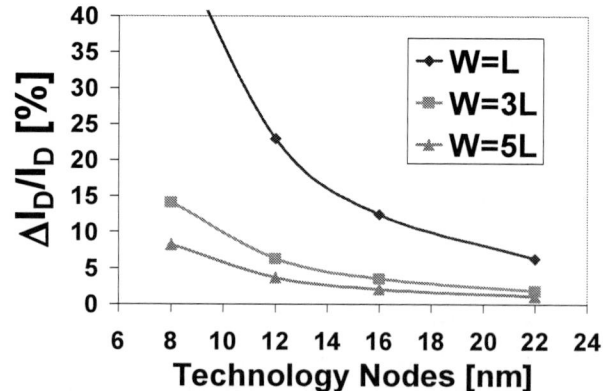

Fig. 6 RTN amplitude at different technology nodes for minimum width ($W = L$) devices as well as wider devices ($W = 3L$, and $W = 5L$).

Note that the extrapolation result in fig. 6 represents the worst case scenario – the defect is right at the center of the active area. This is very likely the tail of the RTN amplitude distribution. To experimentally observe this, the sample size must be large, or one gets lucky as in our case. Since we cannot ignore the tail of the distribution, the results presented in figure 6 are instructional.

Fig. 6 reinforces the conclusion that RTN is a new limiting factor for scaling [10]. Because the RDE has been removed in the data, our conclusion is stronger. It suggests that while going to an undoped channel may alleviate RDE, RTN will remain a huge factor that can cause large V_{TH} or equivalently drain current fluctuations.

Equations (1) and (2), as they are written, are not suitable to assess V_{TH} fluctuation. However, if the "hole in the inversion layer" model is indeed correct, it is not hard to see that the effect is huge at threshold. If we consider that the inversion charge density is two orders of magnitude lower at threshold, the screening length becomes comparable to the minimum device size at 22 nm.

Conclusions

In this work, we show that the "hole in the inversion layer model" together with a percolation path reduced channel width works reasonably well to explain RTN in ultra-small MOSFETs. This establishes the relationship between RTN amplitude and inversion charge density. Using this relation, we show that the RTN amplitude will grow to unacceptable levels in future technology nodes. We also show that V_{TH} fluctuation will continue to grow even when using undoped channels.

References

[1] Ralls, K. S. *et al. Phys. Rev. Lett.* 52(3) 228(1984).
[2] Reimbold, G. *IEEE Trans. Electron Dev.,* 31(9) 1190(1984).
[3] Howard, R. E. *et al. IEEE Trans. Electron Dev.,* **32**(9) 1669(1985).
[4] Gu, S. H. *et al.* Int. Electron Dev. Meeting, 2006, pp1-4.
[5] Kurata, H. *et al. Solid-State Circuits, IEEE Journal of* **42**(6) 1362(2007).
[6] Compagnoni, C. M. *et al. IEEE Electron Dev. Lett.,* 29(8) 941(2008).
[7] Tega, N. *et al.* IEEE Int. Reliab. Phys. Symp., 2008, pp541-546.
[8] Chiu, J. P. *et al.* IEEE Int. Electron Dev. Meeting, 2009, pp1-4.
[9] Ghetti, A. *et al. IEEE Trans. Electron Dev.* **56**(8) 1746(2009).
[10] Tega, N. *et al.* Symp. VLSI Technol., 2009, pp50-51.
[11] Fugazza, D. *et al.* IEEE Int. Reliab. Phys. Symp., 2010, pp743-749.
[12] Ielmini, D. *et al.* Appl. Phys. Lett. **96**: 053503-053503-3(2010).
[13] Roux, O. *et al. Microelectronic Engineering* 15(1-4) 547(1991).
[14] Simoen, E. *et al. IEEE Trans. Electron Dev.,* **39**(2) 422(1992).
[15] Asenov, A. *et al.* Int. Electron Dev. Meeting, 2000, pp279-282.
[16] Vandamme, *et al. Solid-State Electronics* 42(6) 901-905(1998).
[17] Ohata, *et al. J. Appl. Phys.* **68**(1) 200-204(1990).
[18] Debye, P. and E. Huckel, Phys. Z. **24**, 185, 305(1923).
[19] Ando, *et al. Rev. Mod. Phys.* **54**(2) 437 LP - 672(1982).
[20] King, Y.-C. *et al. Semiconductor Sci. & Technol.* **13**(8) 963(1998).

Timing error prevention using elastic clocking

Kwanyeob Chae, Chang-Ho Lee, and Saibal Mukhopadhyay

Georgia Institute of Technology, Atlanta, GA, USA
Email: ky.chae@gatech.edu

Abstract—**"Safety margin" for a logic circuit introduces a performance overhead. But eliminating safety margin makes a system more prone to timing failure, particularly under dynamic operating variations. This paper presents dynamic timing control technique that allows a system to operate without any safety margin. The dynamic control method prevents timing errors utilizing time borrowing and elastic clocking. Time borrowing allows a pipeline to compensate the timing slack by borrowing time from the next pipeline stage and clock stretching pays back the borrowed time to the next pipeline stage. Thus, a system employing such dynamic timing control technique can prevent errors with a small performance penalty and eventually operate without safety margin. The net effect is better power-performance trade-off under voltage scaling i.e. lower power consumption for a target frequency or higher operating frequency for a target power. The proposed technique was validated using a prototype test-chip designed in 180-nm CMOS technology.**

I. INTRODUCTION

Technology scaling leads to higher integration density and faster switching speed but increases power consumption [1]. Rapidly-increasing power consumption contributes to environmental variations in the voltage and temperature, further reducing the reliability [3]-[6]. Increased power consumption also contributes to thermal problems, which raise cooling costs and exacerbate delay variations [3]. Technology scaling also increases the variation in process parameters. Higher process variation results in a significant spread of delay, which causes functional errors and further degrades system reliability [2].

Under increasing process and environmental variations, meeting performance specifications with limited power budget becomes a critical challenge. The traditional worst-case corner based design introduces "safety margin" to tolerate variations for example by operating at lower frequency or higher voltage. Having safety margin, although helps tolerate variations, leads to significant performance loss or power overhead [7]. This has led to investigations in adaptive design methods that compensate for variation by detection and correction of timing related errors. An attractive approach is to utilize replica circuits, which are strongly correlated with the critical path delays of actual circuits [8]-[11]. When delay error is detected in the replica circuits, the supply voltage or the operating frequency can be dynamically changed to prevent chip failure while maintaining higher performance or lower power consumption. However, due to the different geometrical locations on a chip, the actual circuits and replica circuits may experience different process, voltage, and temperature (PVT)

variations [4]. Thus, to compensate for mismatches and prevent chip failure, an adaptive design with replica circuits requires a safety margin as illustrated in Fig. 1.

The alternative approach is to use the in-situ error detection and correction mechanism to remove safety margin. Since the removed safety margin can eliminate power and performance overhead, a system with the in-situ error detection and correction method can lead to lower power consumption, or higher performance [3]-[6]. However, without the safety margin, dynamic environmental variations could cause timing errors during the operation of a system. If the system can handle already-occurred errors, the system can operate without safety margin as shown in Fig. 1. Error handling needs error recovery time and energy. Therefore, at a higher error rate, error recovery methods (e.g. pipeline flushing) incur significant power and performance penalty, which could offset the gains achieved by removing the safety margin [3]. This is particularly challenging problem when the activation probability of critical paths is high. Therefore, circuits with no safety margin for PVT variations require that performance penalty at high activation probability of critical paths should be minimal. This can be achieved if the timing errors can be prevented in advance with minimum performance/power overhead [12].

This paper proposes a method of removing the safety margin with minimal performance penalty. This is achieved by preventing errors in advance. The proposed method is composed of a special flip-flop and a clock shifter. The flip-flop presented in this paper allows time borrowing during a time-borrowing window (T_{BW}) and generates a detection signal in the presence of time borrowing. We also propose a clock shifter that elastically stretches the clock period by T_{BW} when time borrowing is detected to pay back the borrowed time in the next clock cycle. This removes the need for error recovery and minimizes associated performance penalty when time borrowing occurs. In this paper, elastic clocking means

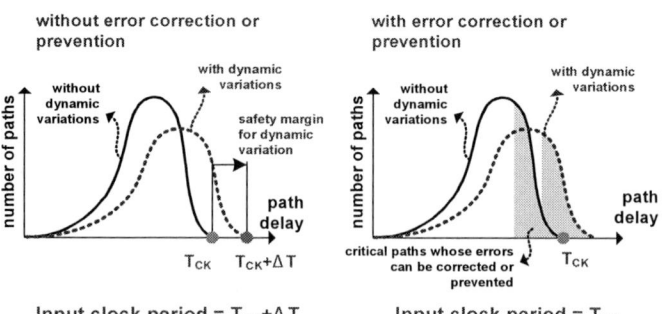

Fig. 1. Clock period considering dynamic variations.

978-1-4244-9019-6/11 $26.00 © 2011 IEEE

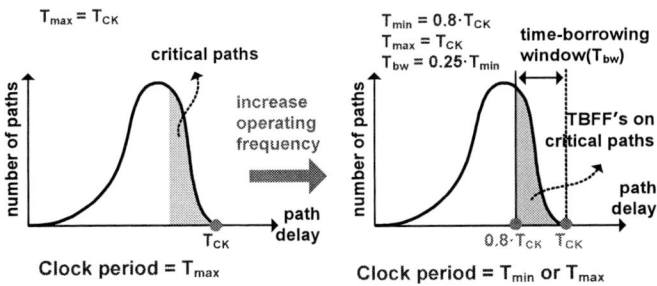

Fig. 2. Basic concept of the proposed dynamic timing control.

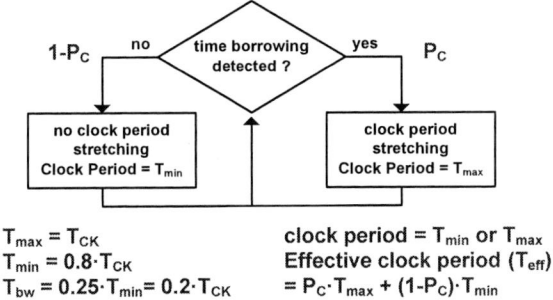

$$T_{max} = T_{CK}$$
$$T_{min} = 0.8 \cdot T_{CK}$$
$$T_{bw} = 0.25 \cdot T_{min} = 0.2 \cdot T_{CK}$$

clock period = T_{min} or T_{max}
Effective clock period (T_{eff})
$= P_c \cdot T_{max} + (1 - P_c) \cdot T_{min}$

Fig. 4. Control flow of the clock stretching scheme.

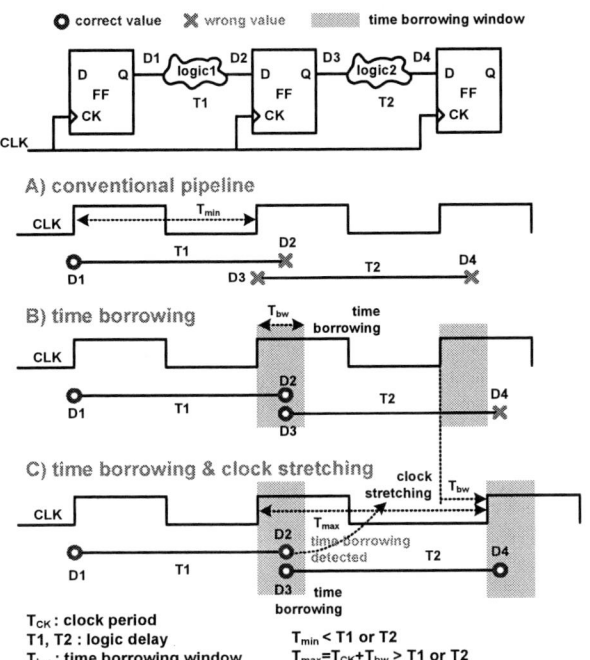

Fig. 3. The operation mechanism of time borrowing and clock stretching.

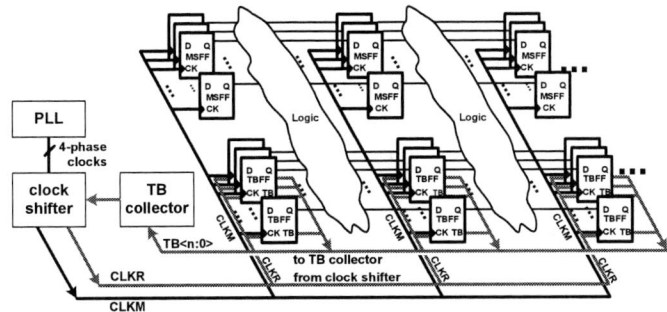

Fig. 5. The proposed pipeline architecture.

dynamic clock period stretching. A design employing the time-borrowing flip-flop (TBFF) in conjunction with the clock shifter can reduce power consumption or increase the clock frequency even at a high activation probability of critical paths.

II. PROPOSED METHODOLOGY

As illustrated in Fig. 2, the maximum operating frequency of a circuit is limited by the maximum delay of its critical paths. In Fig. 2, critical paths whose delay is between $0.8 \cdot T_{CK}$ and T_{CK} will necessarily result in timing failures when the clock period is reduced to $0.8 \cdot T_{CK}$. We propose to use TBFF's with T_{BW} (T_{BW} is defined by a quarter of the input clock period, T_{min}) in critical paths. This helps compensate for time slacks by borrowing time from the next stage as case B in Fig. 3. For the current pipeline stage, time borrowing helps to prevent timing failures [13]. However, it increases the path delay in the next pipeline stage and could end up in a timing failure in the next clock cycle. The possible timing violation in the next pipeline stage can be prevented by stretching the clock period by T_{BW} when time borrowing is detected as shown in case C in Fig. 3. As a result, the design with TBFF's on critical paths and the clock shifter can operate at a clock period of $0.8 \cdot T_{CK}$ when critical paths are not activated. As shown in Fig. 4, in the

detection of time borrowing, the design will operate elastically at the clock period of T_{CK}. In other words, if the activation probability of critical paths is 0, the design with the proposed technique operates at the minimum clock period (T_{min}). On the other hand, if the activation probability of critical paths is 1, the design operates at the maximum clock period (T_{max}). In this way, the design with the proposed scheme can operate at two different clock periods, T_{min} or T_{max}, in response to the activation of critical paths. Since the clock period changes adaptively according to the activation of critical paths, an effective operating frequency (F_{EFF}) is required to estimate the performance. The F_{EFF} for the proposed method is defined as

$$F_{EFF} = (P_c \cdot T_{max} + (1 - P_c) \cdot T_{min})^{-1}, \quad (1)$$

where P_c is the activation probability of critical paths. For the proposed method, the range of F_{EFF} is

$$T_{max}^{-1} \le F_{EFF} \le T_{min}^{-1}. \quad (2)$$

Since T_{max} is equal to $1.25 \cdot T_{min}$ for the proposed method, the F_{EFF} is bounded between $(1.25 \cdot T_{min})^{-1}$ and $(T_{min})^{-1}$. In contrast, for the error detection and recovery method proposed in [3]-[6], T_{max} (T_{max} is clock cycles required to recover and replay in the presence of a timing error) is equal to or more than $2 \cdot T_{min}$. Therefore, as P_c increases, the performance of the proposed method degrades more slowly than that of the error detection and recovery method.

III. IMPLEMENTATION

The pipeline architecture with the proposed method can be implemented as in Fig. 5. TBFF's in critical paths allow time borrowing and detect the occurrence of time borrowing. A collector block is needed to detect the occurrence of any TB from all TBFF's and generate a clock-shift signal that controls the clock shifter. The purpose of the proposed clock shifter is to stretch the clock elastically when time borrowing is detected.

978-1-4244-9019-6/11 $26.00 © 2011 IEEE

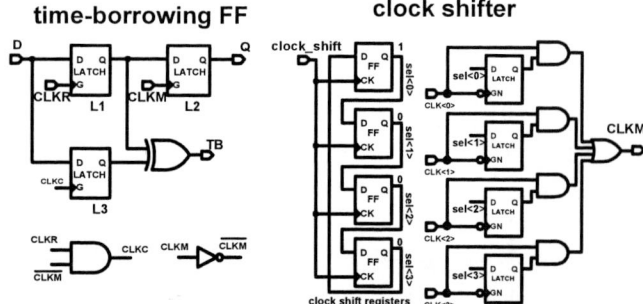

time-borrowing FF **clock shifter**

Fig. 6. Block diagram of the time-borrowing flip-flop and the clock shifter.

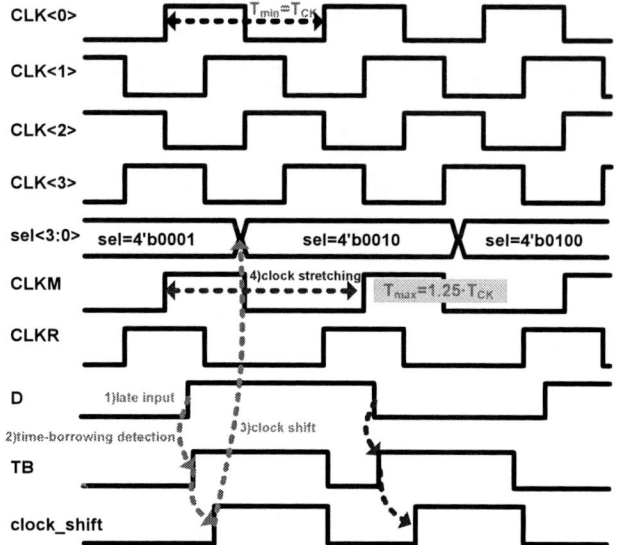

Fig. 7. Timing diagram of time borrowing and clock stretching.

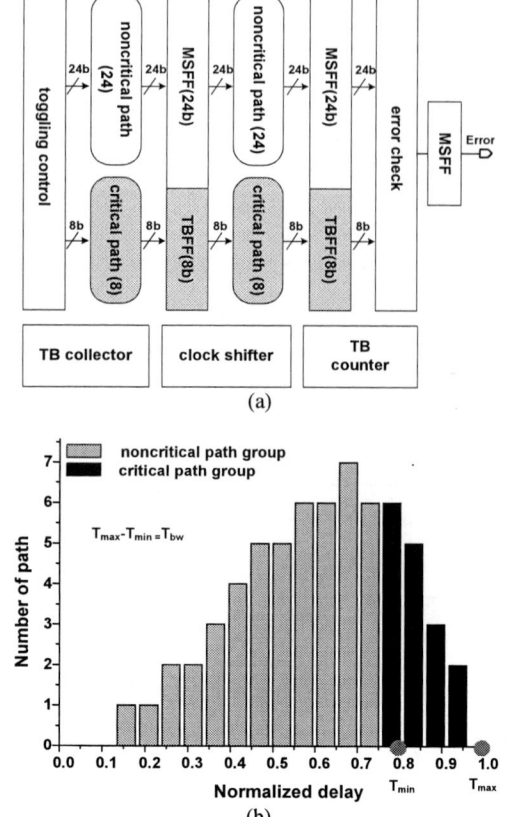

(a)

(b)

Fig. 8. (a) The implemented prototype and (b) the normalized delay distribution of delay paths of the prototype.

To execute clock stretching, the proposed clock shifter requires four-phase clocks. Since a phase-locked loop (PLL), which is embedded in a system for a system clock, can provide multi-phase clock, the clock shifter can be implemented without additional cost.

The TBFF requires a master clock (CLKM) and a reference clock (CLKR) as shown in Fig. 5 and Fig. 6. The phase of the CLKM is $0.25 \cdot T_{min}$ behind that of the CLKR. The CLKM is used as a global clock for both conventional flip-flops in noncritical paths and TBFF's on critical paths. The CLKR is used to define the time-borrowing window ($T_{BW}=0.25 \cdot T_{min}$) of the TBFF. Thus, the TBFF can borrow timing up to T_{BW} from the next pipeline stage. When the data of a latch L3 and L1 do not match, a time-borrowing detection signal (TB) indicating time borrowing, is generated to control the clock shifter circuit to change the clock period. However, if the input arrives before the rising edge of the CLKM, the TB is not set and the TBFF behaves like a conventional flip-flop.

Note that as the TBFF allows time borrowing, this creates more stringent requirements for preventing hold time violations. The hold time of the TBFF increases by T_{BW}, $0.25 \cdot T_{min}$. Therefore, while employing TBFF, it is important to consider the impact in hold time. Inserting buffers in the non-critical paths can to prevent the possibilities of additional hold violations. We expect that the area and power overhead caused by buffer insertion is not critical as TBFF's are used on critical

paths. In addition, design trend towards shallower depth of pipeline stages [4] also relaxes the overhead to fix hold time violations.

A clock shifter that stretches the clock period dynamically, shown in Fig. 7, is proposed. Whenever the clock shift signal transits from low to high, the clock shift registers change clock phase selection signals (sel), which select one clock phase among four-phase clocks with 90° ($0.25 \cdot T_{min}$) phase differences. When the clock moves from one phase to another, the clock period stretches from T_{min} to T_{max}, as shown in Fig. 7. Therefore, the proposed system with TBFF's and the clock shifter can dynamically operate with two different clock periods when

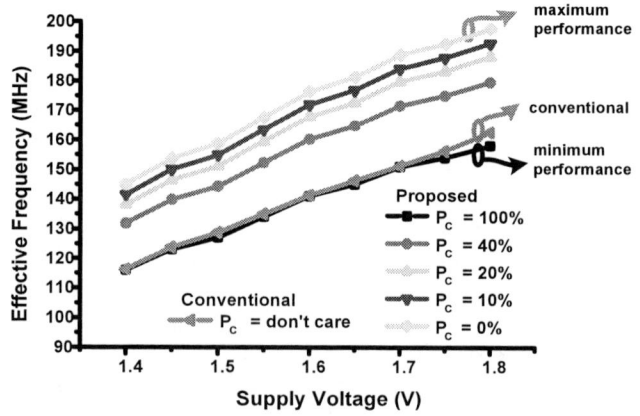

Fig. 9. Measured frequency performance of the conventional design and the proposed prototype.

TABLE I
OPERATION RANGE WITH 162.5 MHZ INPUT CLOCK.

Voltage	Conventional	Proposed
1.80V	Pass	Pass
1.75V	Fail	Pass
1.70V	Fail	Pass
1.65V	Fail	Pass
1.60V	Fail	Pass
1.55V	Fail	Pass
1.50V	Fail	Pass
1.45V	Fail	Fail

time borrowing is present. However, assuming that four-phase clocks with the clock period of T_{CK} are provided, a clock period of T_{CK} or $1.25 \cdot T_{CK}$ and T_{BW} of $0.25 \cdot T_{CK}$ can be generated as shown in Fig. 7. In this case, T_{min} and T_{max} are T_{CK} and $1.25 \cdot T_{CK}$, respectively. In the same way, a clock period of $0.8 \cdot T_{CK}$ or T_{CK} and T_{BW} of $0.2 \cdot T_{CK}$ can be generated from four-phase clocks with the clock period of $0.8 \cdot T_{CK}$. Accordingly, T_{min} and T_{max} are $0.8 \cdot T_{CK}$ and T_{CK}, respectively.

IV. TEST CHIP

A prototype for the proposed method with a three-stage pipeline was implemented in a 180-nm CMOS. In the prototype, the activation probabilities of critical and noncritical paths in pipelined stages are controlled separately by programming registers in a toggling control block in Fig. 8. Inverter chains are used to implement different delay paths for the controllability of the activation probability. The normalized path delay distribution of the prototype is shown in Fig. 8. TBFF's are used on 16 critical paths among 64 delay paths. Master-slave flip-flops (MSFF) are used on noncritical paths. For comparison, a reference pipelined system is also implemented with conventional MSFF's in all paths without any dynamic control. A serial peripheral interface bus (SPI) slave block is integrated to program the activation probability of critical paths and noncritical paths. The purpose of a TB counter in Fig. 8 is to count the number of time borrowing occurrences to calculate the F_{EFF}. The count value and the occurrence of an error can be read through the SPI block.

V. MEASUREMENTS

In the chip measurements, the activation probability of critical paths was set to 0.4, 0.2, or 0.1. In the prototype, measured power overhead caused by the TBFF and the clock shifter was 7.9% at the same operating condition as that of the conventional design. In Fig. 9 the operating frequency was measured changing the input clock frequency until the first error was detected. The measured maximum input frequency for the prototype with the proposed method is when P_c is 0 in Fig. 9. Measured results in different supply voltages show up to a 24.7% increase in the maximum input frequency for the proposed design compared to the conventional design at the same supply voltage. Table I shows that the proposed design can operate over a wider voltage range with the fixed input clock frequency. It implies that the proposed method can tolerate dynamic environmental variations with minimal performance penalty.

VI. CONCLUSION

This paper presented an effective method for preventing timing failures by utilizing time borrowing and elastic clock stretching, thereby eliminating a safety margin. Since an additional recovery or replay operation for the error management is not required, the proposed delay-error tolerant method can minimize the performance penalty. As a result, a design employing the proposed method can tolerate timing variations due to process, voltage, and temperature fluctuations with minimal performance penalty even at high activation probability of critical paths. Hence, the proposed approach allow a system operate over a wider voltage/frequency range.

ACKNOWLEDGMENT

This work was supported in part by National Science Foundation (#CCF-0916083).

REFERENCES

[1] S. Mukhopadhyay, K. Roy, "Modeling and estimation of total leakage current in nano-scaled CMOS devices considering the effect of parameter variation," in *Proc. Int. Symp. Low Power Electronics and Design*, 2003, pp. 172-175.

[2] V. Joshi, David Blaauw, Dennis Sylvester, "Soft-edge flip-flops for improved timing yield: design and optimization," in *Proc. IEEE/ACM Int. Conf. Computer-Aided Design*, Nov. 2007, pp. 667-673.

[3] K. Bowman et al., "Energy-Efficient and Metastability-Immune Resilient Circuits for Dynamic Variation Tolerance," *IEEE J. Solid-State Circuits*, vol. 44, no. 1, pp. 49-63, Jan. 2009.

[4] D. Ernst et al., "Razor: A low-power pipeline based on circuit-level timing speculation," in *Proc. IEEE/ACM Int. Symp. Microarchitecture*, Dec. 2003, pp. 7–18.

[5] S. Das et al., "A Self-Tuning DVS Processor Using Delay-Error Detection and Correction," *IEEE J. Solid-State Circuits*, vol. 41, no. 4, pp.792–804, Apr. 2006.

[6] S. Das et al., "RazorII: In Situ Error Detection and Correction for PVT and SER Tolerance," *IEEE J. Solid-State Circuits*, vol. 44, no. 1, pp.32–48, Jan. 2009.

[7] T. Sato and Y. Kunitake, "A Simple Flip-Flop Circuit for Typical-Case Designs for DFM," in *Proc. Int. Symp. Quality Electronic Design*, 2007, pp.539–544.

[8] J. Tschanz et al., "Adaptive Body Bias for Reducing Impacts of Die-to-Die and Within-Die Parameter Variations on Microprocessor Frequency and Leakage," *IEEE J. Solid-State Circuits*, vol.37, no. 11, pp.1396–1402, Nov. 2002.

[9] A. K. Uht, "Going beyond Worst-case Specs with TEAtime", *IEEE Computer*, vol.37, no.3, pp 51-56, Mar. 2004.

[10] J. Tschanz et al., "Adaptive Frequency and Biasing Techniques for Tolerance to Dynamic Temperature-Voltage Variations and Aging," in *IEEE Int. Solid-States Circuits Conf. Dig. Tech. Papers*, Feb. 2007, pp. 292-293.

[11] A. Drake et al., "A distributed critical-path timing monitor for a 65 nm high-performance microprocessor," in *IEEE Int. Solid-States Circuits Conf. Dig. Tech. Papers*, Feb. 2007, pp. 398–399.

[12] Kwanyeob Chae et al., "A dynamic timing control technique utilizing time borrowing and clock stretching," in *Proc. IEEE Custom Integrated Circuits Conference*, Sept. 2010, pp. 1-4.

[13] K. Bowman et al., "Time-Borrowing Multi-Cycle On-Chip Interconnects for Delay Variation Tolerance," in *Proc. Int. Symp. Low Power Electronics and Design*, Oct. 2006, pp. 79-84.

Time and Workload Dependent Device Variability in Circuit Simulations

D. Rodopoulos[1,2], S. B. Mahato[1,4], V. Valduga de Almeida Camargo[1,5], B. Kaczer[1], F. Catthoor[1,3],
S. Cosemans[1,3], G. Groeseneken[1,3], A. Papanikolaou[2], D. Soudris[2]

[1]IMEC, Leuven, Belgium   [2]NTUA, Greece   [3]KU Leuven, Belgium   [4]TU Munich, Germany & NTU, Singapore   [5]UFRGS, Brazil
email: drodo@microlab.ntua.gr

Abstract—**Simulations of an inverter and a 32-bit SRAM bit slice are performed based on an atomistic approach. The circuits' devices are populated with individual defects, which have realistic carrier-capture and emission behaviour. The wide distribution of defect time scales, accounts for both fast (Random Telegraph Noise – RTN) and near-permanent (Bias Temperature Instability – BTI) defects. The atomistic property of the model allows the detection of workload dependency in the delay of both circuits.**

Index Terms—**Bias-temperature instability (BTI), circuit simulations, parametric reliability, random telegraph noise (RTN), static random access memory (SRAM), workload dependency**

I. INTRODUCTION AND RELATED WORK

Most of the existing BTI simulation approaches, either in terms of modeling or mitigation, employ only the averages of BTI-related parameters. Versions of the reaction-diffusion model appear to be quite popular [5]-[9]. BTI is translated into degradation parameters, which are added to each transistor, thus altering its transient response. The parameters are constant and do not change throughout the simulation of the "aged" circuits [5].

In some cases, averaging of the stress applied on each device is employed [8]. In other approaches, there is poor incorporation of the recoverable part of BTI [9]. This suppression can result in over-estimation of the impact of the BTI mechanism [5]. In a reliability aware design context, this can lead to over-constrained designs.

Devices of older technologies have sufficient size and a large number of defects, thus exhibiting uniform reliability behaviour. As the device dimensions decrease, the stochastic defect behaviour becomes gradually more evident [1], [2].

As previously demonstrated, it is possible to monitor the occupancy of *individual* oxide defects [1], [2]. If this atomistic model is applied to any circuit, it is possible to identify the effect of the defect activity on the operation parameters of the circuit, like the circuit's delay or leakage energy. That way, the circuit's parametric reliability can be studied.

In view of the above, it is increasingly important to shift to a combined deterministic-stochastic view of reliability related phenomena, such as Bias Temperature Instability (BTI) or Random Telegraph Noise (RTN). The deterministic quality refers to the model's workload dependency, which is based on

the gate voltage dependent trap behaviour (not just on the average duty cycle of the ones or zeros). The stochastic component mirrors the probabilistic nature of oxide defect activity. Only that will allow observing and analyzing true workload dependent behaviour. Moreover, that analysis can be the basis of mitigation approaches based on workload tuning. Enabling these important objectives by providing the basic modeling framework is the main differentiator of this paper.

The next section covers the atomistic defect model. The third section contains information on the modules used to illustrate the novel model. Finally, the simulation results are presented along with some brief conclusions.

II. THE ATOMISTIC BTI MODEL

A. Theoretical Model

The atomistic model that is used in the simulations is consistent with the device downscaling trend toward nm dimensions, since it treats each defect of each device separately. Each device is enhanced with a random number of traps, extracted from a Poisson distribution (an average of 10^{12} traps per cm^2 is used). Each trap is characterized by a different threshold voltage impact ($\Delta V_{th} > 0$) and a different set of time constants $\{\tau_{cH}, \tau_{cL}, \tau_{eH}, \tau_{eL}\}$ resembling the capture or emission times for high (H) or low (L) voltages at the device's gate (V_{gs}). Our setup determines each defect's occupancy transitions *during the circuit simulation* using

$$P_{p,v} = \frac{\tau_{\bar{p},v}}{\tau_{e,v} + \tau_{c,v}} \left\{ 1 - \exp\left[-\left(\frac{1}{\tau_{e,v}} + \frac{1}{\tau_{c,v}} \right) \Delta t \right] \right\}, (1)$$

where the process p ($\bar{p}$ being the complement) can be either a capture (c) or an emission (e) event. Δt is the simulation step and v can be either of the $\{H, L\}$ voltage levels.

For every simulation step, a random number is compared to the process probability (1). If the random number is found smaller than the probability, the respective process occurs. When a defect becomes *occupied* (i.e. a carrier is captured in the trap), the respective ΔV_{th} value is added to the runtime V_{th} value of the device. Further details on the model and its assumptions can be found in [2].

978-1-4244-9019-6/11 $26.00 © 2011 IEEE

The distributions of the *defect time scales* are taken from experiments. The interval of time constants is sufficient to account for a wide variety of defect behaviours. Hence, it is possible to observe fast trapping and detrapping events for a single defect (corresponding to *RTN*). It is also possible to observe near-permanent behaviours with larger time constants (corresponding to *BTI*).

B. Netlist Annotation Framework

The simulations presented in the current paper account only for NBTI, but PBTI can also be easily included. For the purpose of the simulations, a framework is required with which to populate the netlist under test with the defects, their time constants and their V_{th} impact. This annotation is performed during pre-processing (Fig. 1). Any sub-circuit is expanded down to device level. Each device is annotated with NBTI parameters alongside the parameters provided by the PTM model [3]. The transient part of the atomistic model is added on top of a Verilog-A implementation of the BSIM4 model.

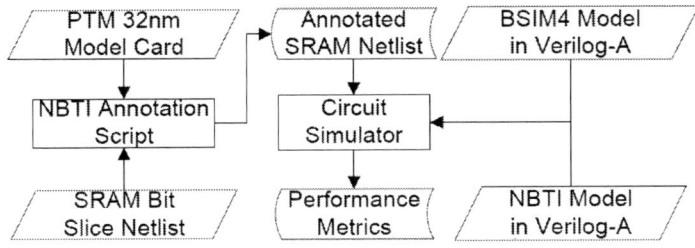

Fig. 1. Generic flowchart of the netlist annotation framework.

C. Workload Dependency

In commercial reliability tools like RelXpert [5], each device is replaced by an aged equivalent, based on the input stress provided to the simulator. Then, a simulation of the "aged" circuit is performed, thus indicating circuit degradation. There is no information provided about the *runtime defect activity*. Such approaches are unable to present a transient view of reliability phenomena. The purpose of this paper is to differentiate the atomistic BTI model from the state-of-the-art BTI approaches. *The main differentiator is the time-dependent workload dependency*. Hence, *all simulations need to assume time and workload as the only independent variables*.

The NBTI model requires a random number per iteration, which is compared to the V_{gs} dependent probability for a capture or emission event. The workload for each device corresponds to the imposed V_{gs} at any given time.

In the default implementation, no correlation is present between the generated random numbers and the imposed V_{gs}.

If we bind the random number generation to the imposed V_{gs} (Fig. 2), we decouple the workload dependency from the model's stochastic component.

The initial seed values for the random number generation should remain the same, irrespective of the imposed workload.

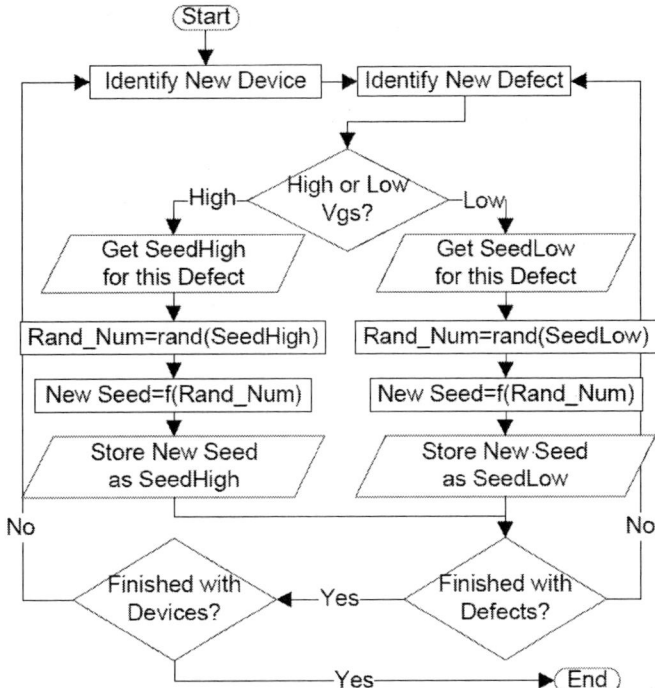

Fig. 2. Customized flowchart of random number generation.

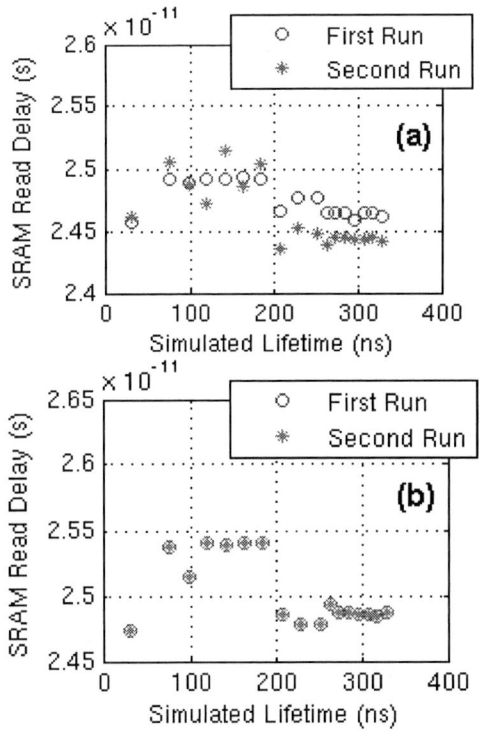

Fig. 3. Two simulations of the same workload: (a) Default implementation with an evident stochastic component and (b) If workload dependency is emphasized, the simulation outputs coincide.

The annotation of the netlist with NBTI parameters has to be performed just once.

In order to clarify the need for the above configurations, we simulate the same SRAM partition workload two consecutive times (Fig. 3). As output of the simulations, we consider the

978-1-4244-9019-6/11 $26.00 © 2011 IEEE

observed delay for the SRAM read operation. If we use the above configurations, we inspect the same output in both runs, since *the workload is the same for both* (Fig. 3b). In the opposite case, the stochastic component is deeply routed in the simulations, so the output is different (Fig. 3a).

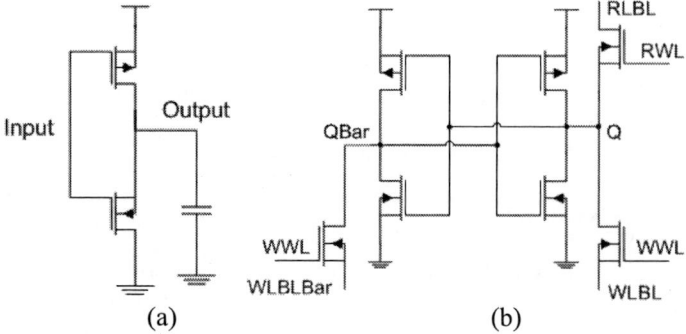

(a)   (b)

Fig. 4. Schematic of the inverter (a) and of an SRAM cell (b).

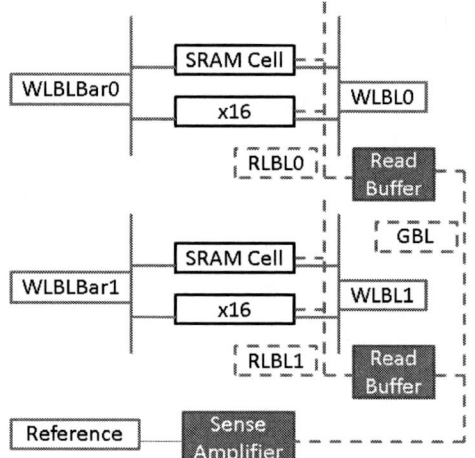

Fig. 5. Schematic of the SRAM Partition (reading path is dashed).

III. CASE STUDIES

A. Inverter

In this case, we use a simple inverter (Fig. 4a) and measure its delay. Since the pull-up branch is affected by NBTI, we expect larger delays for measurements during a 1-0 transition at the input.

B. SRAM Bit Slice

This circuit is an SRAM bit slice of 32 bits, divided in two groups of SRAM cells, based on [4].

The two complementary nodes of each SRAM cell (Fig. 4b) are connected to voltage sources through access transistors. Writing is implemented by applying a word pulse to the respective transistors while maintaining the necessary voltage values at WLBL and WLBLBar. For the reading operation, the two groups of cells have single local read bit lines (RLBL0 and RLBL1) which are connected to a global bit line (GBL) through read buffers (Fig. 5). The GBL is then connected to a sense amplifier (SA). The RLBLs and the GBL are always precharged to a specific voltage and will be discharged to V_{ss}, when a stored 0 is read.

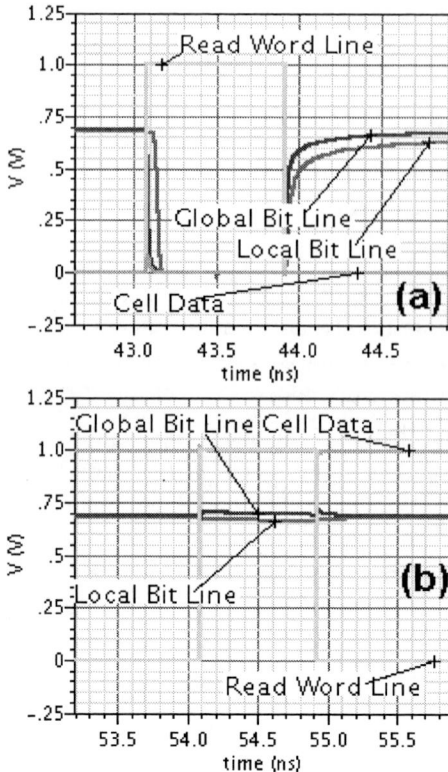

Fig. 6. Instances of the reading operation, either for logic 0 (a) or logic 1 (b). Evidently, only the second case allows a delay measurement.

The performance metric is the *delay of the read operation*. We measure the time from the activation of a word line up to the point where the voltage difference at the inputs of the SA, is enough for the latter to sense (Fig. 5). In the current SRAM organization, this metric can be applied only to cases when the read value is logic 0 (Fig. 6).

```
while(user_defined) {

  /* Write cells consecutively */
  for (i=0; i<=31; i++) {
    write(cell[i]);
  }

  /* Read cells consecutively */
  for (i=0; i<=31; i++) {
    read(cell[i]);

    /* Global Retention in-between */
    for (j=1; j<=10; j++) {
      nop
    }
  }
}
```

Fig. 7. Algorithmic description of the simulated SRAM activity.

C. Stimuli Setup

In the inverter case, we define a specific sequence of bits applied at the input of the inverter as a runtime situation (RTS). The frequency of this sequence is 1GHz.

In case of the SRAM partition, each of the cells can be in one of the Read, Write or Retain operation modes. If a cell is not being read or written, it retains its value. All operation modes are assumed to have the same duration. As long as the consistency of the cells' internal state is maintained (e.g. one cannot read from a cell that has not been previously written), an arbitrary sequence of operation modes can exist, with a user defined frequency of 1GHz. The SRAM activity can be described by a handful of pseudo-code lines (Fig. 7). In the SRAM partition case, a different sequence of values written in the cells indicates a different runtime situation (RTS).

Each control signal or stimulus is defined as a piece-wise linear source, which is set according to the inspected RTS. This setup is performed during pre-processing and is incorporated into the annotation framework presented in section II (Fig. 8).

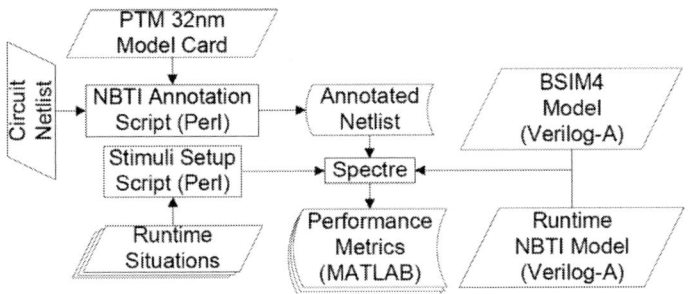

Fig. 8. Customized flowchart of the simulation framework (including automated stimuli setup).

IV. SIMULATIONS

A. Inverter

Based on a bit sequence of 200ns (Reference Workload), we change either the first ("Past") or the last ("Future") 100ns of the sequence (Fig. 9). *The results demonstrate a distinct proof of the model's detailed workload dependency*, which is not averaged out by a pure stochastically based device model. The simulation output of the Changed "Future" is identical to the Reference Workload output up to the point that the bit sequence is changed. In contrast, if we only change the "Past",

the simulation output is entirely different (Fig 10).

What we demonstrate here is that *the atomistic BTI model has a deterministic workload memory. The observed NBTI effect at any time strongly depends on the specific workload that has preceded the current operation state. We can also analyze this on a cycle-by-cycle deterministic basis, which is unique compared to existing circuit-level models.*

B. SRAM Bit Slice

Two entirely different runtime situations are simulated (Fig. 11). Each workload is applied to the NBTI enhanced netlist and to a reference netlist without any oxide defects. No initial stressing of the devices is assumed; all defects are not occupied at the beginning of the simulation. The delay fluctuations differ between the two workloads, in the NBTI Enhanced case. On the contrary, the delay fluctuations are almost identical between the two RTSs, for the Reference netlists.

In another round of simulations, we use the analytical model presented in [2] to stress all the devices of the netlist with an AC signal (80% duty cycle for $-V_{gs}$) for 10^7 seconds. That way, we initialize the trap occupancy in the NBTI Enhanced netlist. A Reference netlist, without defect activity, is also used. Two different RTSs are simulated in both netlists (Fig. 12). We observe that the delay fluctuations in the NBTI Enhanced netlist occupy a much wider interval, than the Reference counterpart. The increased variability of the NBTI Enhanced netlist surpasses the usually accepted 10% criterion.

V. CONCLUSIONS

By monitoring separately every defect of every simulated device, we can see the workload dependent nature of the atomistic BTI model, propagating to the delay of the aged circuit. Even if we change the workload at runtime, we immediately see a difference in the evolution of the circuit's delay. When NBTI is accounted for, the delay measurements

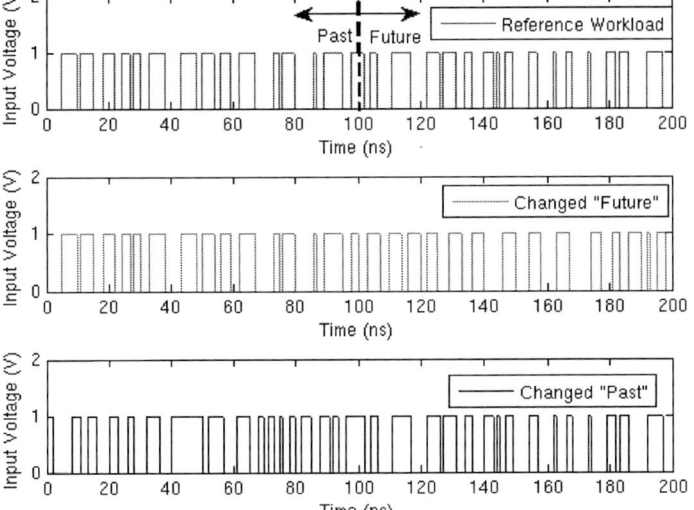

Fig. 9. Runtime changes on the inverter's workload.

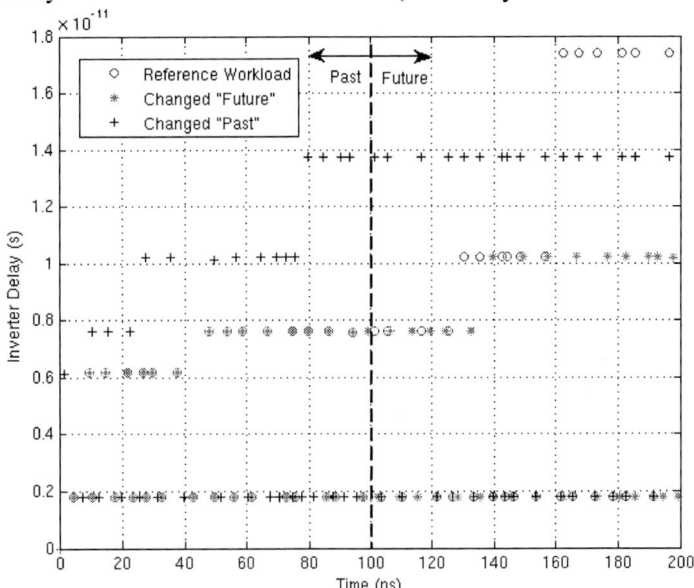

Fig. 10. Delay measurements on the inverter.

978-1-4244-9019-6/11 $26.00 © 2011 IEEE   43

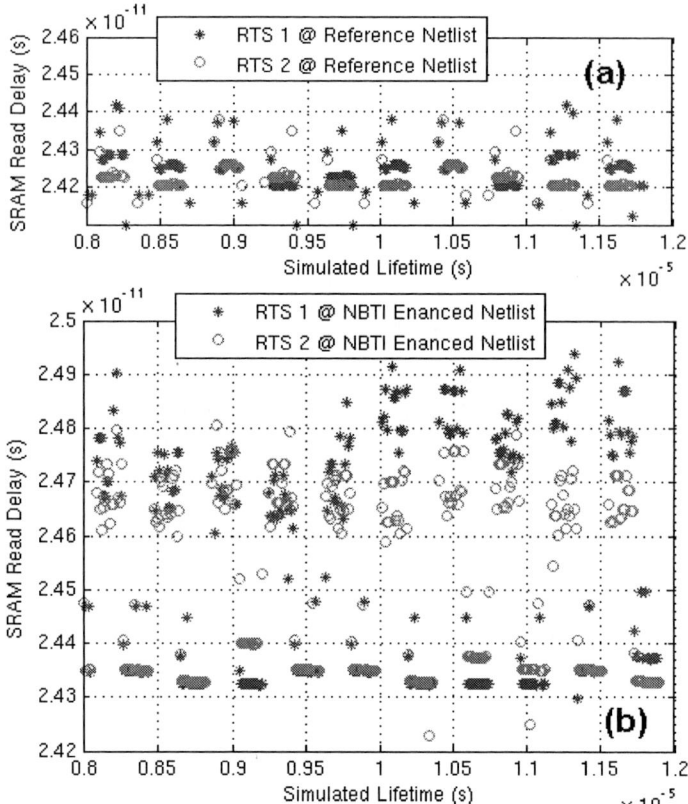

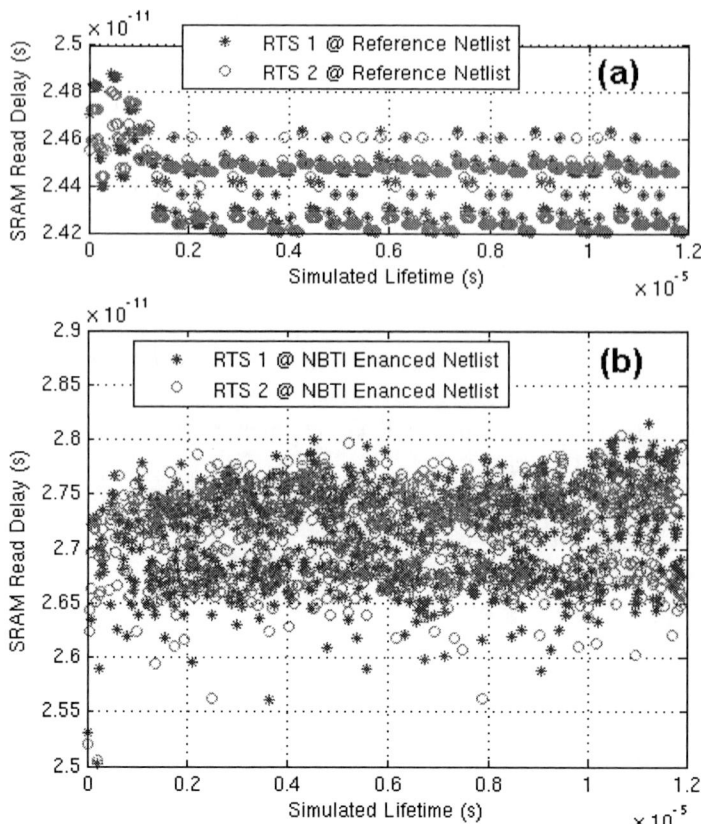

Fig. 11. Runtime delay fluctuations of an SRAM partition for two different RTSs. No initial stressing is assumed (time zero simulation). The delay measurements are almost identical in the Reference netlist (a). For a netlist enhanced with defect activity (b), the fluctuations exhibit greater runtime variability and also differ between the two workloads.

Fig. 12. Runtime delay fluctuations of an SRAM partition for two different RTSs. Initial AC stressing is assumed. When NBTI is not considered (a), the fluctuations are concentrated in a very small interval. In a netlist enhanced with defect activity (b), the delay fluctuations cover a much wider interval.

of the SRAM partition are distributed in a much broader interval.

As a result, the two test cases of this paper illustrate and substantiate the deterministic component of our model. This approach allows the modelling of individual trap behaviour based on the actual input stimuli sequence (and not just on the duty cycle of the ones or zeros as seen in other approaches).

The challenge that lies in view of these conclusions is to start searching for correlations between the imposed workload and its observed impact on performance metrics. That way, a more realistic view of the parametric reliability of larger circuits can be obtained. The current defect model incorporates both the stochastic and the workload dependent nature of oxide traps. Hence, it is a suitable tool with which to explore parametric reliability in a realistic and detailed workload dependent way.

ACKNOWLEDGMENTS

Part of this work was carried out in IMEC's Industrial Affiliation Program funded by IMEC's core partners. V.V.A.C. thanks CNPQ Brazil for financial support.

Discussions with Prof. T. Grasser about the physical aspects of gate oxide charge trapping are gratefully acknowledged.

REFERENCES

[1] Kaczer, B.; Grasser, T.; Roussel, P.J.; Franco, J.; Degraeve, R.; Ragnarsson, L.; Simoen, E.; Groeseneken, G.; Reisinger, H.; , "Origin of NBTI variability in deeply scaled pFETs," *Reliability Physics Symposium (IRPS), 2010 IEEE International* , vol., no., pp.26-32, 2-6 May 2010

[2] Kaczer B. *et al.*, "Atomistic approach to variability of bias-temperature instability in circuit simulations", *Reliability Physics Symposium (IRPS), 2011 IEEE International* (accepted)

[3] http://ptm.asu.edu/ .

[4] Cosemans, S.; Dehaene, W.; Catthoor, F.; , "A 3.6 pJ/Access 480 MHz, 128 kb On-Chip SRAM With 850 MHz Boost Mode in 90 nm CMOS With Tunable Sense Amplifiers," *Solid-State Circuits, IEEE Journal of* , vol.44, no.7, pp.2065-2077, July 2009

[5] Zhihong Liu; McGaughy, B.W.; Ma, J.Z.; , "Design tools for reliability analysis," *Design Automation Conference, 2006 43rd ACM/IEEE* , vol., no., pp.182-187, 0-0 0

[6] Calimera, A.; Macii, E.; Poncino, M.; , "Analysis of NBTI-induced SNM degradation in power-gated SRAM cells," *Circuits and Systems (ISCAS), Proceedings of 2010 IEEE International Symposium on* , vol., no., pp.785-788, May 30 2010-June 2 2010

[7] S. Khan, S. Hamdioui, Temperature Impact on NBTI Modeling in the Framework of Technology Scaling, Digest of the 2nd Design For Reliability (DFR 10), PISA, Italy, Jan 2010.

[8] Kumar, S.V.; Kim, C.H.; Sapatnekar, S.S.; , "NBTI-Aware Synthesis of Digital Circuits," *Design Automation Conference, 2007. DAC '07. 44th ACM/IEEE* , vol., no., pp.370-375, 4-8 June 2007

[9] Paul, B.C.; Kunhyuk Kang; Kufluoglu, H.; Alam, M.A.; Roy, K.; , "Negative Bias Temperature Instability: Estimation and Design for Improved Reliability of Nanoscale Circuits," *Computer-Aided Design of Integrated Circuits and Systems, IEEE Transactions on* , vol.26, no.4, pp.743-751, April 2007

An On-Chip Waveform Capturer for Diagnosing Off-Chip Power Delivery

*Kumpei Yoshikawa, *Takushi Hashida and *,**Makoto Nagata

*Graduate School of System Informatics, Kobe University

1-1 Rokkodai-cho, Nada-ku, Kobe 657-8501, Japan

**CREST, JST

Email: {kumpei,hashida,nagata}@cs26.scitec.kobe-u.ac.jp

Abstract—**In-place diagnosis of off-chip power delivery resonance is demonstrated with on-chip waveform capturer and power delivery network (PDN) exciter that were prototyped in a 65 nm CMOS technology. Oscillatory waveforms are captured after the excitation of PDN, from which an LCR lumped equivalent circuit of PDN seen by on-chip circuits is algorithmically derived. The consistency of component values is confirmed among the demonstrated in-place diagnosis and full-wave analysis.**

I. Introduction

Precise acquisition of on-chip waveforms is deeply applicable to the insight of integrated circuits, diagnosing operation of circuits and interaction with environments. An waveform capture is equipped with digitizing and digital processing features, and expected to coordinate with a host system in capturing and analyzing waveforms in system-on-a-chip (SoC) integration.

This paper demonstrates the usability of such an on-chip waveform capturing function for the diagnosis of a power delivery network (PDN). The resonance frequencies and decay time constants are quantitatively derived from on-chip captured waveforms and used for the calculation of effective parasitic impedances seen from on-chip circuits, and embodying on-the-fly prediction of noise peaks as a function of operating frequency.

General purpose SoC chips, such as automotive, consumer handsets, and factory automation applications, are typically operating at from several 10 MHz to 300 MHz, and inevitably encountering the resonance frequency of PDNs formed in low-cost assembly. This potentially causes operation failures due to the enlarged power noises at these frequencies.

Supply voltage variation due to PDN resonance in a microprocessor was discussed in [1]. A variety of circuit techniques to suppress [2,3] or filter out [4] such voltage variation had been proposed. However, the transfer response of PDN is indeed not accurately provided during the development of general-purpose chips. Therefore, in-place adjustments of operating frequency effectively work for high dependability, with the help of on-chip PDN diagnosis.

The architecture of an on-chip waveform capture will be discussed in Sect. II. On-chip PDN diagnosis will be demonstrated with a 65 nm CMOS prototype chip in Sect. III, and derived impedances are compared with analysis. A brief conclusion will be given in Sect. IV.

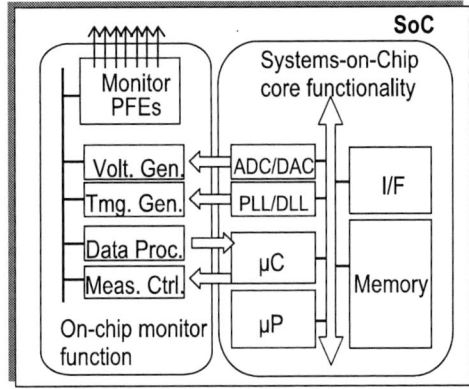

Fig. 1. On-chip waveform capture functionality merged in SoC integration.

II. On-Chip Waveform Capturing

The capability of on-chip waveform capture is merged in SoC integration, shown in a conceptual view of Fig. 1, where general SoC resources are partly involved. An auxiliary micro controller (μC) executes waveform acquisition and locates waveforms in a shared memory space that a main processor (μP) of SoC accesses for subsequent digital processing.

The concrete system diagram of a proposed waveform capturer is given in Fig. 2 [5]. An input channel is provided by probing front-end (PFE) circuitry, that senses and digitizes input voltage, V_{in}, seen at the wiring of interest. Voltage and timing generators, VG and TG, respectively, give reference voltage and sampling timing for the digitization, respectively. A single channel is selected at a time from an array of PFEs that share the VG and TG circuits.

The PFE channel consists of a source follower and a latch comparator [6]. The voltage, V_{in}, continuously sensed by a source follower (SF) is strobed at the timing, T_{strb}, defined by TG, and approximated to the nearest reference voltage, V_{ref}, produced by VG. The nearest search across the space of VG voltage is performed in response to the output of a latch comparator, assuming iterative operations of a target circuit to diagnose.

The PFE channels are prepared for a variety of voltage domains of interest, from the highest voltage of I/O power supply to a common ground voltage. The offset voltage of source followers is tailored for each voltage domains so as to

978-1-4244-9019-6/11 $26.00 © 2011 IEEE

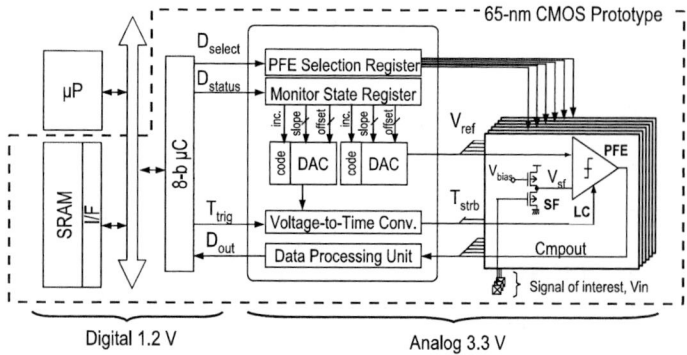

Fig. 2. Block diagram of on-chip waveform capturer.

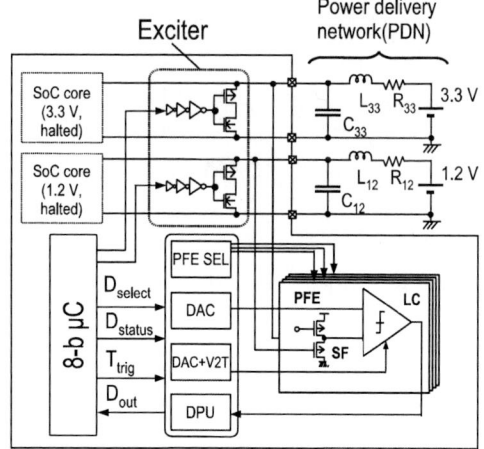

Fig. 3. Power delivery network (PDN) exciter in combination with waveform capturer.

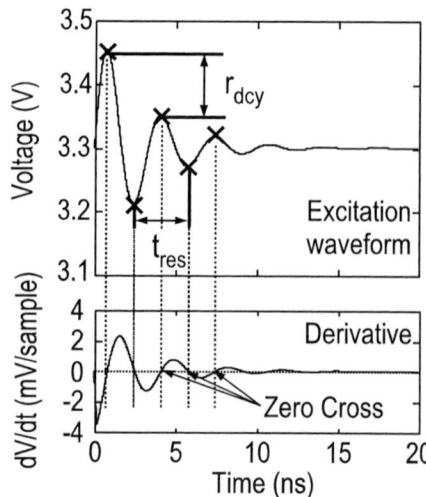

Fig. 4. PDN excitation waveforms and extraction of parameters of resonation.

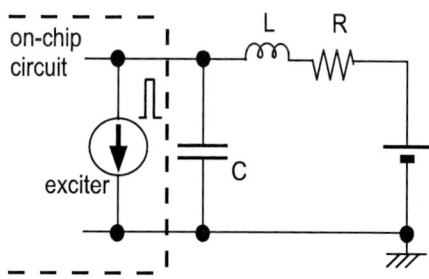

Fig. 5. Off-chip PDN seen by on-chip circuit, resonation equivalently represented by 1st order LCR closed loop.

maintain a constant input DC voltage to the comparator.

A micro controller (μC) communicates with internal operating blocks of a waveform capturer through status registers and executes the search algorithm as the function of output probability of a latch comparator. The output streams from the comparator are measured by a data processing unit (DPU) and accumulated in an output register, D_{out}.

Scalable generation of timing and voltages by TG and VG, respectively, allows to capture fine structures of waveform including such as spikes and coupling noises, while overviewing decaying and oscillations. On-chip digitization reduces the number of pin counts, while digital analysis enables in-place diagnosis of SoC operation and supports run-time adjustments against operation environments. The circuit details and operation of TG and VG are provided in [7].

III. ON-CHIP PDN DIAGNOSIS

A. Power delivery network excitation

Power is delivered to SoC entity through the series connection of on-chip wirings and off-chip traces, forming a power delivery network (PDN). PDN resonance by parasitic inductance (L), capacitance (C), and resistance (R) can seriously degrade the stability of power delivery and emphasize the

voltage bounce by internal circuit operation, namely, power noise.

A power delivery network exciter, given in Fig. 3, intentionally induces LCR resonance to power rails of interest. The resultant waveforms are digitized by an on-chip waveform capturer and stored in a common memory space for subsequent digital processing. The PDN exciter and waveform capturer are managed by system μC in the mode of PDN diagnosis, while core SoC functionality is halted so as to isolate the resonance.

B. Derivation of lumped components

The resonating waveforms are inherent to PDN of a chip in assembly, as shown in Fig. 4. The periods of oscillation, t_{res}, and the decay ratio of amplitude, r_{dcy}, are derived from the waveforms. The series of time locations when the 1st derivative of the waveform crosses the zero border determine t_{res}. Concurrently, the decay ratio is calculated from the adjacent voltage amplitudes that are individually measured at the same set of time locations.

The inductance (L) and capacitance (C) of PDN are related with the derived t_{res} and r_{dcy} according to (1), assuming the equivalency of PDN with the 1st order lumped LCR circuit of Fig. 5.

The resistance (R) needs to be pre-characterized by the circuit simulation of PDN exciter with regard to the oscillation

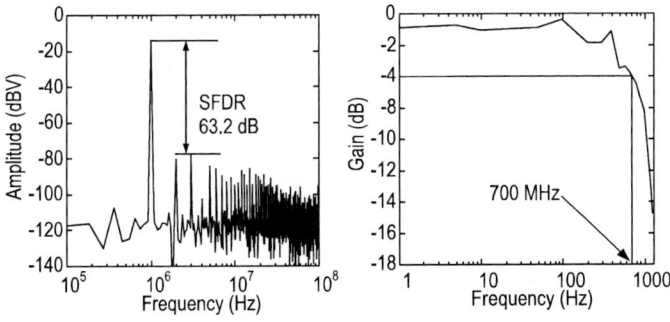

Fig. 6. Measured AC response of on-chip waveform capturer.

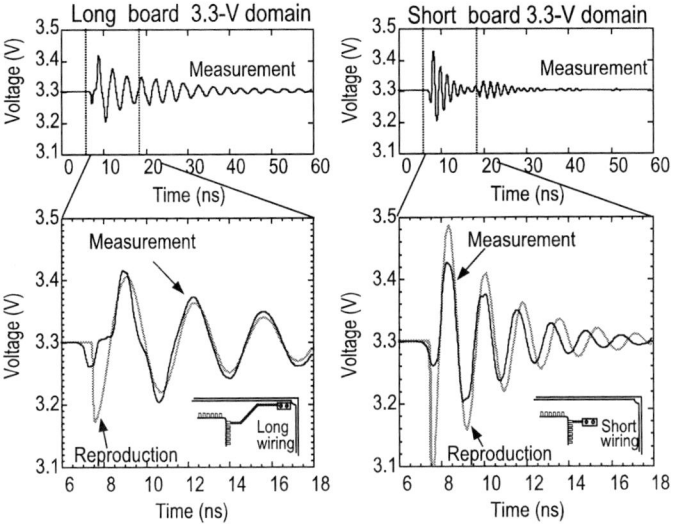

Fig. 7. Measured and reproduced waveforms of PDN excitation.

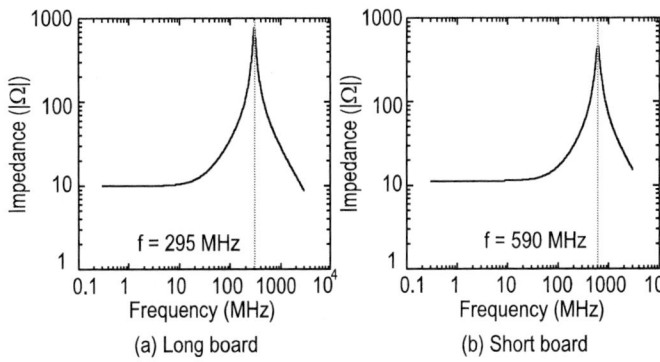

Fig. 8. Simulated off-chip impedance seen from on-chip circuits with PDN exciter.

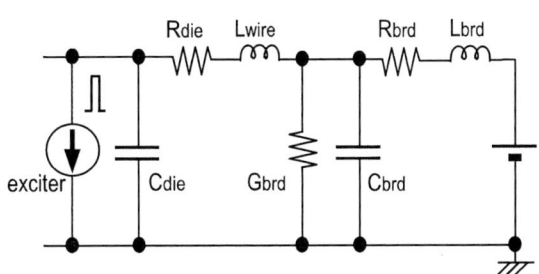

Fig. 9. Entire PDN equivalent circuit.

amplitude, A_0, that is represented by the multiplication of power current and resistance. The 1st peak of oscillation is used to match the pre-characterized A_0 with the acquired waveform.

$$V(t) = A_0 \cdot e^{-t/\tau} \cdot sin(\omega \cdot t + \phi),$$
$$where \ \tau = \frac{t_{res}}{ln(r_{dcy})} = \frac{2L}{R},$$
$$\omega = 2\pi \cdot F_{pdn} = \frac{2\pi}{t_{res}} = \sqrt{\frac{1}{LC} - \frac{1}{\tau^2}} \quad (1)$$

C. Experimental results

On-chip waveform capturer and PDN exciter were prototyped in a 65-nm CMOS technology. Analog circuit blocks including PFE, TG, VG, and selection switches are designed with 3.3 V high voltage (I/O) devices. This is generally needed to cover the whole voltage range in a chip by monitor channels. On the other hand, digital back-end blocks including μC are synthesizable with 1.2 V standard logic cells.

Figure 6 gives AC response of the prototype capturer. The signal purity of 63.2 dB, evaluated by spurious free dynamic range (SFDR), is achieved for capturing sinusoids at 1 MHz.

The effective signal bandwidth of interest is measured as 700 MHz. The resolution of the capturer is programmable, while the finest are 5 ps in timing and 0.2 mV in voltage.

Figure 7 plots captured PDN waveforms of the prototype chip mounted on differently sized FR-4 boards. The oscillatory waveform is naturally generated after a single excitation. The long board obviously exhibits lower oscillation frequency than the short counterpart, reflecting the longer power line traces. In fact, the 1st three oscillations from the excitation dominantly represent the electrical property of a PDN network. The time resolution of 10 ps by TG was needed for extracting electrical parameters with the analysis resolution of 1 MHz, against the resonance frequency of several 100 MHz.

Table I summarizes LCR values calculated from (1) from the captured waveforms, with comparison to full-wave analysis discussed in the next subsection. The off-chip PDN impedance seen from on-chip circuits is simulated as in Fig. 8, with the derived LCR values in the simplified LCR closed loop of Fig. 5. The plot will be useful for the diagnosis of PDN response. On the other hand, a circuit can autonomously avoid the operating frequencies around the integer division of the resonance frequency, that is directly calculated from t_{res} [7].

The waveforms additionally shown in Fig. 7 are reproduced by an equivalent network of Fig. 5 with the derived LCR values. The observed consistency proves the simplification of equivalent PDN circuits.

D. Analysis results

A full-wave solver was applied to simulate the broadband propagation characteristics of power traces on the long and

TABLE I

DERIVED LCR VALUES OF VDD TRACES ON LONG AND SHORT FR-4 BOARDS. ON-CHIP PDN DIAGNOSIS IS COMPARED WITH FULL-WAVE ANALYSIS.

		On-chip diagnosis		Analysis	
Long Board	Fpdn	295 MHz	Fpdn	295 MHz	
	R	10.0 Ω	Rbrd	0.7 Ω	
			Rdie	10.0 Ω	
	L	47.5 nH	Lbrd	34.5 nH	
			Lwire	10 nH	
	C	6.1 pF	Cbrd	2.7 pF	
			Cdie	4.8 pF	
Short Board	Fpdn	590 MHz	Fpdn	590 MHz	
	R	11.3 Ω	Rbrd	0.3 Ω	
			Rdie	10.0 Ω	
	L	18.8 nH	Lbrd	8.4 nH	
			Lwire	10 nH	
	C	3.6 pF	Cbrd	2.0 pF	
			Cdie	3.4 pF	

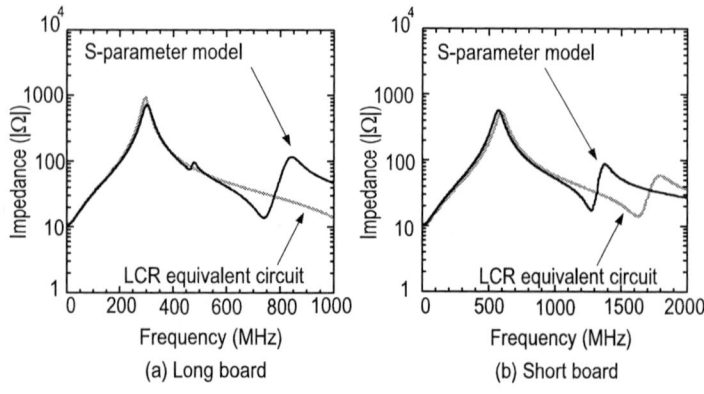

Fig. 10. Impedance comparison of S-parameter model and LCR equivalent circuit.

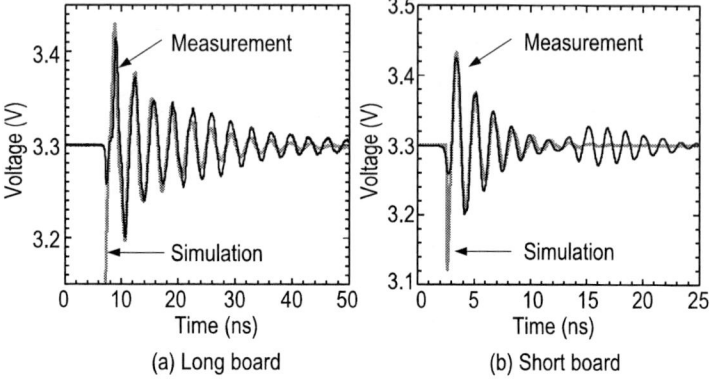

Fig. 11. Simulated PDN noise waveforms and measured waveforms for comparison.

short FR-4 boards, respectively. The derived S-parameter was captured in the 1st order lumped LCR closed loop, within the frequency bandwidth of interest. Then, an PDN model of the entire chip is formed as shown in Fig. 9, including die capacitance, C_{die}, on-chip series resistance, R_{die}, as well as series inductance due to bonding wires, L_{wire}, in addition to the derived LCR closed loop of board traces. We have hand-calculated the values of C_{die}, R_{die}, and L_{wire}, from physical dimensions.

The frequency responce of PDN impedance is simulated for the lumped LCR circuit of Fig. 9 and compared with the original S-parameter model in Fig. 10. It is commonly observed among long and short boards that the lumped LCR model well represents the PDN impedances for the frequency bandwidth of twice as large as the respective 1st resonance frequencies. The component values can now be compared among the full-wave analysis and on-chip PDN diagnosis, as given in Table I. The noise waveforms simulated with the lumped LCR models are also compared with those by on-chip waveform capturing, as shown in Fig. 11. Their results agree consistently with each other.

IV. CONCLUSION

In-place diagnosis of off-chip PDN resonance is demonstrated with on-chip waveform capturer and PDN exciter. It is proven that the 1st order LCR closed loop well represents off-chip PDN seen from on-chip circuits, where the component values computed from on-chip captured oscillatory waveforms are very much consistent with analysis by a full-wave solver.

The in-place diagnosis of off-chip PDN impedance predicts the frequencies at the highest PDN impedance that can vary among diverse assembly options. It will also provide preventive measures against the impacts of dynamic supply voltage variation.

ACKNOWLEDGMENT

This work is in part supported by CREST, JST. The test chip was fabricated by the cooperation of eShuttle, Fujitsu, and VDEC.

REFERENCES

[1] T. Rahal-Arabi, G. Taylor, M. Ma, and C. Webb, "Design and validation of the Pentium III and Pentium 4 processors power delivery," in VLSI Circuits Digest of Technical Papers, pp. 220-223, Jun. 2002.

[2] J. Xu, P. Hazucha, M. Huang, P. Aseron, F. Paillet, G. Schrom, J. Tschanz, C. Zhao, V.De, T. Karnik and G. Taylor "On-die supply-resonance suppression using band-limited active damping," in Intl. Solid-State Circuits Conf., Dig. Tech. Papers, pp. 286-287, Feb. 2007.

[3] S. Pant, and D. Blaauw, "A charge-injection-based active-decoupling technique for inductive-supply-noise suppression," in Intl. Solid-State Circuits Conf., Dig. Tech. Papers, pp. 416-417, Feb. 2008.

[4] J. Gu, H. Eom, and C. Kim, "On-chip supply noise regulation using a low-power digital switched decoupling capacitor circuit," IEEE J. Solid-State Circuits, Vol. 44, No. 4, pp. 1765-1775, Jun. 2009.

[5] T. Hashida, and M. Nagata, "On-chip waveform capture and diagnosis of power delivery in SoC integration," in Symposium on VLSI Circuits Dig. Tech., pp. 121-122, Jun. 2009.

[6] Y. Araga, T. Hashida, and M. Nagata, "An on-chip waveform capturing technique pursuing minimum cost of integration," in Proc. IEEE Int. Symp. Circuits and Systems (ISCAS), pp. 3557-3560 May 2010

[7] T. Hashida, and M. Nagata, "On-chip waveform capture and application to diagnosis of power delivery in SoC integration," IEEE J. Solid-State Circuits, Vol. 46, No. 4, April 2011 to appear.

Interconnect Test for Core-based Designs with Known Circuit Characteristics and Test Patterns

Tung-Hua Yeh and Sying-Jyan Wang
Department of Computer Science and Engineering
National Chung-Hsing University
Taichung 402, Taiwan, ROC
e-mail: {phd9303, sjwang}@cs.nchu.edu.tw

Katherine Shu-Min Li
Department of Computer Science and Engineering
National Sun-Yat Sen University
Kaohsiung, Taiwan, ROC
e-mail: smli@cse.nsysu.edu.tw

Abstract—**System-level interconnect structures become much more complicated and dominate overall performance in multi-core systems. In order to facilitate interconnect test in board-level and system-on-chip (SoC) designs, IEEE standards 1149.1 and 1500 are developed. Dedicated design-for-testability (DFT) architectures for interconnect consisting of through-silicon via (TSV) in future 3-D stacked ICs have also been investigated. Whenever DFTs in such designs are not available due to limits of design constraints or overall cost consideration, testing those inaccessible interconnects becomes a difficult problem and it is rarely discussed in the literature. In this paper, we propose an interconnect test scheme that exploits circuit characteristics, inherent test resources in design, and test patterns of embedded cores to test interconnect. Since chips are often tested before interconnect, our scheme utilizes those good chips to propagate test patterns and observe responses of interconnect.**

I. INTRODUCTION

Modular testing requires various levels of design-for-testability (DFT) supports such as internal scan chain design for cores, IEEE std. 1149.1 [1], and IEEE std. 1500 [2]. The system-in-package (SiP) has used in many cellphone products; this technique utilizes wire bonding to connect stacked dies. Testing SiP also relies on the DFT supports mentioned above. Recently, a new interconnect technology called through-silicon-via (TSV) has been developed to connect stacked dies to form a three-dimensional ICs (3DIC). In order to facilitate testing of 3DIC, various DFT techniques have also been studied [3, 4]. In addition to modules, interconnect can be tested with the help of boundary scan cells and wrappers. A lot of test generation approaches [5-7] were developed assuming boundary scan DFT is available. These algorithms generates small test sets to detect all open and short faults on interconnect. BIST [8] and Oscillation ring [9] provide other alternatives for interconnect test.

Most of previous works requires boundary scan, wrapper, or specific DFT support for interconnect test. However, interconnect test becomes a difficult problem in case those DFTs are not available due to design constraints or overall cost consideration, and this issue is rarely discussed before. Function test is a possible approach for those inaccessible interconnects, but the fault coverage usually cannot be guaranteed. Another method is to generate interconnect tests directly when system behavior is known, but it is too difficult to apply this method to large system as the test generation time can be prohibitive. Besides, this method cannot be applied when circuit architecture is unknown, as

vendors of intellectual property (IP) may only provide test patterns to their customers.

In summary, major problems in interconnect testing are the difficulty of test generation and test generation is impossible when circuit architecture is unknown. To solve these two problems, we address the interconnect test problem at higher level, in which only system connectivity is required while detailed circuit architecture of cores are not necessary. Test patterns for interconnect are derived from known test pattern sets of cores; therefore complicated test generation problem is reduced to a pattern selection problem. In addition, pattern selection still works even if circuit behavior is unknown. The basic idea of the proposed approach is similar to hierarchical testing [10] where test patterns for interconnect faults and their corresponding responses can be produced and propagated by other cores' computation. Pattern selection is still time consuming because there is a large number of pattern combinations. We adopt two strategies to reduce CPU time. (1) All DFTs in designs are fully exploited to help interconnect test. (2) The number of pattern combinations is reduced by calculating *input controllability don't-care set* (CDC_{in}) of cores and pruning decision tree. The major contributions of this paper are summarized as follows:

1. A unified interconnect test method is given in this paper even if the DFT is not supported.
2. The complex test generation problem is reduced to a pattern selection problem. The most important aspect is that this method is still applicable even if circuit architecture or behavior is unknown.
3. Our approach can be used for testability estimation of a system since any possible DFT configuration is acceptable in our program.

II. PRELIMINARIES

A. System Model

In this work, we assume the system is a core-based design described by Verilog-like language hierarchically. Our method can be used for either 3D or 2D designs. For 3D designs, each die is treated as a sub-design in system description. All interconnects connecting to system IO pins or dies are implemented by wire bonding. The code description of the connectivity among dies and the cores inside the dies are similar. Please note that intra-die interconnect is implemented through metal wires, not through wire bonding. In addition to hierarchical description of a system, our system model needs a *core library file* and a *DFT configuration file*. Core library file gives a set of cores that will be carried out in the system. DFT configuration file describes the existing DFT in the system. Those existing DFTs are called *test resources* in this paper, which can make interconnect test easier when they are fully exploited.

B. Test Resource

Scan chain insertion is a necessary step in current design flow to make core test easier; therefore we assume the internal scan chain is an indispensable test resource. Core wrapper (CW) can isolate a core and provides full controllability and observability for core's inputs and outputs. Boundary scan (BS) supports full accessibility for inputs/outputs of dies. In our implementation, the DFT configuration of CW and BS can be set by users freely or according to really existing DFT architectures in the system.

C. Test Environement

In [10], a test environment can compute the test patterns of circuit under test (CUT) from primary inputs and observe the responses of CUT at primary outputs. Similar to [10], the *test environment* defined in this paper is a set of cores that compute interconnect tests from controllable points and then propagate interconnect responses to observable points. The controllable points can be primary inputs of system or test resources with full controllability. System primary outputs and test resources are the observable points providing full observability.

In our implementation, a test environment is divided into two parts: (1) propagation part and (2) activation part. The propagation part propagates fault effect of interconnect from these cores to an observable point. In contrast to propagation part, the activation part of a test environment is used to compute test patterns for target interconnect fault.

D. Pattern Selection

When a test environment is ready, pattern selection is a procedure that selects a set of tests from cores belonging to the test environment. If selected tests can be used to detect the target interconnect fault, the test environment is *useful*. When all pattern combinations of a test environment fail to detect the target interconnect fault, this test environment is *useless*.

III. TEST STRATEGY

When controllability and observability of interconnect are not fully supported, interconnect test is difficult due to lack of direct accessibility. In addition, test generation for interconnect cannot be handled in a reasonable time due to large design size. Instead of test generation, we utilize known test pattern sets of cores and useful test resources to test interconnect. As a result, test generation problem is reduced to pattern selection problem. The target faults are all stuck-at-faults on intra-die interconnect.

A. Assumption

In this paper, two assumptions are needed in our test strategy.
1. All test resources including boundary scan, core wrapper, and internal scan chains are accessible even if the chips are stacked or assembled.
2. The test pattern set to detect all faults on core's pins and the observation points for those faults are known.

The second assumption is made for fault propagation. If that information is unknown, we cannot select a test pattern to propagate interconnect faults to an observable point efficiently. The information of second assumption can be obtained easily by fault simulation.

B. Test Scenario

For each interconnect, it meets one of four possible scenarios shown as follows:
1. Full controllability and full observability.
2. Not full controllability, but full observability.
3. Full controllability, but not full observability.
4. Not full controllability and observability.

In scenario 1, it means the target interconnect is accessible directly by test resources or system IOs; therefore counting sequences [5] can be used directly. For scenario 2-4, we have to create a useful test environment providing full controllability or observability or both. The detail test algorithm is shown in Fig. 1. Test environment extraction and pattern selection for propagation part is executed first (line 3-7). Once propagation part is ready and useful, activation part is then handled (line 8-12). This way can reduce search space and computation time since the problem is divided into two smaller ones. All possible pattern selections will be attempted until the target interconnect fault is detected (line 2-13). If current test environment is useless, the next test environment will be attempted. In case all test environments are useless, the procedure will indicate this fault undetectable and handle next interconnect fault. The details of each step are explained in following subsections.

Procedure: interconnect test algorithm
// f_{inter}: target interconnect fault;
// f_{in} : equivalent core-level input faults;
1.   *while* (existing another test environment) {
2.   *do* {
// handle propagation part
3.   *while* (f_{inter} is not observable) {
4.   model f_{inter} to f_{in};
5.   extract the core as propagation part;
6. P_{pro} = pattern selection to propagate f_{in};
7.   }
// handle activation part
8.   *if* (f_{inter} is observable) {
9.   extract cores as activation part;
10.   pattern selection to activate P_{pro};
11.   *if* (f_{inter} is detectable) *exit*;
12.   }
13. } *until* (no another pattern combination)
14. }

Figure 1.   Interconnect test algorithm.

C. Propagation Part

If the target interconnect fault is observable directly, the steps (line 3-7 in Fig. 1) are ignored. Otherwise, two steps are executed repeatedly until the target fault is observable. The first step models the interconnect fault to an equivalent core-level input fault, while the second step selects test patterns to propagate fault effect to an observable point or to another core.

1) Model Interconnect Fault to One Core-level Fault

An inter-die interconnect communicates signal value to cores eventually, which means the fault effect on inter-die interconnect may affect core behavior. As a result, an interconnect fault can be equivalent to a fault on cores' input pin.

For a two-pin net, it is easy to model an interconnect fault to a core-level fault. An example is given in Fig. 2 in which inter-die interconnect *a* and *d* connect between two dies (Die1 and Die2), and stuck-at-0 (sa-0) fault is assumed in this example. The signal on net *d* will be delivered to pin *G3* of core *u2* eventually so that the fault on net d is modeled as a sa-0 core-level fault on pin $G3$ of core $u2$. For a multi-pin net a in Fig. 2, it has two ways. The first one is to model net fault as a sa-0 core-level fault on $G3$ of core $u3$ and to restrict pin $G1$ of core $u2$ to zero during pattern selection. The other one is to model a sa-0 core-level fault on pin $G1$ of core $u2$ with constraints on pin $G3$ of core $u2$ during pattern selection. In other words, if an inter-die interconnect connects to n core pins, there are n ways to model this interconnect fault.

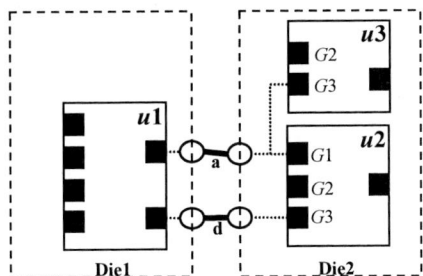

Figure 2.   Interconnect fault to core-level fault

2) Model Interconnect Fault to n-Core-level Fault

When a core-level fault is modeled, the next step is to select a pattern that can propagate the fault to an observable point such as internal scan chains or test resources if possible.

In case the fault cannot be observable after pattern selection, it means the fault has been propagated to other inputs of cores. In this case, this interconnect fault will be modeled as n-core-level faults in next core-level fault modeling (line 4 of Fig. 1) where n is the number of cores the fault goes through. Fig. 3 gives an example. A sa-0 fault on pin $G1$ of core $u1$ is propagated to the output of core $u1$ in the first core-level fault modeling and pattern selection, which is unobservable. In next iteration, this fault will be modeled as 2-core-level faults, sa-0 on pin $G0$ of core $u1$ and sa-0 on pin $G1$ of core $u2$. Please note n-core-level faults needs n test patterns to detect them at the same time during pattern selection, which guarantees the fault effect of inter-die interconnect observable.

When n is larger, those faults are usually hard to be detected simultaneously according to our experimental experiences of pattern selection. Consequently the maximum n is limited to three in this paper. On the other hand, it is preferred to propagate fault to internal scan chains or test resources during pattern selection avoiding appearance of n-core-level faults.

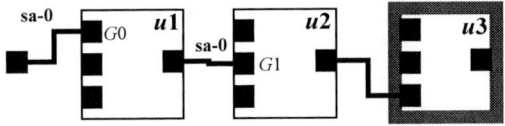

Figure 3.   2-equivalent fault modeling

D. Activation Part

Once propagation part is ready, we perform test environment creation and pattern selection for activation part. A backward depth first search algorithm from cores belonging to propagation part is performed to extract a set of cores as activation part of current test environment. The search will stop when full controllability of test environment is guaranteed. The pattern selection is then executed to activate the patterns selected in propagate part.

An illustrative example is given in Fig. 4. Core $u2$ belong to propagation part of test environment and the test pattern (1, x, 0, 0) is selected to detect sa-0 interconnect fault. Core $u1$ is extracted as activation part and pattern (0, 1, x, 0) with response (1, 0, 0) is selected to activate test pattern selected from core $u2$.

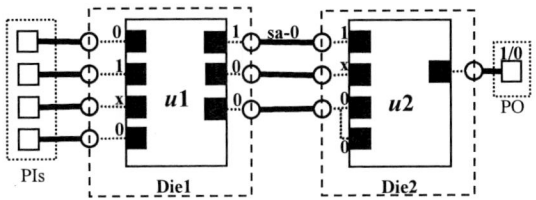

Figure 4.   Test strategy example

E. Time Complexity

The time complexity of our test strategy depends on the number of test environments for each inter-die interconnect faults and the amount of pattern combinations in a test environment. Assume that each inter-die interconnect fault has k test environments. Each test environment has m cores as activation part and n cores belonging to propagation part. In activation part, each core has p test patterns while there are q valid test patterns to detect core-level faults for each core in propagation part.

The amount of pattern combinations for a test environment is $\prod_{i=1 \text{ to } m}(p_i) \times \prod_{j=1 \text{ to } n}(q_j)$. Usually n is small since we try to avoid n-core-level faults during pattern selection of propagation part, and $p \gg q$. As a result, the time complexity is dominated by the term $\prod_{i=1 \text{ to } m}(p_i)$. In next section, two approaches are shown to reduce the search space of pattern selection for activation part.

IV.   PATTERN COMBINATION REDUCTION

A. Input Controllability Don't Care Set

For a Boolean network, its CDC_{in} is all patterns that are never produced by the environment at the network's input. For a test environment, we can calculate CDC_{in} for each core, and then drop the useless test patterns during pattern selection.

Assume that core c_i and c_j belong to the same test environment and some outputs of c_i connect to the inputs of c_j. A test pattern p belongs to CDC_{in} of c_j when it cannot be generated through feeding all test patterns of c_i.

The number of pattern combinations of activation part can be reduced greatly when the CDC_{in} of cores are found out in advance. For example, the activation part of a test environment has three cores $\{c_1, c_2, c_3\}$ and each core has three test patterns. Assume that there is one pattern belonging to CDC_{in} of core c_2. The number of pattern combinations will reduce from 27 (3*3*3) to 18 (3*2*3).

B. Decision Tree Pruning of Pattern Combinations

Another way reducing search space during pattern selection is to prune impossible pattern combinations based on current pattern combination result. The basic concept is similar to decision tree pruning. Assume an order is given for all cores belonging to activation part. A decision tree of pattern combinations can be created based on given order. In our implementation, we adopt the reserve topologic order as core's order and use an array to represent current pattern combination instead of decision tree. An illustrative example is given in Fig. 5, in which current pattern combination is $\{c_0{:}t_1,\ c_1{:}t_1,\ c_2{:}t_1,\ c_3{:}t_1\}$ while $c_0{:}t_1$ indicates test pattern t_1 of core c_0 is selected. Assume that a conflict occurs between $c_0:t_1$ and $c_1:t_1$. In this case, next pattern combination will be $\{c_0:t_1, c_1:t_2, c_2:t_1, c_3:t_1\}$ as five useless pattern combinations are pruned from decision tree.

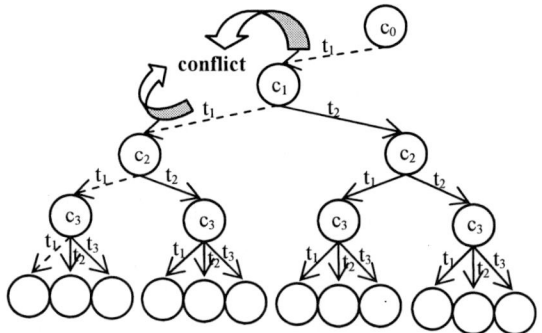

Figure 5. Decision tree example

V. EXPERIMENTAL RESULT

In the experiments, four SiP designs, S1 to S4, with different DFT configurations are generated to validate proposed test methodology. An in-house program is used to generate the related input files for proposed interconnect test algorithm. Three parameters are given to this program: (1) total number of dies in SiP, (2) total number of cores in SiP, and (3) core library in SiP. We adopt some ISCAS benchmark circuits as core library and the statistic of those circuits are shown in Table I. The column under the title #test is the number of known test patterns for each core, and they are a set of patterns that can detect faults on core's pins. Each core with an assigned core library is distributed to a die randomly. The connectivity among cores is also determined randomly and no core's pin is floating. Table II gives SiP structure and Table III gives the DFT configuration of each die and core used in experiments. In the DFT configuration, we set at least one die without support of boundary scan to guarantee the effectiveness of the proposed approach.

TABLE.I   CORE LIBRARY (CLIB)

Circuit	#DFF	#gate	#test	Circuit	#DFF	#gate	#test
s27	3	10	19	s5378	179	2779	756
s208	8	96	47	s13207	638	7951	852
s344	15	160	46	s38417	1636	22179	416
s1423	74	657	79				

TABLE.II   SiP STRUCTURE

SiP	#Die	#Core	Core Distribution
S1	2	3	s27 * 3
S2	3	5	s27 * 2, s208 * 1, s344 *2
S3	2	4	s1423 * 4
S4	2	7	s1423 * 2, s5378 * 1, s13207 * 3, s38417 * 1

Table IV shows the test generation results under different DFT configurations. The total number of interconnect faults is denoted as #F. The column under the title #DF is the number of detectable faults. The fourth column shows the fault coverage (FC) and the last column gives the test generation time. It is clear that proposed approaches can reach at least 87% of fault coverage in all cases by using existing test resources and test patterns of cores, even if a few test resources are given. Another interesting result is that the locality of test resource can affect fault coverage significantly. For example, In SiP S2 and S3, the boundary scan inserted into various die leads to 3.2% and 1.5% fault coverage difference

respectively. System or DFT designer can use our approach to explore possible test solutions to tradeoff test hardware penalty and fault coverage.

TABLE.III   DFT CONFIGURATION

SiP	d1	d2	d3	u1	u2	u3	u4	u5	u6	u7
S1	N	BS	-	N	N	N	-	-	-	-
	BS	N	-	N	N	N	-	-	-	-
	N	N	-	N	N	N	-	-	-	-
S2	BS	N	N	N	N	N	N	N	-	-
	N	BS	N	N	N	N	N	N	-	-
	N	N	N	N	N	N	N	N	-	-
S3	BS	N	-	N	N	N	N	-	-	-
	N	BS	-	N	N	N	N	-	-	-
	N	N	-	N	N	N	N	-	-	-
S4	N	N	-	CW	CW	N	N	CW	N	N
	N	BS	-	N	N	N	N	CW	N	N
	N	N	-	CW	N	N	CW	N	N	N

*BS: Boundary Scan, CW: Core-level Wrapper, N: None DFT

TABLE.IV   TEST GENERATION RESULT

SiP	#F	#DF	FC(%)	CPU(s)
S1	24	24	100	0.14
	24	24	100	0.234
	24	24	100	0.312
S2	94	86	91.49	0.984
	94	89	94.68	1.015
	94	82	87.23	1.328
S3	138	136	98.55	1.062
	138	138	100	0.984
	138	132	95.65	2.187
S4	1640	1627	99,21	98.953
	1640	1427	87.01	14.609
	1640	1599	97.5	464.859

REFERENCE

[1] *IEEE 1149.1 Standard Test Access Port and Boundary-Scan Architecutre*. Abailable: http://grouper.ieee.org/groups/1149/1/

[2] *IEEE 1500 Standard for Embedded Core Test (SECT)*. Available: http://grouper.ieee.org/groups/1500/

[3] E.J. Marinissen, J. Verbree, and M. Konijnenburg, "A structured and scalable test access architecture for TSV-based 3D stacked ICs," in *Proc.VLSI Test Symp.*, pp. 269–274, Apr. 2010.

[4] L. Jiang, Q. Xu, K. Chakrabarty, and T. M. Mak, "Layout-driven test-architecture design and optimization for 3D SoCs under pre-bond test-pin-count constraint," in *Proc. Int. Conf. on Comput.-Aided Design*, pp.191-196, Nov. 2009.

[5] W. H. Kautz, "Testing of Faults in Wiring Networks," *IEEE Transactions on Computers*, vol. 23, no. 4, pp. 358–363, Apr. 1974.

[6] Y. Kim, H. Kim, and S. Kang, "A new maximal diagnosis algorithm for interconnect test, "*IEEE Transactions on VLSI*, vol. 12, no. 5, pp. 532–537, May. 2004.

[7] P. Min, H. Yi, J. Song, S. Baeg and S. Park, "Efficient interconnect test patterns for crosstalk and static faults," *IEEE Transactions on CAD*, vol. 25, no. 11, pp. 2605–2608, Nov. 2006.

[8] C.-H. Chiang, and S.K. Gupta, "BIST TPGs for faults in board level interconnect via boundary scan," in *Proc.VLSI Test Symp.*, pp. 376–382, Apr. 1997.

[9] K. S.-M. Li, C. L. Lee, C. Su, and J.E Chen, "Oscillation ring based interconnect test scheme for SoC," in *Proc. Asia and South Pacific Design Automa. Conf.*, pp. 184–187, Jan, 2005.

[10] B.T. Murray and J.P. Hayes, "Hierarchical test generation using precomputed tests for modules," *IEEE Trans. on CAD*, vol. 9, no. 6, pp. 594-603, Jun. 1990.

978-1-4244-9019-6/11 $26.00 © 2011 IEEE

Architectural-Level Error-Tolerant Techniques for Low Supply Voltage Cache Operation

Shih-Lien Lu, Alaa Alameldeen, Keith Bowman, Zeshan Chishti, Chris Wilkerson and Wei Wu

Intel Labs, Hillsboro Oregon 97124

{shih-lien.l.lu, alaa.r.alameldeen, keith.a.bowman, zeshan.chishti, chris.wilkerson, wei.a.wu}@intel.com

Abstract—Supply voltage (V_{CC}) scaling is the most effective technique for reducing the energy consumption of microprocessors. Since V_{CC} scaling increases the impact of parameter variations on circuit performance and functionality, circuits eventually fall out of specification, thus limiting the minimum operating supply voltage (V_{CCMIN}) for the microprocessor. The last-level cache (LLC) often determines V_{CCMIN}. To maximize cache capacity, the LLC memory cell consists of near minimum-sized transistors, which are highly sensitive to process variations. For a tradition LLC, a small fraction of memory cells with large variations limit the V_{CCMIN} for the entire microprocessor. In this paper, error-tolerant techniques dynamically reconfigure the cache to either disable or correct these failing memory cells to enable a lower V_{CCMIN} at the cost of lower cache capacity, thus enhancing the microprocessor energy efficiency. At the high-V_{CC} operating mode, the cache operates at full capacity to satisfy the high-performance target. At the low-V_{CC} operating mode, energy consumption is the primary concern, and the cache is dynamically reconfigured with lower capacity to mitigate the impact of the failing memory cells on reliability. Since the clock frequency significantly reduces for the low-V_{CC} mode as compared to the high-V_{CC} mode, the reduction in cache capacity has a smaller effect on performance. In comparison to a traditional LLC design, simulation results indicate that the error-tolerant cache techniques decrease V_{CCMIN} by 13-28%, corresponding to a 20-42% reduction in energy per instruction. Adding these techniques only incurs a 5-10% performance penalty at the low-V_{CC} operating mode in comparison with a hypothetical idea cache.

Index Terms—Resilient design, error-tolerant cache, error-correcting codes, ECC, reliability, low voltage operation.

I. INTRODUCTION

Today's microprocessors provide multiple supply voltage (V_{CC}) and clock frequency (F_{CLK}) operating modes to satisfy a wide range of performance demands with low energy consumption. A high-V_{CC} operating mode enables the maximum F_{CLK} for the microprocessor to execute high-performance applications. In contrast, a low-V_{CC} operating mode reduces the V_{CC} and F_{CLK} to lower the energy consumption for low-performance applications. This dynamic operating range is critical to the microprocessor performance and energy efficiency.

For the low-V_{CC} operating mode, V_{CC} scaling is the most effective way to reduce total energy since both dynamic and leakage components decrease. Reducing V_{CC}, however, increases the sensitivity of circuit performance and functionality to parameter variations. Eventually, circuits start to fail with V_{CC} scaling, thus limiting the minimum operating supply voltage (V_{CCMIN}) for the microprocessor. The last-level

cache (LLC) in today's microprocessors often determines the V_{CCMIN}. To maximize cache capacity, the memory cell design typically consists of near minimum-sized transistors. Since the uncorrelated parameter variations (e.g., random-dopant fluctuations) are inversely proportional to the square-root of the transistor gate area, the memory cell with near minimum-sized transistors is highly sensitive these sources of process variations. Thus, wide distributions exist for the cache memory cell read, write, and retention voltages while satisfying a target F_{CLK} requirement. For a tradition LLC, a small fraction of memory cells with large variations limits the V_{CCMIN} for the entire microprocessor.

The traditional approach toward reducing the cache V_{CCMIN} is to upsize the memory cell design by employing transistors with larger gate length and width dimensions. Since uncorrelated parameter variations are inversely proportional to the square-root of the transistor gate area, the larger transistor gate area narrows the distributions for the memory cell read, write, and retention voltages. In contrast to simply upsizing the 6-transistor (6-T) cell design, recent work has demonstrated V_{CCMIN} benefits with an 8-T or even a 10-T cell design [1]-[2]. Each of these approaches trade-off a better V_{CCMIN} at the cost of lower cache density. At the high-V_{CC} operating mode with the maximum F_{CLK}, performance is tightly coupled with the cache capacity. Thus, improving V_{CCMIN} at the cost of an overall lower cache capacity, significantly affects the microprocessor performance at the high-V_{CC} mode.

In this paper, we explore architectural-level error-tolerant techniques to dynamically reconfigure on-chip caches in modern microprocessors to improve V_{CCMIN} at the low-V_{CC} mode while not affecting the performance at the high-V_{CC} mode. Section II provides a description of the memory cell failures, and Section III describes the primary variation sources. In Section IV, the error-tolerant techniques are presented to dynamically reconfigure the cache to mitigate the impact of the error sources on V_{CCMIN}. A simulation framework is described in Section V to evaluate the benefits of the error-tolerant techniques. Finally, Section V concludes by summarizing the key insights.

II. MEMORY CELL FAILURES

Fig. 1 describes a traditional 6-T SRAM cell. The cell contains two cross-coupled inverters to either store a logic "0" or a logic "1" and two access transistors to enable read and write operations. The word-line select (WL) signal allows the access transistors to either read or write the cell by connecting the bit-lines to the cell storage nodes. In a read operation, bit-lines (Bit0 and Bit1) are initially pre-charged to V_{CC}. Then, the WL is asserted to activate the access transistors, resulting in a differential voltage across the two bit-lines. Then, a shared sense-amplifier is coupled to the selected bit-lines and is activated to amplify the bit-line differential voltage to read the cell logic value. In a write operation, the bit-lines are oppositely driven to the desired logic value, and then the WL is asserted to pass data from the bit-lines through the access transistors to the cell.

A memory cell failure can occur during a memory read or write operation or during the retention of the cell bit value. These failure mechanisms are described in detail below [2].

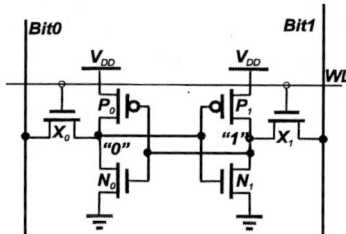

Figure 1. A 6-transistor SRAM cell.

1) Read Failure: A read failure occurs when the stored value flips during a read operation. This happens when the noise developed on the node storing a "0" from charge sharing with the highly-capacitive pre-charged Bit0 is larger than the trip point of inverter (P1/N1). A memory cell variation that strengthens the access transistor X_0 and weakens the pull-down transistor N_0 makes the cell more susceptible to a read failure. In addition, a read failure can occur when the differential voltage across the bit-lines during a read operation is not sufficient for the sense amplifier to identify the correct value.

2) Write Failure: A write failure occurs when the cell contents cannot be over written to the opposite value. To toggle the contents of the cell in Fig. 1, Bit1 is driven to V_{SS} while Bit0 remains pre-charged to V_{CC}. In an operational cell, transistor N_1 starts discharging the node storing a "1" and the positive feedback completes the operation by driving the "0" to "1". A memory cell variation that strengthens the pull-up transistor P_1 and weakens the access transistor X_1 makes the cell more susceptible to a write failure.

3) Retention Failure: A retention failure occurs when the stored value in the cell is lost during standby [3]. This results from a strong mismatch between the trip points of the two crossed-coupled inverters. In modern cache designs, read and write failures limit V_{CCMIN} as compared to retention failures.

III. VARIATION SOURCES

The variation sources for memory bit failures are classified into two categories: persistent and non-persistent. Persistent memory bit failures result from static process variations (e.g., random-dopant fluctuations). The persistent variation contributes to the majority of bit failures at low V_{CC}. As V_{CC} scales, the persistent memory bit failures increase exponentially [1], [4]. Persistent bit failures are detected through traditional cache testing.

Non-persistent memory bit failures exhibit sporadic behavior. Example variation sources include radiation-induced soft errors and erratic bit failures [5]. Radiation-induced soft errors typically results from neutrons in cosmic rays. The neutron strike generates enough charge on a circuit node to flip the logic state. Erratic memory bit failures result from temporary degradation and recovery in the gate dielectric of a memory cell transistor [5]. Consequently, the SRAM cache V_{CCMIN} erratically fluctuates in time across different measurements [5]. Due to the random nature, erratic cells may escape standard cache testing, resulting in potential failures in the field. Since non-persistent failures occur randomly across the cache during operation, traditional memory testing cannot adequately address this source of variation. As a result, these failures do not directly contribute to memory yield loss. Rather, these sources of variation directly impact the failures-in-time (FIT) for a microprocessor.

Fig. 2 compares the probability of a memory bit failure (P_{FAIL}) across V_{CC} for different variation sources. The P_{FAIL} from persistent variations is based on circuit simulations that are validated with measured data [1]. In Fig. 2, the P_{FAIL} from persistent variations increases exponentially as V_{CC} reduces [1], [4]. The P_{FAIL} from SER is estimated from previous measured data [6], which is extrapolated to lower V_{CC} values. Since the data in [6] is measured by inducing neutrons from a nuclear reactor, the SER is scaled by a factor of one billion to estimate the SER under normal conditions at sea-level [7]. The SER is assumed to increase by ~3X for every 500mV reduction in V_{CC}, resulting from the SER exponential dependency on the charge stored at a particular node, which changes linearly with V_{CC}. Although V_{CC} scaling increases the P_{FAIL} from SER [7]-[10], Fig. 2 illustrates that the P_{FAIL} from SER increases at a dramatically slower rate as compared to the P_{FAIL} from persistent variations.

Rigorous models to capture the erratic bit behavior are unknown. Given this lack of a detailed analysis, a simple model is applied to estimate erratic bit failures for two hypothetical process technologies with high and low rates of erratic behavior [11]. In Fig. 2, the P_{FAIL} from erratic behavior is modeled as a fixed proportion of the P_{FAIL} from persistent variations.

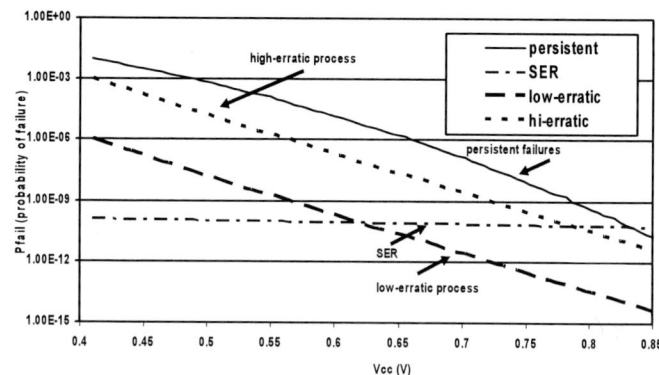

Figure 2. Probability of memory cell failure (P_{FAIL}) versus supply voltage for persistent and non-persistent variations.

IV. ARCHITECTURAL-LEVEL ERROR-TOLERANT TECHNIQUES

Persistent failures are reliably identified using standard memory testing methodologies, which are referred as testable bit failures. In contrast, non-persistent failures are non-

978-1-4244-9019-6/11 $26.00 © 2011 IEEE

testable bit failures since these bits cannot be identified with traditional cache testing. Two architectural-level error-tolerant techniques are presented to address testable and non-testable bit failures: (i) Dynamic cache reconfiguration with fine-grain disabling and (ii) Dynamic cache reconfiguration with error-correcting codes (ECC). The following subsections describe these techniques in detail.

A. Dynamic Cache Reconfiguration with Fine-Grain Disabling

We examined two architectural mechanisms to dynamically reconfigure the cache with fine grain disabling: (i) word disable and (ii) bit-fix [12]. Word disable identifies and disables defective words in the cache. Bit fix identifies broken bit pairs and maintains patches to repair the cache line. Both mechanisms leverage memory tests to identify portions of the cache that are defective at low V_{CC}, which would otherwise limit the cache V_{CCMIN}. At the high-V_{CC} mode, these defective portions of the cache operate correctly, so the cache remains at full capacity to satisfy the high-performance demands. At the low-V_{CC} mode, these defection portions of the cache are disabled to ensure correct cache operation at low V_{CC} [13], thus further scaling V_{CCMIN} at the most energy-efficient mode. The word-disable mechanism isolates the defects on a word-level granularity and then disables words containing the defective bits. Each tag of a cache line keeps a defect map with one bit per word that represents whether the word is defective or valid. For each cache set, two physical lines in the same set (e.g., line 0 and line 1) combine to form one logical line. After disabling the defective words, the two physical lines store the contents of one logical line. This cuts both the cache size and associativity in half.

The bit-fix mechanism differs from the word-disable mechanism in three respects. First, instead of disabling at word-level granularity, the bit-fix mechanism disables groups of 2 bits. These groups are defective pairs in which at least one bit is defective. Second, for each defective pair, the bit-fix mechanism maintains a 2-bit patch that can be used to correct the defective pair. Third, the bit-fix mechanism requires no additional storage for repair patterns. Instead, repair patterns are stored in selected cache lines in the data array. This eliminates the need for the additional tag bits required to store the defect map in the word-disable mechanism. It does, however, require storage for repair pointers elsewhere in the system (e.g., main memory) while not in use (i.e., high-V_{CC} mode).

B. Dynamic Cache Reconfiguration with Error-Correcting Codes (ECC)

ECC utilizes information redundancy to protect bit failures. Hamming and Hsiao codes [14]-[15] are commonly applied in commercial memory chips and on-chip caches to provide single-error correction, double-error detection (SECDED) due to the code simplicity. With continued transistor scaling, large on-chip SRAM arrays now incorporate multi-bit ECCs. As an example, double-error correction, triple-error detection (DECTED) is designed from a binary BCH code [9]. Table 1 gives an example of the BCH ECC logic and latency overheads for different numbers of correction bits with latency minimization as the primary goal

for a common cache line. In Table 1, n is the number of bits in the code word, k is the number of data bits, d is the Hamming distance, and t is the number of correctable bits. Codes with a small number of correction bits provide sufficient redundancy to deal with non-persistent bit failures at the high-V_{CC} operating mode. In order to push the cache V_{CCMIN} to ultra-low values (e.g., 500 mV) at the low-V_{CC} mode, addition redundancy bits are desired. Codes with a large number of correction bits, however, require significant logic overhead.

Table 1. Comparison of BCH ECC logic and latency for a parallel implementation.

	SEC-DED (266, 256, 4)	DEC-TED (275, 256, 6)	TEC-QED (284, 256, 8)	tEC-(t+1)ED (n, k, d)
Logic				
Encoding	1030 XOR	2325 XOR	3637 XOR	O(kr/2) XOR
Decoding	1040 XOR	2351 XOR	3640 XOR	O(nr/2) XOR
Correction	256 XOR 4608 AND/OR	5530 XOR 2387 AND/OR	14018 XOR 2486 AND/OR	Step 1 (BM algorithm) (4mt+4) FF (3t) GF(m)-Multiplier Step 2(Per-bit correction) O(km²t/2)XOR O(km) AND/OR
Latency				
Encoding	7 XOR	8 XOR	8 XOR	Log2(n) XOR
Decoding	7 XOR	9 XOR	9 XOR	Log2(n)+1 XOR
Correction	1 XOR 5 AND (1 cycle)	8 XOR 5 AND/OR (1 cycle)	19 XOR 5 AND/OR (~2 cycles)	Step 1 (BM algorithm) * (2t-2) cycles Step 2(Per-bit correction) O(log₂(m/2)+1) XOR O(log₂(m/2)+3) AND/OR

From Table 1, BCH ECC logic and latency does not scale well to correct a large number of bits in the cache line. To address this deficiency, the majority logic decodable codes provide an alternative. Orthogonal Latin square codes (OLSC) [15] is an example. While OLSC requires more check bits than BCH ECC, the OLSC implementation reduces the logic and latency overheads as compared to traditional BCH ECC. Moreover, the orthogonal property of the code enables a modular construction of the parity H matrix (encoder and decoder). Thus, each additional module provides extra correction ability without disturbing the existing modules.

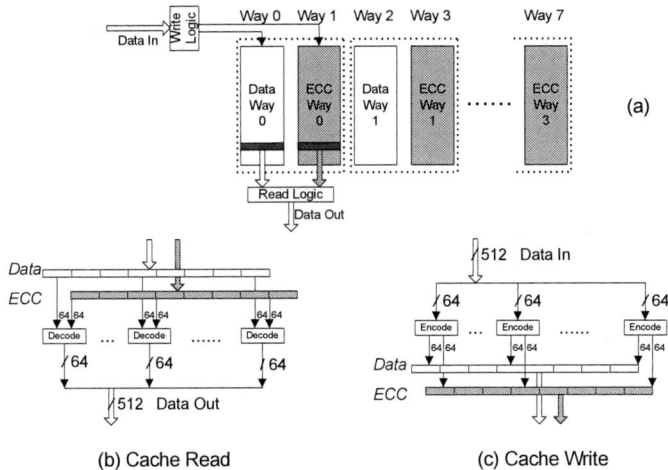

Figure 3. Example of multi-bit segmented ECC with eight 64-bit segments.

We propose a multi-bit segmented ECC (MS-ECC) to achieve a high-level of correction with low overhead. The purpose of MS-ECC is to correct multi-bit errors by implementing error correction at finer granularity segments within a cache line. In the high-V_{CC} mode, the entire cache capacity is available for the microprocessor with conventional ECC to mitigate the impact of radiation-induced soft errors. In the low-V_{CC} mode, a portion of the cache stores additional ECC information on granularities finer than a cache line, thereby enabling more error correction per cache line. Because the size of each segment is smaller than the entire cache line, the logic complexity and latency for MS-ECC is significantly less than conventional (un-segmented) ECC. As an example, assume a 2MB 8-way L2 cache with 64-byte lines. In the low-V_{CC} mode, we divide the eight physical ways in each set among (i) data ways and (ii) ECC ways. The ratio of data ways to ECC ways depends on the desired reliability level, which in turn depends on the target V_{CCMIN}. Decreasing this ratio increases the redundancy, and thus the reliability, at a cost of lower cache capacity during low-V_{CC} operation. We analyze the impact of this ratio on V_{CCMIN} and performance in Section V. For this example, assume there is one ECC way for every data way (i.e., 50% cache capacity available in low-V_{CC} mode). We apply a fixed mapping to associate data ways with their corresponding ECC ways (Fig. 3(a)). Physical way 1 stores the ECC for physical way 0, physical way 3 stores the ECC for physical way 2, and so on. Thus, in the low-V_{CC} mode, the cache effectively reconfigures to a 1MB 4-way set-associative cache.

V. EVALUATION

A. Simulation Methodology for Performance

We use a cycle-accurate, execution-driven simulator running IA32 binaries to evaluate performance. The micro-operation (uOp) based simulator executes both user and kernel instructions and models a detailed memory subsystem. Table 2 provides the parameters for the baseline out-of-order processor. For the baseline processor, both the L1 and L2 caches use SECDED ECC on a per-line granularity to tolerate persistent as well as non-persistent bit failures.

Table 2. Baseline processor parameters.

Voltage Dependent Parameters		
	High VCC	Low VCC.
Processor frequency	3 GHz	500 MHz
Memory latency	300 cycles	50 cycles
Voltage	1.3v	0.5v
Voltage Independent Parameters		
ROB size	256	
Register file	256 fp, 256 int	
Fetch/schedule/retire width	6/5/5	
Scheduling window size	32 FP, 32 Int, 32 Mem	
Load buffer size	32	
Store buffer size	32	
Branch predictor	16k bytes TAGE	
Hardware data prefetcher	Stream-based (16 streams)	
Cache line size	64 bytes	
L1 Instruction/Data cache	32k bytes, 8-way, 3 cycles	
L2 Unified cache	2M bytes, 8-way, 20 cycles	

To quantify the performance overhead from the loss in cache capacity and the ECC encoding/decoding latency changes of the proposed techniques, we also simulate an ideal, defect-free, low-V_{CC} baseline that has reliable caches without any capacity loss or latency overheads. Since transistor switching delay depends on V_{CC}, we perform circuit simulations to predict the F_{CLK} at different V_{CC} values. Table 3 provides the benchmark programs that are executed by the simulator.

Table 3. Benchmark suite.

Category	# of traces	Example benchmarks
Digital home (DH)	60	H264 decode/encode, flash
FSPEC2K (FP2K)	25	www.spec.org
ISPEC2K (INT2K)	26	www.spec.org
Games (GM)	49	Doom, quake
Multimedia (MM)	77	Photoshop, raytracer
Office (OF)	52	Excel, outlook
Productivity (PROD)	43	File compression, Winstone
Server (SERV)	48	SQL, TPCC
Workstation (WS)	82	CAD, bioinformatics
ALL	**462**	

B. Reliability Analysis

The V_{CCMIN} values are compared for the MS-ECC and the fine-grain disabling technique with bit-fix [12]. The MS-ECC design assumes a segment size of 64 bits capable of correcting 4 error bits per segment. Since the basic bit-fix design does not tolerate non-persistent failures, SECDED ECC is incorporated into the bit-fix design, which is referred as BFXECC. Fig. 4 plots the P_{FAIL} versus V_{CC} for MS-ECC and BFXECC. For a yield loss of 10^{-3}, simulations indicate a V_{CCMIN} of 475mV and 490mV for BFXECC and MS-ECC, respectively, when only considering persistent failures. When non-persistent failures are included, both techniques provide a V_{CCMIN} of 520mV for a 10^{-3} FIT for a low-erratic process. For a high erratic process, however, the MS-ECC technique does significantly better than BFXECC as the high frequency of erratic bit failures at low-V_{CC} overwhelms the SECDED ECC that augments BFXECC.

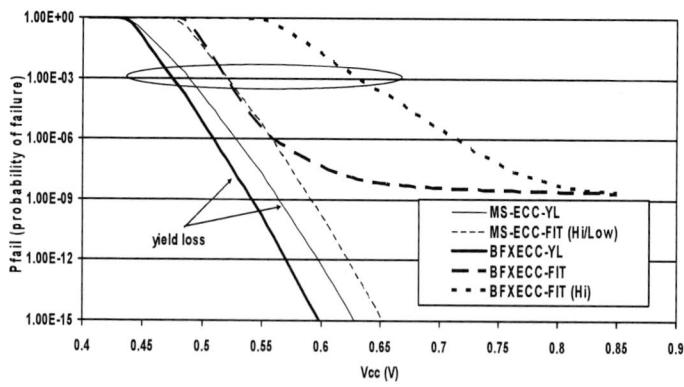

Figure 4. P_{FAIL} versus V_{CC} for MS-ECC and BFXECC with persistent and non-persistent variations.

978-1-4244-9019-6/11 $26.00 © 2011 IEEE

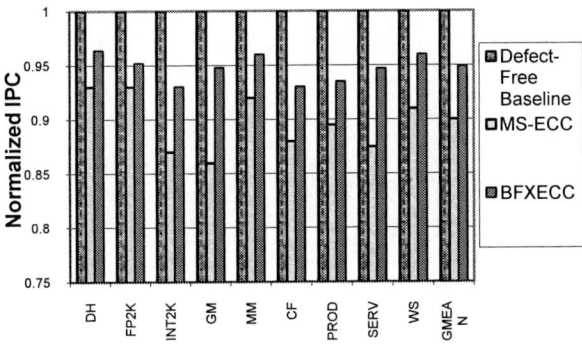

Figure 5. Performance impact of MS-ECC and BFXECC at the low-V_{CC} operation mode.

C. Performance and Energy Efficiency

To quantify the performance overhead of using MS-ECC and BFXECC in the low-V_{CC} mode, Fig. 5 plots the normalized IPC relative to the defect-free low-V_{CC} baseline when applying either MS-ECC or BFXECC to both the L1 and L2 caches with a V_{CC} of 500mV. For this comparison, the defect-free low-V_{CC} baseline contains a fully functional cache at the low-V_{CC} mode. Averaging across all the workloads, MS-ECC and BFXECC techniques result in a 10% and a 5% IPC degradation, respectively, as compared to the defect-free baseline. Although these techniques degrade performance at the low-V_{CC} operating mode, Table 4 highlights the benefits of these techniques to lower V_{CCMIN} as compared to the baseline design. From Table 4, MS-ECC and BFXECC decrease the V_{CCMIN} by 28% and 13%, respectively, as compared to the baseline design, corresponding to a reduction in the energy per instruction of 42% and 20%, respectively.

Table 4. Comparison of V_{CCMIN}, F_{CLK}, normalized power, and normalized energy per instruction between the baseline processor, BFXECC, and MS-ECC.

Scheme	Vccmin (mV)	Frequency (MHz)	Norm. Power	Norm. EPI
Baseline	725	1400	1	1
BFXECC	630	1000	0.57	0.8
MS-ECC	520	700	0.29	0.58

VI. CONCLUSION

Two architectural-level error-tolerant techniques are described to dynamically reconfigure the cache to either disable or correct failing memory cells to enable a lower V_{CCMIN} to enhance the microprocessor energy efficiency. One technique combines fine-grain disabling with ECC to mitigate both persistent and non-persistent cache cell failures. The other technique employs a modular ECC scheme to protect segmented cache lines. At the high-V_{CC} operating mode, the cache operates at full capacity to meet the high-performance demands. At the low-V_{CC} operating mode, the cache is dynamically reconfigured with lower capacity to mitigate the impact of the failing cache cells on reliability. Since the clock frequency significantly reduces with V_{CC} scaling, the reduced cache capacity has a minor effect on performance. In comparison to a traditional LLC design, simulation results indicate that the error-tolerant cache techniques reduce V_{CCMIN} by 13-28%, resulting in a 20-42% savings in energy per instruction. At the same time these error-tolerant techniques only incurs a 5-10% performance penalty at the low-V_{CC} operating mode in comparison with a hypothetical idea cache.

REFERENCES

[1] J. P. Kulkarni, K. Kim and K. Roy, "A 160 mV Robust Schmitt Trigger Based Subthreshold SRAM," *IEEE J. of Solid-State Circuits*, Vol. 42, no. 10, pp. 2303-2313, Oct. 2007.

[2] M. M. Khellah, A. Keshavarzi, D. Somasekhar, T. Karnik, and V. De, "Read and Write Circuit Assist Techniques for Improving Vccmin of Dense 6T SRAM Cell," in *Proc. IEEE Intl. Conf. Integrated Circuit Design Tech. (ICICDT)*, June 2008, pp. 185-188.

[3] H. Qin, Y. Cao, D. Markovic, A. Vladimirescu, and J. Rabaey, "Standby supply voltage minimization for deep sub-micron SRAM", *IEEE Microelectronics Journal*, Aug 2005, vol. 36, pp. 789-800

[4] A. Kumar, J. Rabaey and K. Ramchandran, "SRAM Supply Voltage Scaling: A Reliability Perspective," in *Intl. Symp. Quality Electronic Design (ISQED)*, Mar. 2009, pp. 782-787.

[5] M. Agostinelli, et. al, "Erratic fluctuations of SRAM cache Vmin at the 90nm process technology node," in *IEDM Tech. Dig.*, Dec. 2005, pp. 655–658.

[6] K. Ünlü, et al., "Neutron-induced Soft Error Rate Measurements in Semiconductor Memories," *Nuclear Instruments and Methods in Physics Research*, Section A, Volume 579, Issue 1, pp. 252-255, 2007.

[7] J.F. Ziegler, et al., "Accelerated Testing for Cosmic Soft-Error Rate," *IBM Journal of Research and Development*, Vol. 40, No. 1, pp. 51-72, Jan. 1996.

[8] S. Hareland, et al., "Impact of CMOS Scaling and SOI on Soft Error Rates of Logic Processes," in *Symp. VLSI Technology Dig. Tech. Papers*, June 2001, pp. 73-74.

[9] P. Shivakumar, et al., "Modeling the Effect of Technology Trends on the Soft Error Rate of Combinational Logic," in *Proc. International Conference on Dependable Systems and Networks*, June 2002, pp. 389-398.

[10] X. Li, et al., "Scaling of Architecture Level Soft Error Rates for Superscalar Processors," in *Proc. 1st Workshop on the System Effects of Logic Soft Errors (SELSE)*, Apr. 2005.

[11] Z. Chishti, A. R. Alameldeen, C. Wilkerson, W. Wu and S.-L. Lu, "Improving Cache Lifetime Reliability at Ultra-Low Voltages," in *42nd Intl. Symp. Microarchitecture (MICRO-42)*, Dec. 2009.

[12] C. Wilkerson, et al., "Trading Off Cache Capacity for Low Voltage Operation," *IEEE Micro*, Jan./Feb. 2009.

[13] J. Chang et. al., "The 65-nm 16-MB Shared On-Die L3 Cache for the Dual-Core Intel Xeon Processor 7100 Series," *IEEE J. Solid-State Circuits*, Vol 42, Is. 4, pp. 846-852, Apr. 2007.

[14] S. Lin and D. Costello, *Error Control Coding*, (2nd Edition), Prentice Hall, 2004.

[15] H. Y. Hsiao et al., "Orthogonal Latin Square Codes," *IBM Journal of Research and Development*, Vol. 14, No. 4, pp. 390-394, July 1970.

Special Considerations for 3DIC Circuit Design and Modeling

Sally Liu, Yung-Chow Peng, and Fu-Lung Hsueh
Taiwan Semiconductor Manufacturing Company, 9, Creation Road 1, Hsinchu Science Park,
Hsinchu, Taiwan 300-77, ROC

Abstract—In this paper, the new elements in 3DIC are examined for enabling optimal 3D products: including 3D interconnect which maybe the limiting factor to achievable speed; 3D chip design strategy (partition and implementation) to achive optimal performance; wireless testing to address the challenges in testing a partial system / chip before stacking and with limited observation points after stacking.

Index Terms—3D interconnect, design strategy, wireless testing

I. INTRODUCTION

AS 3D integration technologies advance in recent years [1], it has become a viable alternative to increase SOC integration density [2]. Most critically it enables tighter "hetero-technology 3D integration" and holds great promise for new product innovation and cost effectiveness. Die/wafer stacking with vertical connectivity formed with micro-bumps and through chip via (TSV) or contactless through chip interconnect (TCI) realize denser integration. The 3rd dimension of chip integration not only invites designers to rethink their design strategy, potentially a paradigm shift in partition and implementation of subsystems but also brings about new risk factors and the daunting challenges in testing a partial system / chip before stacking and with limited observation points after stacking. These critical facets are examined in the following sections.

II. 3D INTERCONNECT

Chips in a 3DIC stack can be placed side-by-side on top of a 3rd chip which may contain only passives, as the interposer substrate, suitable for integration of high-performance chips with memories, both of which tend to be power hungry and require good heat dissipation management; or a stacking of IC chips. The later case can be either face-to-face or face-to-back. The chips at lower stacks most likely are low power chips. The overall goal is to achive smaller form factor, higher interconnect density and lower energy consumption than that in the SiP or PoP packages.

The horizontal connectivity on interposer usually requires traces of a few milli-meter without active repeater due to

Manuscript received April 12, 2011..
Sally Liu, Yung-Chow Peng, and Fu-Lung Hsueh are with Taiwan Semiconductor Manufacturing Company, 9, Creation Road 1, Hsinchu Science Park, Hsinchu, Taiwan 300-77, ROC (e-mail: s_liu@tsmc.com).

physical constrains. At such length a complete wire model should predict correct effective loading at driving and receiving points but also correct trace delay, including both RC delay and flight time. Futhermore, impact from reflection waves should also be counted for a more accurate jitter prediction if the line is not terminated.

The vertical connectivity in face-to-face stacking can be formed through micro-bumps while in face-to-back stacking requires TSV through thinned wafers. Today 25um wafer thickness, or TSV height, has been demonstrated. While reaping the benefits in speed and power, stress induced by the formation of TSV and stacking on top of micro-bumps must be managed in design which requires linking stress analysis tools to the mainstream design tools with the underlying stress model characterized. Thin wafer will enhance coupling between top and bottom metal layers through the substrate. The traditional BEOL modeling methodology normally assumes substrate as conductor, instead of semiconductor. This assumption needs to be examined, particularly for interposer substrate in which may only sparsely populated with passive devices. Proper shielding is needed to protect high-speed and sensitive signals from coupling noises. This also holds for top-bottom die coupling with insulator in between.

The 3rd dimension of chip integration requires significant extension / modification to the work flow and tool set to achieve overall covergence of chip stacking designs. In addition to the inclusion of new elements, *e.g.* TSV and micro-bump representations and 3D parasitic coupling, the major challenges are 1) simulation allowing different circuit portions from different IC technologies which is counter intuitive to the IC design flow, 2) seamless flow to enable chip interface optimization which requires a smooth data exchange flow to share data among 3DIC design tools at different stages, 3) interdependence among electrical, thermal and mechanical analyses, not only each of which needs a solid underlying model / analysis but also a smooth data exchange flow to share data. These facets are just started to be addressed by the EDA industry and research institutions.

III. 3D DESIGN STRATEGIES

When adopting 3D solution several design angles are changing and moving to wide parallel interface, circuit repartition and module base parallel-processing for achieving better system performance, lower power and noise isolation.

Wide parallel interfaces will gradually replace almost all of the conventional interfaces among dies or chips when using 3D or 2.5D design because it could achieve order of BW (bandwidth) improvement under limited interface power budget. The interface power consumption is close to [DC power (through terminator) + AC power ($C_{loading}*V_{IO}^2*BW$)]. Here, $C_{loading}$: total channel loading; V_{IO}: IO supply voltage. When through 3D design, it could be DC power free due to no terminator needed for ISI concern. And AC power could be cut to < (1/10) because $C_{loading}$ and IO supply voltage could be ~1/3 and ~1/4 of conventional solution, respectively. Another consideration in design is how to increase BW without increasing the channel data rate to simplify the design complexity. This could also be achieved by expanding interface to wide through 3D benefits because pitch of TSV or micro-bump is much smaller than that of normal C4 bump. Tables I and II compares loading capacitance and power consumptions of 3D, 2.5D and PCB solutions.

Capacitance (pF)	3D	2.5D	PCB
TSV/micro-bump/IO pad	0.2	0.2	1
Channel Routing	0	~1	~10
Memory dies + controller	5	5	2
C total	1	~2	~12

Table I: C-loading comparison of 3D/2.5D/PCB solutions

Power Consumption (mW/channel)	3D	2.5D	PCB
DC	0	0	28.8
AC @ 1Gbps	1.44	2.88	17.28
Total	1.44	2.88	46.08

Table II: Power consumption comparison of 3D/2.5D/PCB solutions

Circuit repartition could achieve better PPA (performance, power, area) on noise isolation, thermal balance and lower cost. On noise isolation pure analog part of a circuit could be protected in a die with clean power supply, like analog part of a PLL could be separated into another die from main chip. Then noise injection from main chip to it could be reduced around 20dB. When designing a 3D chip another killer is thermal dissipation, since it is worse than conventional approach for dies in the middle of a stack. To overcome this drawback thermal balance handling is one of possible ways. Like for digital application, an ALU could be repartitioned into different dies based on high, middle and low bit-groups with balance toggle rate to achieve lowest thermal self-heating. Regarding to lower cost, the way is to separate the RF, IO or ESD on a SoC chip from main die with most advance technology to dies with best match non-advance technology.

Module base parallel-processing is another important area. 3D structure opens the possibility to easily link individual processor of a die with related DVFS (dynamic voltage frequency scaling) controller and memory banks on another dies. Through this system power consumption can be minimized. And the memory access latency and BW could be improved due to shorten access path and wider bus.

IV. 3DIC INTER-CHIP SOC TREND

Reducing power consumption is one of the major reasons driving for 3DIC. For inter-chip data communication between stacking chips, output drivers can be powered by the same core-device power supply due to reduced connection loading. This points to the trend of two kinds of chip partition in 3DIC: "core-device only" SOC for inter-chip data communications, and "core-IO mixed" SOC for driving peripheral devices.

Low-cost and low-power consumption are two big advantages for "core-device only" SOC. Without I/O devices, the 3-5 silicon fabrication process steps are eliminated and cost saving is about 5% and chip size is reduced by using high-density core devices in circuits. For "core-device only" chips, each chip will choose its process technology for PPA (power, performance, and area) optimization. So, different "core-device only" chips in 3DIC might use different core-device power supply voltages (for example, 0.9V, 1.0V, 1.2V, 1.5V, etc) in different technology nodes. For inter-chip data communication compatibility, each chip must have level-shift circuit at the output driver side and input receiver side. The level-shift circuit must not degrade speed performance and energy efficiency.

For "core-IO mixed" chips, their main function is to drive and to interface with peripheral devices, such as displays, speakers, off-chip high-power amplifiers, sensors, actuators, etc. The technology choice is broad and is toward the mainstream high-voltage technologies. The chip also needs high-speed core devices for interfacing with "core-device only" chips. The technology consideration for "core-IO mixed" chip is focusing on mostly speed and high-V, rather in low-power consumption.

V. WIRELESS TESTING IN 3DIC

In 3DIC, each chip must be a known-good-die (KGD) before and after packed with other chips. It is critically important to leverage BIST (built-in-self-test) capability to assure the robust function in each chip before stacking. The BIST strategy can also be extended to cover inter-chip connections. Wireless testing [3] can be deployed to ease the constrains of limited observation points after stacking. In 3DIC environment, each chip is thinned to less than 50um. The inductive coupling and capacitive coupling become suitable for inter-stacking chips data communications and non-contact wafer functional testing.

Inductive coupling uses the magnetic near-field induced by micro-coils for data transmission. The coils can be arranged in a dense array and have been analyzed with very small crosstalk. The inductive coupling link can be applied to both the face-up face-down chip stacking, and more than two chips stacking. The effectiveness of inductive coupling between receiver for transmitter does not degrade with the silicon substrate resistivity. However, in the capacitive coupling link, the coupling coefficient between the transmitter and receiver is a strong linear function of silicon substrate resistivity. The displacement current under the metal plate shields the electric field in the capacitive coupling links. So, the capacitive coupling link can only be applied to face-to-face chip stacking of two chips. It concludes that inductive coupling is more suitable as a wireless interface than the capacitive coupling.

VI. CONCLUSION

3DIC enables further scaling of product form factors and promises plenty opportunities of product innovation along the way as 3DIC technology matures. The 3rd dimension of chip integration not only invites designers to rethink their design strategy, potentially a paradigm shift in partition and implementation of subsystems but also brings about new risk factors. It is time for both model developers and circuit designers togather with EDA vendors and stardardization institutions needs to collaborate to enrich the 3DIC eco-system, leveraging the latest development in design technologies and software technologies.

REFERENCES

[1] J.C. Lin etc, "High density 3D integration using CMOS foundry technologies for 28 nm node and beyond", *2010 IEDM*

[2] N. Yu, "The coming of age of 2.5D and 3D IC's: Interface standardization for product realization"., 2010 SEMICON

[3] H.Ishikuro etc, "Wireless proximity interface with a pulse-based inductive coupling technique", IEEE Communication Magzine, October 2010.

A Single TSV-rail 3D Quasi Delay Insensitive Asynchronous Signaling

M. Belleville, *Senior Member, IEEE*, E. Beigne, A. Valentian

Abstract—**Asynchronous communications are foreseen as mandatory for implementing 3D multiple tiers circuits. The drawback of asynchronous rails compared to synchronous ones is the higher number of interconnects. This number needs to be decreased when horizontal interconnects are replaced by Through Silicon Vias (TSV) because of their big silicon footprint. A circuit using only one TSV for asynchronous, quasi delay insensitive 3D signal propagation is proposed. This achieves to save two TSVs out of three, while offering 1Gbits/s capability.**

Index Terms—**3D Integration, Globally Asynchronous Locally Synchronous, Quasi Delay Insensitive, Through Silicon Via**

I. INTRODUCTION

To cope with the market requirements of more functionalities and performances, while keeping reasonable power consumption, the microelectronic industry has always extensively relied on 2D technology scaling. However, with the technical and economic challenges increasing dramatically with the very advanced nodes, 3D integration is now recognized as a very attractive alternative solution to sustain increased system integration. In particular, mixing different technologies in the 3D stacking, like i.e. memory on logic or logic on silicon interposer, is an opportunity to achieve lower cost and higher performances. However, in addition to the technological and economical issues, 3D IC designers have to face complex new challenges.

In advanced CMOS technologies, process variations, and especially their local contributions, are increasing with scaling leading to over-conservative design margins and very difficult timing closures in complex digital System on Chips. To cope with this issue, Globally Asynchronous Locally Synchronous (GALS) architectures have been proposed.[1]. In such a case, timing closure has only to be achieved locally, while global data transfers are based on asynchronous signaling. Another advantage of this approach is to ease synchronization between independent frequency islands and thus enable local DVFS (Dynamic Voltage Frequency Scaling) implementation [2].

Although some preliminary demonstration of 3D clock distribution has been demonstrated [3], it is expected that 3D timing closure will be even more challenging than 2D one. Indeed, process variations may be uncorrelated between two tiers, if they come from different technologies, lots or foundries [4]. In addition, thermal coupling between tiers will

have a major impact on local environmental variations and has to be considered at the design stage [5].

Thus, GALS architectures will be very helpful for the development of 3D ICs. However, asynchronous signaling is based on request / acknowledge exchanges which require more interconnect lines than synchronous transfers [6]. For instance, a single bit transfer using the asynchronous Quasi Delay Insensitive (QDI) protocol needs three interconnect lines.

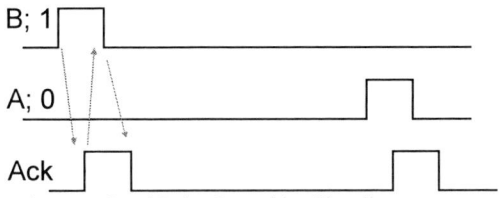

Fig. 1. Asynchronous Quasi Delay Insensitive Signaling.

This point, already sensitive in 2D ICs, is a big issue for 3D interconnects. Indeed, the area of a single Through Silicon Via (TSV) (e.g. 15µm diameter for medium density ones) is equivalent to the one of several thousands of transistors (e.g. in 45nm CMOS). Thus, minimizing the number of TSVs is a key challenge to be able to benefit from the GALS architectures advantages. To achieve this, a first proposal is to serialize the transmitted data [7].

This paper proposes a complementary approach, based on a new circuit able to transmit QDI signals, using only one TSV instead of three. It is organized as follows: Section II will introduce the signaling principle; transmitter and receiver schematics will be described next, respectively in Sections III and IV; finally, complete link results will be presented and commented in Section V.

II. ASYNCHRONOUS QUASI-DELAY-INSENSITIVE SIGNALING AND PROPOSED 3D, SINGLE TSV IMPLEMENTATION

As illustrated in Fig. 1, a basic QDI data transfer requires three signals: a "1" is transmitted when line B rises up; a "0" is transmitted when line A rises up; next, the receiver sends an acknowledge (ACK) when it has stored this information; upon the reception of this acknowledgement, the transmitter resets A or B.

In a first attempt to minimize the number of interconnect lines, using ternary signals was proposed [8]. The principle is presented in Fig. 2. A and B signals are merged into a single ternary signal. Idle state (A=B=0) correspond to the intermediate voltage value (Vmed). A "1" is transmitted when

---

M. Belleville, E. Beigne and A. Valentian are with CEA, LETI, MINATEC Campus, Grenoble, FRANCE (e-mail firstname.lastname@cea.fr).

978-1-4244-9019-6/11 $26.00 © 2011 IEEE

the ternary signal switches from Vmed to Vhigh, and a "0" when it switches from Vmed to Vlow. One line is still dedicated to the acknowledge signal.

One drawback of this proposal is the need of two supply voltages. However, in complex digital circuits, such power supplies are quite often available: Vi/o and Vcore in every digital circuit in advanced technology, or Vcore1 and Vcore2 like in Vdd Hopping implementations [2].

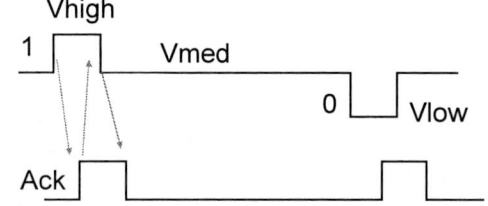

Fig. 2. Ternary Asynchronous Quasi Delay Insensitive Signaling.

This work extends this proposal by using only one ternary line to support all the QDI protocol. The basic principle is described in Fig. 3. Two new interface blocks are introduced between the standard asynchronous transmitter (TR) and receiver (RE): IFT and IFR. Only those two blocks will be detailed in this paper. Standard QDI signals (Ai, Bi, Acki, and Ao, Bo, Acko) connect IFT to TR and IFR to RE.

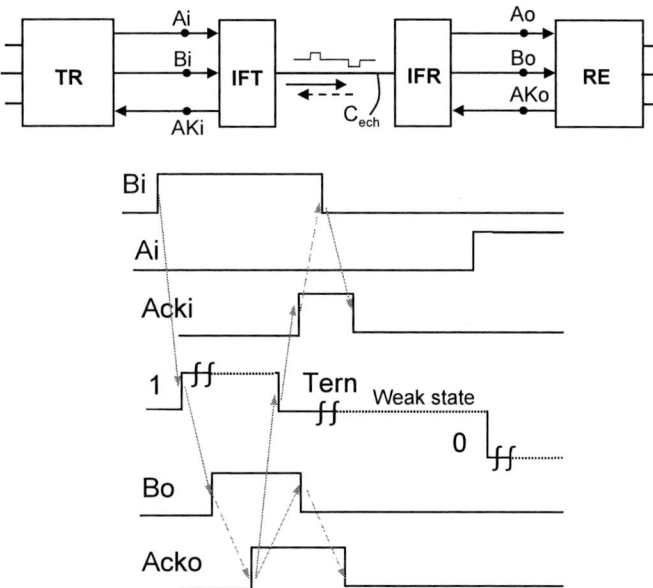

Fig. 3. Proposed single conductor Quasi Delay Insensitive Signaling.

When a "1" is transmitted, IFT sets the ternary signal (Tern) to Vhigh. Internally, IFT monitors Tern and detects the completion of the transition to Vhigh. Upon completion, IFT stops driving Tern, and preferably maintains the Vhigh level with a weak driver. The IFR block detects this transition on Tern and activate Bo. Upon reception of Acko, IFR resets Bo, and simultaneously drives Tern to Vmed. Internally, IFR monitors Tern and detects the completion of the transition to Vmed. At this completion, IFR stops driving Tern. Then, a weak driver keeps Tern Vmed level. On the transmit side, IFT detects Tern transition to Vmed, and consequently sets Acki. Finally, when Bi is reset, IFT resets Acki. An equivalent sequence occurs when Ai is activated, Tern switching from

Vmed to Vlow.

III. TRANSMITTER

This section details the IFT block architecture. It is based on a set of flip-flops detecting transitions on Ai (rising), Bi (rising), or Tern (rising and falling). Two specific buffers, with logic thresholds adjusted accordingly, are used to switch, on Tern transitions between Vmed and Vhigh (Th2), as well as between Vmed and Vlow (Th1). Two switches are driving Tern either to Vhigh or Vlow. Level shifters may be required to drive efficiently the two switches connected to Vhigh and Vlow. This will be used if Vhigh is for instance higher than Vdd. An acknowledge (Acki) is generated on a Tern transition either between Vhigh and Vmed (Th2), or between Vlow and Vmed (Th1).

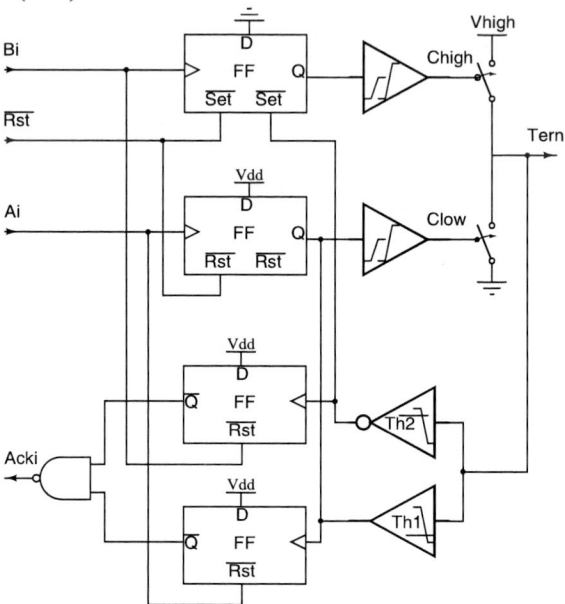

Fig. 4. Binary to ternary transmitter architecture.

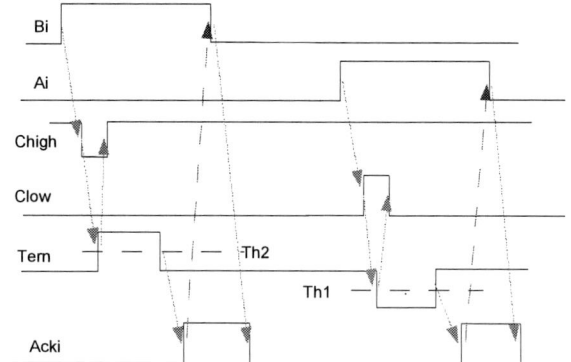

Fig. 5. Binary to ternary transmitter timing diagram

This architecture has been implemented in a low power 65nm CMOS technology. In this case, Vhigh is set to Vcore (Vdd=1.2V nominal), and Vmed is set to Vcc=0.6V. Corresponding schematics are detailed in Fig. 6 and Fig. 7. Th1 and Th2 are designed to have the appropriate logic thresholds and to have no static power consumption, whatever Tern state. Two weak transistors are used to maintain Tern into its state after completion of the transition. A reset signal is

978-1-4244-9019-6/11 $26.00 © 2011 IEEE   62

used to guarantee a good startup.

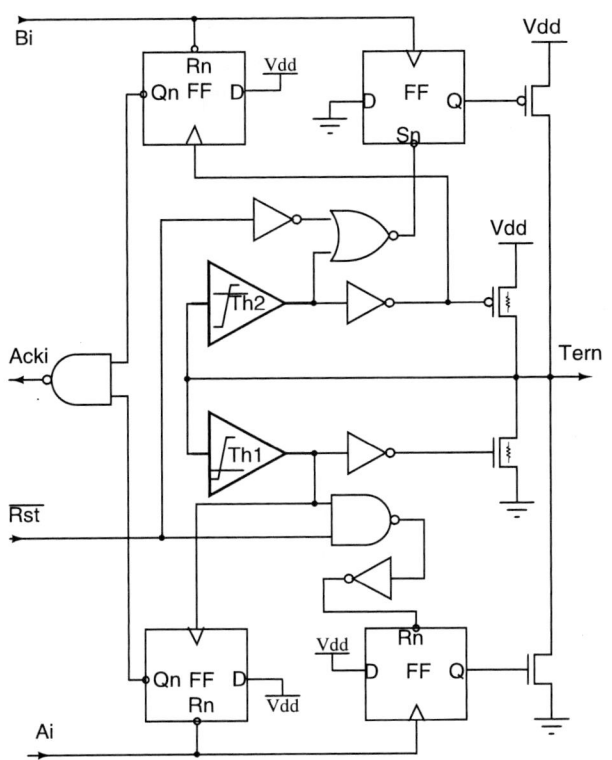

Fig. 6. Binary to ternary transmitter 65nm schematic (Vdd=1.2V, Vcc=0.6V)

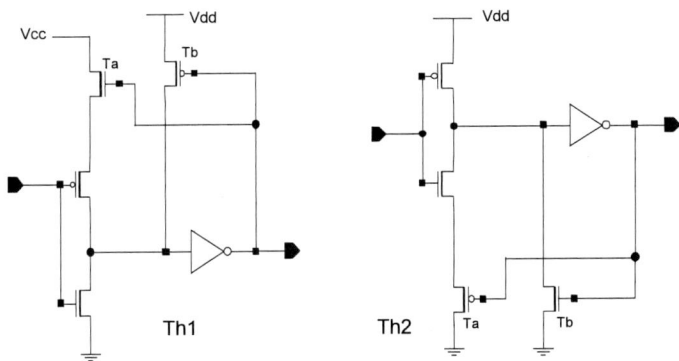

Fig. 7. Th1 and Th2 electrical schematics

IV. RECEIVER

This section details the IFR block architecture. It is based on a set of flip-flops detecting transitions on Acko (rising) and Tern (rising and falling). Again, the same specific buffers are used to switch, on Tern transitions between Vmed and Vhigh (Th2), as well as between Vmed and Vlow (Th1). A switch can drive Tern to Vmed. A level shifter may be required to drive efficiently this switch. Upon detection of a Tern transition between Vhigh and Vmed (respectively between Vlow and Vmed) Bo (respectively Ao) are set. Then the detection of the acknowledge signal (Acko) resets Bo or Ao, and drives the Tern line to Vmed. This strong drive stops at the completion of the transition. The corresponding 65nm implementation is detailed Fig. 10.

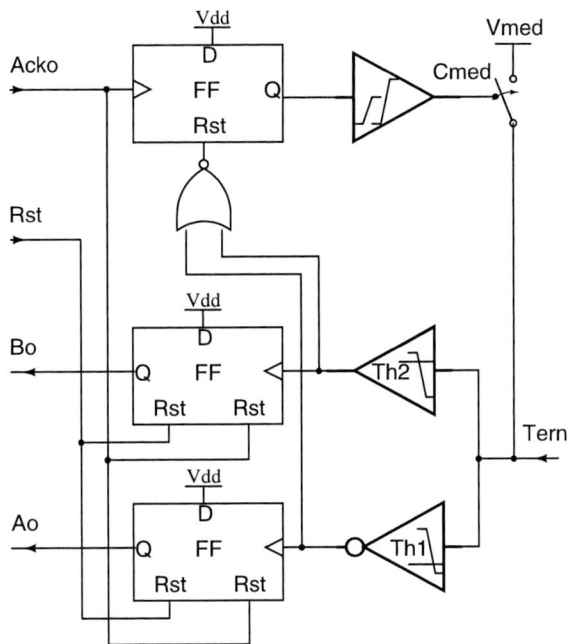

Fig. 8. Ternary to Binary receiver architecture.

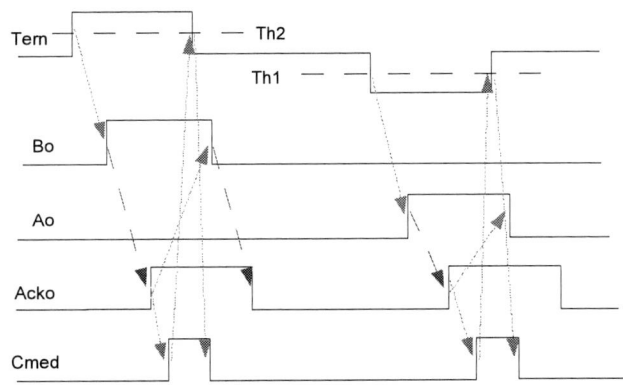

Fig. 9. Ternary to Binary receiver timing diagram.

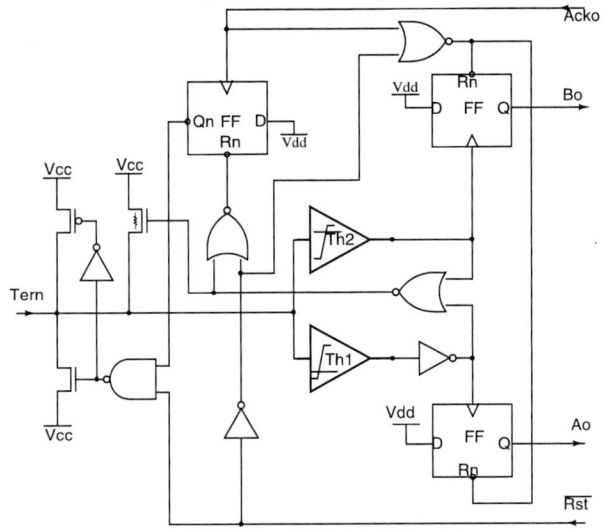

Fig. 10. Ternary to Binary receiver 65nm schematic (Vdd=1.2V, Vcc=0.6V)

Again, a reset signal is used to guarantee a proper startup, and a weak transistor is driving Tern to Vmed during the idle

978-1-4244-9019-6/11 $26.00 © 2011 IEEE

state. A full CMOS switch has been chosen to efficiently drive Tern to Vmed.

V. TOTAL 3D ASYNCHRONOUS LINK RESULTS

Fig. 11 presents simulation results of the total 3D asynchronous link, under typical conditions (Vdd=1.2V, Vcc=0.6V). Typical TSV capacitance was assumed [9], and the switches sized accordingly.

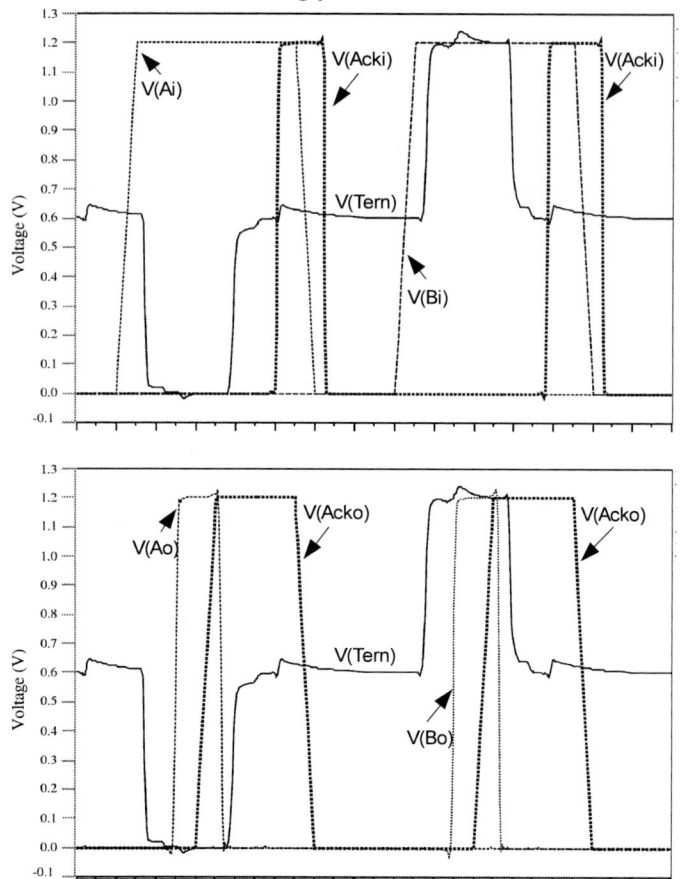

Fig. 11. Total link electrical waveforms (Vdd=1.2V, Vcc=0.6V)

This back and forth conversion to single TSV ternary adds two delays to a regular QDI data transfer: 250ps from Ai to Ao (or Bi to Bo) and 360ps from Acko to Acki, in typical conditions. Thus, typical 1Gbit/s data transfers are achievable. Note that one key advantage of asynchronous communications is that the speed is always optimal, whatever the process and environmental conditions.

The IFT block is made of 138 transistors, plus the two PMOS and NMOS driving Tern to Vdd and Gnd which may have multi-fingers according to the load. The IFR block consists of 87 transistors, plus the NMOS/PMOS driver to Vcc, which may also have multi-fingers. Thus, considering this small added complexity, using this proposed circuit to reduce the number of TSVs will significantly save silicon area, while achieving a data rate compatible with portable applications.

Due to the weak impedance state that is used in this circuit, a particular attention has to be paid at the layout stage. Indeed,

this weak state induces sensitivity to capacitive coupling or inductive effects. Thus the Tern connection has to be quite small (in addition to the 3D TSV). In this work, we propose to build dedicated library cells, embedding IFT or IFR layout and one TSV, therefore mastering the connection length as well as the TSV stress region[10]. Equivalent cells, without the TSV, are used in the non-crossed tier, and constraints minimizing the final inter-tier routing are given to the Place&Route tool.

VI. CONCLUSION

Asynchronous connections between tiers are very good candidates to mitigate synchronization issues of future complex digital 3D ICs, which may mix different technologies. This paper presents a circuit solution to use only one TSV, instead of three, in Quasi Delay Insensitive asynchronous 3D connections. The silicon area of the added circuit is largely compensated by the reduction of the number of TSVs. In addition, 1Gbits/s transfer capabilities are demonstrated in a 65nm Low Power CMOS technology, making this solution suitable for many applications.

REFERENCES

[1] M. Krstic, E. Grass, F. K. Gürkaynak and P. Vivet, "Globally Asynchronous, Locally Synchronous Circuits: Overview and Outlook", *IEEE Design & Test of Computers*, vol 24, n°5, 2007, pp 430-441.

[2] E.Beigne, F.Clermidy, H.Lhermet, S.Miermont, Y.Thonnart, Xuan-Tu Tran; A.Valentian, D.Varreau, P.Vivet, X.Popon and H.Lebreton, "An Asynchronous Power Aware and Adaptive NoC Based Circuit", *IEEE Journ. of Solid-State Circuits*, Vol.: 44 , Issue: 4, 2009 , pp. 1167-1177

[3] V. F. Pavlidis, I. Savidis and E. G.Friedman, "Clock Distribution Networks in 3-D Integrated Systems", *IEEE Trans. on VLSI Systems*, vol. PP, Issue: 99,: 2010 , pp. 1-11

[4] S.Garg and D.Marculescu, "3D-GCP: An analytical model for the impact of process variations on the critical path delay distribution of 3D ICs", *ISQED*, 2009., pp. 147-155.

[5] A.K Coskun,. J.L Ayala, D Atienza, T.S.Rosing and Y Leblebici, "Dynamic thermal management in 3D multicore architectures", *DATE '09*, 2009 , pp. 1410-1415

[6] A.J.Martin and M.Nystrom, "Asynchronous Techniques for System-on-Chip Design", *Proceedings of the IEEE*, Volume: 94 , Issue: 6, 2006 , pp. 1089-1120

[7] P. Vivet, and F. Darve, "An Asynchronous Serial Link for a 3D Network-on-Chip '", *DATE 2010 – 3D Integration Workshop*.

[8] T. Felicijan and S. B. Furber, "An Asynchronous Ternary Logic Signaling System", *IEEE Trans. on VLSI Systems*, Vol. 11, N°. 6, December 2003, pp.1114-1119

[9] L. Cadix, M. Rousseau, C. Fuchs, P. Leduc, A. Thuaire, R. El Farhane, H. Chaabouni, R. Anciant, J.-L. Huguenin, P. Coudrain, A. Farcy, C. Bermond, N. Sillon, B. Flechet and P. Ancey, "Integration and frequency dependent electrical modeling of Through Silicon Vias (TSV) for high density 3DICs", *International Interconnect Technology Conference (IITC)*,: 2010 , pp. 1-3

[10] T. Dao and V.Adams, "Through-Silicon-Via stress 3D modeling and design", *IEEE International Conference on IC Design and Technology (ICICDT)*, 2010 , pp. 114-117

Smart Stacking[TM] technology: an industrial solution for 3D layer stacking

C. Lagahe Blanchard[1], I. Radu[1], M. Sadaka[2] and K. Landry[1]

[1]SOITEC, Parc Technologique des Fontaines, Bernin 38926 Crolles Cedex, France
[2] SOITEC USA Inc., 1010 Land Creek Cv, Austin, Texas, 78746, USA

Abstract— Smart Stacking™ is a wafer-to-wafer stacking technology of partially or fully processed wafers. This technology enables transferring very thin layers in a high volume manufacturing environment. The core technologies are surface conditioning, low temperature direct bonding and wafer thinning (figure 1). This technology is adapted for advanced semiconductor applications such as Back Side Illumination (BSI) CMOS Image Sensors (CIS) as well as 3D integration approaches [1,2].

I. SMART STACKING[TM] TECHNOLOGY FOR BSI CIS APPLICATIONS

In traditional front side illuminated image sensors the photodiode quantum efficiency (QE) has been suffering with attempts to decrease pixel size and increase the number of backend metal layers and image processing circuits. By flipping the processed layer top down, light can now be detected through the backside of the sensor with significantly higher QE than with the front side method. The industry has now identified BSI image sensors as the primary path to continue increasing pixel density, resolution and speed without sacrificing QE.

This application segment requires a defect free bonding process with high bonding strengths and no induced distortion of bonded layer, to stand the post transfer integration steps and to fulfill color filter array (CFA) lithography requirements.

Because of the patterned surface of the CMOS wafer, bonding requires a capping oxide deposition followed by chemical mechanical planarization. Adequate deposited oxide nature, surface flatness, micro-roughness (< 3Å r.m.s) and cleanliness are needed for direct oxide bonding in order to maximize the post annealing bonding strengths. For IC's stacking, the thermal budget is limited to 400°C due to the presence of metal interconnects. Because of this temperature limitation, specific surface preparations are essential to reach high bonding strengths after low temperature annealing.

A minimum bonding energy level is needed to withstand the most stressing steps such as wafer thinning, edge treatment, post processing metal deposition, etching, thermal treatments... and to allow a wide process window at these steps (Figure 2).

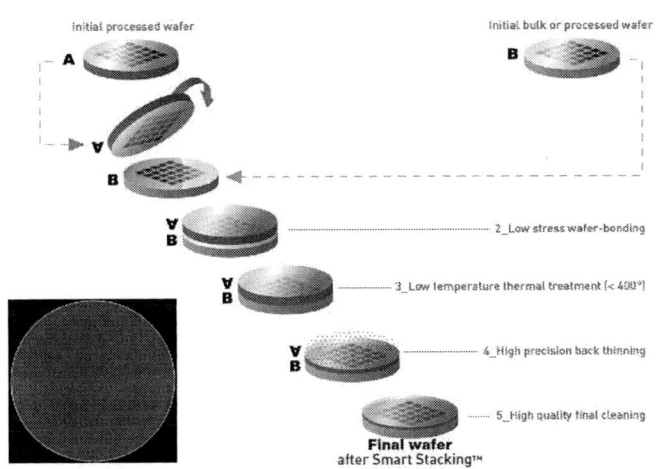

Figure 1: Schematics of Smart Stacking[TM] technology for 3D layer stacking.

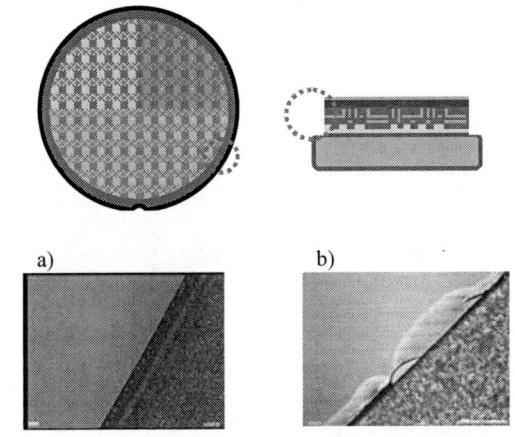

Figure 2: Comparison of layer edge quality after standard edge treatment on transferred processed film.
a) No peel off for structure with 1,5 J/m[2] bonding energy
b) Peel off for structure with 1,2 J/m[2] bonding energy

It has been evidenced that increasing the bonding energy leads to the creation of peripheral bonding voids if standard direct bonding process is used (Figure 3). Such bonding voids are in the range of 10-200 µm in diameter and they are located few mm away from the bonding edge limit. Therefore the challenge is to reduce these edge peripheral voids without loosing bonding energy [3]. Morphology of the surface of wafer is of primary importance to avoid bonding defects. Residual topography by lack of CMP removal or CMP scratches can cause bonding voids. Although scratches are not detectable on standard defectivity tools (threshold 0,13 µm), the scratches become visible after the bonding, especially at the wafer edge.

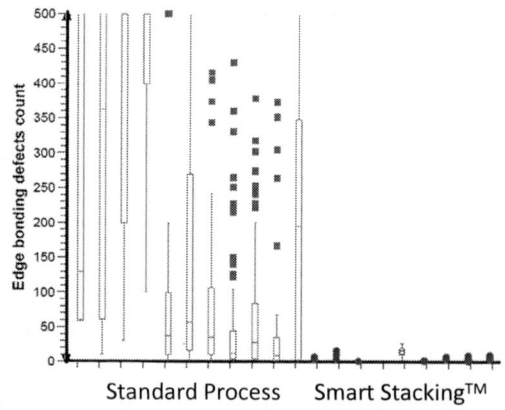

Standard process

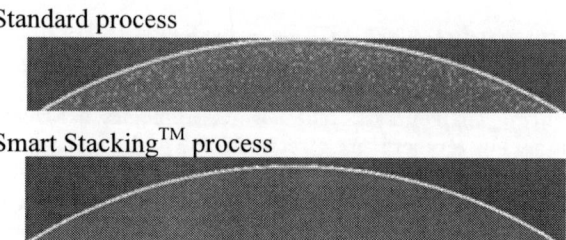

Smart Stacking™ process

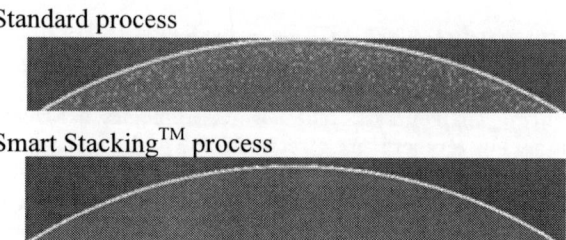

Figure 4: Edge bonding voids count for a BSI CIS bonding production line before and after introduction of Smart Stacking™ bonding process (acoustic microscopy monitoring). Acoustic microscopy images of edge quality depending on bonding process.

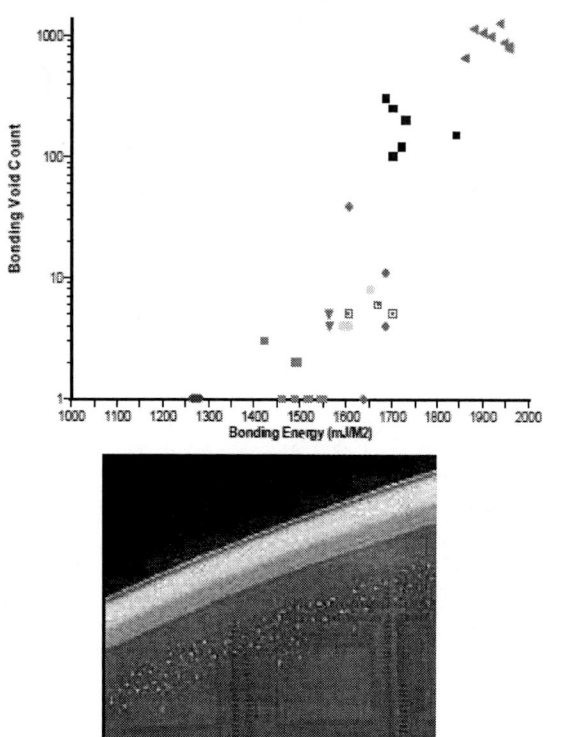

Figure 3: Edge bonding voids count as a function of bonding energy with standard direct oxide bonding process (log10 scale). Different symbols and colors represent different surface preparation conditions.

To eliminate these edge bonding defects, various parameters such as surface morphology, edge roll off, surface chemistry and adsorbed water molecule structure on surface to be bonded have to be adequately set up during the bonding process. The bonding process implemented in Smart Stacking™ technology successfully avoids edge bonding defects formation as shown by process monitoring (Figure 4), without decreasing the bonding energy (2 J/m^2).

The bonding interface stability was assessed after thermal annealing around 400°C and did not show any change in the acoustic microscopy scan.

In parallel, the distortion printed during bonding step into the bonded layers must be minimized to fulfill requirements imposed by later processing of color filters array : final overlay must be less than 100nm for pixel size >1,1µm and even lower for smaller pixels. Smart Stacking™ bonding process has been optimized in order to reach those requirements but also in order to be as independent as possible from the initial processed wafer. The current performance has proven its compatibility with the current pixel node of BSI CIS image sensors in 200mm and 300mm.

II. SMART STACKING™ TECHNOLOGY FOR 3D INTEGRATION APPLICATIONS

The emerging field of 3D integration aims at providing highly integrated systems by vertically stacking and connecting various materials, technologies, and functional components together [4]. Due to stringent reliability standards, a fundamental motivation is to achieve a mechanical bond between the face-to-face bonded wafers without intermediate materials. Achieving a robust mechanical bond without introducing new materials is a promising 3D IC path required for both manufacturability and reliability.

The wafer stacking technology for 3D integration requires high quality bonding interfaces with uniform bonding films. Therefore, the within wafer non uniformity (WIWNU) of the oxide layers needs to be minimized. The CMP planarization processes have been optimized to reduce the surface topography with limited non uniformity. To address the thermal budget constrain imposed by stacking of processed wafers (i.e. < 400°C), specific pre-bonding surface conditioning adapted to bonded materials nature and post-bonding thermal treatment were defined to control and increase the bonding strengths [5].

Precision wafer-to-wafer alignment represents a key requirement for the wafer scale vertical integration of 3D ICs. The alignment error has a direct influence on the size of the metal pad and the pitch of the vertical interconnects. The alignment accuracy is dependent on the tool capability and on the process conditions. Based on optimization and process control the alignment of 300mm bonded wafers is currently reaching sub micron accuracy with best in class data reported at <0.25μm (all wafer sites) (figure 5) [6].

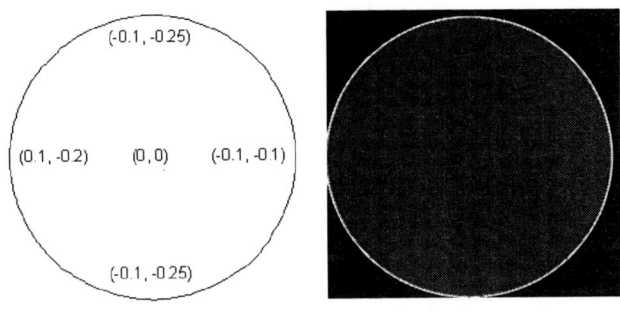

(a)   (b)

Figure 5: (a) Wafer-to-wafer alignment accuracy of bonded 300mm wafers (x,y) ; (b) Acoustic microscopy image of bonded 300mm wafers.

III. CONCLUSION

Smart Stacking[TM] is a competitive technology for BSI image sensors applications, specifically aimed to reach high performance bonding (defect free, low induced distortion, proven reliability insured by high bonding strengths, small edge exclusion) and to fulfil key industrial requirements (throughput, yield, manufacturability). Additional needs linked to 3D integration applications such as WIWNU of bonded oxide layers and wafer-to-wafer alignment are also addressed by specific flavours of this generic stacking technology.

ACKNOWLEDGMENT

Authors would like to acknowledge the Process Development team of Tracit BU Platform and the R&D team in charge of 3D integration developments in Soitec Group and CEA/LETI.

REFERENCES

[1] C. Lagahe et al, Wafer stacking : key technology for 3D integration, IEEE International SOI conference, San Francisco, October 2009

[2] M. Sadaka et al, 3D integration: Advantages, Enabling Technologies and Challenges, presented at IEEE International Conference on Integrated Circuit Design&Technology (ICICDT), Grenoble, June2-4, 2010

[3] A.Castex et al, ECS, Montreal 2011 (to be published)

[4] C. S. Tan et al, Wafer Level 3-D ICs Process Technology, Springer, 2008

[5] A. Castex et al, "Low Temperature Direct Wafer Bonding Process for Back Side Illumination Image Sensors and 3D Stacking ", IEEE conference Low Temperature Bonding for 3D Integration, Tokyo, January 2010

[6] G. Gaudin et al., « Low temperature direct wafer to wafer bonding for 3D integration", presented at IEEE 3D-IC conference, Munich 2010.

TSV Number Minimization Using Alternative Paths

Chun-Hua Cheng, Chih-Hsien Kuo, Shih-Hsu Huang[a)]

Department of Electronic Engineering,
Chung Yuan Christian University
Chung Li, Taiwan, R.O.C.
a) Email: shhuang@cycu.edu.tw

Abstract—In a three-dimensional integrated circuit (3D IC) design, through-silicon-vias (TSVs) are used for data transfer across layers. However, TSVs act as obstacles during the stage of placement and routing and have a negative impact on chip yield. Therefore, TSV number minimization is an important topic for 3D IC design. In this paper, we point out that there often exist idle functional units and idle TSVs at each control step. If we use idle functional units and idle TSVs to form an alternative path to replace direct TSVs for data transfer, the number of TSVs can be reduced. Based on that observation, we present an ILP (integer linear programming) approach to formally draw up our problem. Given a high-level synthesis result and a clock period constraint, we perform post-processing to fully utilize alternative paths for TSV number minimization. Compared with previous work that minimizes the TSV number without considering alternative paths, experimental results show that our approach can further reduce 16.92% TSV number without affecting the circuit performances.

I. INTRODUCTION

In deep sub-micron CMOS process technology, interconnect width and transistor feature size will gradually face the photolithography of physical limits [1]. Interconnect delay has increased significantly compared to gate delay and become a bottleneck in the design of SOC (System-on-a-chip). Three-dimensional integrated circuits (3D ICs) instead of SOC designs is a trend to overcome the bottleneck in the interconnect scaling. In 3D ICs design, multiple chip layers are stacked together and communicate with direct vertical interconnects. These direct vertical connections are called through-silicon-vias (TSVs). Long 2D planar interconnects can be replaced by shorter direct TSVs in 3D ICs. However, TSVs act as obstacles during the stage of placement and routing [2-3]. Besides, TSVs also has a negative impact on chip yield. Therefore, minimizing the TSV number becomes a very important topic in 3D IC design.

Several research efforts have been paid to the minimization of TSV number in the high-level synthesis stage. Previous works. Mukherjee et al. [4] presented the first literature to integrate operation scheduling, resource binding and layer assignment in one ILP formulation. Their objective [4] is to maximize the number of data transfers in the same layer. As a result, the TSV number can be approximately minimized. Further, Lee et al. [5] accurately formulate the TSV number minimization problem in the high-level synthesis stage. Based on their ILP formulation [5], the number of TSVs can be minimized in the high-level synthesis stage.

On the other hand, recently, Kim and Liu [6] demonstrate that idle functional units can be used as a pass-logic (an alternative path) for data transfer to reduce the global interconnect requirement in 2D planar designs. However, in 3D IC designs, to form a pass-logic (an alternative path) for data transfer, we not only require idle functional units but also require idle TSVs. Since Kim and Liu [6] do not consider TSVs, their work [6] cannot be applied to 3D IC designs.

Actually, there often exist idle functional units and idle TSVs at each

control step in a high-level synthesis result that includes scheduling, binding and layer assignment. Therefore, it is possible and practical to use idle functional units and idle TSVs to form a pass-logic (an alternative path) to replace direct TSVs for data transfer. Obviously, if we can fully utilize these alternative paths to replace direct TSVs, the number of TSVs can be further reduced.

Based on that observation, in this paper, we study the TSV number minimization by fully utilizing alternative paths. Our inputs include a high-level synthesis result and a clock period constraint. We perform post-processing to eliminate those TSVs whose data transfers can be replaced by alternative paths. Our objective is to maximize the number of eliminated TSVs. An ILP (integer linear programming) formulation is proposed to formally draw up our problem.

Compared with previous work [5] that already minimizes TSV number without considering alternative paths, experimental results show that our approach can further reduce 16.92% TSV number. Note that our solution does not degrade the circuit performance. It should also be mentioned that, in each benchmark circuit, the required CPU time of our approach is less than one second. Therefore, our approach is effective and efficient.

The rest of this paper is organized below. Section II describes our motivation. Section III presents our ILP approach to minimize the number of TSVs by using alternative paths. In Section IV, we report our experimental results and comparisons with previous works. Finally, in Section V, we make some concluding remarks.

II. PRELIMINARIES

In high-level synthesis [7], a behavior-level description is represented by a data flow graph (DFG), in which each node corresponds to an operation, and each directed edge corresponds to a dependency relation. Take the DFG ex1 shown in Fig. 1 for illustration. This DFG has three operations, including operations O_1, O_2, and O_3. This DFG has three dependency relations, including $O_1 \rightarrow O_2$, $O_1 \rightarrow O_3$, and $O_2 \rightarrow O_3$.

Given a DFG, the high-level synthesis of 3D IC includes three main tasks: operation scheduling, resource binding, and layer assignment. Operation scheduling [8-10] is to assign each operation to a control step to start its execution (under dependency relation constraints and resource constraints). Resource binding [11-12] is to assign each operation to a functional unit that can execute it (under operation lifetime constraints and resource constraints). Layer assignment [4-5][13] is to assign each functional unit to a layer that can accommodate it (under the constraint on the number of active device layers and the constraint on footprint area).

Lee et al. [5] integrate operation scheduling, resource binding and layer assignment in one ILP formulation to minimize the TSV number. We use the DFG ex1 shown in Fig. 1 as an example. By applying the approach proposed in [5], we have the following high-level synthesis result: operation O_1 is scheduled at control step 1, operation O_2 is scheduled at control step 2, operation O_3 is scheduled at control step 3, operation O_1 is assigned to multiplier M1, operation O_2 is assigned to

978-1-4244-9019-6/11 $26.00 © 2011 IEEE

adder A1, operation O_3 is assigned to divider D1, adder A1 is placed at layer 1, multiplier M1 is placed at layer 1, and divider D1 is placed at layer 2. The scheduled DFG is displayed in Fig. 1, and the resource binding and layer assignment result is displayed in Fig. 2 (a). With an analysis, we find that this high-level synthesis result requires two TSVs: one TSV (denoted as TSV_1) is from adder A1 to divider D1, and one TSV (denoted as TSV_2) is from multiplier M1 to divider D1. Note that the approach proposed in [5] guarantees minimizing the number of TSVs (if alternative paths are not considered). For the convenience of readers, we provide the detailed circuit graph in Fig. 2(b).

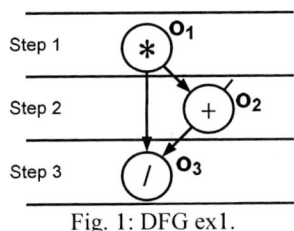

Fig. 1: DFG ex1.

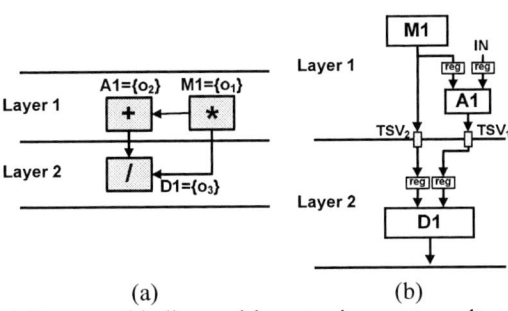

(a)   (b)

Fig. 2: (a) Resource binding and layer assignment results of [5]. (b) Corresponding detailed circuit graph.

III. MOTIVATION

Kim and Liu [6] demonstrate that idle functional units can be used as a pass-logic (an alternative path) for data transfer to reduce the global interconnect requirement in 2D planar designs. However, their work [6] cannot be directly applied to 3D IC designs. Here, we borrow their concept [6] to define the alternative path of a direct TSV data transfer below.

Consider a direct TSV data transfer from the output of functional unit k_1 to the input of functional unit k_2 at control step c. A path p from the output of functional unit k_1 to the input of functional unit k_2 can be an alternative path of this direct TSV data transfer, unless the following two conditions are met:

(1) All functional units (exclude k_1 and k_2) in the path p should be idle at control step c.

(2) To prevent timing violation, the delay of the path p should be less than or equal to the clock period.

Let's use Fig. 2(a) as an example. Suppose that the delays of multiplier M1, adder A1, and divider D1 are 10 ns, 5 ns, and 20 ns (denoted as $D_{M1} = 10$ ns, $D_{A1} = 5$ ns, $D_{D1} = 20$ ns), respectively. Thus, the clock period is at least 20 ns. We consider the data transfer from the output of operation O_1 to the input of operation O_3, (denoted as $O_1 \rightarrow O_3$). In Fig. 2(a), this data transfer is implemented by a direct TSV data transfer from the output of multiplier M1 to the input of divider D1 at control step 1. With an analysis, we find that the data transfer $O_1 \rightarrow O_3$ can be replaced by the alternative path $M1 \rightarrow A1 \rightarrow D1$ instead of the direct TSV data transfer $M1 \rightarrow D1$. We analyze the reasons below.

(1) Adder A1 is idle at control step 1. Therefore, we can use adder A1 as the pass logic at control step 1.

(2) The delay of alternative path $M1 \rightarrow A1 \rightarrow D1$ is less than one clock cycle ($D_{M1} + D_{A1} = 10$ ns +5 ns = 15 ns < 20 ns).

Based on the above analysis, we can use the alternative path $M1 \rightarrow A1 \rightarrow D1$ to implement the data transfer $O_1 \rightarrow O_3$. Note that, in Fig. 2(a), there is only one data transfer (i.e., the data transfer $O_1 \rightarrow O_3$) from the TSV from the output of multiplier M1 to the input of divider D1. Since the data transfer $O_1 \rightarrow O_3$ has been replaced by an alternative path, the TSV from the output of multiplier M1 to the input of divider D1 can be eliminated. As a result, we have the result displayed in Fig. 3(a).

For the convenience of readers, we provide the detailed circuit graph of our post-processing result in Fig. 3(b). Compared with Fig. 2(b), one TSV is eliminated. In addition, the global wiring from the output of multiplier M1 to the input of divider D1 can also be saved. On the other hand, it should also be mentioned that, compared with Fig. 2(b), two multiplexers and one de-multiplexer are added. However, since TSVs have many negative impacts and the global wiring often occupies a large area, our post-processing is worthy.

Finally, we explain the operations of multiplexers and de-multiplexer in Fig. 3(b) below. At control step 1, the left side of each multiplexer and the left side of de-multiplexer are selected; at control step 2, the right side of each multiplexer and the right side of de-multiplexer are selected. As a result, the circuit can work correctly with only one TSV.

At the end of this section, we make a summary. Since the approach proposed in [5] does not make use of alternative paths, their high-level synthesis result can be further improved. In this example, we find that the TSV number can be reduced from 2 to 1 by utilizing alternative paths.

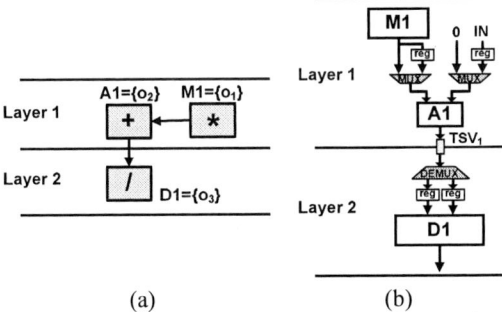

(a)   (b)

Fig. 3: (a) Our post-processing result. (b) Corresponding detailed circuit graph.

IV. OUR ILP APPROACH

In this section, we present an ILP approach to formally draw up the problem of TSV number minimization using alternative paths. Given a high-level synthesis result (including the results of operation scheduling, resource binding, and layer assignment) and a clock period constraint, we perform a post-processing to reduce the TSV number by fully utilize alternative paths. Note that, even if we have use the approach proposed in [5] to minimize the TSV number, we still can further reduce the TSV number by using alternative paths.

First, Table 1 defines the notations used in our ILP formulation.

Our optimization goal is to minimize the the number of TSVs. Therefore, the objective function is

$$\text{Minimize} \sum_{i=1}^{m} V_i$$

Each data transfer requires a path to implement. Therefore, for each cross-layer data transfer $O_a \rightarrow O_b$, which is performed at control step c, we have the following constraint:

$$\sum_{\forall j \in N(Oa \rightarrow Ob)} P_{j,c} = 1. \qquad \text{(Formula 1)}$$

If path j at control step c is used, all the TSVs in path j are used. Therefore, for each TSV $i \in U_j$, we have following constraint:

$$P_{j,c} \le V_i. \qquad \text{(Formula 2)}$$

All functional units in path j should be idle; otherwise, path j at control step c cannot be used as the pass-logic. Therefore, for each functional unit $k \in F(j)$, we have following constraint:

$$P_{j,c} \leq I_{k,c}. \qquad \text{(Formula 3)}$$

If path j_1 and path j_2 have an overlap, only one of them can be used at each control step. Therefore, for each control step c, we have following constraint:

$$P_{j1,c} + P_{j2,c} \leq 1. \qquad \text{(Formula 4)}$$

Each path must satisfy timing constraint (i.e., each path delay can not exceed clock period T_{clock}). Therefore, for each path j, we have the following constraint:

$$\left(T_{clock} - \sum_{k \in F(j)} D_k\right) \times P_{j,c} \geq 0. \qquad \text{(Formula 5)}$$

Note that the clock period T_{clock} and the delay of each functional unit k, denoted as D_k, are constant values. Therefore, $T_{clock} - \sum_{k \in F(j)} D_k$ is also a constant.

Table I. List of the notations used in our ILP formulation.

Notation	Description
V_i	The notation Vi is a binary variable. Suppose that there are m TSVs (numbered from 1 to m) in the given high-level synthesis result. Then, there are m corresponding varibales. If the number i TSV is used in our post-processing result, then $V_i = 1$; otherwise, $V_i = 0$.
$N(O_a \rightarrow O_b)$	The notation $N(O_a \rightarrow O_b)$ denotes the set that includes all paths that can implement the data transfer from the output of operation O_a to the input of operation O_b.
U_j	The notation U_j denotes the set that includes all TSVs in the path j.
$F(j)$	The notation $F(j)$ denotes the set that includes all the functional units in the path j.
$I_{k,c}$	The notation $I_{k,c}$ is a constant derived from the high-level synthesis result. Accrding to the high-level synthesis result, if the functional unit k at control step c is idle, then $I_{k,c} = 1$; otherwise, $I_{k,c} = 0$.
$P_{j,c}$	The notation $P_{j,c}$ is a binary variable. If the path j at control step c, then $P_{j,c} = 1$; otherwise, $P_{j,c} = 0$.
D_k	The notation D_k is a constant that denotes the delay of functional unit k.
T_{clock}	The notation T_{clock} is a constant that denotes the clock period constraint.

Take the DFG ex2 shown in Fig. 4 as an example. We use the approach proposed in [5] to derive the high-level synthesis result as the input. The high-level synthesis result is shown in Fig. 4 and Fig. 5(a): Fig. 4 gives the operation scheduling result, Fig. 5(a) gives resource binding and layer assignment results. Suppose that we are given two multipliers M1 and M2, one adder A1, and one divider D1. The delay of multiplier M1, multiplier M2, adder A1, and divider D1 are 10 ns, 10 ns, 5 ns, and 20 ns (denoted as $D_{M1} = D_{M2} = 10$ ns, $D_{A1} = 5$ ns, $D_{D1} = 20$ ns), respectively. The clock period constraint is 20ns. From Figure 5(a), we know the number of TSVs is 3 in the given high-level synthesis result.

The objective function is to minimize the number of TSVs:
Minimize = $\{V_1 + V_2 + V_3\}$.
The constraints are below.

Formula 1. Using the cross-layer data transfer O_1 to O_4 at control step 1 an example. There are two paths M1→D1 and M1→A1→D1 to implement this cross-layer data transfer. However, only one path can be used. Thus, we have the constraint $P_{M1 \rightarrow D1,1} + P_{M1 \rightarrow A1 \rightarrow D1,1} = 1$. All the constraints due to Formula 1 are listed in the following.

$$P_{M1 \rightarrow D1,1} + P_{M1 \rightarrow A1 \rightarrow D1,1} = 1, \qquad P_{M2 \rightarrow D1,1} + P_{M2 \rightarrow A1 \rightarrow D1,1} = 1,$$
$$P_{A1 \rightarrow D1,2} = 1.$$

Formula 2. If the path M1→D1 in used at control step 1, the TSV V_1 in the path M1→D1 is used. Thus, we have the constraint $P_{M1 \rightarrow D1,1} \leq V_1$. All the constraints due to Formula 2 are listed in the following.

$$P_{M1 \rightarrow D1,1} \leq V_1, \qquad P_{M1 \rightarrow D1,2} \leq V_1, \qquad P_{M1 \rightarrow A1 \rightarrow D1,1} \leq V_2,$$
$$P_{M2 \rightarrow A1 \rightarrow D1,1} \leq V_2, \qquad P_{A1 \rightarrow D1,1} \leq V_2, \qquad P_{M2 \rightarrow D1,1} \leq V_3.$$

Formula 3. The functional unit A1 should be idle at control step 1; otherwise, the path M1→A1→D1 cannot be used at control step 1. Thus, we have the constraint $P_{M1 \rightarrow A1 \rightarrow D1,1} \leq I_{A1,1}$. All the constraints due to Formula 3 are listed in the following.

$$P_{M1 \rightarrow A1 \rightarrow D1,1} \leq I_{A1,1}, \qquad P_{M2 \rightarrow A1 \rightarrow D1,1} \leq I_{A1,1}.$$

Formula 4. Path M1→A1→D1 and M2→A1→D1 have an overlap because of adder A1 and divider D1. Therefore, path M1→A1→D1 and M2→A1→D1 cannot be used at the same control step. Thus, for contrl step 1, we have the constraint $P_{M1 \rightarrow A1 \rightarrow D1,1} + P_{M2 \rightarrow A1 \rightarrow D1,1} \leq 1$.

Formula 5. Each path must satisfy the timing constraint. Thus, we have the constraint $(20 - 10 - 5) \times P_{M1 \rightarrow A1 \rightarrow D1,1} \geq 0$. All the constraints due to Formula 5 are listed in the following.

$$(20 - 10 - 5) \times P_{M1 \rightarrow A1 \rightarrow D1,1} \geq 0,$$
$$(20 - 10 - 5) \times P_{M2 \rightarrow A1 \rightarrow D1,1} \geq 0.$$

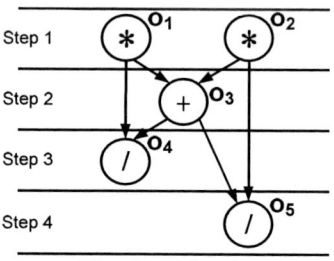

Fig. 4: DFG ex2.

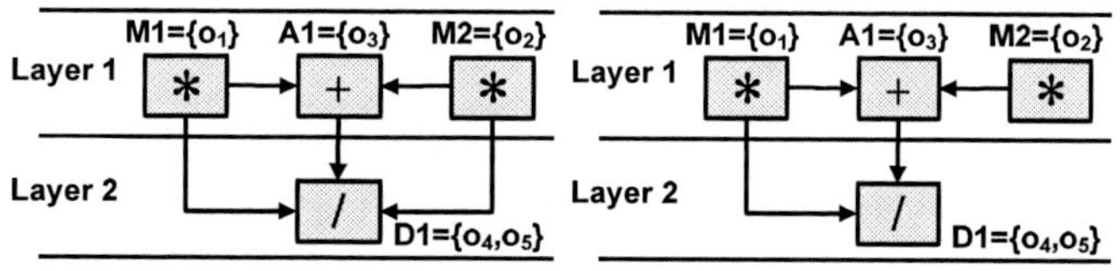

Fig. 5: (a) Resource binding and layer assignment results of [5]. (b) Our post-processing result.

After solving the ILP formulation, we find that: $P_{M1 \rightarrow D1,1} = P_{M2 \rightarrow A1 \rightarrow D1,1} = P_{A1 \rightarrow D1,2} = I_{A1,1} = V_1 = 1$, $V_2 = 1$, and the values of other binary variables are 0. Thus, we have the post-processing result as shown in Fig. 5(b). Note that the number of TSVs is only 2. Therefore, our approach reduce the TSV number from 3 to 2.

V. EXPERIMENTAL RESULTS

We use Extended LINGO Release 11.0 as the ILP solver. The platform is Windows 7 x64 running on Intel Xeon E5355 CPU with 4GB RAM. We adopt the modules in Synopsys DesignWare to implement functional units, including the adder, subtractor, multiplier, divisor, and selector. These functional units are 8-bit designs and targeted to UMC 0.13μm cell library. The area of adder, subtractor, multiplier, divisor, and selector are 252 μm², 842 μm², 4193 μm², 4434 μm², and 221 μm², respectively. The maximum delay of adder, subtractor, multiplier, divisor, and selector are 1.64 ns, 1.66 ns, 3.52 ns, 10.13 ns, and 1.05 ns, respectively.

Six benchmark circuits are used to test the effectiveness of our approach. Circuit TGFF14 is a benchmark adopted from [4], circuits BF, G2, and G5 are popular DSP applications used in the high-level

978-1-4244-9019-6/11 $26.00 © 2011 IEEE

synthesis community [14-15], and circuits IDCT1, IDCT2 and are the representative functions adopted from the MediaBench suite [16]. We use the approach proposed in [5] to derive their high-level synthesis results (including operation scheduling, resource binding and layer assignment results) for comparisons.

Table II tabulates the characteristics of benchmark circuits. The columns *Operations* and *Resources* give 5-tuple (#add, #sub, #mul, #div, #sel), where #add, #sub, #mul, #div, and #sel denote the numbers of additions (adders), subtractions (subtractors), multiplications (multipliers), dividers (divisors), and selectors (selections), respectively. The column *Steps* gives the number of control steps. The column *Clock* gives the clock period.

Table II. Characteristics of benchmark circuits.

Circuits	Operations	Steps	Clock	Resources
TGFF14	(7,3,3,1,0)	8	10.13 ns	(4,2,1,1,0)
BF	(12,6,11,0,0)	10	3.52 ns	(7,5,6,0,0)
G2	(9,0,9,0,6)	10	3.52 ns	(7,0,4,0,6)
G5	(24,0,0,0,4)	10	1.64 ns	(8,0,0,0,2)
IDCT1	(16,10,20,2,0)	14	10.13 ns	(7,6,6,2,0)
IDCT2	(32,20,40,4,0)	25	10.13 ns	(9,8,3,2,0)

Table III shows our experimental results. The column *Layer* gives the the number of layers. The column *Area* gives the footprint area. The column *#vias* gives the number of TSVs. We use the high-level synthesis results of previous work [5] as our inputs. The column *Imp* (in percentage) gives our improvement over previous work [5]. Compare with previous work [5], our approach can reduce 16.92% TSV number. It should also be mentioned that, no matter of the design constraint, the required CPU time of each benchmark circuit is less than one second. Therefore, our approach is effective and efficient.

Table III. Comparisons between previous work and our approach.

Circuits	Design Constraint		#vias		
	Layer	Area (μm^2)	[5]	Ours	Imp (%)
TGFF14	3	5033	5	4	20.00%
	4	4788	8	7	12.50%
	5	4434	12	8	33.33%
BF	3	15452	15	13	13.33%
	4	11589	20	17	15.00%
	5	9627	15	13	13.33%
G2	3	8386	11	9	18.18%
	4	7677	11	9	18.18%
	5	6142	36	28	22.22%
G5	3	1277	18	14	22.22%
	4	776	5	4	20.00%
	5	755	34	27	20.59%
IDCT1	3	16337	8	7	12.50%
	4	12253	21	17	19.05%
	5	9874	25	22	12.00%
IDCT2	3	11724	23	20	13.04%
	4	9057	53	48	9.43%
	5	7309	31	28	9.68%

VI. CONCLUSIONS

In this paper, we study TSV number minimization by using alternative paths. Given a high-level synthesis result and a clock period constraint, we perform post-processing to eliminate those TSVs whose data transfers can be replaced by alternative paths. We use an ILP approach

to formally draw up our problem. Experimental result show that compared with previous approach, our approach can reduce 16.92% TSV number without affecting the circuit performance. Furthermore, the required CPU time of each benchmark circuit is less than one second. Therefore, our approach is effective and efficient.

VII. REFERENCES

[1] T. Vucurevich, "3-D Semiconductor's: More from Moore", Proc. of IEEE/ACM Design Automation Conference, pp. 664, 2008.

[2] T.Y. Chiang, S. J. Souri, C. O. Chui, and K. C. Saraswat, "Thermal Analysis of Heterogeneous 3-D ICs with Various Integration Scenarios," IEEE International Electron Devices Meeting, pp. 681—684, 2001.

[3] D.H. Kim, S. Mukhopadhyay, and S. K. Lim, "Through-Silicon-Via Aware Interconnect Prediction and Optimization for 3D Stacked ICs," Proc. of ACM/IEEE International Workshop on System Level Interconnect Prediction, pp. 85—92, 2009.

[4] M. Mukherjee and R. Vemuri, "Simultaneous Scheduling, Binding and Layer Assignment for Synthesis of Vertically Integrated 3D Systems," Proc. of IEEE/ACM International Conference on Computer Design, pp. 222—227, 2004.

[5] C.H. Lee, S. H. Huang, and C.H. Cheng, "High-Level Synthesis of 3D IC Designs for TSV Number Minimization", Proc. of Synthesis And System Integration of Mixed Information technologies, pp. 260—265, 2010.

[6] T. Kim and X. Liu, "A Global Interconnect Reduction Technique during High Level Synthesis", Proc. of ACM/IEEE Asia and South Pacific Design Automation Conference, pp. 695—700, 2010.

[7] G.D. Micheli, "Synthesis and Optimization of Digital Circuits", Kluwer Academic Publishers, 1994.

[8] C.T. Hwang, J.H. Lee, and Y.C. Hsu, "A Formal Approach to the Scheduling Problem in High-Level Synthesis", IEEE Transactions On Computer-Aided Design of Integrated Circuits and Systems, vol. 10, no. 4, pp. 464—475, 1991.

[9] S.H. Huang and C.H. Cheng, "A Formal Approach to the Slack Driven Scheduling Problem in High-Level Synthesis", Proc. of IEEE International Symposium on Circuits and Systems, pp.5633—5636, 2005.

[10] S.H. Huang and C.H. Cheng, "An ILP Approach to the Simultaneous Application of Operation Scheduling and Power Management", IEICE Transactions on Fundamentals of Electronics, Communications, and Computer Sciences, vol. E91-A, no. 1, pp. 375—382, 2008.

[11] C. Tseng and D.P. Siewiorek, "Automatic Synthesis of Data Paths in Digital Systems", IEEE Trans. on Computer-Aided Design", vol. 5, no. 3, pp. 379—395, 1986.

[12] C.H. Cheng, S.H. Huang, and W.P. Tu, "Module Binding for Low Power Clock Gating", IEICE Electronics Express, vol. 5, no. 18, pp. 762—768, 2008.

[13] M. Mukherjee and R. Vemuri, "On Physical-Aware Synthesis of Vertically Integrated 3D Systems", Proc. of International Conference on VLSI Design, pp. 647—647, 2005.

[14] S.H. Huang and C.H. Cheng, "An ILP Approach to the Simultaneous Application of Operation Scheduling and Power Management", IEICE Transactions on Fundamentals of Electronics, Communications, and Computer Sciences, vol. E91-A, no. 1, pp. 375—382, 2008.

[15] S.H. Huang and C.H. Cheng, "Timing Driven Power Gating in High-Level Synthesis", Proc. of IEEE Asia and South Pacific Design Automation Conference, pp. 173—178, 2009.

[16] C. Lee, M. Potkonjak, and W. H. Maggione-Smith, "MediaBench: A tool for evaluating and synthesizing multimedia and communications systems," Proc. of IEEE International Symposia Microarchitecture, pp. 330—335, 1997.

Through Silicon Via Technology using Tungsten Metallization

G. Parès, N. Bresson, S. Minoret, V. Lapras, P. Brianceau, J.F. Lugand, R. Anciant, N. Sillon

Abstract— Through Silicon Vias (TSV) is a very promising technology in advanced packaging, for the replacement of wire bonding. This technology is becoming mandatory for fully integrated products such as SiP, SoP, 3D components integration (e.g memory stacking), or MEMS structure packaging. Different alternatives are currently investigated such as via-first or via-last. Into the via-first family two different approaches can be considered. The TSV's can be done before the FEOL (pre-process approach) or in-between the FEOL and the BEOL (mid process approach). Each solution has advantages and drawbacks depending on the final application in particular.

In a first part of this paper the tungsten mid-process TSV technology will be presented and briefly compared to the copper mid-process approaches.

Then, the process of the tungsten TSV fabrication will be detailed and morphological characterizations will be presented. We will focus on two specific parts of the process which have been specifically optimized for the tungsten TSV technology: the low temperature insulation oxide and the tungsten deposition-etch back sequence to fill the vias. The results of those optimizations will be presented and discussed.

Last, we will introduce the electrical test vehicle used in this work and present the main results regarding via resistances. Some specific recommendations will by proposed in term of design and integration rules in relation with the process constraints.

Index Terms— 3D integration, DRIE, SACVD, TSV, Tungsten, Via-mid.

I. INTRODUCTION

AS miniaturization of the CMOS components are becoming more and more difficult and costly new fields for improving the ICs performances are actively developed. One of the most promising ways is certainly the 3D integration

of stacked chips. Through Silicon Vias (TSV) or through wafer interconnections are most likely the solution to go to 3D device stacking [1], [2]. Different integration schemes have been proposed, but no single solution has emerged as yet. The choice to make the TSVs before or during the devices fabrication (via-first approach) or after (via-last approach), the nature of the filling material (polysilicon, tungsten or copper), will mainly depend on the constraints of the final product. In the following of this paper we will focus on the via-mid technology which is yet in development stage. This option consists in fabricating the TSVs just after the Front End of the Line (FEOL) operations implying thermal budget limitation in the range of 400° C to avoid any degradation of the CMOS performances.

Annular shape of the vias [2], [3], as well as narrow trenches array [4] have already been shown to be a good compromise to achieve low resistance while using relatively thin deposited films and is thus the more suitable designs for the W TSV approach. However, in this scheme, the aspect ratio (AR) of the TSVs needs to be higher than 20:1 for vias in the range of 100μm of depth. The Deep Reactive Ion Etching (DRIE) Bosch etch process is used close to its best capability to achieve such very deep structures.

Mid-process tungsten TSV offers an interesting compromise to the copper via-mid TSV mainly due to its well established compatibility with the CMOS transistors as well as its very good step coverage of high aspect ratio via with commonly used WF_6 based CVD chemistry. Also W has a Coefficient of Thermal Expansion (CTE) that matches better the one of the silicon than the copper.

The main disadvantage of using tungsten as the filling conductive material is the high stress of the film that limits drastically the allowed deposited thickness [3]. In table 1 we have summarized the main characteristics related to W and Cu mid-process TSV.

In particular the specific surface resistances of the two TSV types are reported considering the resistivity of the metals and the silicon surface utilized and for a common given depth of 100 μm. For that we have considered that the maximum width of the W TSV is limited at 3 μm as discussed later in this paper irrespectively of the design geometry (annular or bars) meanwhile we have considered a typical 10 μm of diameter via hole in the case of the Cu TSV.

We found near a decade in favor of the copper technology indicating a determinant advantage for this option in case of

---

Manuscript received xxxx. This work was supported in part by ST Microelectronic and by the PACA council area in France.

Gabriel Parès is with CEA/LETI –Minatec, France in charge of 3D integration development projects involving TSV technologies (phone: 33438786099; fax: 33438785012; e-mail: gabriel.pares@cea.fr.

Nicolas Bresson is with SOITEC France. nicolas.bresson@soitec.fr

Stéphane Minoret, Valérie Lapras, Pierre Brianceau, Jean-François Lugand are with the CEA/LETI –Minatec, France, they are respectively in charge of process development activities like metal deposition, process integration, etching, CMP.

R. Anciant is with CEA/LETI –Minatec, France in charge of electrical tests for 3D components.

N. Sillon is with CEA/LETI –Minatec, France is leading the Laboratory of 3D integration and packaging. nicolas.sillon@cea.fr

978-1-4244-9019-6/11 $26.00 © 2011 IEEE

high density or high current interconnection needs.

II. TECHNOLOGY

A. Test vehicle features

TABLE I
COMPARISON BETWEEN W AND CU FOR TSV FILLING

Symbol	Tungsten	Copper [a]
Resistivity (μOhm.cm)	12[1]	3
CTE (ppm.K^{-1})	4.6	17
Deposition temperature (°C)	440-470[1]	25
Maximum aspect ratio	30:1	12 :1
Intrinsic stress as-deposited (MPa)	1400[1]	20
Specific surface resistance for 100 μm deep TSV (Ohm.μm^2)	~28[1]	~3

[1] this work

Here we used a specially designed test vehicle with annular TSVs of different features; 2 rings and 3 rings with 4 μm or 5 μm ring width [6]. The complete integration of the mid-process W TSV module has been realized using this test vehicle including the front side and the back side interconnections to obtain electrical characteristics of the TSV structures as illustrated in Fig. 1.

 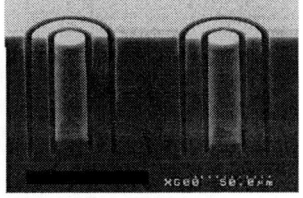

(a) TSV test vehicle architecture   (b) TSV 2 rings design
Fig. 1. Illustrations of the test vehicle used TSV W-filled technology.

B. DRIE etching of the TSVs trenches

A hard mask is needed in the process flow for CMP of the top W layer to stop with a good selectivity. A sacrificial oxide or nitride can be used for this purpose. Following this top layer etching an optimized DRIE Bosch process has been used to fabricate trenches with a depth of 80 μm. The requirement is to have a regular little tapered profile allowing a good filling in the next stages and to minimize the scalloping effect for easier filling as well as to reduce breakdown field [7]. With this kind of narrow high aspect ratio structures tuning of the slope is very limited, for example with 1.5° of slope the trench is closed at a depth of 80μm for a trench of 4μm of width. Practically a slope around 0.5° is obtained leading to a width reduction of about 1 μm at the bottom of 80μm trenches.

Very good uniformity is achieved within a wafer with less than 2% depth variation between the centre and edge of the wafer.

C. Insulation oxide

In the range of temperature acceptable for the mid-process technology there are two available deposition techniques that gather the expected criteria of good film conformity as well as good electrical properties: TEOS/Ozone Sub Atmospheric CVD (SACVD) [8]or silane High Density Plasma CVD (HDP)

[9]. Both of these techniques have been evaluated in our structures. In the case of HDP deposition a standard gap fill chamber has been used with a process setup in a temperature range of 700°C tuned to obtain good step coverage. We obtain conformity on the walls of the structure of less than 10% which was not good enough for our application.

In the case of the SACVD two different temperature regimes were also evaluated: 480°C or 540°C. We found that at a higher deposition temperature the film has a better coverage of the scalloping (Fig. 2.). The conformity at the bottom of the trenches is as good as 70% with this deposition technique.

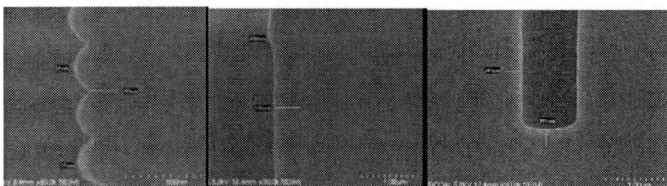

a) Top at 480°C   b) Top at 540°C   c) Bottom at 540°C
Fig. 2. SEM cross sections showing the step coverage of the SACVD deposition inside the trenches.

D. W TSV filling

Tungsten Chemical Vapor Deposited (CVD) as been widely utilized in VLSI CMOS technology for a long time [10] in particular for contact plugs to connect the silicided gates and S/D of the transistors to the BEOL interconnections. CVD tungsten is the metal of choice for this application because it is fully compatible with the middle end brick in terms of resistance to electromigration, metal diffusion, thermal budget, it has a relatively low resistivity (in the range of 8 to 15 μOhm.cm) and it is very conformal to fill the very narrow contact structures.

However, as a first constraint, CVD tungsten needs an adhesion layer especially when deposited on a dielectric. With the high aspect ratio of our structures a very conformal barrier process is required to ensure a good adhesion of the W inside the cavity. In this study we used a 20 nm MOCVD (TDMAT) TiN film applied in four consecutive densified layers.

The second issue with the tungsten is the very high stress of the CVD deposition process particularly when a conformal layer is needed. In the TSV application the process requirements appear to be much more difficult than for the contact plugs because both the deposited thickness and the AR is one order of magnitude higher (respectively few microns of deposition and AR of 20:1 or higher). For instance, with our structures the space to be filled with W after deposition of the isolation oxide is equivalent to 2.5μm for the 4μm design and 3.8 μm for the 5μm design. It is not possible to deposit such a thick conformal tungsten layer in one time otherwise the stress of the wafer will be too high. Different ways have been investigated in order to obtain a good filling of the structures while keeping the stress at an acceptable value. First we have selected a lower temperature of deposition of 440°C in order to improve the film conformity in a more surface reaction-controlled regime.

978-1-4244-9019-6/11 $26.00 © 2011 IEEE

In a first approach we used the deposition of a tungsten layer of 1 μm with very good film conformity, named Via-Fill. The stress of this film is of the order of 1.4 GPa tensile for 800 nm of deposition and 1 μm is the maximum value to keep the wafer bow acceptable for the subsequent process steps. With this film we found that the stress in the trenches was likely to be too high and end up with the delamination of the tungsten layer inside the structures (Fig. 3.).

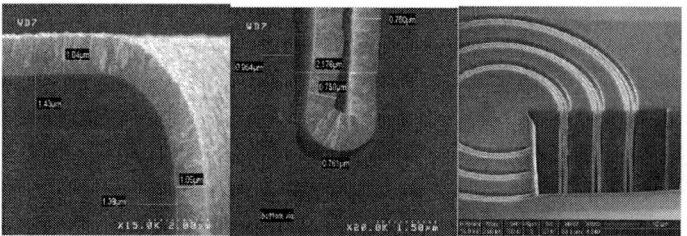

a) top deposition   b) bottom deposition   c)delamination
Fig. 3. High conformal tungsten deposition SEM pictures.

Therefore a second process has been developed consisting in the deposition of a less stressed film with a slightly lower conformity. With these conditions of process the goal is to deposit a thicker film, less stressed and without delamination. The process recipe has been modified essentially by significantly reducing the WF6 flow while keeping the H2 flow constant and with more dilution in Ar in order to decrease the deposition rate leading to shift the process toward a less surface reaction-limited regime. By this way the stress of the film is reduced down to 800 MPa for a film of 1.7μm while the film conformity is little degraded (-20% approximately).

In addition we used a multi steps deposition/etch back sequence to obtain a better filling of the trenches. RIE etch back process has been also optimized to obtain a very anisotropic behavior in order to remove the tungsten layer only on the top surface and to keep as much as possible the material on the trench walls as shown in Fig. 4. With this process all the W is removed from the front side, stopping on the TiN layer meanwhile only few tens of nanometers are removed from the top trench walls and nothing in the bottom of the cavity.

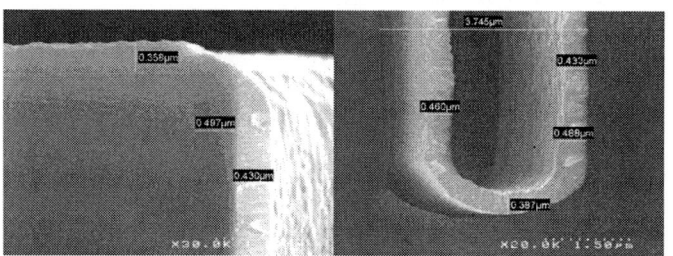

Fig. 4. Anisotropic W etch back prevent etching of the trench side walls.

Finally the optimized sequence is as followed: a first 500 nm layer of W Via-Fill + Etch back, a second 500 nm layer of W Via-Fill + Etch back and a last 1.8 μm of Low Stress W + CMP.

E. Front side metallization

The front side metallization is made with W plugs and AlCu metal lines to connect the TSVs. This brick is completely compatible with the CMOS interconnection process flow.

F. Bonding, back-lapping and back side CMP and metallization

After having made the front side interconnections the wafers are flipped and bonded on a silicon substrate receiver to perform the back-lapping and the RDL metallization. We used temporary HT10.10^{TM} glue from Brewer Science for the bonding [11].

Then back lapping was done to thin the wafer down and to uncover the TSVs.

This step has been found to be one of the more critical of the back side process since the presence of the hard tungsten material leads to high defectivity in the neighboring silicon especially during the fine grinding step. This issue is amplified with the presence of voids inside the bottom of the trench due to the incomplete tungsten filling. Indeed ripping out of some W pieces is more likely to occur with the presence of voids inside the trenches. In this work only the 4μm structures were almost free of defect since they were better filled with the W as shown in Fig. 6.

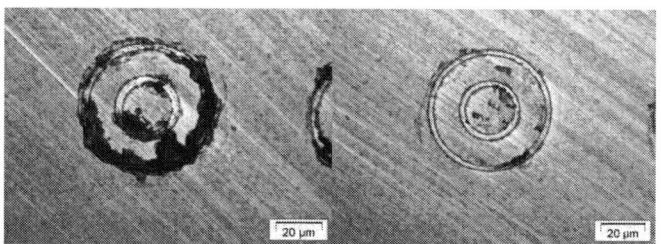

a) High defectivity with 5 μm trenches b) Low defectivity with 4 μm trenches
Fig. 6. Back side views after silicon back grinding and TSV revelation.

The subsequent CMP step is designed to remove the hardened silicon layer and obtain a low roughness final surface. If the defectivity generated during the grinding sequence is too large particularly in the depth of the silicon substrate the CMP step will not remove it.

The defectivity caused by the presence of the hard tungsten material during the thinning of the Si substrate needs further optimizations for the grinding as well as the CMP processes. This issue can also be overcome by applying a different approach that avoids opening the trenches during these steps. In this case the TSVs can be revealed by an additional selective etching step done by RIE or by wet chemistry [3] that leaves the vias in protrusion relatively to the silicon surface (W nails approach).

Contact and Redistribution Layer (RdL) levels are aligned directly on the specific fiducials transferred from the front side to the back side at the same time than the TSVs.

The backside contacts were realized through a low temperature PECVD oxide (LTO). Then a classic Ti/AlCu metallization has been realized for RDL to directly measure the test structures.

978-1-4244-9019-6/11 $26.00 © 2011 IEEE

III. ELECTRICAL RESULTS

Specific test structures are designed to characterize all the different process steps and to allow ultimately the choice of the optimum TSV design. Pseudo Kelvin structures [8] were used to measure precisely the final TSV resistances as well as Daisy chains with 8, 100, 200 and 400 TSV to give an overview of the yields.

Electrical data that are reported here have been obtained with a non optimized sequence of W filling detailed above as shown by SEM cross section of the structure in Fig. 7. Only a liner of tungsten of approximately 1 µm is present on the side walls of the trenches in the bottom part of the trenches.

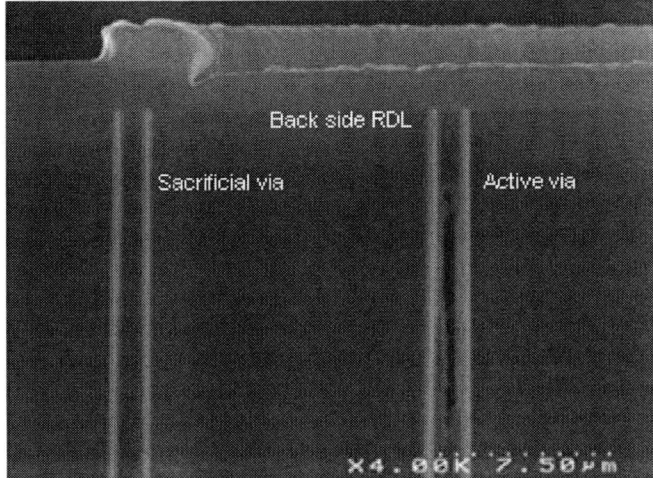

Fig. 7. Bottom of a trench partially filled with CVD tungsten.

The first observation is that the 5µm TSV rings give equivalent or slightly higher Kelvin resistances than the 4µm ones (0.17 Ohms against 0.15 Ohms) [3]. This is a clear indication of the partial filling of the tungsten inside the trenches: via resistance is determined by the thickness of the tungsten layer inside the cavity and not by the size of the cavity itself.

The second observation is that the 5µm rings give high spread chain resistances confirming the presence of defective TSVs inside the chains caused by the W rip out during TSV opening.

Regarding the yield of the TVSs we used the Daisy chains to estimate the process robustness as illustrated in Fig. 8. The distribution is fairly tight and the estimated yield is higher than 90% for both 2 and 3 ring designs on the 4 µm ring design.

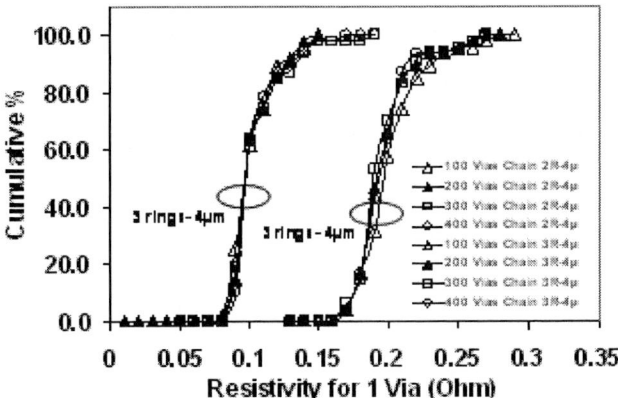

Fig. 8. 4µm TSV resistances with Daisy chains of 100, 200, 300 and 400 vias.

IV. CONCLUSION

Mid-process high aspect ratio annular TSVs with CVD tungsten filling were demonstrated. A specific process development has been presented to obtain a better filling inside the trenches using very conformal and not too stressed tungsten film. Comparing 4µm and 5µm wide trenches it has been highlighted that only the small ones can be sufficiently filled. In particular, the presence of a void inside the trenches results in very large defectivity at the TSV opening step and causes high resistances and low chain yield. With 4µm design we obtain resistances as low as 0.1 Ohm/via with near 90% chain yield over a wafer. A 3 µm width TSV design seems to be an optimum to fill completely the trenches using the CVD tungsten deposition. This last requirement leads to the relatively high specific surface resistance of the W-TSV in comparison to the copper metallization. Then W-TSV seems to be less interesting in particular for high density vertical interconnections.

ACKNOWLEDGMENT

This work is part as been realized conjointly in the STMicroelectronics facility at Rousset (FRANCE 13) and in the CEA-LETI Minatec facilities in Grenoble (FRANCE 38).

We would like to acknowledge the engineers in CEA-LETI D. Bouchu and C Brunet-Manquat for the SEM analysis.

REFERENCES

[1] Knickerbocker J. et al, "3D Silicon Integration", Proc 58th Electronic Components and Technology Conf, Lake Buena Vista, FL, May 2008, pp. 538 - 543.

[2] K. Takahashi et Al, "Process integration of 3D Chip stack with vertical interconnection", 54th Electronic Components and Technology Conf, Las Vegas, Nevada, May 2004, pp 601-609.

[3] Tsang C.K. et al, "CMOS-Compatible Trough Silicon Vias for 3D Process Integration", Mater. Res. Soc. Symp. Proc. Vol 970, 2007, pp 145 – 150.

[4] Wright S. et al, "Reliability Testing of Through-Silicon Vias for High-Current 3D Applications", Proc 58th Electronic Components and Technology Conf, Lake Buena Vista, FL, May 2008, pp. 879 – 883

[5] Parès G. et al, "Mid-Process Through Silicon Vias Technology Using Tungsten Metallization: Process Optimization And Electrical Results",

978-1-4244-9019-6/11 $26.00 © 2011 IEEE

Proc 12th Electronics Packaging Technology Conf, Singapore, Dec. 2009, pp. 772 - 777.

[6] Henry D. et al, "Via First Technology Development Based on High Aspect Ratio Trenches Filled with Doped Polysilicon", Proc 57th Electronic Components and Technology Conf , Reno, NV, May 2007, pp. 830 – 835.

[7] Laviron C. et al, "Via First Approach Optimisation for Through Silicon Via Applications", Proc. Electronic Components and Technology Conf, 2009, 59th , pp. 14 - 19.

[8] Kikuchi H. et al, "Tungsten through silicon via technology for Three-Dimensional LSIs", Jpn. J. Appl. Phys. Vol. 47, No. 4 (2008), pp. 2801 - 2806.

[9] Liu F. et al, "A 300-mm Wafer-Level Three-Dimensional Integration Scheme Using Tungsten Through-Silicon Via and Hybrid Cu-Adhesive Bonding", IEDM 2008.

[10] I-Shun Chang et al, "Growth characteristics and electrical resistivity of chemical vapor deposited tungsten film", Thin Solid Films 333 (1998) 108-113.

[11] A. Jouve & Al, "Facilitating Ultrathin Wafer Handling for TSV Processing", EPTC 2008, 9-12 december 2008, Singapour.

Statistical Delay Calculation with Multiple Input Simultaneous Switching

Qin Tang, Amir Zjajo, Michel Berkelaar and Nick van der Meijs

Circuits and Systems Group, Delft University of Technology

Q.tang@tudelft.nl

Abstract—**The increasing process variations which goes along with the continuing CMOS technology shrinking necessitate accurate statistical timing analysis. Multiple Input Simultaneous Switching (MISS) is simplified to Single Input Switching (SIS) in most of the recent approaches, which introduces significant errors in Statistical Static Timing Analysis (SSTA). Hence, we propose a new modeling and statistical analysis method to capture statistical gate delay variations, able to accurately handle MISS. Experiment results obtained with a 45nm technology show that our approach accurately obtains not only mean and standard deviation, but also the third moment, skewness.**

I. INTRODUCTION

Static Timing Analysis (STA) tools are widely used for performance verification due to their ability to perform efficient timing checks on large chips. However, STA faces a number of accuracy issues related to false paths and MISS. Since process variations do not shrink at the same ratio as the process geometries, they have an increasing impact with every new technology generation. Small changes in transistor channel length and doping density cause more dramatic changes in transistor behavior compared to older technologies. The traditional corner-based STA is not able to accurately and efficiently model delay variability. A better way to estimate the variability of timing is to perform Statistical Static Timing Analysis (SSTA), which models path or circuit delay variations as a function of all process variations of interest using statistical techniques. SSTA comes in two flavors: path-based SSTA and block-based SSTA. Although block-based SSTA does not have the tough task to select critical paths considering process variations like path-based SSTA, it requires a solution to the basic statistical sum and maximum operations for the propagation of arrival times from the source node to the sink node. Since the maximum is a nonlinear function, the maximum of two normal (Gaussian distributed) arrival times at the inputs of a gate will result in a non-normal arrival time at its output, typically with positive skewness. The error caused by ignoring skewness in the maximum operation is larger if the input arrival times have similar means but dissimilar variances [1]. Approximations for the statistical maximum operator have been proposed for both Gaussian [2] and non-Gaussian random variables [3], but they use an assumption of statistical independence between the input signals. The computation of the exact statistical maximum also needs

This research is sponsored by the European Union and the Dutch government as part of the ENIAC MODERN project

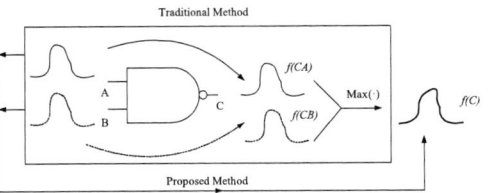

Fig. 1. Exact statistical simulation for MISS

exact correlation information among the input arrival time variables, which requires extensive computation and storage of dependence. In [6] and [7] discrete probability density functions (pdf) or cumulative density functions (cdf) are used to propagate statistical information, avoiding the maximum operation, however they also assume independent arrival times and circuit delay.

New methods for handling reconvergence and spatial correlations have been proposed in block-based SSTA, for example [4], [5]. In most of these approaches, there is a focus on the correct computation of the statistical maximum, taking proper correlation into consideration. However, an essential source of calculation error is due to MISS, which can not be handled accurately by the above methods, as they all use black-box gate delay models, in which the electrical effect of multiple input signals switching (near-)simultaneously is not modeled. These black-box models only model SIS. The approach presented in this paper uses gate models built out of statistical transistor models, and is hence fundamentally capable of modeling the effects of MISS.

MISS arises when multiple inputs of a gate switch in close proximity in time. Fig. 1 explains the SIS assumption for statistical delay analysis of a NAND2. In a SIS approach, it is always assumed that only one input is switching while the others are deterministic stable (V_{dd} for NAND). The output arrival time distributions $f(CA)$ and $f(CB)$ are calculated by propagating input distributions through the gate separately based on SIS. The final distribution $f(C)$ is found after statistical maximum operation of $f(CA)$ and $f(CB)$. In traditional timing analysis, MISS is a significant problem in both STA and SSTA. It has been reported that not modeling MISS can result in as much as 100% error in STA [8]. [9] shows SIS underestimates the mean delay of a stage by up to 20% and overestimates the standard deviation up to 26%. The existing SSTA approaches considering MISS mainly model delay as a function (linear function, orthogonal polynomial or high dimensional function) of the absolute input arrival times [9]–[13]. Fixed distributions are required in [9]–[11] without

978-1-4244-9019-6/11 $26.00 © 2011 IEEE

77

considering varying input slope in SSTA, which can also cause significant errors [12], [13].

In this paper, we propose a stochastic waveform model (SWM) and delay calculation method that include MISS effects and process variations. As illustrated in Fig. 1, we propose to consider all the inputs together and directly calculate output statistical information avoiding the maximum operation. The SWM has both varying crossing times and input slopes. In our delay calculation algorithm, the discrete *pdf* of crossing time was used. We tested our approach in circuits with MISS up to four inputs. The first three moments—mean (μ), standard deviation (σ) and skewness (γ) values were compared to transistor-level Monte Carlo simulation. Experimental results show that our approach gives accurate results for these three moments even under MISS conditions.

II. MODELING AND STATISTICAL ANALYSIS CONSIDERING MISS

A. Stochastic Waveform Model (SWM)

In traditional SSTA, the arrival time and gate delay are represented as statistical variables while the slope variability is neglected and treated as a deterministic value in the simplest case. As only the arrival time distribution and deterministic slope are known, the statistical waveform is represented by a set of ramp signals shown in Fig. 2. However, the slope is also an important factor which has direct effect on gate delay [14]. [12] shows that large errors occur without considering the varying slope of input signals. Fig. 2 also shows the realistic stochastic waveforms of a buffer (left) and the output waveforms of a 3-stage inverter chain (right) from Spectre MC simulations. Clearly the stochastic waveforms are not exactly symmetric with respect to the time axis. Similar to [15] and [16], the variational voltage waveform is represented by a time domain stochastic variable:

$$v(t) = v_0(t) + \sum_{k=1}^{M} \alpha_k(t) \cdot \xi_k \tag{1}$$

where $\alpha_k(t)$ is the sensitivity of voltage $v(t)$ to the corresponding process variation ξ_k. Therefore in our SWM, the voltage, rather than crossing time, is modeled as a stochastic variable. The sensitivity matrix $\alpha(t)$ is the parameter which must be calculated during delay calculation. Given the statistical information of process variations, the moments of and covariance between voltages are easily obtained.

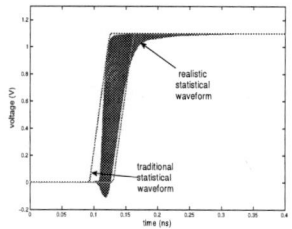 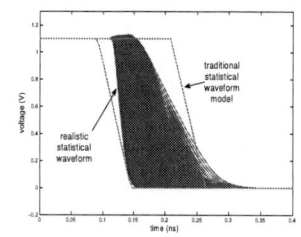

Fig. 2. Stochastic waveform modeling

B. RDE-based Statistical Simulator

As technology scales to 45nm and below, the process variations have greater impact on transistor behavior. To make our gate model as physical (and hence accurate) as possible, we create a gate model at the transistor level. For this paper, a table-based statistical transistor model (STM) is used for gate modeling [17]. Every transistor is modeled as a current source I_{ds} and five capacitors (C_{gs}, C_{gb}, C_{gd}, C_{sb} and C_{db}) as shown in Fig. 3. All elements in the STM are represented as a linear function of process variations of interest.

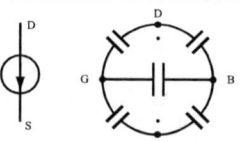

Fig. 3. Transistor model

Monte Carlo is too CPU time intensive for statistical timing analysis since normally at least 1000 runs are required and the number grows when more random variables are considered. As stated in Section II-A, the nominal voltages $v_0(t)$ and sensitivity matrix $\alpha(t)$ must be calculated for SWM. In our RDE-based statistical simulation algorithm [18], the random circuit equation is processed directly to compute $v_0(t)$ and $\alpha(t)$ in (1).

By introducing process variations, the random circuit equation can be expressed in compact form:

$$F(x', x, t, \xi) = C(t, \xi)x'(t, \xi) + G(t, \xi)x(t, \xi) = J(t, \xi) \tag{2}$$

where x is the nodal voltage vector including all input and output voltages and x' is its time derivative. G and C are the conductance and capacitance matrices and J is a current source value vector. The equation can be further simplified by eliminating and moving V_{dd} and input voltages to the right side since they are known. If there are no process variations ($\xi=0$), the equation is a typical transient analysis equation $F(x', x, t) = 0$ and the solution is denoted as $x_s(t)$ in this paper. In order to manage the random nonlinear equation (2), Taylor expansion is used to linearize (2) at the nominal value $x'_s(t)$, $x_s(t)$ and the mean values of the process variables P_0.

After linearization, the random equation is converted to a linear random differential equation (RDE). Denoting the voltage variation as $y(t) = x(t) - x_s(t)$, the RDE equation can be written as:

$$y'(t) = R(t)y(t) + Q(t)\xi \tag{3}$$
$$R(t) = -C^{-1}(G - \partial J/\partial x) \tag{4}$$
$$Q(t) = C^{-1}\partial J/\partial p \tag{5}$$

If the $\partial C/\partial p$ and $\partial G/\partial p$ are comparable to $\partial J/\partial p$, they must be included in (5). If C is singular, it stays in the left of (3).

According to the mean square integral theorem, the solution of (3) is proportional to ξ assuming the initial value is not random [19]. Using $\alpha(t)$ as the coefficient of proportionality and substituting $y = \alpha\xi$ in (3), the equation for $\alpha(t)$ turns out to be an ordinary differential equation:

$$\alpha'(t) = R(t)\alpha(t) + Q(t) \tag{6}$$

which can be solved by fast numerical methods. After solving $x_s(t)$ and (6), the stochastic output waveform model is obtained in (1). Based on the moments and correlations of process variations, the moments of voltage can be calculated by using common statistical operations.

C. Delay Moments Calculation

For timing analysis, the problem of interest is to compute the moments of arrival time, gate delay or in general the crossing time. The crossing time t_η is defined as the first time for voltages to cross the threshold voltage $V_\eta = \eta\% \cdot V_{dd}$. By using a numerical integral method, e.g, backward Euler or the trapezoidal rule, the solution of x_s and α at a specific time point are calculated from that at the previous time point, making the output $x(t)$ a Markovian process. During the period when the nominal voltage is in transition, the calculation of crossing time cdf (F_n in (7)) starts and for a rising transition this is expressed as:

$$F_n = P(t_\eta \le t_n) = 1 - P(t_\eta > t_n) = 1 - G_n \quad (7)$$

$$G_n = P(v_1 \le V_\eta \cap v_2 \le V_\eta \cap \ldots \cap v_n \le V_\eta) \quad (8)$$

$$= P(v_n \le V_\eta | v_{n-1} \le V_\eta, \ldots, v_1 \le V_\eta) \cdot G_{n-1} \quad (9)$$

$$= P(v_n \le V_\eta | v_{n-1} \le V_\eta) \cdot G_{n-1}(n = 2 : N) \quad (10)$$

$$= \frac{P(v_n \le V_\eta \cap v_{n-1} \le V_\eta)}{P(v_{n-1} \le V_\eta))} \cdot G_{n-1} \quad (11)$$

where v_i is the voltage of interest at time t_i. According to the properties of a Markovian process $v(t_n)$, (9) is rewritten in (10). Based on (7) to (11) an iteration method is used to calculate the cdf of the corresponding crossing time with initial condition $G_1=1$. Given the moments and covariances calculated in the RDE-based statistical simulator, the joint probability and single probability in (11) are easy to obtain.

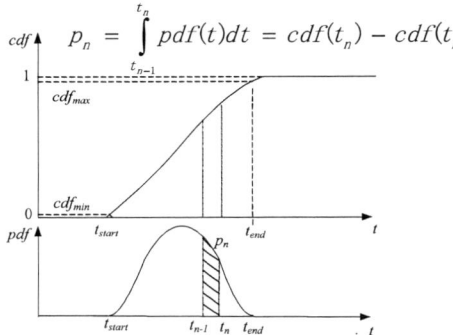

Fig. 4. Cumulative density function and discrete probability density function

To simplify the calculations, the cdfs and pdfs have these properties: i) $cdf = 1$ if $cdf \ge cdf_{max}$ and $cdf = 0$ if $cdf \le cdf_{min}$. The time t_{start} and t_{end} are the time corresponding to cdf_{min} and cdf_{max} respectively shown in Fig. 4; ii) the pdf is calculated during the period $[t_{start} \; t_{end}]$. At any time t_n, the discrete pdf is approximated by $p_n = \int_{t_{n-1}}^{t_n} pdf(t)dt$ with $p(t_{start}) = 0$.

The next step is to restrict the cdf within t_{start} and t_{end} (denoted $cdfr$ in this paper), shown in Fig. 4. Since the simulation uses a dynamic time step algorithm for efficiency, the $cdfr$ needs to be uniformly sampled for pdf computation. After

uniformly sampling $cdfr$ with N_s samples and interpolating $cdfr$, the resulting time vector T_1 and $cdfu$ vectors are used to calculate the pdf vector Λ with element $\Lambda_k = cdfu_k - cdfu_{k-1}$ ($\Lambda_1 = 0$).

The last step is to calculate the moments of crossing time. Denoting Λ^T as the transposition of the row vector Λ, the calculation method can be explained as following:

$$\mu = T_1 \Lambda^T \quad (12)$$

$$\sigma = T_2 \Lambda^T - \mu^2 \quad (T_{Nk} = T_{1k}^N \quad k = 1 : N_s) \quad (13)$$

$$\gamma = (\Gamma - 3\mu\sigma^2 - \mu^3)/(\sigma^3) \quad (\Gamma = T_3 \Lambda^T) \quad (14)$$

The calculation method for a falling transition is similar to the above methods with the only difference in (8). By replacing v_i by $V_{dd} - v_i$, (8) to (11) are still used in the same way. If the waveform is non-monotonic and crosses V_η multiple times, the method above can be used to iteratively find all crossing times.

III. EXPERIMENTAL RESULTS

The proposed model and algorithm were implemented and tested on a set of gates and circuits in the 45nm PTMVTG technology [20]. Transistor-level Monte Carlo(MC) simulation results (2500 samples) are regarded as the golden reference. Fig. 5 shows the discrete pdf with 50 samples and the histogram of MC simulation in Spectre of a NAND2 with falling output, AOI21 and AOI22 with rising output. All inputs of each gate have the exact same mean value of arrival times. The discrete pdf was scaled to provide a straightforward shape comparison. In the beginning, the statistical input signal at every input of a multi-input gate was modeled as ramp signals of $100ps$ transition time with voltage variations. The σ of voltages and arrival time differences among input signals are varied to obtain results at diverse scenarios.

Table I[1] lists the average error of mean (μ), standard deviation (σ) and skewness (γ) of delay in gates with different levels of complexity. It shows that the worst average σ and γ error occur in NAND4 and NAND3 respectively. These gates have the most transistors stacked among the gates with corresponding same number of inputs. Fig. 6 illustrates the errors of all the experiments. The μ errors are within 1%. All the σ errors are within 5% except two biggest σ cases (6.02% and 6.42%) coming from NAND4 with rising output and falling output respectively. All of the skewness errors are within 7%. We also simulated two combinational circuits (COM1 and COM2 in Table I) with identical paths to the output gate. The process variables are taken to be length and width with 3σ of 20% and 43% of the mean value. By using identical input signals switching at the same time, the inputs of the output gate has MISS and realistic waveforms produced by process variations like the curves in Fig. 2. The results of COM1 and COM2 are compared to 10000× MC results in Spectre. It is observed that the smaller the arrival time difference, the larger skewness, so the skewness should not

[1]importantance sampling-based MC was used for all senarios of NAND2 and NOR2 as comparison references

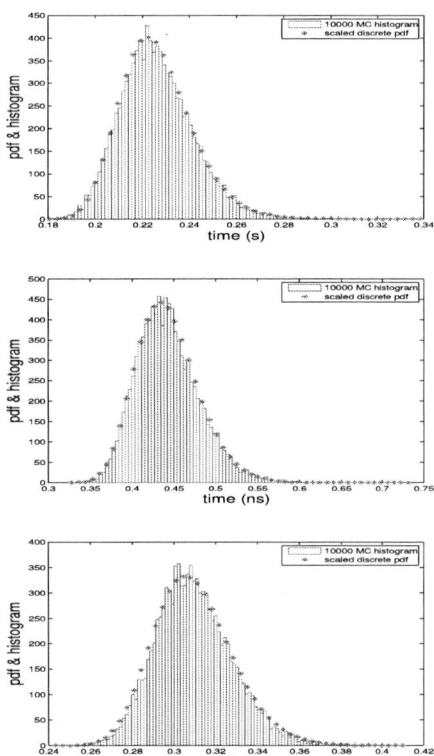

Fig. 5. pdf and histogram comparison of NAND2, AOI21 and AOI22

be ignored. The experiment results indicate the necessity of skewness estimation and the ability of the proposed method to accurately calculate three moments.

Compared to 1000 MC runs, our method achieves 62×speedup on average. Although our method only needs to simulate once for statistical output voltages, the equations of the RDE-based statistical simulation have an extra set of equations for α computation in (1) and need sensitivity calculation for (3), which slows down the simulation. We are working on using a faster differential equation solver, even higher speedup is expected.

TABLE I

ACCURACY COMPARISON OF THREE MOMENTS FOR MULTI-INPUT GATES

cell	errors of rising output			errors of falling output		
	μ	σ	γ	μ	σ	γ
NAND2	0.52%	2.00%	2.52%	0.20%	1.14%	2.20%
NOR2	0.38%	1.03%	3.07%	0.15%	1.73%	2.91%
NOR3	0.55%	0.82%	3.26%	0.15%	1.90%	2.82%
NAND3	0.70%	2.38%	5.52%	0.41%	2.04%	3.05%
AOI21	0.04%	1.56%	2.00%	0.09%	1.34%	1.71%
AOI211	0.13%	3.06%	2.62%	0.04%	1.84%	4.33%
AOI22	0.04%	0.65%	1.60%	0.19%	1.68%	3.06%
NAND4	0.75%	4.48%	2.70%	0.33%	4.54%	3.27%
COM1	0.93%	1.99%	4.00%	0.37%	3.89%	4.26%
COM2	0.64%	4.39%	5.10%	0.41%	5.58%	5.75%

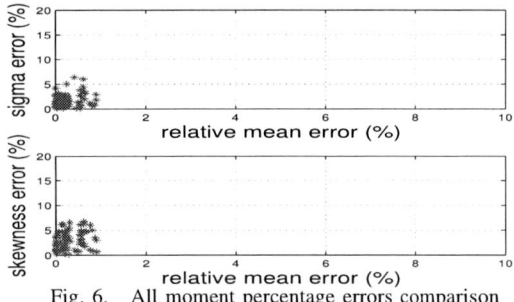

Fig. 6. All moment percentage errors comparison

IV. CONCLUSION

The errors introduced by the SIS assumption and statistical maximum operation motivate us to propose a novel modeling and simulation method to capture process variations and MISS, avoiding the maximum operation. We represent variational waveforms of any shape in time domain statistical variables including the influence of slope. The transistor-level gate modeling and RDE-based statistical simulation method provides variational output waveforms, from which the moments of crossing times are computed. Due to the increasing process variations and continuing shrinking technology, the μ and σ are not always enough to represent variational gate delay. Additionally in this paper we computed the skewness of gate delay. Experimental results indicated the high accuracy of our approach compared to Monte Carlo simulations.

REFERENCES

[1] C. Clark, "The greatest of a finite set of random variables," *J. Oper. Res.*, vol. 9, no. 2, pp. 145–162, Mar./Apr. 1961.

[2] C. Visweswariah, K. Ravindran, S. W. K. Kalafala, and S. Narayan, "First-order incremental block-based statistical timing analysis," in *DAC*, 2004, pp. 331–336.

[3] J. Singh and S. Sapatnekar, "Statistical timing analysis with correlated non-Gaussian parameters using independent component analysis," in *DAC*, 2006, pp. 155–160.

[4] L. Zhang, W. Chen, Y. Hu, J. A. Gubner, and C. C. Chen, "Correlation-preserved statistical timing with a quadratic form of Gaussian variables," *Trans. Comput.-Aided Des. of Integr. Circuits Syst.*, vol. 25, no. 11, pp. 2437–2449, 2006.

[5] V. Khandelwal and A. Srivastava, "A quadratic modeling-based framework for accurate statistical timing analysis considering correlations," *Trans. on VLSI Syst.*, vol. 15, no. 2, p. 206-215, 2007.

[6] J.-J. Liou, K.-T. Cheng, S. Kundu, and A. Krstic, "Fast statistical timing analysis by probabilistic event propagation," in *DAC*, 2001, pp. 661–666.

[7] A. Devgan and C. Kashyap, "Block-based static timing analysis with uncertainty," in *ICCAD*, 2003, pp. 607–614.

[8] C. Amin, C. Kashyap, N. Menezes, k. Killpack, and E. Chiprout, "A multi-port current source model for multiple-input switching effects in CMOS library cells," in *In DAC 2006*, 2006, pp. 247–252.

[9] A. Agarwal, F. Dartu, and D. Blaauw, "Statistical gate delay model considering multiple input switching," in *DAC*, 2004, pp. 658–663.

[10] K. Y. Satish, J. Li, C. Talarico, and J. Wang, "A probabilistic collocation method based statistical gate delay model considering process variations and multiple input switching," in *DATE*, 2005, p. 770-773.

[11] S. Yanamanamanda, J. Li, and J. Wang, "Uncertainty modeling of gate delay considering multiple input switching," in *ISCAS*, 2005, pp. 2457–2460.

[12] J. Sridharan and T. Chen, "Modeling multiple input switching of CMOS gates in DSM technology using HDMR," in *DATE*, vol. 1, 2006, pp. 1–6.

[13] J. Sridharan and T. Chen, "Gate delay modeling with multiple input switching for static (statistical) timing analysis," in *VLSID*, 2006, pp. 323–328.

[14] T. Kouno and H. Onodera, "Consideration of transition-time variability in statistical timing analysis," in *SOCC*, 2006, pp. 207–210.

[15] H. Fatemi, S. Nazarian, and M. Pedram, "Statistical logic cell delay analysis using a current-based model," in *DAC*, 2006, pp. 253–256.

[16] B. Liu and A. B. Kahng, "Statistical gate level simulation via voltage controlled current source models," in *BMAS Workshop*, 2006, p. 23.

[17] Q. Tang, A. Zjajo, M. Berkelaar, and N. van der Meijs, "Transistor level waveform evaluation for timing analysis," in *VARI*, 2010, pp. 1–6.

[18] Q. Tang, A. Zjajo, M. Berkelaar, and N. van der Meijs, "RDE-based transistor-level gate simulation for statistical static timing analysis," in *DAC*, 2010, pp. 787–792.

[19] T.T. Soong, "Random differential equations in science and engineering," New York: Academic Press, 1973.

[20] Arizona State University, "Predictive Technology Model (PTM)", http://ptm.asu.edu.

Balanced Truncation of a Stable Non-Minimal Deep-Submicron CMOS Interconnect

Amir Zjajo, Qin Tang, Michel Berkelaar, Nick van der Meijs

Abstract—As the widening of process variability in submicron CMOS technology calls for accurate timing models, their deployment requires well-controlled characterization techniques to cope with the complexity and scalability. In this context, model order reduction techniques have been used extensively to reduce the complexity of extracted interconnect circuits and to expedite fast and accurate circuit simulation. In the interconnect modeling, solving large-scale Lyapunov equations arises as a necessity in model order reduction techniques based on Balanced Truncation. In this paper, within this framework, dominant eigensubspaces of the product of the system Gramians are approximated directly. We construct orthogonal basis sets for the dominant subspaces of controllability and observability Gramians and perform eigenvalue decomposition to reduce the cost of singular value decomposition. As the experimental results indicate, the proposed approach can significantly reduce the complexity of interconnect, while retaining high accuracy in comparison to the original model.

Index Terms—interconnect model, model order reduction, balanced truncation.

I. INTRODUCTION

GATE and interconnect delay are critical issues in present day low power VLSI circuit design. As we are moving towards nanometer technology, variations in process, voltage, and temperature are increasing, causing significant uncertainty in the delay estimation [1] and greatly impacting the yield [2]. As a consequence, various statistical static timing analysis (SSTA) algorithms [3]-[5] have been proposed to compute the statistical variations of timing performance due to the underlying process parameters. Deriving an efficient characterization methodology and model order reduction (MOR) techniques that can provide parameterized interconnects and facilitate efficient logic stage delay calculation is one of the critical tasks. In an asymptotic waveform evaluation (AWE) algorithm [6] explicit moment matching was used to compute the dominant poles via Padé approximation. As the AWE method is numerically unstable

This research was sponsored by the European Union and the Dutch government as part of the ENIAC/MODERN project.

The authors are with Circuits and Systems Group, Delft University of Technology, Mekelweg 4, 2628 CD, Delft, The Netherlands. (e-mail: amir.zjajo@ieee.org).

for higher-order moment approximation, a more elegant solution to the numerical problem of AWE is to use projection-based MOR methods. In the Padé via Lanczos (PVL) method [7], the Lanczos process, which is a numerically stable method for computing eigenvalues of a matrix, was used to compute the Krylov subspace. In PRIMA [8] the Krylov subspace vectors are used to form the projector for the congruence transformation, which leads to passive models with the matched moments in the rational approximation paradigm. However, these methods are not efficient for circuits with many inputs and output terminals as the reducing cost are tied to the number of terminals; the number of poles of reduced models is also proportional to the number of terminals. Additionally, PRIMA-like methods do not preserve structure properties like reciprocity of a network.

Another approach to circuit-complexity reduction is to reduce the number of nodes in the circuits and approximate the newly added elements in the circuit matrix in reduced rational forms by approximate Gaussian elimination for RC circuits [9]. Alternatively, model order reduction can be performed by means of singular-value-decomposition (SVD) based approaches such as control-theoretical-based truncated balance realization (TBR) methods, where the weakly uncontrollable and unobservable state variables are truncated to achieve the reduced models [10]-[16]. The major advantage of SVD-based approaches over Krylov subspace methods lies in their ability to ensure the errors satisfying an a-priori upper bound [14]. Also, SVD-based methods typically lead to optimal or near optimal reduction results as the errors are controlled in a global way, although, for large scale problems, iterative methods have to be used to find an adequate balanced approximation (truncation). In this respect, ideas based on balanced reduction methods are significant since they offer the possibility to perform order selection during the computation of the projection spaces and not in advance. Typically in balanced reduction methods, there is a rapid decay in the Gramians eigenvalues. As a consequence these Gramians can be well approximated using low-rank approximations, which are used instead of the original. Accordingly, several SVD approaches approximate the dominant Cholesky factors (dominant eigensubspaces) of controllability and observability Gramians [11],[15]-[16] to compute the reduced model.

In this paper, we adjust the dominant subspaces projection model reduction (DSPMR) [11] and provide an approximate balancing transformation for circuits whose coefficient matrices are large and sparse such as in interconnect. The approach presented here produces orthogonal basis sets for the dominant singular subspace of the controllability and observability Gramians significantly reducing the complexity and computational costs of singular value decomposition, while preserving model order reduction accuracy and the quality of the approximations of the TBR procedure.

II. Adjusted Approximated Truncated Balance Realization Method

In the analysis of delay or noise in on-chip interconnect we study the propagation of signals in the wires that connect logic gates. These wires may have numerous features: bends, crossings, vias, etc., and are modeled by circuit extractors in terms of a large number of connected circuit elements: capacitors, resistors and more recently inductors. Given a state-space formulation of the interconnect model,

$$C(dx/dt) = Gx(t) + Bu(t)$$
$$y(t) = E^T x(t) \tag{1}$$

where $C, G \in \mathcal{R}^{n \times n}$ are matrices describing the reactive and dissipative parts of the interconnect, respectively, $B \in \mathcal{R}^{n \times p}$ is a matrix that defines the input ports, $E \in \mathcal{R}^{p \times n}$ is matrix that defines the outputs, and $y(t) \in \mathcal{R}^q$ and $u(t) \in \mathcal{R}^p$, are the vectors of outputs and inputs, respectively, the model reduction algorithm seek to produce a similar system

$$\widehat{C} d\widehat{x}/dt = \widehat{G}\widehat{x}(t) + \widehat{B}u(t)$$
$$\widehat{y}(t) = \widehat{E}^T \widehat{x}(t) \tag{2}$$

where $\widehat{C}, \widehat{G} \in \mathcal{R}^{k \times k}$, $\widehat{B} \in \mathcal{R}^{k \times m}$, $\widehat{E} \in \mathcal{R}^{p \times k}$, of order k much smaller than the original order n, but for which the outputs $y(t)$ and $\widehat{y}(t)$ are approximately equal for inputs $u(t)$ of interest. The Laplace transforms of the input output transfer functions

$$H(s) = E^T (G + sC)^{-1} B$$
$$\widehat{H}(s) = \widehat{E}^T (\widehat{G} + s\widehat{C})^{-1} \widehat{B} \tag{3}$$

are used as a metric for approximation accuracy if

$$\left\| H(s) - \widehat{H}(s) \right\| < \varepsilon \tag{4}$$

for a given allowable error ε and an allowed domain of the complex frequency variable s, the reduced model is accepted as accurate.

Balanced truncation [10],[16], singular perturbation approximation [18], and frequency weighted balanced truncation [19] are model reduction methods for stable systems. Except for modal truncation each of the above methods is based either explicitly or implicitly on balanced realizations, the computation of which involves the solutions of Lyapunov equations

$$GXC^T + CXG^T = -BB^T$$
$$G^T YC + C^T YG = -E^T E \tag{5}$$

where the solution matrices X and Y are controllability and observability Gramians.

The original implementation of balanced truncation [10] involves the explicit balancing of the realization (1). This procedure is dangerous from the numerical point of view because the balancing transformation matrix T tends to be highly ill-conditioned.

The square root method [16] is an attempt to cope with this problem by avoiding explicit balancing of the system. The method is based on the Cholesky factors of the Gramians instead of the Gramians themselves. In [20] the use of the Hammarling method was proposed to compute these factors. Recently, in [11] and [15] it has been observed that solutions to Lyapunov equations often have low numerical rank, which means that there is a rapid decay in the eigenvalues of the Gramians. Indeed, the idea of low-rank methods is to take advantage of this low-rank structure to obtain approximate solutions in a low-rank factored form. The principal outcome of these approaches is that the complexity and the storage are reduced from $O(N^3)$ flops and $O(N^2)$ words of memory to $O(N^2 r)$ flops and $O(Nr)$ words of memory, respectively, where r is the approximate rank of the Gramian ($r \ll N$). Moreover, approximating the Cholesky factors of the Gramians directly and using these approximations to provide a reduced model, has a comparable cost to that of the popular moment matching methods. It requires only matrix-vector products and linear solvers.

For large systems with a structured transition matrix, this method is an attractive alternative because the Hammarling method can generally not benefit from such structures. In the original implementation this step is the computation of exact Cholesky factors, which may have full rank. We formally replace these (exact) factors by (approximating) low rank Cholesky factors [11],[15]. The iterative procedure approximates the low rank Cholesky factors Z_X and Z_Y with $r_X, r_Y \ll n$, such that $Z_X Z_X^H \approx X$ and $Z_Y Z_Y^H \approx Y$, where H is Hermitian (complex-conjugate) matrix. Note that the number of iteration steps i_{max} needs not be fixed a priori. However, if the Lyapunov equation should be solved as accurate as possible, correct results are usually achieved for low values of stopping criteria that are slightly larger than the machine precision. Let

$$Z_Y^H Z_X = U_Y \Sigma U_X^H \tag{6}$$

be SVD of $Z_Y^H Z_X$ of dimension $N \times m$. The cost of this decomposition including the construction of U is $14Nm^2 + O(m^3)$ [21].

To avoid this, in this paper we perform eigenvalue decomposition

$$(Z_Y^H Z_X)^H Z_Y^H Z_X = U_Y \Lambda U_X^H \tag{7}$$

Comparing (7) with (6) shows that the same matrix U_X is constructed and that

$$(Z_Y^H Z_X U_X)^H Z_Y^H Z_X U_Y = \Lambda = \Sigma^H \Sigma \tag{8}$$

This algorithm requires Nm^2 operations to construct $(Z_Y^H Z_X)^H Z_Y^H Z_X$ and $Nmn + O(m^3)$ operations to obtain $Z_Y^H Z_X U_X \Sigma^{-1}$ for $n \times n$ Σ.

978-1-4244-9019-6/11 $26.00 © 2011 IEEE

The balancing transformation matrix T is used to define the matrices $S_X = T_{(1:k)}$ and $S_Y = T^T_{(1:k)}$. If $\sigma_k \neq \sigma_{k+1}$, the reduced order realization is minimal, stable, and balanced, and its Gramians are equal to $diag(\sigma_1, \ldots, \sigma_k)$. The balancing transformation matrix can be obtained as

$$S_X = Z_X U_X \Sigma^{-1/2} \quad S_Y = Z_Y U_Y \Sigma^{-1/2} \qquad (9)$$

then, under a similarity transformation of the state-space model, both parts can be treated simultaneously after a transformation of the system (C, G, B, E) with a nonsingular matrix $T \in \mathcal{R}^{n \times n}$ into a balanced system

$$\hat{C} = S_X C S_Y^H \quad \hat{G} = S_X G S_Y^H \quad \hat{B} = S_Y^H B \quad \hat{E} = E S_X \qquad (10)$$

In this algorithm we assume that $k \leq r$ ($rank\ Z_Y^H Z_X$). Note that SVDs are arranged so that the diagonal matrix containing the singular values has the same dimensions as the factorized matrix and the singular values appear in non-increasing order.

III. EXPERIMENTAL RESULTS

The proposed method and all sparse techniques have been implemented in Matlab. All the experimental results are carried out on a PC with an Intel Core 2 Duo CPU running at 2.66 GHz and with 3 GB of memory. To characterize the timing behavior, a lookup table-based library is employed which represents the gate delay and output transition time as a function of input arrival time, output capacitive load, and several independent random source of variation for each electrical parameter (i.e., R and C). In each case, both driver and interconnect are included for the stage delay characterizations. The analytical delay distribution obtained using the quadratic interconnect model in 45 nm CMOS technology is illustrated in Figure 1. The nominal value of the total resistance of the load and the total capacitance is chosen from the set 0.15kΩ-1kΩ and 0.4pF-1.4pF, respectively.

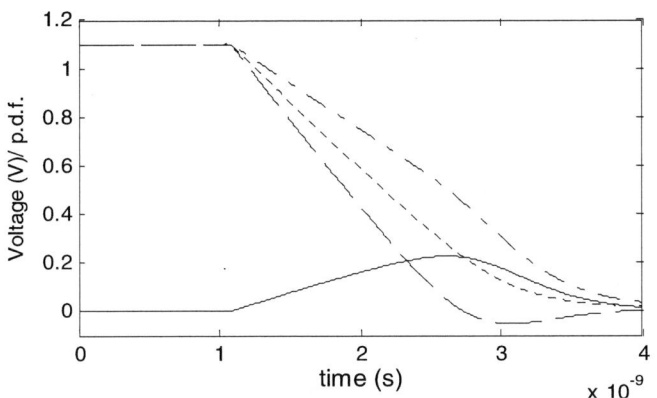

Figure 1: Analytical delay distribution in 45 nm CMOS technology. Solid line illustrates delay variance.

The sensitivity of each given data to the sources of variation is chosen randomly, while the total σ variation for each data is chosen in the range of 10% to 30% of their nominal value. The scaled distribution of the sources of variation is considered to have a skewness of 0.5, 0.75, and 1.

For model order reduction we consider a RC-chain with 2002 capacitors and 2003 resistors. In Figure 2 and Figure 3 the convergence history with respect to the number of iteration steps for solving the Lyapunov equation is plotted. For the tolerances at a residual norm of about the same order of magnitude, convergence is obtained after 40 and 45 iterations, respectively. The cpu-time needed to solve the Lyapunov equations according to the related tolerance for solving the shifted systems inside the iteration is 2.7 seconds.

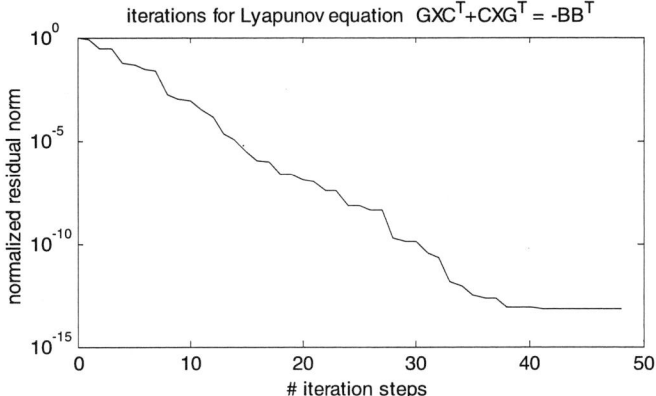

Figure 2: Convergence history of residual forms. The convergence is obtained after 40 iterations.

Note further that saving iteration steps means that we save large amounts of memory-especially in the case of multiple input and multiple output systems where the factors are growing by p columns in every iteration step. When very accurate Gramians (e.g. low rank approximations to the solutions) are selected, the approximation error of reduced system as illustrated in Figure 4 is very small compared to the Bode magnitude function of the original system. The lower two curves correspond to the highly accurate reduced system; the proposed model order reduction technique delivers a system of lower order, and the upper two denote $k=20$ reduced orders. The frequency response plot is obtained by computing the singular values of the transfer function $H(j\omega)$, which is the frequency response (4) evaluated on the imaginary axis (Figure 5). The error plot is the frequency response plot of the singular values of the error system as a function of ω.

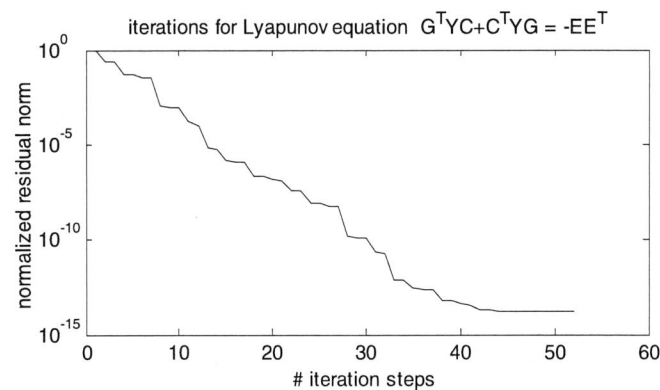

Figure 3: Convergence history of residual forms. The convergence is obtained after 45 iterations.

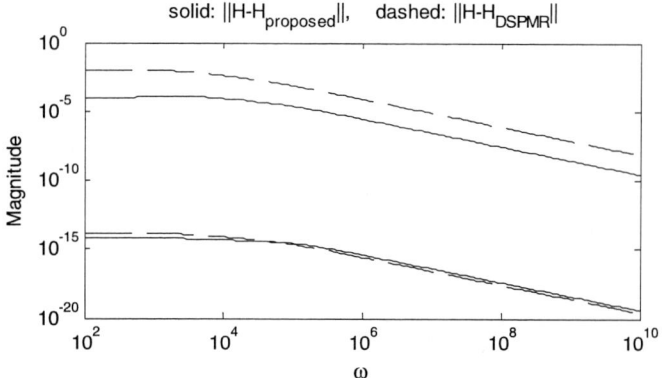

Figure 4: The Bode magnitude plot of the approximation errors.

The reduced order is chosen in dependence of the descending ordered singular values $\sigma_1, \sigma_2, \ldots \sigma_r$, where r is the rank of factors which approximate the system Gramians. For n variation sources and l reduced parameter sets, the full parameter model requires $O(n^2)$ simulation samples and thus has a $O(n^6)$ fitting cost. On the other hand, the proposed parameter reduction technique has a main computational cost attributable to the $O(n+l^2)$ simulations for sample data collection and $O(l^6)$ fitting cost significantly reducing the required sample size and the fitting cost.

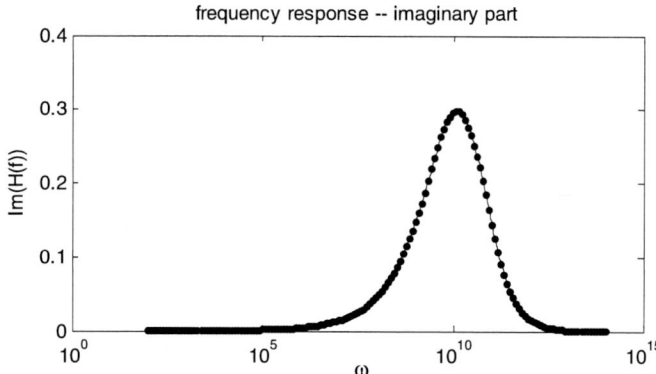

Figure 5: Frequency response of the interconnect model.

IV. CONCLUSION

This paper presents an efficient methodology for interconnect model reduction based on adjusted dominant subspaces projection. By adopting the parameter dimension reduction techniques, interconnect model extraction can be performed in the reduced parameter space, thus provide significant reductions on the required simulation samples for constructing accurate models. Extensive experiments are conducted on a large set of random test cases, showing very accurate results.

REFERENCES

[1] C. Forzan, D. Pandini, "Statistical static timing analysis: A survey," *Integration, The VLSI Journal,* vol. 42, no. 3, pp. 409-435, 2009

[2] S.R. Nassif, "Modeling and analysis of manufacturing variations," *Proceedings of IEEE Custom Integrated Circuit Conference,* pp. 223-228, 2001

[3] L. Zhang, W. Chen, Y. Hu, A. Gubner, C. Chen, "Correlation-preserved non-Gaussian statistical timing analysis with quadratic timing model," *Proceedings of IEEE Design Automation Conference,* pp. 83-88, 2005

[4] V. Veetil, D. Sylvester, D. Blaauw, "Efficient Monte Carlo based incremental statistical timing analysis," *Proceedings of IEEE Design Automation Conference,* pp. 676-681, 2008

[5] Q. Tang, A. Zjajo, M. Berkelaar, N. van der Meijs, "RDE-based transistor-level gate simulation for statistical static timing analysis," *Proceedings of IEEE Design Automation Conference,* pp. 787-792, 2010

[6] L. T. Pillage, R. A. Rohrer, "Asymptotic waveform evaluation for timing analysis," *IEEE Transaction on Computer-Aided Design of Integrated Circuits and Systems,* vol. 4, pp. 352-366, 1990

[7] P. Feldmann, R.W. Freund, "Efficient linear circuit analysis by Pade approximation via the Lanczos process," *IEEE Transaction on Computer-Aided Design of Integrated Circuits and Systems,* vol. 14, pp. 639-649, 1995

[8] A. Odabasioglu, M. Celik, L. Pileggi, "PRIMA: Passive reduced-order interconnect macromodeling algorithm," *IEEE Transactions on Computer-Aided Design of Integrated Circuits and Systems,* pp. 645-654, 1998

[9] P. Elias, N. van der Meijs, "Including higher-order moments of RC interconnections in layout-to-circuit extraction," *Proceedings of IEEE Design, Automation and Test in Europe Conference,* pp. 362-366, 1996

[10] B. C. Moore, "Principal component analysis in linear systems: controllability, observability, and model reduction," *IEEE Transaction on Automatic Control,* vol. 26, pp. 17-31, 1981

[11] J. Li, J. White, "Efficient model reduction of interconnect via approximate system Grammians," *Proceedings of IEEE International Conference on Computer Aided Design,* pp. 380-384, 1999

[12] J. R. Phillips, L. Daniel, L. M. Silveira, "Guaranteed passive balancing transformations for model order reduction," *Proceedings of IEEE Design Automation Conference,* pp. 52-57, 2002

[13] J. R. Phillips, L. M. Silveira, "Poor man's TBR: a simple model reduction scheme," *Proceedings of IEEE Design, Automation and Test in Europe Conference,* pp. 938-943, 2004

[14] W.F. Arnold, A.J. Laub, "Generalized eigenproblem algorithms and software for algebraic Riccati equation," *Proceedings of IEEE,* vol. 72, pp. 1764-1754, 1984

[15] T. Penzl, "A cyclic low-rank Smith method for large sparse Lyapunov equations," *SIAM Journal on Scientific Computing,* vol. 21, pp. 1401-1418, 2000

[16] M.G. Safonov, R.Y. Chiang, "A Schur method for balanced-truncation model reduction," *IEEE Transactions on Automatic Control,* vol. 34, pp. 729-733, 1989

[17] A.J. Laub, M.T. Heath, C.C. Paige, R.C. Ward, "Computation of system balancing transformations and other applications of simultaneous diagonalization algorithms," *IEEE Transactions on Automatic Control,* vol. 32, pp. 115-122, 1987

[18] K.V. Fernando, H. Nicholson, "Singular perturbational model reduction of balanced systems," *IEEE Transactions on Automatic Control,* vol. 27, pp. 466-468, 1982

[19] D. Enns, "Model reduction with balanced realizations: an error bound and a frequency weighted generalization," *Proceedings of IEEE Conference on Decision and Control,* pp. 127-132, 1984

[20] M.S. Tombs, I. Postlethwaite, "Truncated balanced realization of stable, non-minimal state-space systems," *International Journal of Control,* vol. 46, pp. 1319-1330, 1987

[21] G. Golub, C. van Loan, *Matrix computations,* Johns Hopkins University Press, Baltimore MD, 1996

Enabling TLM-2.0 Interface on QEMU and SystemC-based Virtual Platform

Tse-Chen Yeh, Zin-Yuan Lin, and Ming-Chao Chiang
Department of Computer Science and Engineering
National Sun Yat-sen University
70 Lienhai Road, Kaohsiung 80424, Taiwan, ROC
sdgp03@ms18.hinet.net, m983040031@student.nsysu.edu.tw, mcchiang@cse.nsysu.edu.tw

Abstract—**This paper presents a QEMU and SystemC-based virtual platform that is capable of hardware modeling using TLM-2.0 interface. The proposed virtual platform is not only capable of running an operating system, but it is also capable of using such an interface to connect hardware models, such as the instruction set simulator to a bus model. We verify the functionality of such a platform by using it to boot up a full-fledged Linux while at the same time estimating its performance at the instruction-accurate level. Furthermore, TLM-2.0 interface makes our framework more compatible with other models using TLM-2.0 and more suitable for modeling at the early stage of ESL design flow.**

Keywords—**TLM; bus functional model; QEMU; SystemC; virtual platform; ESL**

I. INTRODUCTION

In order to overcome the complexity of System-on-Chip (SoC), most of the Electronic System Level (ESL) [1] tools provide the virtual platform for co-simulating hardware/software at the early stage of the design flow. In order to unify the interfaces of modeling the transactions between diverse hardware models, TLM-2.0 standard [2] was proposed. The hardware/software emulation framework QEMU-SystemC [3], proposed in 2007 for the development of software and device drivers, has been upgraded to use TLM-2.0 as its modeling interface in 2009 [4]. Although the TLM-2.0 standard can assist to model the memory-mapped bus model, the transactions of the bus model are incomplete because it lacks transactions provided by the processor model. Another virtual platform [5], [6] that combines QEMU-SystemC with CoWare's Platform Architect was proposed in 2009. This framework utilizes lots of models provided by off-the-shelf Model Library [7] modeled with TLM-2.0 interface.

To show that the QEMU and SystemC based virtual platform we proposed is capable of the TLM-2.0 interface, we connect the instruction-accurate instruction set simulator (IA-ISS) with the direct memory access controller (DMAC) model using the simple bus modeled in TLM-2.0 interface. Certain adaptions on interfaces of simple bus are needed because some properties of transaction modeling are maintained in the QEMU internally.

II. RELATED WORK

In this section, we begin with a brief introduction to SystemC and TLM-2.0 standard. Then, we turn our discussions to QEMU-SystemC and the QEMU and SystemC-based virtual platform we propose, which uses QEMU as the processor model.

A. TLM-2.0

TLM-2.0 is an modeling interface focused on modeling the memory-mapped bus model. It is capable of modeling from the temporal decoupling (TD) down to the approximately-timed (AT). However, modeling at the cycle-accurate level is beyond the scope of TLM-2.0. The capability of mixed level modeling is demonstrated by the simple bus, which is modeled using AT coding style, connected with diverse initiator and target models using TD, loosely-timed (LT), and AT coding style. We can demonstrate that our virtual platform is capable of TLM-2.0 by using the simple bus it provided to connect our ISS and DMAC model.

B. QEMU-SystemC

QEMU-SystemC [3] is an open source hardware/software emulation framework for the SoC development. It allows devices to be inserted into specific addresses of QEMU and communicates by means of the PCI/AMBA bus interface. Although the waveform of the AMBA on-chip bus of the QEMU-SystemC framework can be used to trace the operations performed on the slave device, no transactions about the processor and master device are provided by QEMU-SystemC or the TLM-2.0 appending version [4]. For instance, the instructions executed, the memory accessed, and so on, which are so valuable to the system designers, are unfortunately not provided.

C. QEMU and SystemC-based Virtual Platform

In order to provide all the information necessary for modeling the bus transactions, QEMU has to be converted from a virtual machine to an ISS that is capable of simulating all the hardware models down to the cycle-accurate level [8].

The virtual platform described herein is implemented as two threads running in a single process and communicating with a unidirectional FIFO, as shown in Fig. 1. In other words, the communication between QEMU and SystemC is one way so that the relative order of the instructions executed, the memory accessed, and the I/O write operations is retained by the packet receiver within the ISS wrapper and infrastructure interface. Thus, the interface can simulate different bus transactions for modeling the Bus Functional Model (BFM) by means

978-1-4244-9019-6/11 $26.00 © 2011 IEEE

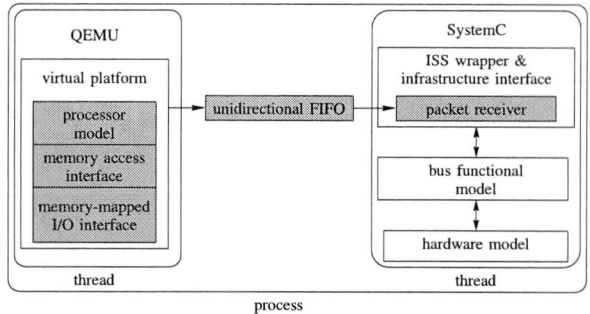

Figure 1. The inter-process communication (IPC) mechanism used by the ISS wrapper & infrastructure interface described herein.

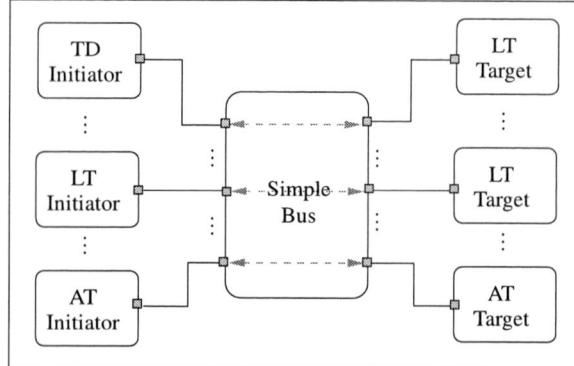

Mixed Level Simple Bus Connections

Figure 2. Simple bus provided by TLM-2.0. All of the ports use simple sockets for connecting with each other. Also notice that the TD initiator is connected with LT target, and both of them are modeled by using the same LT coding style.

of the information in the received packets. In addition, the synchronization between QEMU and SystemC is only needed by the I/O read operations, which can be achieved by having QEMU call the I/O read function—which will pass the pointer to the data to be read to SystemC—and then block until SystemC returns.

III. Virtual Platform Integration

In this section, we give a more detailed description about how the simple bus is integrated into the virtual platform described herein. Because certain attributes of transaction modeling are maintained in QEMU, there is no need to re-model them.

A. Organization of Simple Bus

As described in Section II-A, the simple bus using AT coding style can be used to connect with different level of models. The simple bus is constructed by initiator and target sockets for connecting the external target and initiator models. The number of initiator and target sockets can be parameterized by using the template parameters at the time of model instantiation. All of the paths for propagating the transactions are similar no matter the blocking or non-blocking interface is used, as shown in Fig. 2. Furthermore, the mapped interface and utilities for modeling at different levels are summarized in Table I.

The simple bus is built on the base protocol pre-defined by TLM-2.0 instead of any specific bus protocol. The LT coding style supports two timing points for modeling the start and end of a single transaction, while the AT coding style provides four timing points for modeling the start and end of the request phase and the response phase.

B. Simple Bus Integration

In order to integrate the simple bus into our virtual platform, the first thing that has to be taken into consideration is what type of the simple socket is to be used. To model the connections between the ISS and DMAC model, we have to take into account four master ports, i.e., the processor instruction port, the processor data port, DMAC M1 master port, and DMAC M2 master port. Moreover, the processor data port can be divided into data port for the memory access

and data port for the memory-mapped I/O access. For the simulation speed, we choose the LT style simple socket for modeling the initiator and target ports of our models.

As Fig. 2 shows, most of the initiator models are paired with the target models the propagating path built by the simple bus. However, two types of transactions initiated by the master ports need not to be transferred to the target models in our virtual platform. In this paper, we use the prefix "pseudo-" to indicate them, i.e., the pseudo-initiator, the pseudo-target, and the pseudo-transaction. The first type of the pseudo-transaction is initiated by the QEMU and launched by means of the pseudo-initiator socket of ISS wrapper. Moreover, the transaction sequence and the address translation are maintained by the transaction FIFO and the software MMU of QEMU, respectively.

The second type of the pseudo-transaction is initiated by the master ports of DMAC model and used to access the internal memory of QEMU. Because the pseudo-transactions to the internal memory model are managed by QEMU, all we need to do is to send these pseudo-transactions to the bus model to pretend that the memory read/write operations are actually performed on the bus.

No matter what type the transaction belongs to, the read-/write commands of both types are only a hint, not a real operation applied on the hardware models. Thus, the pseudo-sockets mentioned above are not required to be paired with the target models except the data port for memory-mapped I/O access. Therefore, we preserve the primary template parameters of simple bus for the sockets which need to be paired and add new template parameters for the pseudo-initiator sockets which need not be paired with the target models. The new simple bus template is as given below.

```
1  template <int NR_OF_INITIATORS, int NR_OF_TARGETS, int
              NR_OF_IBUS, int NR_OF_DBUS>
2  class SimpleBusAT_QSC2 : public sc_core::sc_module
3  {
4  public:
5      typedef tlm::tlm_generic_payload transaction_type;
6      typedef tlm::tlm_phase phase_type;
```

TABLE I. THE MAPPED INTERFACE AND UTILITIES OF DIFFERENT LEVEL MODELING.

Coding style	Interface		Utility	
			quantum keeper	payload event queue
Temporal Decoupling	`b_transport()`		Yes	–
Loosely-Timed	`b_transport()`		–	–
Approximately-Timed	initiator	`nb_transport_fw()`	–	Yes
	target	`nb_transport_bw()`		

```
 7    typedef tlm::tlm_sync_enum sync_enum_type;
 8    typedef tlm_utils::simple_target_socket_tagged<SimpleBusAT_QSC2>
          target_socket_type;
 9    typedef tlm_utils::simple_initiator_socket_tagged<SimpleBusAT_QSC2
          > initiator_socket_type;
10   public:
11    target_socket_type target_socket[NR_OF_INITIATORS];
12    initiator_socket_type initiator_socket[NR_OF_TARGETS];
13    target_socket_type ibus_socket[NR_OF_IBUS];
14    target_socket_type dbus_socket[NR_OF_DBUS];
15    ...
16    uint32_t insn_cnt[NR_OF_IBUS];
17    uint32_t ld_cnt[NR_OF_DBUS];
18    uint32_t st_cnt[NR_OF_DBUS];
19    ...
20   }
```

The pseudo-target socket of the modified simple bus, `ibus_socket` and `dbus_socket`, has the same type of `target_socket` but is parameterized by the template parameters `NR_OF_IBUS` and `NR_OF_DBUS`. We use `ibus_socket` to receive the transactions on the instruction bus and `dbus_socket` to receive the transactions on the data bus. Also, the performance counters `insn_cnt`, `ld_cnt` and `st_cnt` are used to record, respectively, the number of instructions executed, the number of load and store operations performed. In essence, `ibus_socket` and `dbus_socket` both use the transport interface except that the performance counters are different.

Although the simple bus is modeled using AT coding style, the transport function `NBTransport_fw()` will actually call the LT transport function `BTransport()`. Finally, the virtual platform that integrates the IA-ISS and DMAC model, i.e., the `model.arm_n` and `model.dmac` instance, with the adapted simple bus, i.e., the `SimpleBusAT_QSC2<1,1,1,3>`. The model instantiation is as given below.

```
 1    ...
 2    soc_model model("model");
 3    SimpleBusAT_QSC2<1,1,1,3> bus("bus");
 4    ...
 5    model.arm_busio−>socket(bus.target_socket[0]);
 6    model.arm_ibus−>socket(bus.ibus_socket[0]);
 7    model.arm_dbus−>socket(bus.dbus_socket[0]);
 8    model.dmacm1−>socket(bus.dbus_socket[1]);
 9    model.dmacm2−>socket(bus.dbus_socket[2]);
10    bus.initiator_socket[0](model.busio_dmac−>socket);
11    ...
```

The initiator sockets of IA-ISS and DMAC model need only call `b_transport()` the implementation of which resides in the target socket of the modified simple bus. Also notice that the only path that will transfer the transaction from the initiator model to the target model via the bus model is the following sequence: `model.arm_busio->socket` ⇌ `bus.target_socket[0]` ⇌ `bus.initiator_socket[0]` ⇌ `model.busio_dmac->socket`. The double arrows between the sockets indicate the direction of the forward and the backward interface used to transfer the transactions.

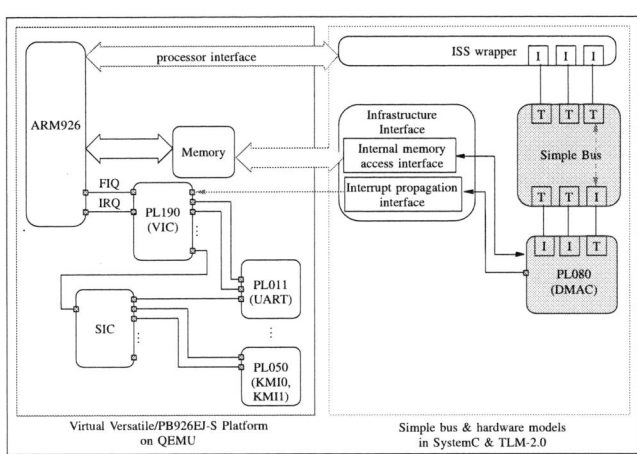

Figure 3. The block diagram of the virtual platform Versatile/PB926EJ-S of QEMU combined with the models written in SystemC and TLM-2.0.

TABLE II. NOTATIONS USED IN TABLE III

min	The best-case co-simulation time of 30 runs.
max	The worst-case co-simulation time of 30 runs.
μ	The mean of co-simulation time of 30 runs.
σ	The standard deviation of co-simulation time of 30 runs.
N_{TI}	The number of target instructions simulated.
N_{LD}	The number of load operations of the virtual processor.
N_{ST}	The number of store operations of the virtual processor.
N_{DMAR}	The number of read operations of DMAC.
N_{DMAW}	The number of write operations of DMAC.
N_{TX}	The total number of transactions.

IV. EXPERIMENTAL RESULTS

In this section, we turn our discussion to the experimental results of using the modified simple bus as the interconnect of the virtual Versatile/PB926EJ-S platform and as the interconnect of the IA-ISS and ARM PrimeCell PL080 DMAC [9] modeled in SystemC, the details of which are as shown in Fig. 3. Because PL080 has two master ports, M1 and M2, we use M1 to read and M2 to write data in turn.

Because the arbitration, the burst transfer, and the split transfer are not modeled in the simple bus, we only model the signal pins corresponding to the attributes of a transaction. In addition, the processor model is based on the ARM9 processor without cache. Because Linux kernel dose not provide the device driver for PrimeCell PL080 of the Versatile/PB926EJ-S platform, we also develop our own DMAC device driver as a char device for the purpose of testing, and its behavior can be controlled by an application program by calling the `ioctl()` function defined within the driver.

We provide two different metrics to measure the performance of the proposed virtual platform: (1) the simulation

978-1-4244-9019-6/11 $26.00 © 2011 IEEE   87

TABLE III. SIMULATION TIME OF BOOTING UP THE LINUX KERNEL WITH DMA TEST BENCH USING TLM-2.0 SIMPLE BUS FOR 30 TIMES.

Statistics	Co-simulation time	N_{TI}	N_{LD}	N_{ST}	N_{DMAR}	N_{DMAW}	N_{TX}
min	40m50.756s	744,864,386.00	254,566,331.00	135,886,958.00	1,024,000.00	1,024,000.00	1,137,365,675.00
		(65.49%)	(22.38%)	(11.95%)	(0.09%)	(0.09%)	(100.00%)
max	44m20.429s	789,182,764.00	271,736,320.00	147,344,322.00	1,024,000.00	1,024,000.00	1,210,311,406.00
		(65.20%)	(22.45%)	(12.17%)	(0.08%)	(0.08%)	(100.00%)
μ	42m05.767s	759,358,111.80	260,203,215.90	139,598,175.97	1,024,000.00	1,024,000.00	1,161,207,503.67
		(65.39%)	(22.41%)	(12.02%)	(0.09%)	(0.09%)	(100.00%)
σ	00m49.962s	10,957,387.63	4,197,037.64	2,815,094.29	0.00	0.00	17,969,519.57

time it takes to boot up a full-fledged Linux kernel and to run the DMAC test bench, and (2) the statistics it can collect at instruction-accurate level while the system is being boot up. For all the experimental results given in this section, a 2.13GHz Intel Core 2 6420 processor machine with 3.2GB of memory is used as the host, and the target OS is built using the BuildRoot package [10]. The Linux distribution is Gentoo, and the kernel is Linux version 2.6.30. QEMU version 0.11.0-rc1, SystemC version 2.2.0 (including the reference simulator provided by OSCI), and TLM-2.0 (released on July 15, 2009) are all compiled by gcc version 4.3.2.

A. Time to Boot up Linux

In order to gather the statistics, the initial shell script is modified to enable the option of executing the DMAC test bench and rebooting the virtual machine automatically as soon as the booting sequence is completed. Furthermore, the pre-defined no-reboot option of QEMU will catch the reboot signal once the OS completed the DMAC test bench and started the reboot command to shut the QEMU down. Thus, the test bench can easily estimate the co-simulation time of QEMU and SystemC at the OS level.

Our test bench uses DMA to move data between memory allocated from the kernel space. The amount of data moved is 2,048,000 words, one half of which are read and the other half of which are write. Moreover, each word occupies 4 bytes. The simulation results of the simple bus is shown in Table III, and the notations used in Table III can be found in Table II.

As Table III show, the time taken to boot up a full-fledged Linux kernel is as shown in the column labeled "Co-simulation time." The column labeled "N_{TX}" gives the total number of target instructions executed and load and store operations performed. Because the number of read/write operations of the slave port of DMAC (PL080 in this case) has been counted as the load and store operations of the virtual processor, only the number of read/write operations of the master ports has to be counted. That is,

$$N_{TX} = N_{TI} + N_{LD} + N_{ST} + N_{DMAR} + N_{DMAW}.$$

The percentages given in parentheses are defined by

$$\frac{N_\alpha}{N_{TX}} \times 100\%$$

where the subscript α is either TI, LD, ST, DMAR, DMAW or TX.

V. CONCLUSION AND FUTURE WORK

This paper presents a QEMU and SystemC-based virtual platform that is capable of using TLM-2.0 interface and estimating its performance with an OS up and running. The IA-ISS is used as the processor model and is connected with DMAC model via the modified simple bus. In addition, using TLM-2.0 as the interface can provide the compatibility of other hardware models with the same interface. Moreover, the proposed virtual platform can provide statistics—such as the number of instructions executed, memory accessed, and DMAC read/write operations—that are valuable for the system designers at the early stage of ESL design flow. Although the co-simulation takes 44m20.429s to boot up a full-fledged kernel is still acceptable in the worst case, one of our goals is to speedup the simulation speed of TLM-2.0 interface in the future.

ACKNOWLEDGMENT

This work was supported in part by National Science Council, Taiwan, ROC, under Contract No. NSC99-2221-E-110-052.

REFERENCES

[1] M. Creamer, "Nine reasons to adopt SystemC ESL design," http://www.eetimes.com/news/design/columns/eda/showArticle.jhtml?articl%eID=47212187.

[2] J. Aynsley, OSCI TLM-2.0 Language Reference Manual, 2009, http://www.systemc.org/members/download_files/check_file?agreement=tlm_%2-0_080606.

[3] M. Montón, A. Portero, M. Moreno, B. Martínez, and J. Carrabina, "Mixed SW/SystemC SoC Emulation Framework," in Proceedings of IEEE International Symposium on Industrial Electronics, June 2007, pp. 2338–2341.

[4] M. Montón, J. Carrabina, and M. Burton, "Mixed Simulation Kernels for High Performance Virtual Platforms," in Proceedings of Forum on Specification and Design Languages, September 2009, pp. 1–6.

[5] C.-C. Wang, R.-P. Wong, J.-W. Lin, and C.-H. Chen, "System-level development and verification framework for high-performance system accelerator," in Proceedings of International Symposium on VLSI Design, Automation and Test, Apr. 2009, pp. 359–362.

[6] J.-W. Lin, C.-C. Wang, C.-Y. Chang, C.-H. Chen, K.-J. Lee, Y.-H. Chu, J.-C. Yeh, and Y.-C. Hsiao, "Full system simulation and verification framework," in Proceedings of Fifth International Conference on Information Assurance and Security, Aug. 2009, pp. 165–168.

[7] Synopsys Inc., "Platform Architect Models," http://www.synopsys.com/Tools/SLD/VirtualPrototyping/SLLibraries/Pages/%PlatformArchitect.aspx.

[8] T.-C. Yeh, G.-F. Tseng, and M.-C. Chiang, "A fast cycle-accurate instruction set simulator based on QEMU and SystemC for SoC development," in Proceedings of the 15th IEEE Mediterranean Electrotechnical Conference, Apr. 2010, pp. 1033–1038.

[9] ARM Ltd., PrimeCell DMA Controller (PL080) Technical Reference Manual, 2003, http://infocenter.arm.com/help/index.jsp.

[10] P. Korsgaard, "BuildRoot," http://buildroot.uclibc.org/.

A Fast Custom Network Topology Generation with Floorplanning for NoC-based Systems

Katherine Shu-Min Li, *Member, IEEE*, Shu-Yu Chen, *Student Member, IEEE*, Liang-Bi Chen, *Member, IEEE*, and Ruei-Ting Gu

Abstract—**This paper proposes a fast full-chip synthesis methodology which can be built a custom Network-on-Chip (NoC) topology for NoC-based systems. The processors and their communications are synthesized simultaneously in the system-level floorplanning process. The proposed method leads to accurate area estimation, which makes an algorithm much more efficient than previous approaches. Moreover, the wirelength-aware floorplanning is carried out to optimize circuit size as well as wire length. As a result, experimental results show that the proposed approach produces custom NoCs with better performance than previous methods while the computation time is significantly shorter. This method is also more scalable, which makes it ideal for complicated NoC-based systems.**

Index Terms—**Custom NoC, Floorplanning, Network-on-Chip (NoC), Topology Generation.**

I. INTRODUCTION

THE advance of process technology keeps on reducing device size. As a result, a system consists of multiple processing elements (PEs), which can be integrated into a single chip, called System-on-a-Chip (SoC). Compared with the traditional approach, the SoC technology offers many advantages such as higher performance and lower power consumption. The requirement for communication would be very complicated in an SoC with a large number of PEs. The multiple-to-multiple connection may also be required in complicated SoC designs. The conventional on-chip communication architecture, which consists of point-to-point connection and bus infrastructure, may not be able to provide sufficient communication support for an SoC.

In the point-to-point connection structure, a direct link is assigned to any two modules that need communication. As a result, a large number of links are required, which dramatically inflates chip size. Besides, long communication wires may be required under this approach, which leads to excessive signal delay. The bus-based communication infrastructure is both

Manuscript received February 13, 2011; revised April 7, 2011. This work was supported in part by the National Science Council (NSC), Taiwan, R.O.C. under Grants NSC-97-2220-E-110-009 and NSC-98-2221-E-110-090-MY2, and by the Ministry of Economic Affairs, Taiwan, R.O.C. under Grant 99-EC-17-A-01-S1-104.

The authors are with Department of Computer Science and Engineering, National Sun Yat-Sen University, Kaohsiung 80424, Taiwan, R.O.C. (e-mail: smli@cse.nsysu.edu.tw; liangbi.chen@gmail.com).

efficient and flexible. On the flip side, the communication channel can only be utilized by a pair of modules at a time; thus, the bus may become a bottleneck of overall system performance in applications requiring heavy communication. A promising solution to these problems is to deploy the packet-switched networking infrastructure for on-chip communication. This approach is referred to as the Network-on-Chip (NoC) [1]-[4].

The NoC greatly improves the scalability of SoCs and achieves higher power efficiency compared to other types of communication structures. In an NoC-based system, which consists of network components including data links, network interfaces (NI), and routers. Links provide communication media on which data are transmitted. An NI is the interface between a PE and a router, and it is responsible for transforming data into packets or vice versa. The physical transmission path for a packet is set up by routers. The topology of the NoC is determined by the physical connections of links and routers, while connection paths for packets are selected by the routing algorithms.

NoC topologies can be classified into two categories: regular, and irregular. If each PE is mapped a router, the PE and its assigned router are combined as a tile. A router is usually connected to other routers in neighboring tiles in addition to the PE in the same tile. There are two types of irregular topologies. The first type is constructed mainly to deal with the placement of PEs with diversified sizes and aspect ratios. In the second type, a custom topology is synthesized specially for given design goals. A custom NoC no longer consists of tiles, as routers are allocated according to the communication demand. In this paper, we will focus on the problem of irregular topology generation for custom NoC-based systems.

II. RELATED WORK

Regular topologies are better choices for general-purpose systems where traffic conditions are not predictable, and they are also easy to design. On the flip side, they involve more network components than necessary for ASICs with special communication demands. As a result, the chip becomes larger with higher power consumption and longer latency. These problems become more prominent as chips become larger. The custom NoC provides a better solution to the above problems in ASICs, as this approach produces smaller circuits. Specially, it was reported that power consumption in an NoC is highly correlated to the link wirelength. Many custom NoC design methods have been discussed and developed in [5]-[10].

Previous studies on custom NoC designs treat topology generation and message routing separately, and route allocation has to be carried out for every demand on the given topology. Although this approach can generate good network topology, the computation time for route allocation can be a problem, especially when the number of PEs and communication data volume increase. Besides, network components are usually added after floorplan for PEs being decided, which may render the estimation of overall chip size inaccurate.

Therefore, this paper proposes a custom NoC synthesis methodology for ASICs with special communication demands. The goal is to synthesize an irregular custom NoC floorplan that requires fewer network components and consumes less power. The proposed algorithm is also very fast, and can achieve accurate area estimation for the synthesized circuits. The proposed method consists of two steps: a topology generation based on communication analysis and co-floorplanning for PEs and routers. This scheme provides some advantages over previous methods. Since the proposed methodology can take the communication demands into account in the topology synthesis procedure so it is not necessary to allocate a route for each message. As a result, routing in NoCs synthesized by the proposed methodology is more efficient. The area estimation is more accurate with the proposed method, as PEs and network components are floorplanned at the same time. Besides, it provides a complete system-level floorplan where circuit size and overall network link length are optimized.

III. THE PROPOSED METHODOLOGY

The proposed custom NoC design flow is shown in Fig. 1. The goal is to develop a fast synthesis scheme that minimizes power consumption and overall wire length in a custom NoC-based system. A *Communication Trace Graph (CTG)* is a graph $CTG = (V_G, E_G)$, where each $v \in V_G$ represents a block and each $e \in E_G$ is an edge in the corresponding block diagram. In a *CTG*, $w(e)$ stands for the communication demand of edge e, and η_G is an edgemap showing the source node u and target node v where $\eta_G(e) = (u, v)$, $u, v \in V_G$, $e \in E_G$.

Given a *CTG* and the input/output port limits of routers, our synthesis method generates a custom NoC topology with the corresponding system-level floorplan. The custom NoC topology is generated in the first two phases, where the communication requirement is analyzed (**Phase I**) and then *Router Sharing Groups (RSGs)* are formed (**Phase II**) accordingly. In **Phase III**, a wire-length driven floorplanning that takes into account PEs and shapes of the routers is carried out, while routing paths are allocated as well.

A. Phase I: Communication Analysis

The optimal NoC structure is mainly determined by the communication requirement. In order to facilitate the computation, first a *Communication Trace Bipartite Graphics*

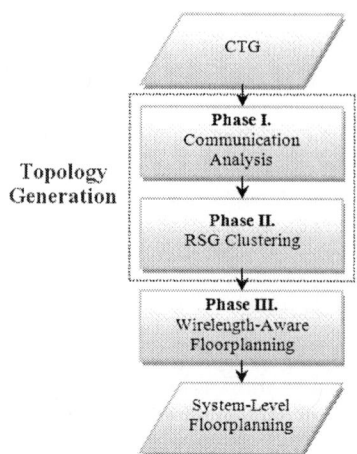

Fig. 1. Flow chart of the proposed methodology.

(CTB) is pruned by removing trivial arcs and vertices. The simplified bipartite graph is then analyzed so that the *RSGs* can be constructed in the next phase.

1) Isolated Arcs Removal (IAR)

Let $d_{in}(v)$ and $d_{out}(v)$ be the in-degree and out-degree of a vertex v, respectively. In a bipartite graph, $d_{in}(v)=0$ if v is a source vertex while $d_{out}(v)=0$ if v is a target vertex. An arc is said to be *isolated* if it is the only communication link connected to the source and target vertices.

2) Isolated Vertices Removal (IVR)

A source vertex s is *isolated* if $d_{out}(s)=0$; similarly, a target vertex t is *isolated* if $d_{in}(t)=0$. In a *CTB*, such vertices are not connected to any arc.

3) Successors Subset Tree Construction (STC)

In order to improve the router utility, it is imperative to put a set of vertices with heavy communication traffic in the same RSG. In other words, it is desirable to put target vertices sharing the same source vertex in the same group. Similarly, source vertices sharing the same target vertices can be put in the same group.

Definition 1: Successors of a source vertex v are those target vertices that are connected to v through arcs in the *CTB*. The set of vertices are denoted as $Suc(v)$.

Definition 2: Predecessors of a target vertex v are those source vertices that connect to v through arcs in the *CTB*. The set of vertices are denoted as $Pre(v)$.

Source vertices sharing the same successors are partitioned into the same group, and this partition is achieved through a subset tree construction. The root of the tree contains all source vertices (S_B) in the *reduced CTB (RCTB)*, while each of the remaining tree nodes contains a subset of S_B. Each source vertex s_i is contained in exactly one non-root tree node. This tree is constructed according to the following rules. (1) If $Suc(s_i) = Suc(s_j)$, source vertices s_i and s_j belong to the same tree node. (2) $Suc(s_i) \subset Suc(s_j)$, then the tree node containing s_i is a descendant of the tree node containing s_j. (3) If $Suc(s_i) \not\subset Suc(s_j)$ and $Suc(s_j) \not\subset Suc(s_i)$, then the only ancestor of tree nodes containing s_i and s_j is the root. The tree construction algorithm is given in Fig. 2.

Algorithm: *STC* (Subset Tree Construction)
Input: Reduced *CTB''*(*S''*, *T''*, *A''*)
Output: Subset Tree
1. *L* ← sorted *S''* by |Suc(*s*)| in descending order , *s* ∈ *S''*
2. set *root node of ST* as *S''*
3. **for** (all *v* ∈ *L*) **do**
4.   insert_subset_tree (*v*, *root node of ST*)
5. **end for**
function insert_subset_tree (*v*, *root node of ST*)
1. *q*.enqueue(*root node of ST*)
2. **while** (*q* is not empty) **do**
3.   node *current* ← *q*.front
4.   **if** (Suc(*v*) = Suc(*current*)) **then**
5.   insert *v* to *current*
6.   **else if** (Suc(*v*) ⊂ Suc(*current*)) **then**
7.   **if** (*current* is leaf node) **then**
8.   insert *v* as child of *current*
9.   **else if**
10.   *counter* ← 0
11.   **while** (*current* is not null) **then**
12.   **if** (Suc(*v*) = Suc(*current*) **or** Suc(*v*) ⊂ Suc(*current*)) **then**
13.   *q*.enqueue(*current*)
14.   *counter* ← *counter* + 1
15.   **end if**
16.   node *prev* ← *current*
17.   *current* ← *current*.sibling
18.   **end while**
19.   **if** (*counter* = 0)
20.   insert *v* as sibling of *prev*
21.   **end if**
22.   **else**
23.   insert *v* as sibling of *current*
24.   **end if**
25. **end while**
end function

Fig. 2. Pseudo code of *STC*.

4) Target Vertices Partitioning (TVP)

When Step 3 (*STC*) in Phase I is finished, the set of all source vertices (S_B) has been partitioned into several disjoint subsets, with vertices in the same subset sharing exactly the same subset of target vertices (T_B). Each of these disjoint subsets of source vertices is called a *Source Component* (*SC*), and it is a minimal set of source vertices that can be put into an *RSG*.

B. Phase II: RSG Clustering

Once we have all the source and target components, the next step is to search for components that can use the same router. A set of such components become an *RSG*. The goal is to construct larger *RSG*s that maximize router utility under the given I/O port constraints. The rationale for larger routers is that fewer routers are created, so that the number of hops for each communication is reduced. As a result, communication delay can be improved, so is the power consumption.

C. Wirelength-Aware Floorplanning

A fast simulated-annealing based algorithm [11] is used to generate a floorplan in the proposed scheme. In order to optimize circuit size as well as overall wire length, the cost function is defined as follows:

$$Cost(fp) = 0.5 \times Area(fp) + 0.5 \times WL(fp), \qquad (1)$$

where *fp* is a floorplan, while *Area*(*fp*) and *WL*(*fp*) are the estimated area and wirelength of floorplan *fp*, respectively.

IV. EXPERIMENTAL RESULTS

We conduct the experiments on Ubuntu 8.0.4, with Xeon 2.00GHz CPU, 3GB memory. The benchmark circuits are application-specific SoC designs, including video applications: MPEG-4 decoder, Multi-Window Display (MWD), Picture-in-Picture (PIP), and Video Object Plane Decoder (VOPD); and multimedia applications: H.263 video encoder, H.263 video decoder, MP3 audio encoder, and MP3 audio decoder. In order to make a fair comparison to results obtained from the mesh topology optimized for performance, the upper bounds on router inputs (r_i) and outputs (r_o) are set to 5, and the bus width is limited to 128. The custom NoC designs are synthesized with the 70 nm technology file provided by CosiNoC, while the area and power consumption of routers are estimated by Orion [12].

The statistics of benchmark SoCs are shown in Table I. Also shown in Table I are the NoCs synthesized by the proposed method as well as results from mesh-based implementation. The indices and circuit descriptions are given in Columns One and Two, while the numbers of cores (#Core) and communications (#Comm.) of each circuit are listed in Columns Three and Four. The next column gives the average amount of communication in each benchmark (#Comm./#Core), which gives an indication of the communication traffic in each example. Columns Six and Seven give the number of routers required in mesh and the proposed method, while the last two columns are the number of links for mesh and the proposed method. Synthesis results of the proposed methodology will be compared to a custom NoC synthesis scheme: CosiNoC [10]. Since the source code of CosiNoC is available, a complete comparison between CosiNoC and this work will be provided in Table II.

V. CONCLUSIONS

A custom NoC topology generation method with floorplanning is proposed in this paper. The topology generation is based on communication analysis, which leads to efficient designs in shorter computation time. Experimental results show that the proposed methodology can achieve lower wirelength and power consumption than previous methods while the runtime is significantly shorter. Since both network components and PEs are planned simultaneously, an accurate estimation of the synthesized circuit can be made. Besides, this approach is also helpful to restrict runtime, as the floorplanning process can be done in one pass. In contrast, previous methods try to add the on-chip network in a floorplan where PEs have been processed, which may cause design violation and require more iterative loops of the process.

The proposed method is also scalable, as it is capable of providing good results in reasonable time for larger circuits. Hence, the proposed method is also particularly useful for large

TABLE I
BENCHMARKS STATISTICS WITH SYNTHESIS RESULTS OF THIS WORK AND MESH

SoC	Description	# Core	# Comm.	Avg. Comm. (# Comm. / # Core)	# Router		# Link	
					n × n MESH	This Work	n × n MESH	This Work
G1	PIP	8	8	1.00	9	6	30	14
G2	H.263 enc MP3 dec	12	12	1.00	16	7	56	21
G3	MWD	12	12	1.00	16	9	56	23
G4	MP3 enc MP3 dec	13	13	1.00	16	8	56	21
G5	H.263 dec MP3 dec	14	15	1.07	16	8	56	24
G6	VOPD	12	14	1.17	16	9	56	24
G7	VOPD + MPEG-4 + MWD	36	52	1.44	36	24	240	81
G8	VOPD + MPEG-4	24	40	1.67	25	15	182	58
G9	MPEG-4 + PIP	20	34	1.70	25	12	132	46
G10	MPEG-4	12	26	2.17	16	6	132	32
G11	IMP	27	96	3.56	36	12	380	74
Average				Data	20.64	10.55	125.09	38.00
				Normalized	1.96	1.00	3.29	1.00

TABLE II.
THE PROPOSED METHODOLOGY SYNTHESIS RESULTS AND COMPARISON TO CosiNoC [10]

SoC	Method	Size (mm²)	#Hop	Network				Power (mW)	Dynamic Energy (μJ)	CPU Time (sec)
				#Router	A_R (mm²)	#Link	W_L (mm)			
G1	CosiNoC	2.473	2	6	0.007	14	7.786	5.06		0.04
	This Work	2.7	2	6	0.007	14	6.167	4.79	4.18	0.001
G2	CosiNoC	2.912	2.17	7	0.014	19	15.686	8.92		0.08
	The Work	2.934	2.17	7	0.014	21	6.381	7.35	3.3	0.001
G3	CosiNoC	2.706	2.17	7	0.013	18	16.46	8.9		0.06
	This Work	2.816	2.17	9	0.013	23	8.293	7.67	10.42	0.001
G4	CosiNoC	2.32	2	8	0.012	21	16.358	8.41		0.09
	This Work	2.436	2	8	0.012	21	7.24	6.89	0.12	0
G5	CosiNoC	4.666	2.13	8	0.017	24	26.029	12.59		0.12
	This Work	4.54	2.07	8	0.015	24	11.264	8.95	0.2	0.001
G6	CosiNoC	8.025	2.43	9	0.019	25	34.006	15.8		0.1
	This Work	8.26	2.07	9	0.015	24	15.773	10.23	39.81	0.001
G7	CosiNoC	12.906	3.69	26	0.074	81	97.519	58.5		5.37
	This Work	13.737	2.25	24	0.054	81	49.708	35.89	253.18	0.01
G8	CosiNoC	10.252	2.98	14	0.058	56	52.959	39.78		0.59
	This Work	10.834	2.28	15	0.041	58	33.276	26.31	242.77	0.001
G9	CosiNoC	5.17	2.5	10	0.037	42	56.596	30.12		1.25
	This Work	4.98	2.24	12	0.032	46	20.278	19.58	199.37	0.001
G10	CosiNoC	2.511	2.81	6	0.028	29	19.517	18.51		0.37
	This Work	2.499	2.31	6	0.024	32	12.48	14.45	195.18	0.001
G11	CosiNoC	6.298	4.74	22	0.079	78	66.277	50.84		16.74
	This Work	6.29	2.68	12	0.073	74	50.43	46.8	370.2	0.01
Avg.	CosiNoC	5.476	2.69	11.18	0.0325	37.0	37.199	23.40		2.2555
	This Work	5.639	2.20	10.55	0.0273	38.0	20.117	17.17		0.0025
Improvement		-2.98%	18.22%	5.69%	16.20%	-2.70%	45.92%	26.62%		886.1X

systems where a complicated NoC-based system is required.

REFERENCES

[1] W. J. Dally and B. P. Towles, *Principles and Practices of Interconnection Networks*. Morgan Kaufmann, Jan. 2004.

[2] J. Nurmi, "Network-on-chip:a new paradigm for system-on-chip design," in *Proc. IEEE System-on-Chip Conf.*, pp. 2–6, 2005.

[3] B. S. Feero and P. P. Pande, "Networks-on-chip in a three-dimensional environment: A performance evaluation," *IEEE Trans. Computers*, vol. 58, pp. 32–45, Jan. 2009.

[4] R. Marculescu, U. Y. Ogras, L.-S. Peh, N. E. Jerger, and Y. Hoskote, "Outstanding research problems in noc design: System, microarchitecture, and circuit perspectives," *IEEE Trans. Computer-Aided Design of Integrated Circuits and Systems*, vol. 28, pp. 3–21, Jan. 2009.

[5] K. Srinivasan, K. S. Chatha, and G. Konjevod, "Linear-programming based techniques for synthesis of network-on-chip architectures," *IEEE Trans. VLSI Systems*, vol. 14, pp. 407–420, Apr. 2006.

[6] K. Srinivasan, K. S. Chatha, and G. Konjevod, "Linear-programming-based techniques for synthesis of network-on-chip architectures," *IEEE Trans. VLSI Systems*, vol.14, pp. 407-420, 2006.

[7] K. Srinivasan and K. S. Chatha, "A methodology for layout aware design and optimization of custom network-on-chip architectures," in *Proc. International Symp. Quality Electronic Design*, pp. 352–357, 2006.

[8] K. Srinivasan, K. S. Chatha, and G. Konjevod, "An automated technique for topology and route generation of application specific on-chip interconnection networks," in *Proc. IEEE/ACM International Conf. Computer-Aided Design*, pp. 231–237, 2005

[9] L. Benini, "Application specific NoC design," in *Proc. IEEE/ACM Design, Automation, and Test in Europe*, pp. 491–495, 2006

[10] J. A. Roy, S. N. Adya, D. A. Papa, and I. L. Markov, "Min-cut floorplacement," *IEEE Trans. Computer-Aided Design of Integrated Circuits and Systems*, vol. 25, pp. 1313–1326, Jul. 2006.

[11] T.-C. Chen, Y.-W. Chang, and S.-C. Lin, "A new multilevel framework for large-scale interconnect-driven floorplanning," *IEEE Trans. Computer-Aided Design of Integrated Circuits and Systems*, vol. 27, pp. 286–294, Feb. 2008.

[12] A. B. Kahng, B. Li, L.-S. Peh, and K. Samadi, "Orion 2.0: A fast and accurate NoC power and area model for early-stage design space exploration," in *Proc. IEEE/ACM Design, Automation and Test in Europe*, pp. 423–428, 2009.

Evolution of Embedded Flash Memory Technology for MCU

Hideto Hidaka, *Member, IEEE*

Abstract—**Embedded flash memory technology has undergone tremendous growth of demands with various performance requirements driven by expanded applications of MCU (Micro Controller Unit) products. High temperature operations with highest reliability for auto-motive applications, very low power embedded EEPROM functions for smart-cards, and ultra low-voltage operations for medical applications are driving factors in developing embedded flash technologies. Together with evolving memory cell technology, resolving performance/power trade-offs by developing dedicated design platforms with optimized eFlash technology, memory interface & bus designs, and the whole chip design methodologies, has realized advanced MCU products line-ups by split-gate MONOS flash technology with a wide range of applied products including auto-motive and security applications.**

Index Terms—**embedded flash memory, split-gate flash memory cell, charge-trapping flash memory cell**

I. INTRODUCTION

Flash-MCU, Micro-Controller Unit with embedded flash memory storage (eFlash), has made a leaping progress in the market acceptance according to the expansion of real-time control applications in 2000's. The programmable code storage by eFlash in place of on-chip mask-ROM has triggered rapid expansion of adaptive control and data stream applications. Together with over-all production and inventory cost reduction, this development has realized an innovation with remarkable cost/value advantage over MCU with fixed ROM or MCU with stand-alone flash memory.

Diversified eFlash technologies for flash-MCU products have challenged new market drivers such as auto- motive, smart-IC card, and medical applications and have expanded the MCU market, while alternately eFlash has become the most successful, largest business in embedded memory technology only second to CMOS-inclusive embedded SRAM. In 1990's Flash-MCU was mainly used in proto-types for debugging systems. With the advancement of Flash-MCU innovation, almost all the MCU market has been focused onto flash-MCU solutions (Fig. 1), where eFlash technology evolution represents the most important factor.

Manuscript received March 18, 2011.
Hideto Hidaka is with Renesas Electronics Corp. Itami, Hyogo, 664-0005 JAPAN (e-mail: hideto.hidaka.pz@renesas.com).

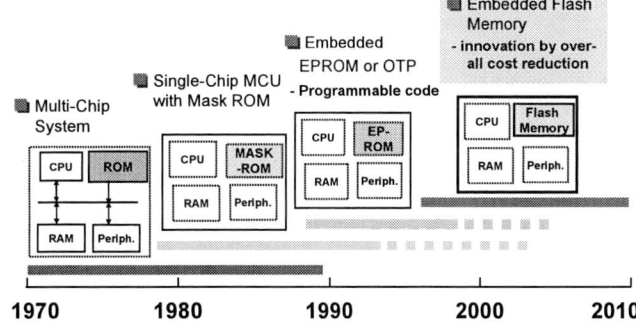

Fig.1. Evolution of MCU products in a memory-centric view.

This paper reviews the past and current status and explore the future directions of eFlash technology for MCU, from the viewpoints of process, circuit, and applications. Innovative factors different from stand-alone flash memory and a future MCU concept in the memory-centric view are also presented.

II. EVOLUTION OF EMBEDDED FLASH TECHNOLOGY

In the flash memory technology tree in Fig. 2, those suitable for embedded uses have been quite selective because of requirements specific to embedded uses. Because high-density flash technologies for stand-alone data memory don't meet the requirements for embedded uses in reliability and performance, eFlash technologies have evolved on its own.

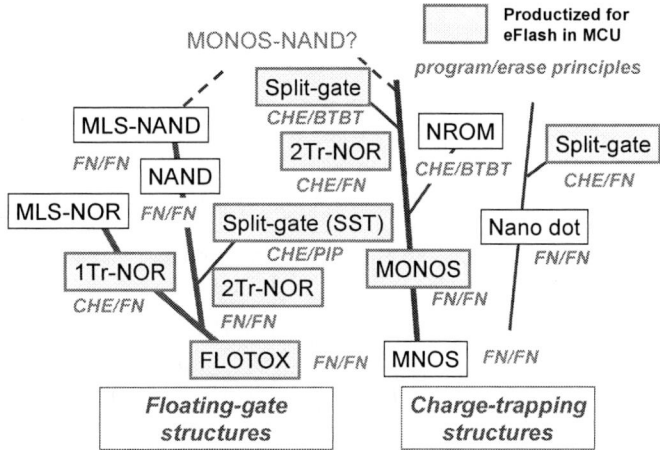

Fig.2. Evolution of Flash and eFlash memory technologies.

Type	1Tr NOR cell	1.5Tr cell (SuperFlash™)[1]	2Tr cell	1Tr SONOS (NROM™) [2]	1.5Tr Nano-dot [3]	2Tr SONOS (PMOS) [4]
Program	CHE	SSI	FN	CHE	SSI	CHE
Erase	FN (poly-sub)	FN (poly-poly)	FN (poly-sub)	HH	FN	FN
Device structure						
Advantage	High density	Fast program	Low power P/E	2bits/cell	Fast, low-power program	Low power P/E

Fig.3. Embedded flash memory technologies. 1Tr-NOR cells for high density are giving way to 1.5Tr/2Tr cell for performance. Charge-trapping cells are emerging in some applications for reliability.

Two technology transitions specific to eFlash technologies have been remarkable (Fig.3):
(1) 1T cell to 1.5T (split-gate structure) and 2T cells for high-performance/low power, and
(2) Discrete charge-trapping cell technology (MONOS and Nano-dot[5]) for higher reliability.

These trends prove a deviation from the standardized stand-alone flash memory products such as NAND-flash memories. Although the conventional floating-gate NOR structure will survive in some of the high-density embedded uses, pervasive use of split-gate and charge-trapping storage structures are expected according to the requirements by diversified MCU market segments.

Considering the structural compatibility of the split-gate with advanced underlying CMOS logic transistor, Access-Gate first (Control-Gate last), which is only realized by a thin-film charge-trapping storage layer, not by stacked floating gate structure, is preferred. This may indicate that the technological convergence point lies in the split-gate, charge-trapping cell structure in the future. The advantage of this choice has been proved by the implementation in the CMOS logic platforms at 90nm [6],[7].

III. STATE-OF-THE-ART MCU DESIGN

In addition to the over-all cost reduction through design, production and inventory control by programmability in Flash-MCU, the "embedded-ness" is favorably utilized for high performance and data security properties. The advantages and possible drawbacks of embedded flash memory are listed in Table.1. Product design should consider utilizing embedded flash properties with the overall cost and value advantages over-coming drawbacks of higher wafer process cost to incorporate flash memory. In general much diversified MCU product line-ups are efficiently supported only by a unified design platform to employ optimized eFlash macros.

By scaling the device and by circuit developments, a steady scaling trend has been realized in the auto-motive applications of MCUs (Figs.4 and 5), with x20 CPU performance growth by 10 years and x8 ROM capacity growth by 10 years. The techno-logy node development for MCU is not so aggressive as most-advanced SOCs because of the smaller system on the chip and somewhat tailored technology for required higher temperature operations.

Advantages of embedded flash memory:
 - **By internal access path,**
   (1) Fast, low-power
   (2) High data security
   (3) High-reliability, low EMI, low system cost
   (4) Higher design freedom
      memory capacity, interface, designed functions, and operating voltage.
 - **By low activation rate, act as thermal cooler.**
 - **Contributes to optimization of LSI functions and cost.**

Possible drawbacks in embedded flash memory:
   (1) Higher cost in large and small capacity of memory (owing to low density, higher process cost)
   (2) Cost by non-standardization
   (3) Single source
   (4) Difficult in integrating multiple types of memory.

Table 1. Advantages and possible drawbacks of embedded flash memory, as compared with stand-alone flash memory.

Example requirements to current state-of-the-art MCU designs are shown in Table.2. Because eFlash in MCU is inherently required a high-speed access to meet the CPU execution speed, performance-oriented design is required in many applications, which is quite different from stand-alone flash memory products. Also data reliability as well as high-temperature and low-leakage product strategy are important factors in most of MCU applications.

Requirements here describe the natures of current MCU market segments as well as eFlash specifications.
(1) Auto-motive applications such as power-train require high-density, high-temperature and highly reliable eFlash designs. A random access as fast as 10ns at 160□(max) in eFlash is achieved by the current state-of-the-art design with

978-1-4244-9019-6/11 $26.00 © 2011 IEEE

hierarchical sensing, optimized memory mat division, and highly reliable memory cell technology to fit this application.

(2) Security functions against attacks account for a large portion of the MCU design for smart-IC card applications. Embedded EEPROM for data manipulation functions with quite fast and low-power program/erase operation is essential in non-contact smart-cards, which makes this field of application very selective in the choice of eFlash technologies.

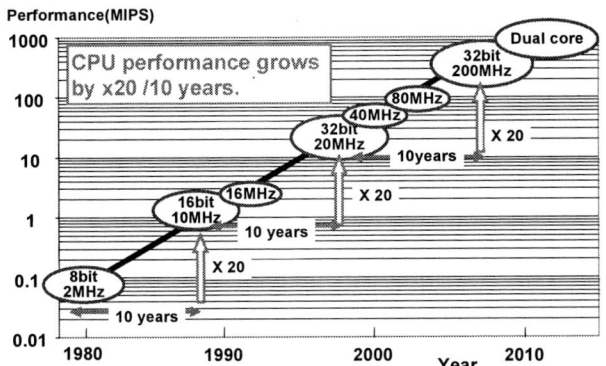

Fig.4. Performance trend in CPU and eFlash for high-end auto-motive market.

(3) Very low-voltage/low-power MCU products are required for emerging medical/health-care applications with battery operations. eFlash program/erase is necessary in the future data collection and storage for data streaming operations in the advanced health-care environments.

These sample requirements indicate much diversified MCU product line-ups, suggesting very wide expectations to eFlash performance and functions, much to be explored and exploited in technology/ circuit/system developments.

Fig. 6 depicts key architecture and circuit technologies in eFlash designs. In frequent read operations in the code storage applications, non-boosted word-line utilizing a split-gate cell structure is promising to enhance the random read access speed and for low-power consumptions. Lowe-power design in eFlash is strongly affected by the charge-pumping circuitry in program/ erase operations. Because flash memory technologies have inherent difficulty in scaling down the program/ erase voltage, energy-efficient program/erase algorithms and optimally generated high-voltage waveforms to mitigate excessive power consumption are important in power-aware designs.

Fig. 7 shows a current MCU product line-up employing a split-gate MONOS eFlash technology, with up to 100MHz read access at code storage with 500K program/erase capability in EEPROM on the chip, all under Tj=160□(max). The operating voltage ranges down to 1.62V for battery operations.

Fig.5. Trend in eFlash capacity for high-end auto-motive market.

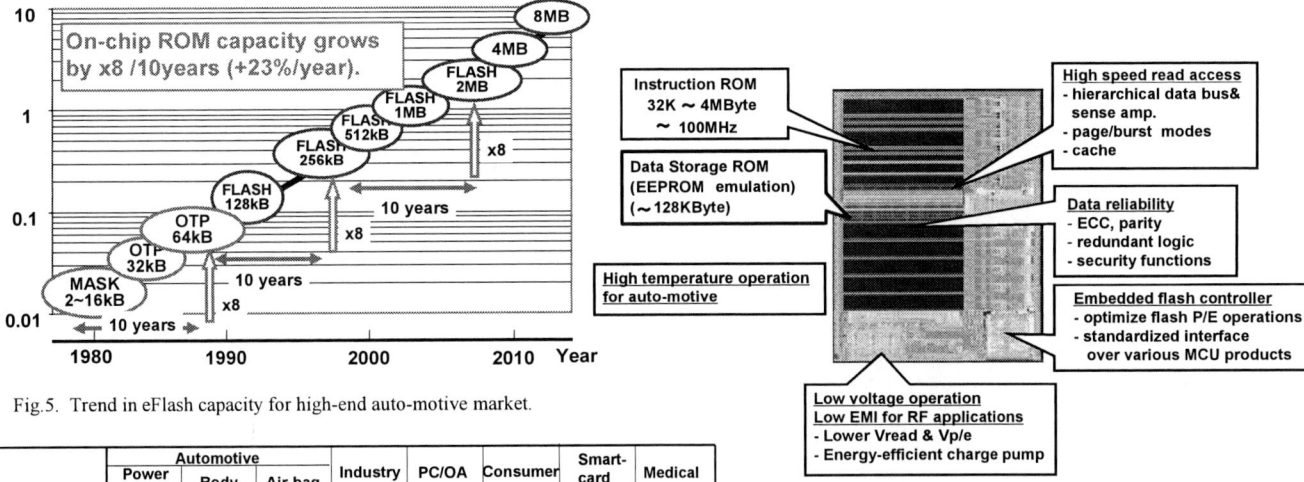

Fig.6. Key considerations in eFlash design [8],[9].

- 90nm CMOS w/i MONOS eFlash
- CPU : 240MHz(max)
- eFlash(split-gate MONOS)
  - Code Flash : 4M Byte(max)
    (freq. =100MHz @ random read)
  - EEPROM : 128K Byte(max)
- RAM:128kB max, A/D:12bit x 37ch etc.
- Applications: auto-motive, industry, consumer, PC/OA, smart-card etc.

		Automotive			Industry	PC/OA	Consumer	Smart-card	Medical
		Power Train	Body	Air-bag					
MCU	Performance (frequency)	~300MHz	150~ 200MHz	100MHz	~300MHz	25~ 50MHz	20~ 100MHz	15~ 50MHz	1~ 10MHz
	Power	0.5mA /MHz	0.5mA /MHz	0.25mA /MHz	1mA /MHz	0.5mA /MHz	0.25mA /MHz	0.2mA /MHz	0.1mA /MHz
	Temp.(Ta)	- 40 ~ 125 C			max 85 C		- 20 ~ 85 C		
	Density	8MB	2MB	2MB	1MB	2MB	1MB	512KB	256KB
FLASH	P/E cycle	Code : 1K-10K cyc. / Data : 100K cyc.(EEPROM)						-500K (EEPROM)	100K (EEPROM)
	Small Cell	✓							
	Small Macro			✓		✓	✓	✓	✓
	Fast Access	✓			✓				

Table.2. Requirements for embedded flash memory in MCU applications.

Fig.7. 90nm flash-MCU products.

978-1-4244-9019-6/11 $26.00 © 2011 IEEE   95

IV. IMPACT OF ENERGY-EFFICIENT NV- MEMORY

Energy-efficient system approaches require low-energy frequent re-write performance of non-volatile memories. As the stand-by leakage problem has emerged as a critical factor in the advanced LSI systems, embedded non-volatile memory will play more important roles in reducing the system power by intermittent power switching schemes (Fig. 8), to frequently switch idling states into power-off states. Frequent power On/Off necessitates frequent non-volatile store and retrieval of circuits states and data. Fast and energy-efficient re-write is required for non-volatile memories in this context.

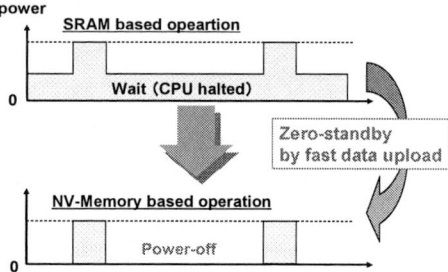

Fig. 8. Intermittent system operations with normally off schemes.

Fig. 9 describes re-write speed and energy in various eFlash technologies compared with Magnetic-RAM. Because system requirements are diverse in power control schemes, all the current eFlash and next-generation NV-memories are good for uses in intermittent power control applications. However, efficient power-down schemes will store and retrieve circuit states instantly, favoring distributed fast non-volatile storage on the chip. One remarkable feature realized by emerging NV memory is orders-of-magnitude lower energy per bit re-write than existing eFlash by fast and low-voltage re-write capabilities (Fig. 10), which can be exploited effectively [10].

Programmability provided on the chip has been an important factor in the design and cost structure of LSI (Table 3). Beginning with the ROM-based logic operated by stored instructions in the CPU, alterable/reconfigurable logic organizations have emerged in the 2nd stage, where the main players are Flash-MCU and re-configurable logic products. The 3-rd stage realized by NV-RAM will see a much broader possibility of innovation by energy-efficient NV memories.

V. CONCLUUSIONS

Flash-MCU has achieved a rapid market penetration attributed to the leap in value/cost for innovation by virtue of evolving eFlash technologies. eFlash cell and circuit technologies will see some convergence points quite different from stand-alone flash memory, according to diversified market requirements. Designs of energy-efficient MCU and of MCU for energy-efficient systems will be the next focus of technology, circuit, and system co-development in emerging applications, where non-volatile memory technologies will play important roles in the scaled LSI environments.

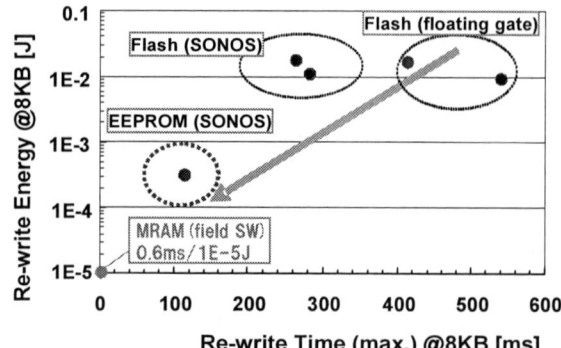

Fig.9. Re-write performance by energy/bit in various eFlash and MRAM. Re-write energy and time by memory cell and periphery circuitry.

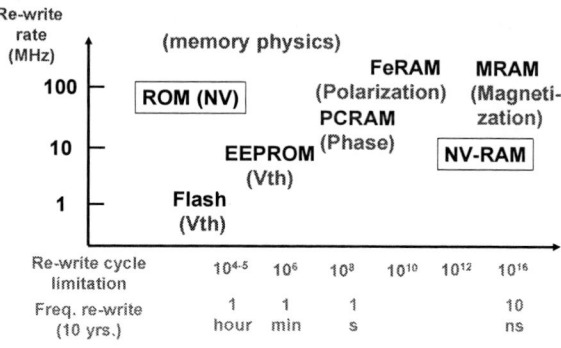

Fig. 10. Re-write performance by non-volatile memories. Re-write frequency requirement ranges over power On/Off schemes in the system.

	Innovation	Main Enabler	Product	Effect
1	Memory-based Logic	ROM Program Register-based comput.	MPU, MCU	Programmable Logic
2	Alterable Logic	SRAM/Flash	Flash-MCU, FPGA	Re-configurable Production, Inventory, Delivery Efficiency
3	Universal Memory	NV-RAM w/i energy-efficient re-write	Unified-Memory/ MCU, SOC	Instant ON Intermittent operation Re-usable logic

Table 3. Evolution of on-chip programmability [9].

ACKNOWLEDGEMENT

The author would like to express sincere thanks to all the members of Embedded Memory Core Development Division, Renesas Electronics Corp. for their supports.

REFERENCES

[1] S. Kianian et al., Symp. VLSI Tech. Dig. Tech. Papers, pp.71-72 (1994).
[2] B. Eitan et al., Proc. Int. Conf. Solid State Devices and Materials, pp. 522-524 (1999).
[3] J. A. Yater et al., NVSMW, pp. 77-78 (2007).
[4] H. M. Lee et al., Dig. NVSMW, pp. 15-16 (2006).
[5] K. Baker, Proc. ICICDT, pp.185-189 (2009).
[6] J. A. Yater et al, International Memory Workshop (2009).
[7] W. Stenzl and J. Hupper, in EE-Times Europe, March 17 (2008).
[8] M. Hatanaka and H. Hidaka, Proc. Tech. Papers, ASSCC, pp.38-41 (2007).
[9] H.Hidaka, in chap. 7, ed. K. Zhang, "Embedded memories for nano scale VLSI", Springer (2009).
[10] M. Zwerg et al., Dig. Tech. Papers, ISSCC, pp. 334-335 (2011).

978-1-4244-9019-6/11 $26.00 © 2011 IEEE

Impacts of Intrinsic Device Variations on the Stability of FinFET Subthreshold SRAMs

Yin-Nien Chen, Chien-Yu Hsieh, Ming-Long Fan, Vita Pi-Ho Hu, Pin Su and Ching-Te Chuang

Abstract—In this work, we investigate the impacts of intrinsic device variations on FinFET subthreshold SRAMs, including the conventional tied-gate 6T SRAM, tied-gate 10T Schmitt Trigger based SRAMs, and recently proposed independent-gate controlled 8T Schmitt Trigger based SRAMs. The impacts of intrinsic random device variations, including Fin Line-Edge Roughness (LER) and Work Function Variation (WFV), on the device threshold voltage V_{th} , Subthreshold Swing (S.S.) and stability of FinFET SRAMs operating in subthreshold region are assessed using 3D atomistic mixed-mode Monte-Carlo simulations. The results indicate that Fin LER is the dominant factor limiting the stability of FinFET subthreshold SRAMs, since Fin LER degrades both V_{th} fluctuation and S.S., while WFV mainly affects only V_{th} fluctuation. The independent-gate controlled Schmitt Trigger SRAMs are shown to offer adequate stability for the intended subthreshold applications even considering intrinsic device variations.

Index Terms— Subthreshold SRAM, Schmitt Trigger, Line-Edge Roughness (LER), Work Function Variation (WFV)

I. INTRODUCTION

As the scaling of bulk CMOS devices reaches the physical limit, FinFET device emerges as a promising candidate to extend scaling due to its superior EI (Electro-Integrity), better SCE (Short-Channel-Effect) and subthreshold slope, reduced leakage, and immunity to RDF (Random-Dopant Fluctuation) [1] by using lightly-doped or undoped silicon fin. For ultra-low-power applications such as portable devices, implanted medical instruments, and wireless body sensing networks, operating at below threshold voltage is an effective solution [2] to reduce both static and dynamic power consumption. However, the stability of conventional tied-gate 6T FinFET SRAM (Fig. 1(a)) deteriorates significantly in subthreshold region [3]. Schmitt Trigger based SRAMs have been proposed to enhance Read Static Noise Margin (RSNM), Write-ability, and the tolerance to process and device variations, and shown to be advantageous for subthreshold operation [4, 5]. As shown in Fig. 1(b) [4] and Fig. 1(c) [5], these 10T Schmitt Trigger SRAM cells (designated as ST1 and ST2, respectively) add stacking transistors (NL1 and NR1) and feedback transistors (NFL/NFR in Fig. 1(b), and AXL2/AXR2 in Fig. 1(c)) to provide the feedback mechanism for conditioning the intermediate node to raise the cell-inverter trip voltage for rising input, thus improving RSNM. Recently, several independent-gate controlled FinFET Schmitt Trigger SRAM cells (shown as

IG_ST1, IG_ST2, and IG_ST3 in Fig. 2(a), 2(b), and 2(c), respectively) have been proposed [6]. By splitting the front-gate and back-gate of NL1 (NR1), where the front-gate is used as the stacking device, and the back-gate as the intermediate node conditioning device to provide built-in feedback mechanism for Schmitt Trigger action, these 8T independent-gate controlled FinFET Schmitt Trigger SRAM cells reduce the cell transistor count/area while achieving improved RSNM and better tolerance to process and device variations.

On the other hand, with technology scaling down to deca-nanometer scale, the local random intrinsic device variations such as Random Dopant Fluctuation (RDF), Line-Edge Roughness (LER), Work Function Variation (WFV) and nonuniformity of interface become especially important [7, 8]. Among these local random variations, Fin LER and WFV have become the dominant factors for FinFET device variability as LER, which does not scale with technology, approaches device critical dimensions; and the grain size of the metal-gate becomes significant with respect to the channel length and thus WFV induced by the grain surface orientations becomes increasingly important for the state-of-the-art high-k/metal-gate MOSFETs.

In this work, we evaluate the impacts of Fin LER and WFV on threshold voltage V_{th} and S.S. of FinFET devices. The stability and variability of FinFET SRAM cells, including the conventional tied-gate 6T cell, tied-gate 10T Schmitt Trigger SRAM cells, and independent-gate controlled 8T Schmitt Trigger SRAM cells are evaluated for subthreshold operation.

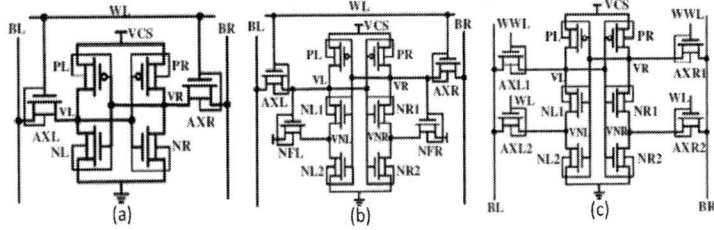

Fig. 1. Schematic of various FinFET cells: (a) conventional tied-gate 6T, (b) tied-gate 10T Schmitt Trigger (ST1) [4] , (c) tied-gate 10T Schmitt Trigger (ST2) [5] .

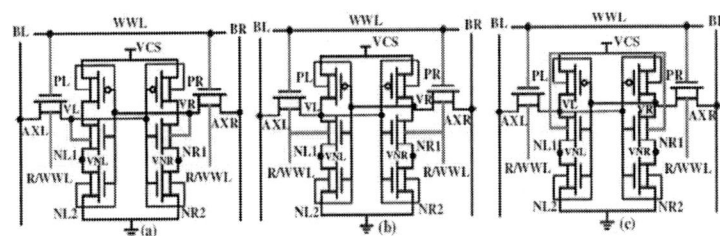

Fig. 2. Schematic of independent-gate controlled 8T Schmitt Trigger FinFET cells: (a) IG_ST1, (b) IG_ST2, (c) IG_ST3.

This work was supported in part by the Ministry of Economic Affairs in Taiwan under Contract 98-EC-17-A-01-S1-124, and in part by the Ministry of Education in Taiwan under ATU Programs. The authors are with the Department of Electronics Engineering, National Chiao Tung University, Hsinchu 300, Taiwan (e-mail: snoopyfairy@gmail.com; chingte.chuang@gmail.com)

978-1-4244-9019-6/11 $26.00 © 2011 IEEE

II. INDEPENDENT-GATE CONTROLLED SCHMITT TRIGGER SRAM CELLS

A. Operations

The proposed IG_ST1 cell (Fig. 2(a)) forms Schmitt Trigger feedback path by connecting the back-gate of NR1 (NL1) to cell storage node VR (VL). During Read operation (assume VL=0 VR=1), the feedback mechanism is enabled with VNR conditioned to higher voltage by the back-gate of NR1, thus increasing trip voltage of the cell inverter (PR-NR1-NR2) and improving RSNM. Notice that as VL rises and VR falls, the feedback mechanism becomes weaker and the switching slope (steepness) of IG_ST1 cell would degrade. For access pass-transistor AXL (AXR), the split-gate topology is used, so only one gate is enabled during Read to reduce Read disturb, while both gates are enabled during Write to improve Write-ability and performance. During Write operation (assume VL=0 VR=1), due to reduced NL1 strength with its back-gate connected to VL (= 0), and the series NL1-NL2 pull-down configuration, the trip voltage of the left cell inverter (PL-NL1-NL2) is raised, thus further improving the Write-ability. In IG_ST2 (Fig. 2(b)), the back-gates of NR1 (NL1) and AXR (AXL) are connected to the R/WWL thus providing a firmer intermediate node conditioning action, and a steeper switching transition than IG_ST1 cell for the back-gate of NR1 is always "High" during Read and Write, but this may slightly degrades the Write-ability with respect to IG_ST1 cell. In IG_ST3 cell (Fig. 2(c)), the back-gates of NR1 (NL1) are connected to V_{CS}. Therefore, IG_ST3 would have better HSNM, and the same RSNM and WSNM compared with IG_ST2 cell.

B. Stability

Our following analyses are based on FinFET device as shown in Fig. 3 with $N_A=1\times10^{17}$cm-3, $L_{eff}=25$nm, $W_{fin}=7$nm, $H_{fin}=20$nm and EOT=0.65nm. The threshold voltage of the devices V_{TN} and V_{TP} are both near 0.41V. The Read Static Noise Margin (RSNM), Write Static Noise Margin (WSNM), and Hold Static Noise Margin (HSNM) as defined in Fig. 4 are compared in Fig. 5(a), 5(b), and 5(c), respectively.

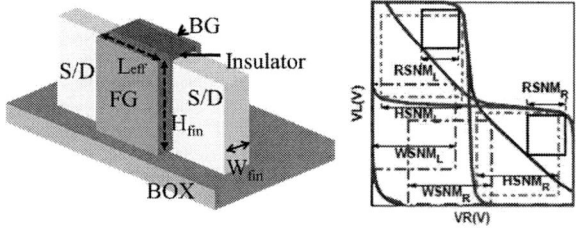

Fig. 3. FinFET device structure. & Fig. 4. Voltage transfer characteristic curves used to calculate SNM: Read, Write, and Hold

RSNM – In Fig. 5(a), the normalized nominal RSNM of different cells are compared in subthreshold region ($V_{CS} = 0.4$V). As shown in the figure, IG_ST2 and IG_ST3 have the most significant improvement over the conventional tied-gate FinFET 6T SRAM in the nominal RSNM (~81%) than other Schmitt Trigger based cells due to their stronger intermediate node conditioning action (steeper switching transition) and reduced Read-disturb since only the back-gate of the access pass-transistor is on during Read operation.

WSNM – In Write mode (Fig. 5(b)), Schmitt Trigger based cells also show better nominal WSNM (1% to 33%) over the conventional tied-gate 6T SRAM. The improvement is most significant for ST2 cell due to its two parallel discharging paths for

cell internal nodes and tied-gate pass-transistor configuration.

HSNM – In Hold mode (Fig. 5(c)), IG_ST1 and IG_ST2 have slightly lower nominal HSNM due to the split-gate configuration of NL1 (NR1) which slightly degrades the switching slope (steepness). On the other hand, IG_ST3 exhibits HSNM comparable to (1% better) tied-gate 6T cell since it preserves the Schmitt Trigger feedback mechanism in Hold mode as discussed in part A.

Notice that WSNM and HSNM do not appear to be limiting factors on SRAM stability, while RSNM does. Thus, we will focus on the RSNM when considering the impacts of intrinsic device variations in section III.

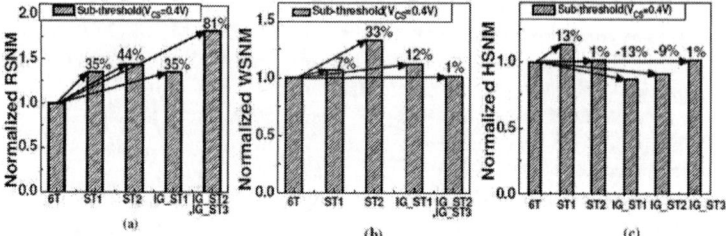

Fig. 5. Comparison of normalized nominal (a) RSNM, (b) WSNM, and (c) HSNM for different cells.

C. Area

Based on published design rules of 32 nm technologies [9] and scaling factor from ITRS Roadmap, the cell area of various FinFET SRAM cells are estimated and compared in Fig. 6. For the tied-gate 6T thin-cell layout [10], the area is 0.09 μm^2. For ST1 and ST2 cells, extra feedback (NFL/NFR and AXL2/AXR2) and stacking (NL1/NR1) transistors result in increase of 69% and 50% in horizontal and vertical dimension, respectively. Furthermore, extra Metal-2 track is required to connect the internal nodes. In contrast, the independent-gate controlled Schmitt Trigger SRAM cells reduce the areas occupied by the two feedback transistors (horizontal dimension) and the contacts at NL2/NR2 drain side (vertical dimension). As shown in Fig. 6, IG_ST1, IG_ST2/IG_ST3 cells can save 30% - 39% area compared with ST1 and ST2 cells.

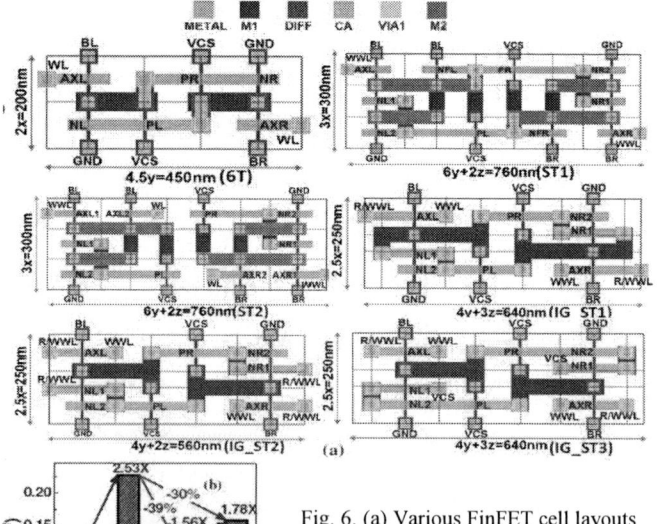

Fig. 6. (a) Various FinFET cell layouts and (b) comparison of cell areas of different FinFET cells.

III. IMPACTS OF FIN LER AND WFV ON SUBTHRESHOLD FINFET SRAM CELLS

In this section, we first introduce the methodology for assessing the impacts of WFV on the stability of FinFET SRAM cells. We then compare the impacts of Fin LER and WFV on the variability of the threshold voltage and subthreshold swing of FinFET devices. Finally, we analyze the impacts of Fin LER and WFV on the stability of FinFET subthreshold SRAMs using 3D atomistic TCAD mixed-mode Monte-Carlo simulations with 100 samples for each case.

A. Methodology for assessing WFV on subthreshold FinFET SRAM cells

The WFV depends on the metal-gate material [10] used as each material has its own distinct grain size and orientation dependency. For example, MoN has the <110> (probability 60%) and <112> (probability 40%) grain orientations with the associated work function values of 5.0eV and 4.4eV, respectively; while TiN has the grain orientations <200> (probability 60%) and <111> (probability 40%), but with more close associated work function values of 4.6eV and 4.4eV, respectively. Thus, TiN has the intrinsic superiority (over MoN) with lower WFV if the grain size is assumed to be the same. The other factor influencing the WFV is the manufacture processing of gate formation that may change the grain size of the gate material. Here, as shown in Table-1, we choose TiN as the gate material for both N/PMOS to emphasize the physically achievable intrinsic limit that WFV may have on the stability of the analyzed cells.

We construct the physical FinFET structures with the metal-gates randomly populated with two different grain orientations <200> and <111> with the probability of 60% and 40%, as shown in Fig. 7(a) designated by the yellow colors and orange colors, respectively. With 200 samples, from the corresponding Id-Vg dispersion curves (Fig. 7(b)), the spreading of the V_{th} (Fig. 7(c)) can be obtained and translated into the distribution of effective work function for N/PMOS (Fig. 7(d)) respectively, due to the parallelism of the subthreshold slopes. For the follow-on SRAM-level mixed-mode Monte-Carlo simulations, the effective work function values are sampled from the library of distribution of work function built (Fig. 7(d)).

Grain Size	3nm
Gate Material	TiN
Grain Orientation	<200> 60% <111> 40%

Table-1: Parameters of the gate material used in the analysis.

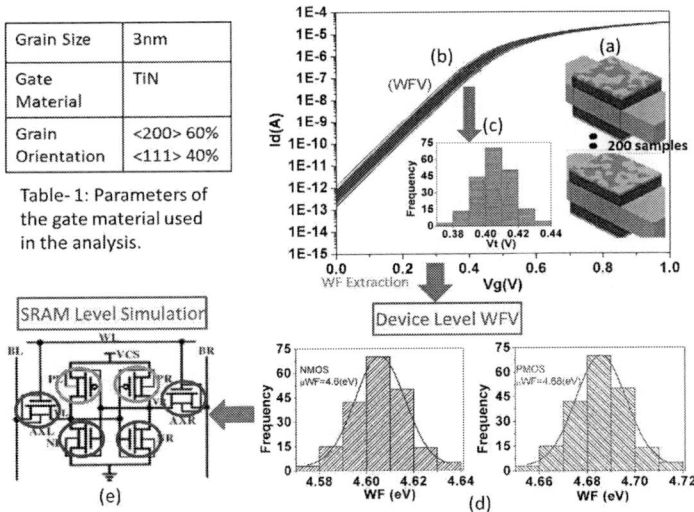

Fig. 7. Methodology flow for modeling WFV.

B. Device level impacts of Fin LER and WFV

Once the metal-gate material is chosen, the distribution of the work function corresponds to the translational shift of Id-Vg dispersion curves (Fig. 7(b)) in the subthreshold region. The Subthreshold Swing (S.S.) are almost the same (Fig. 7(b)) for these samples with the same effective work function for both the front-gate and the back-gate. On the other hand, since Fin LER affects the effective fin width and therefore electrostatic integrity, it impacts not only ΔV_{th} but also $\Delta S.S$. We analyze the impacts on device characteristics between Fin LER and WFV based on ΔV_{th} and $\Delta S.S$ to differentiate their major differences.

In order to assess the contribution of V_{th} and S.S. within the operation range separately, we use constant current method at high enough current level to derive the V_{th} near the operation voltage ~0.4V to isolate V_{th} from S.S.. In Fig. 8(a) and 8(b), the σV_{th} and $\sigma S.S$. due to Fin LER and WFV are compared. As can be seen, the $\sigma V_{th,WFV}$ is comparable to $\sigma V_{th,FinLER}$, while $\sigma S.S_{,WFV}$ is one order of magnitude smaller than $\sigma S.S_{,FinLER}$.

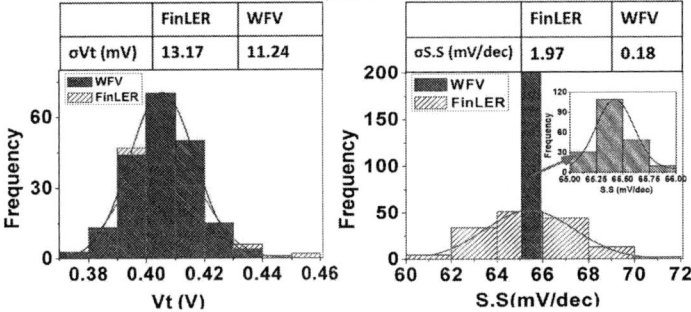

Fig. 8. (a) The distribution of Vt caused by Fin LER and WFV, and (b) The distribution of S.S. caused by Fin LER and WFV.

The sub-threshold drain current can be expressed as [12]

$$I_D \propto \exp\left(\frac{V_{GS} - V_{th}}{S.S}\right) \quad (1)$$

Thus the σV_{th} and $\sigma S.S$. can be presented by the percentage of I_D offset from the nominal drain current as shown in the vertical axis of Fig. 9. From equation (1), the corresponding I_D offset percentage due to the variance of S.S. depends on the gate voltage overdrive (V_{GS}) as indicated in the horizontal axis of Fig. 9. We can see that at V_{GS} ~0.4V, the I_D offset percentage due to Fin LER is comparable to that due to WFV since their σV_{th} values are comparable. As the gate voltage overdrive reduces, the contribution of the Fin LER increases significantly, far beyond that of the WFV, because $\sigma S.S_{,FinLER}$ becomes much more important to the I_D offset percentage under lower gate voltage.

Since the transition region in Voltage Transfer Curve (VTC) near half V_{CS} is critical for SRAM stability, it is clear from Fig. 9 that the additional $\Delta I_{D,S.S}$ caused by Fin LER will cause serious degradation of SRAM stability as discussed below.

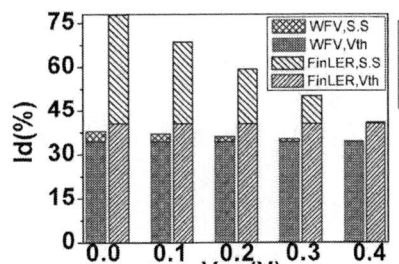

Fig. 9. Percentage of I_D offset (composed of $\Delta I_{D,S.S}$ and $\Delta I_{D,Vth}$) from the nominal drain current caused by Fin LER and WFV versus gate overdrive voltage.

C. Impact of Fin LER and WFV on the stability of subthreshold FinFET SRAM cells

Fig. 10 and Fig. 11 illustrate the variations of butterfly curves (at $V_{CS} = 0.4V$) caused by Fin LER and WFV for different cells. 3D atomistic TCAD mixed-mode Monte-Carlo simulations with 100 samples for each case are analyzed. The probability distribution of the RSNM (at $V_{CS} = 0.4V$) of various cell structures considering Fin LER and WFV are compared in Fig. 12(a) and 12(b), respectively. From the butterfly curves and the σ from the tables in Fig. 12, the noticeable difference between Fin LER and WFV is the spreading near the transition region as discussed previously. For Fin LER, ΔI_D is much more severe due to its large $\Delta I_{D,S,S}$, thus causing more serious mismatch of the driving capability of N/PMOS. It is worthwhile to mention that the deviation caused by WFV for IG_ST cells is smaller than that for the conventional tied-gate 6T cell since WFV mainly affect V_{th} and has less influence on EI (Electrostatic Integrity). It can be seen that Fin LER represents the dominant factor for RSNM variation (its μ/σ ratio is almost half of that of WFV for the same cell). And we can see that except for IG_ST1 cell (μ/σ ratio = 5.45), the IG_ST2 and IG_ST3 cells (μ/σ ratio = 9.52) can provide significantly better margin than the conventional tied-gate 6T cell (μ/σ ratio = 5.83), and comparable to that of ST1 (μ/σ ratio = 8.62) and ST2 (μ/σ ratio = 11.06). This is because IG_ST1 cell operating in independent-gate mode has worse electrostatic integrity than the tied-gate mode [13], and its nominal RSNM improvement (over 6T cell) is less than IG_ST2 and IG_ST3 cells due to its softer (less steep) switching characteristics.

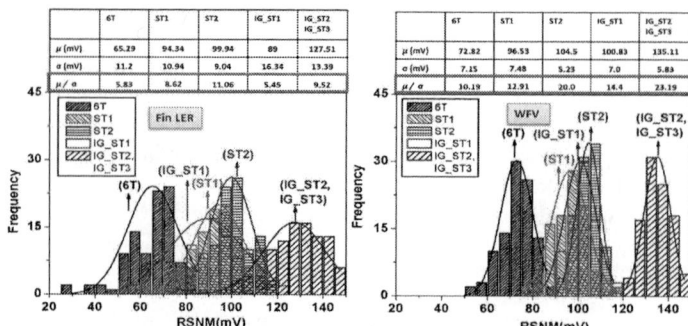

	6T	ST1	ST2	IG_ST1	IG_ST2 / IG_ST3
μ (mV)	65.29	94.34	99.94	89	127.51
σ (mV)	11.2	10.94	9.04	16.34	13.39
μ / σ	5.83	8.62	11.06	5.45	9.52

	6T	ST1	ST2	IG_ST1	IG_ST2 / IG_ST3
μ (mV)	72.82	96.53	104.5	100.83	135.11
σ (mV)	7.15	7.48	5.23	7.0	5.83
μ / σ	10.19	12.91	20.0	14.4	23.19

Fig. 12. Probability distribution of RSNM at Vcs=0.4V considering (a) Fin LER (b) WFV for different SRAM cell structures from 3D atomistic mixed-mode Monte-Carlo simulations.

IV. CONCLUSION

The impacts of intrinsic device variations on the stability of FinFET subthreshold SRAMs including the conventional tied-gate 6T SRAM, tied-gate 10T Schmitt Trigger based SRAMs, and recently proposed independent-gate controlled 8T Schmitt Trigger based SRAMs were investigated and compared. The 8T IG_ST2 and IG_ST3 cells were shown to have superior RSNM (~81% improvement) and comparable WSNM compared with conventional tied-gate 6T cell while significantly reducing the cell area (30% to 39%) compared with the 10T ST1 and ST2 cells. 3D atomistic mixed-mode Monte-Carlo simulations were performed to evaluate the impacts of Fin LER and WFV. Due to its impact on the subthreshold swing and the resulting I_D variation ($\Delta I_{D,S,S}$), Fin LER was shown to be the dominant factor limiting the stability of subthreshold FinFET SRAM cells. The independent-gate controlled Schmitt Trigger SRAMs were shown to offer adequate stability for the intended subthreshold applications even considering intrinsic device variations.

Acknowledgment

This work was supported in part by the Ministry of Economic Affairs in Taiwan under Contract 98-EC-17-A-01-S1-124, and in part by the Ministry of Education in Taiwan under ATU Programs.

References

[1] E. Baravelli, M. Jurczak, N. Speciale, K. D. Meyer, and A. Dixit, "Impact of LER and Random Dopant Fluctuations on FinFET Matching Performance," *IEEE Transactions on Nanotechnology*, vol. 7, no. 3, pp.291-298, May. 2008.

[2] S. Hanson, M. Seok, D. Sylvester and D. Blaauw, "Nanometer Device Scaling in Subthreshold Logic and SRAM," *IEEE Transaction on Electron Devices*, vol. 55, no. 1, pp. 175-185, Jan. 2008.

[3] T. H. Kim, J. Liu, J. Keane and C. H. Kim, "A High-Density Subthreshold SRAM with Data-Independent Bitline Leakage and Virtual Ground Replica Scheme," in *Dig. Tech. Papers, ISSCC*, pp. 330, Feb. 2007.

[4] J. P. Kulkarni, K. Kim, and K. Roy, "A 160 mV Robust Schmitt Trigger Based Subthreshold SRAM," *IEEE J. Solid-State Circuits*, vol. 42, no.10, pp. 2303-2313, Oct. 2007.

[5] J. P. Kulkarni, K. Kim, and K. Roy, "Process Variation Tolerant SRAM Array for Ultra Low Voltage Applications," *Proc. Design Automation Conference*, pp. 108-113, Jun. 2008.

[6] Chien-Yu Hsieh, Ming-Long Fan, Vita Pi-Ho Hu, Pin Su, and Ching-Te Chuang, "Independently-Controlled-Gate FinFET Schmitt Trigger Sub-threshold SRAMs, *Proc. 2010 IEEE International SOI Conference*, pp. 133-134, Oct. 2010.

[7] E. Baravelli, A. Dixit, R. Rooyackers, M. Jurczak, N. Speciale, and K. D. Meyer, "Impact of Line-Edge Roughness on FinFET Matching Performance," *IEEE Transactions on Electron Devices, vol. 54, no. 9, pp. 2466-2474, Sep. 2007.

[8] K. Ohmori et.al, "Impact of Additional Factors in Threshold Voltage Variability of Metal/High-k Gate Stacks," *IEEE IEDM.*, pp.1-4, 2008.

[9] S. Natarajan et.al, "A 32nm logic technology featuring 2nd-generation high-k + metal-gate transistors, enhanced channel strain and 0.171μm² SRAM cell size in a 291Mb array" *IEDM*, pp. 1-3, Dec. 2008.

[10] F. Bauer et.al, "Layout Options for Stability Tuning of SRAM Cells in Multi-Gate-FET Technologies," *ESSCIRC*, pp.392-395, Sep. 2007

[11] H. Dadgour, K. Endo, V. De, K. Banerjee, "Modeling and Analysis of Grain-Orientation Effects in Emerging Metal-Gate Devices and Implications for SRAM Reliability," IEDM, pp.1-4, 2008.

[12] Y. Tsividis, *Operation and Modeling of the MOS Transistor*. London, U.K.: Oxford Univ. Press, 1999.

[13] Z. Lu and J. G. Fossum, "Short-Channel Effects in Independent-Gate FinFETs," *IEEE Electron Devices Letter*, vol. 28, no. 2, pp. 145-147, Feb.2007.

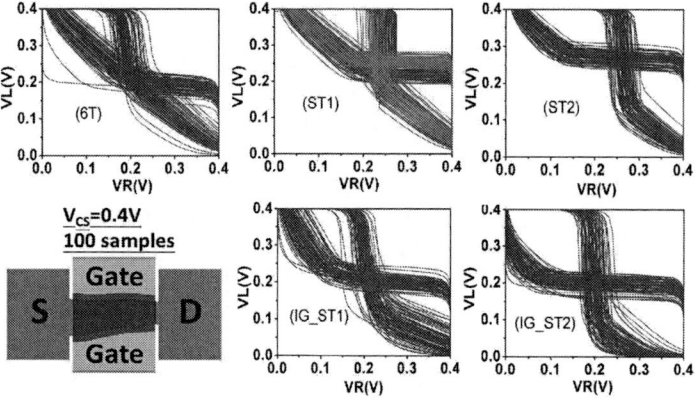

Fig. 10. Butterfly curves of various cells considering Fin LER from 3D atomistic mixed-mode Monte-Carlo simulations.

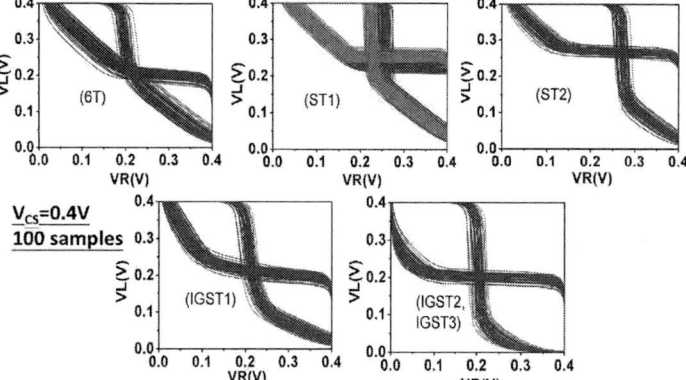

Fig. 11. Butterfly curves of various cells considering WFV from 3D atomistic mixed-mode Monte-Carlo simulations.

Low-Cost Embedded Flash Memory Technology

Wein-Town Sun[*], Cheng-Jye Liu[*], Chun-Yuan Lo[*], Yun-Jen Ting[*], Ying-Je Chen[*], Tai-Yi Wu[*], Eng-Huat Toh[**], Xiao-Hong Yuan[**], Ko-Li Low[**], Qiu Han[**], Young-Seon You[**], Ying-Keung Leung[**], and Swee-Tuck Woo[**]

Abstract—A simple and low cost logic based single poly Flash memory technology, NeoFlash®, with fast programming and high reliability is demonstrated in this paper. Programming with channel hot-hole-induced hot-electron injection and erasure with uniform channel Fowler-Nordheim tunneling are utilized to achieve fast programming, high endurance and good reliability characteristics. Owing to its simple cell structure and operation schemes, only 3 additional non-critical masks are needed, and the complexity of process integration and device tuning is much reduced. The SONOS based technology has been successfully embedded into 0.35 μ m ~ 65nm CMOS logic process. Because of electrons stored in nitride layer of ONO film, no tail bit during endurance and retention test is observed. As a result, NeoFlash® is a promising embedded Flash technology for SoC applications.

Index Terms—Nonvolatile memory, SONOS devices, Flash memory

I. INTRODUCTION

IN order to meet the slim and light demand of portable electronics, embedded non-volatile memory (NVM) becomes more and more indispensable. As a result, a logic process compatible NVM is a basic requirement to meet the portable electronics. Because of the additionally complex process, it is more difficult for embedded floating gate NVM to be implemented quickly into logic and its derivative processes. As a result, Silicon-Oxide -Nitride-Oxide-Silicon (SONOS) devices would be a good solution due to less additional mask process and introduced thermal cycles. On the other hand, the charges for SONOS device are stored in the isolation traps of nitride layer. If there is a stress induced defect generated in the tunneling oxide during operation, only few charges near the defects are lost. On the contrary, charge loss due to stress induced leakage current is more serious for floating gate NVM. NeoFlash®, a kind of SONOS technology, is a truly logic-based single-poly p-channel embedded SONOS Flash as shown in Fig. 1[1]. Owing to its simple cell structure, the complexity of process integration and device tuning is much reduced. At most, three non-critical masking layers are added. Moreover, because NeoFlash is a SONOS type NVM, no tail bit during endurance test is observed. In comparison with other embedded Flash technologies with complicated structure, expensive process and time-consuming development, NeoFlash is quite promising in response to embedded Flash demands in SoC

The authors marked with "*" are with Ememory Technology Inc., Hsinchu County 30265 Taiwan, R.O.C. (e-mail: wtsun@ ememory.com.tw).
The authors marked with "**" are with the GLOBALFOUNDRIES Singapore Pte. Ltd., 60 Woodlands Industrial Park D Street 2, Singapore 738406.

applications. In the paper, we provide an overview of NeoFlash technology, including cell operation, cell and array characteristics, and reliability. The major content is based on the result of 0.18 μ m NeoFlash technology. Furthermore, 65nm NeoFlash cell performance is also demonstrated at the first time.

II. CELL OPERATION AND CHARACTERISTICS

A. Cell Operation

The NeoFlash programming (PGM) mechanism is the "channel-hot-hole induced hot-electron injection (CHHIHE)". The CHHIHE PGM methodology was first reported in 1992 [1]. As presented in that work, figure 2 shows drain current and gate current characteristics of PMOS transistors at various gate voltages. The CHHIHE PGM method applies a voltage near the threshold voltage (Vt) of erase state of SONOS cell on the control-line (CL). At such a PGM condition, PMOS device is operated at low current conduction mode with maximum gate current. It was reported that the charge injection efficiency is as high as 10^{-4} [3], which is at least three orders larger than that of N-channel NVM with channel hot electron PGM mechanism.

During a short PGM period, the characteristics of gate current and drain current versus gate voltage shifts. The peak of gate current is gradually moved to the positive value as PGM is proceeding. Vt of memory cell also becomes more positive and then the gate current moves from electron-favored region toward hole-favored region. It is believed that part of reliability degradation is caused by hot-hole injection [4] and become a major concern of charge losses in advanced technology. That is, if we would not change the bias on CL, VCL, the PGM efficiency and film reliability would be depressed by the large channel hole current.

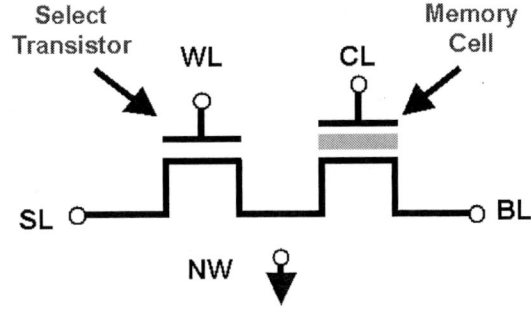

Figure 1.   Schematic circuit of a NeoFlash bitcell.

To keep the best efficiency and have good reliability, we adjust the pulse applied on the control-line [5]. Figure 3(a) illustrates the bias setting during PGM operation. The VCL is continually increased to trace Vt of memory cell during PGM.

The gate current is kept almost at the maximum, and then PGM efficiency is pretty good during the PGM period. As a result, the NeoFlash PGM speed is very fast without hot-hole damage.

As for ERS mechanism, NeoFlash adopts the F-N tunneling mechanism. Figure 3(b) is the bias configuration during ERS operation. The selector transistor (ST) is turned on and the source side positive potential (6.5V) can be transmitted through select transistor to the middle node between ST and memory cell. On the other hand, a negative voltage (-6V) is applied on the gate of memory cell. Therefore, there is a strong electrical field across ONO film to push the electrons out of nitride into Si substrate through the bottom oxide. Because ERS bias is divided between memory cell's gate and other terminals, the highest bias level does not exceed 6.5V. Consequently, there is no high-voltage (HV) device adopted for charge pumping circuit and peripheral HV system no matter of PGM or ERS operations.

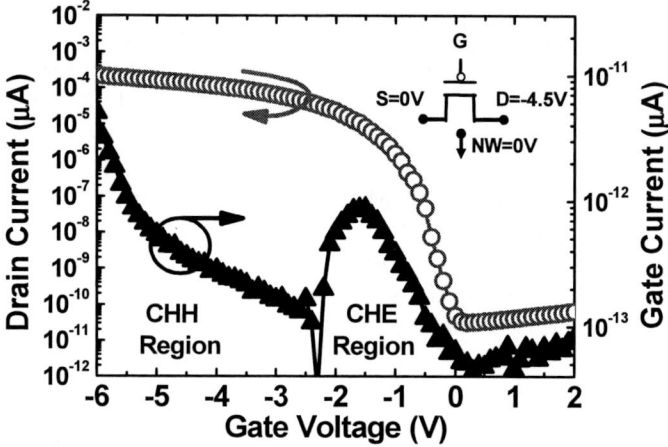

Figure 2. Drain current and gate current characteristics of PMOS transistors at various gate voltages.

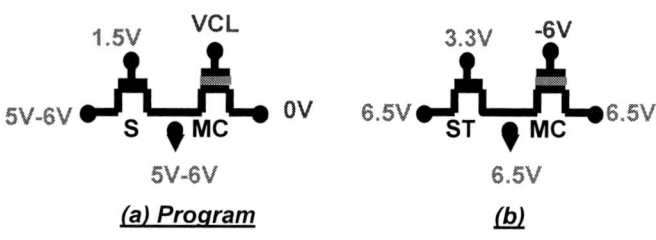

Figure 3. NeoFlash bias setting for (a) PGM; (b) ERS.

B. Cell characteristics

Figure 4 demonstrates PGM and ERS trend of a 0.18 μ m NeoFlash cell. The PGM trend indicates that the PGM speed is much fast. The PGM procedure would be accomplished with 3.5V PGM/ERS window within 5 μ s. That is because the PGM procedure is operated at the maximum hot-electrons injection point. The PGM current can be also kept low during PGM procedure. As for the ERS characteristics, the bitcell can be erased to be smaller than 0V in 200ms in Fig. 4.

While electron-hole pairs are generated at Bit-Line (BL) side in the beginning of PGM, hot electrons surmount the bottom oxide and are trapped in the nitride film. The trapped electrons attract holes in the channel. It seems like that the drain of memory cell extends into the channel. Then, electron-hole pairs are generated inwards at a new position. Some electrons inject into nitride again and then attract holes inwards into the channel. This process continues and the electrons quickly distribute to the nitride film of the whole channel. As a result, it is believed that the trapped charge distribution along the channel is almost uniform. The Vt distribution along the channel length direction was extracted by the charge-pumping method [6]. With lower VBL, the charge distribution correlated with the Vt change in Fig. 5 proves the model.

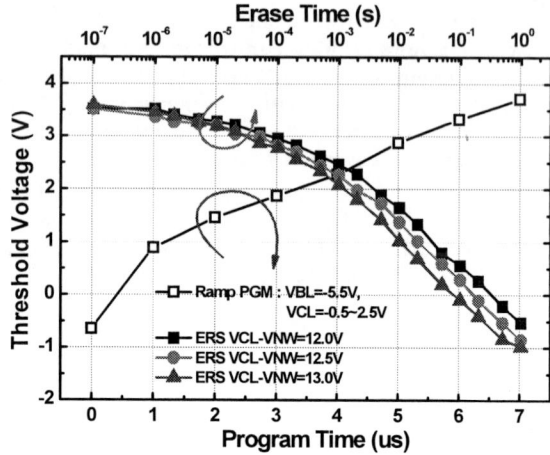

Figure 4. PGM and ERS trend of a 0.18 μ m NeoFlash cell.

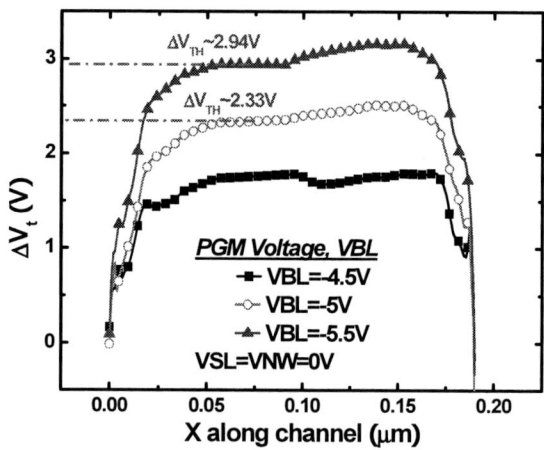

Figure 5. Vt distribution along the channel length direction with various BL bias.

III. RELIABILITY

One important feature of non-volatile flash memory is PGM/ERS endurance. It means how many cycling times the memory cell could be programmed and erased. Figure 6 shows the endurance characteristic of 32k8 bits of NeoFlash array with different PGM/ERS cycling count. It shows that the cell Vt distribution is almost the same from 1 cycle to 3k cycles. The

978-1-4244-9019-6/11 $26.00 © 2011 IEEE

cell Vt distribution after 10k PGM/ERS cycle is about 0.3V lower than the initial one.

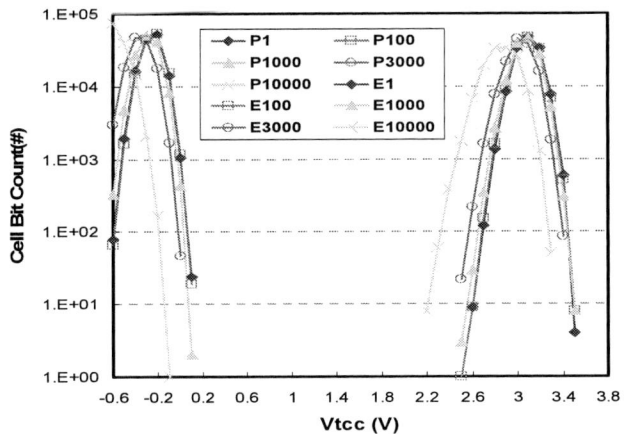

Figure 6.   Cell Vt distribution of a 32k8 bits of testarray with different PGM/ERS cycling count.

To understand the bitcell degradation during cycling, the PGM/ERS endurance test with a bitcell was also executed. The cycling trend of Vt and maximum transconductance, Gm_max, are illustrated in Fig. 7. Figure 7(a) shows the Vt cycling trend is very similar to that gotten in Fig. 6 by array level. The Vt trend hardly changes till 2k cycling. The Gm_max cycling trend at ERS state in Fig. 7(b) reveals the slight Vt degradation after 2k PGM/ERS endurance. That is because the quality of the interface between Si substrate and bottom oxide would be damaged since 2k cycling [8]. Furthermore, the interface traps would increase, and then the characteristics of NeoFlash devices would change. The Vt decreases a little after 10k cycles. The more obvious Gm_max degradation also means that the interface traps increased after 10k cycles. The Gm_max degradation accounted for the mobility degradation due to the increased interface trap density at Si surface including channel and source/drain regions of memory cells.

To evaluate whether NeoFlash can sustain 10 year of retention criterion at 85°C, the energy required for charge loss, activation energy (Ea), is extracted. In order to extract Ea, charge loss characteristics after 1k cycling were tested under different baking temperatures, including 200°C, 225°C, 250°C, and 275°C. The mean time to failure for each temperature is obtained and used for Ea extraction as shown in Fig. 8. It shows the relationship between mean time to failure and reciprocal of temperature. By using Arrehenius equation, Ea=1.27eV is obtained from the slope of the fitted line in Fig. 8. Based on the Ea value, it ensures that NeoFlash can guarantee 10 years of retention @85°C after 1k PGM/ERS endurance.

In order to further confirm whether NeoFlash can pass 10 years of retention @85°C after 1k endurance or not, the baking experiment with 30 samples were really performed at 85°C up to 3.8 hours. Figure 9 shows charge retention characteristics of the 30 samples after 1k cycling at 85°C and baked at 85°C up to 3.8 hours. By the extrapolation to 10 years, there is still an enough margin to pass the criterion, the normal read level with a speed margin. The cell current distribution of one of the 30 samples during baking is illustrated in Fig. 10. It is a representative of all 30 samples. Contrast to floating-gate NVM,

NeoFlash doesn't have tail bits even if baking at 85°C up to 3.8 hours.

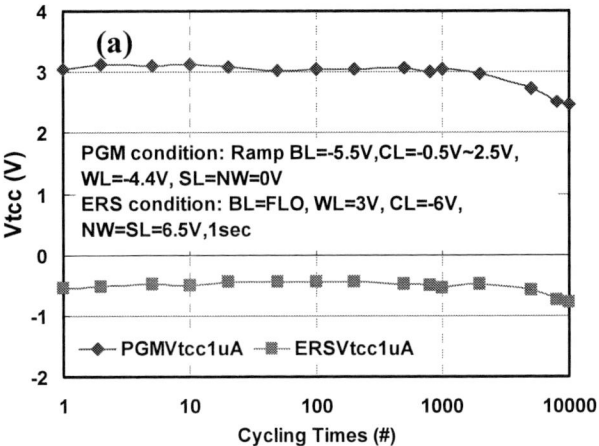

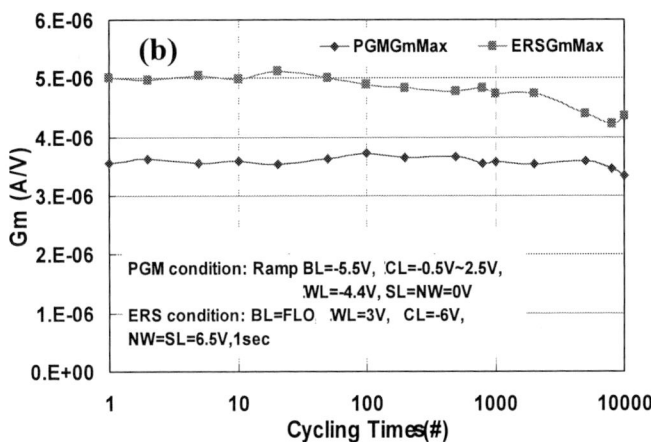

Figure 7.   (a) Threshold voltage trend of a 0.18 μ m SONOS cell as cycling. (b) Transconductance trend of a 0.18 μ m SONOS cell as cycling.

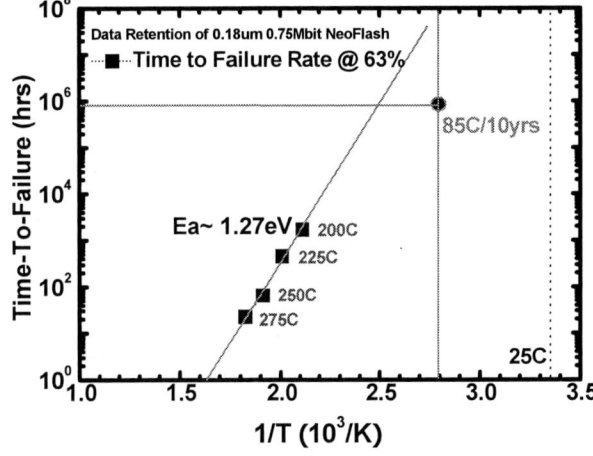

Figure 8.   The relationship between data retention lifetime and reciprocal of temperature.

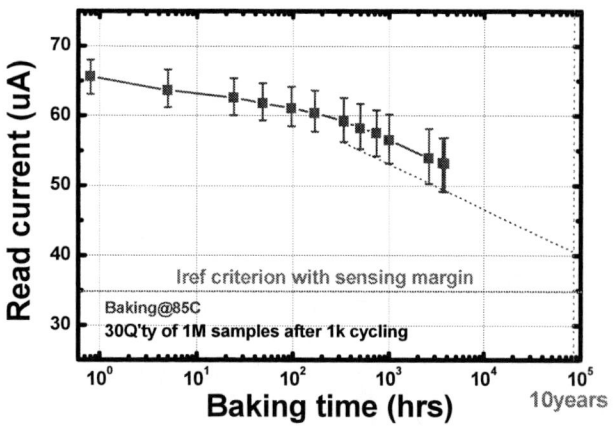

Figure 9. Charge retention characteristics of the 30 samples after 1k cycling at 85°C and baked at 85°C up to 3.8 hours.

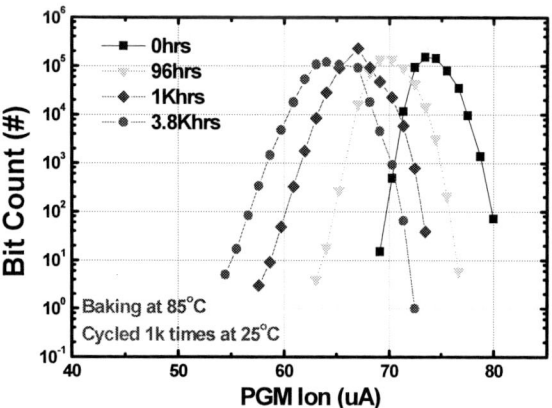

Figure 10. Bit cell current distribution of a 1M bits of NeoFlash baked at 85°C with different readout time.

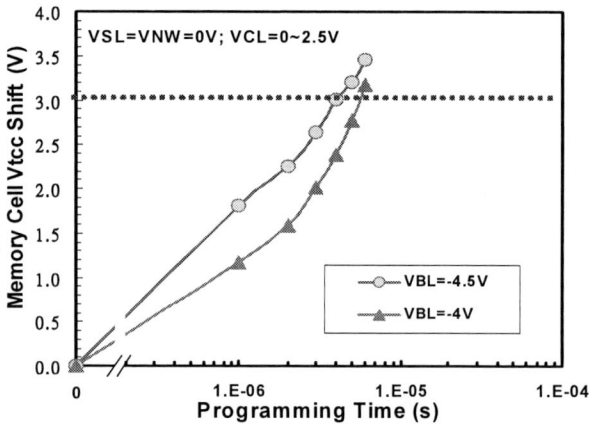

Figure 11. PGM trend of a 65nm NeoFlash cell with different BL bias.

IV. 65NM NEOFLASH PERFORMANCE

The PGM trend of 65nm NeoFlash cell is shown in Figure 11. Like the 0.18 μm NeoFlash shown in Fig. 4, the PGM speed is very fast. In 4μs, the cell can be programmed with a 3V

PGM/ERS window with VBL=-4.5V. Even if VBL=-4V, the bitcell can have a 3V PGM/ERS window in 6 μ s. The endurance performance is tested as shown in Fig. 12. It shows that only slight Vt degradation occurs during 1k cycling.

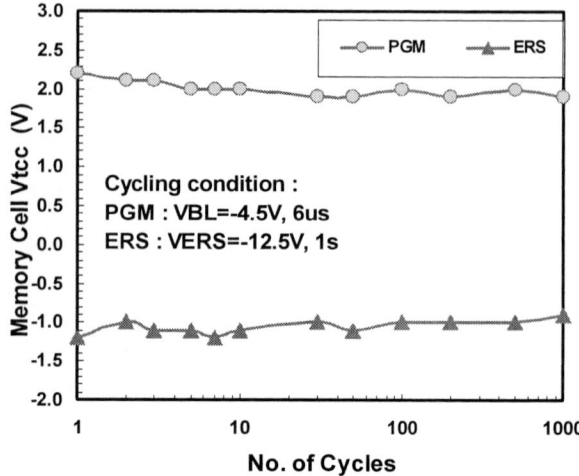

Figure 12. Endurance performance of a 65nm NeoFlash cell.

V. CONCLUSION

In this paper, NeoFlash technology with fast PGM, uniform ERS and high reliability is demonstrated. Only 3 masks are required in addition to baseline logic process. Tracing Vt hot electron injection can shorten the PGM time to 5~7 μ s for emerging high speed and high density applications. Hot-hole-free CHHIHE PGM leads to high reliability. Charge retention performance is sufficient for 10 years at 85°C based on the Ea extraction and the extrapolation result of the real baking data of 30 dies samples up to 3.8 hours. No tail bit during endurance and retention test is observed. Furthermore, 65nm NeoFlash cell performance is demonstrated at the first time. As a result, NeoFlash is a very promising embedded NVM solution due to its low simple fabrication process, design and low cost.

REFERENCES

[1] H.M. Lee, L. Lim, S. M. Jung, S.T. Woo, H. M. Chen, C. Y. Lin, R. Shen, C. D. Wang, C. C.-H. Hsu, and S.C. Sun, "NeoFlash – True logic based 0.18um single poly embedded SONOS flash," SSDM, IEEE, 2005, p.196.

[2] C.C.-H. Hsu, A. Acovic, L. Dori, B. Wu, T. Lii, D. Quinlan, D. DiMaria, Y. Taur, M. Wordeman, and T. Ning, "A high speed, low power p-channel flash EEPROM using silicon rich oxide as tunneling dielectric," SSDM, IEEE, 1992, pp.140.

[3] S.S. Chung, S.N. Kuo, C.M. Yih, and TS. Chao, "Performance and Reliability Evaluation of P-channel Flash Memories with Different Programming Schemes", IEDM Tech. Dig., pp.295-298, 1997.

[4] A. Bravaix, D. Goguenheim, N. Revil, and E. Vincent, "Comparison of low leakage and high speed deep submicron PMOSFETs submitted to hole injcetions," IRW Final Report, IEEE, 2002, p.14-20

[5] Ying-Je Chen, Cheng-Jye Liu, Chun-Yuan Lo, Yun-Jen Ting, T. H. Hsu, and Wein-Town Sun, "A Highly Reliable Embedded P-Channel SONOS Memory using Dynamic Programming method", to be puslished in IRPS, IEEE, 2011.

[6] C.E. Weintraub, E. Vogel, J.R. Hauser, N. Yang, V. Misra, J.J. Wortman, J. Ganem, and P. Masson, "Studying of low-frequency charge pumping

on thin stacked dielectrics", IEEE Trans. Electron Devices, vol. 48, no. 12, pp. 2754-2762, Dec. 2001.

[7] Ying-Je Chen, Yun-Jen Ting, Cheng-Jye Liu, Wein-Town Sun, and Rick Shen, "Precision programming power control in embedded p-channel SONOS flash using transient-IV method," SSDM, IEEE, 2009

[8] Sung-Rae Kim, Kyung Joon Han, Kin-Sing Lee, Pavan Singaraju, Rophina Li, Patty Liu, Yingbo Jia, Ben Schmid, Yu Wang, Fethi Dhaoui, Frank Hawley, and Huan-Chung Tseng, "Cycling impact on the gm degradation and GIDL current of 65nm 2T-embedded flash memory," NVMTS, IEEE, 2009, p.77-79.

[9] A. Arreghini, N. Akil, F. Driussi, D. Esseni, L. Selmi, and M.J van Duuren, "Long term charge retention dynamics of SONOS cells," Solid-State Electronics, 2008, p.1460-1466.

Wein-Town Sun was born in Kaohsiung, Taiwan, R.O.C., in 1970. He received the B.S. and Ph.D. degrees in electrical engineering from National Tsing-Hua University, Hsinchu, Taiwan, in 1992 and 1998, respectively.
He was with Process Integration Department, eMemory Technology Inc., Hsinchu, in 2006, working in the area of SONOS Flash non-volatile memory. His research interests are device physics of sub-micron MOSFETs and TFTs, and process development of gate and drain engineering for sub-micron MOSFETs. Currently, his research involves in developing high PGM/ERS endurance SONOS cells and next-generation embedded SONOS Flash.

Crystallization Technique of Epitaxial HfO$_2$ Thin Films on Si Substrates and their Potential for Advanced High-k Gate Stack Technology

Shinji Migita and Hiroyuki Ota

Abstract—Crystalline phase high-k films are promising gate stack structure for the advanced CMOS technology because they are thermodynamically stable and have higher dielectric constant when compared with amorphous phase high-k films. A disadvantage of crystalline high-k films, however, is the large leakage current, which is sometimes caused by grain boundaries and non-crystallized region in ultra-thin crystalline high-k films. We developed a unique crystallization technique that realizes epitaxial growth of HfO$_2$ films on Si substrates. MOS capacitors of closely packed epitaxial HfO$_2$ films achieved extremely small EOT with suppressed leakage current. It demonstrates that crystallization process is the key for the application of high-k crystal films.

Index Terms—Gate leakage, Grain boundaries, High-K gate dielectrics, MOSFETs,

I. INTRODUCTION

SCALING of gate dielectric thickness in advanced CMOS technology has been progressed by integration of high-k materials [1]. The thickness of gate dielectric films should be designed to be thin in order to enhance the controllability of channel potential by the gate electrode and suppress the short channel effect of MOSFETs. However, when the physical thickness of dielectric thin film is reduced to be less than 2 nm, a direct tunneling current through the film increases exponentially and increases the power consumption of the VLSI chips drastically. Thus SiO$_2$ and SiON gate dielectrics cannot be scaled to less than 2 nm (Fig. 1(a)). High-k materials such as HfO$_2$ and HfO$_2$-based compounds have much larger dielectric constant than SiO$_2$. A gate dielectric structure which composed of high-k film with SiO$_2$ interfacial layer can suppress the leakage current owing to the physical thickness of the stacked layers. At the same time, the equivalent oxide thickness (EOT) can be scaled to 1 nm because of the large dielectric constant of high-k layers (Fig. 1(b)). In the future, EOT scaling toward 0.6 nm is expected. In order to achieve 0.6

nm EOT, crystalline high-k film with absence of SiO$_2$ interfacial layer is demanded (Fig. 1(c)). SiO$_2$ interfacial layer must be eliminated because it usually occupies 0.4 to 0.6 nm thickness in EOT. Crystalline phase high-k films are promising because they are thermodynamically stable, and their dielectric constants are larger than those of amorphous phase high-k films. Crystalline high-k films are also attractive because their dielectric constants can be controlled by modification of crystalline structures [2-4]. However, crystalline high-k films have been avoided in the history of high-k gate stack development because they sometimes induce harmful events in electrical performances, such as and increments of leakage currents, hysteresis, and variability. These phenomena are always linked to the existence of grain boundaries and defects in the crystalline films.

In this paper, we report preparation of epitaxial HfO$_2$ thin films on Si substrates and their excellent electrical properties. Different from previous works of epitaxial high-k film growth on Si by molecular beam epitaxy (MBE) technique [5-7], we made epitaxial HfO$_2$ film growth through deposition of amorphous phase film by atomic-layer deposition (ALD) followed by a rapid thermal crystallization (RTC) treatment [8]. Satisfactory small leakage currents of epitaxial HfO$_2$ MOSCAPs infer their potential for advanced high-k gate stack technology. Based on the crystallization behavior of amorphous HfO$_2$ films, the importance of temperature management before and during crystallization is discussed as a guiding principle in the development of ultra thin crystalline high-k film.

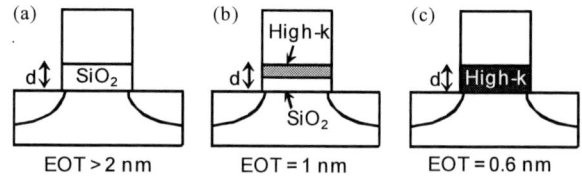

Fig. 1. Comparison of gate dielectric materials and structures for advanced MOSFETs. (a) SiO$_2$ or SiON gate dielectric. Because of the direct tunneling leakage current, the thickness d cannot be scaled to less than 2 nm. (b) A gate dielectric structure composed of high-k film with SiO$_2$ interfacial layer. Owing to the large dielectric constant of high-k material, the equivalent oxide thickness (EOT) can be scaled to 1 nm while the physical thickness d is maintained to 2 nm or larger. (c) A future gate dielectric structure which consists of crystalline high-k film with absence of SiO$_2$ interfacial layer. This structure is indispensable for achieving 0.6 nm EOT.

Manuscript received March 22, 2011. This work was supported by the "Next-generation Semiconductor Materials and Process Technology (MIRAI) Project of the New Energy and the Industrial Technology Organization (NEDO).

Shinji Migita and Hiroyuki Ota are with the Nanodevice Innovation Research Center, National Institute of Advanced Industrial Science and Technology, Onogawa 16-1, Tsukuba 305-8569, Ibaraki, Japan (e-mails: s-migita@aist.go.jp and hi-ota@aist.go.jp).

II. Process Concept and Experimental

A. Temperature Management for HfO₂ Crystal Film Growth

Crystallization technique in this work was constructed through the consideration of transformation behavior of amorphous phase HfO_2 film into crystalline phase. There are several works concerning the crystallization of HfO_2 films. In a study using chemical solution deposition, the onset of crystallization was observed at 500°C. The volume of crystal phase increased with the increment of process temperature [9]. In another study using metal-organic chemical vapor deposition (MOCVD), it is observed that crystal nuclei come into existence even at a deposition temperature of 350°C [10]. The number of nuclei increases at a higher deposition temperature, but the size of crystal grains is maintained to be less than 10 nm in diameter. These studies suggest that transformation of amorphous film into crystal phase does not progress at a specific temperature, but it progresses gradually in the transition temperature range (Fig. 2(a)). The deposition temperature of MBE is high enough to form crystalline films directly. The CVD produces films which consist of crystal grains in amorphous film because the deposition temperature exists in the transition temperature range. The deposition temperature of ALD may low enough to produce amorphous HfO_2 films.

Following to the preparation of amorphous HfO_2 films by ALD, crystallization was processed by rapid thermal annealing (RTA) equipment. In contrast to the conventional RTA process which is designed with an attention of the peak temperature, the RTC process in this work is designed with the consideration of the transition temperature range of HfO_2 films (Fig. 2(b)) [8]. A fast heating up commences the crystallization of HfO_2 film from the Si interface concurrently.

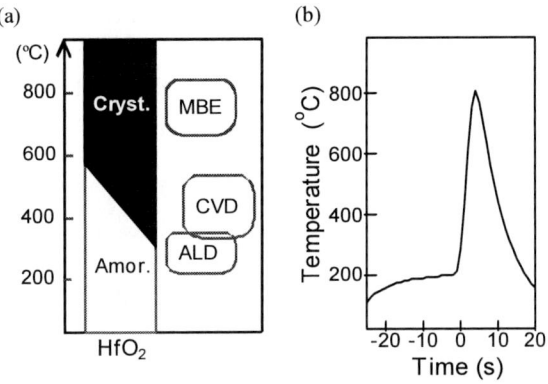

Fig. 2. (a) A schematic diagram showing transformation temperature of amorphous phase HfO_2 film into crystalline phase, and the deposition temperatures of typical growth techniques for high-k films. (b) Temperature-time program of the developed rapid thermal crystallization (RTC) process after deposition of amorphous phase HfO_2 films on Si substrates using ALD. The process is featured by that the film passes the onset temperature of crystallization with a high heating rate.

B. Experimental

Native oxides on Si (100) substrates were removed by an HF solution treatment. Amorphous phase HfO_2 films were deposited directly on Si surfaces by ALD at 250°C using a cyclic supply of tetrakis(diethylamino)hafnium and H_2O as hafnium and oxygen sources. The RTC was performed with a heating rate of 200°C/s at the transition temperature range. The peak temperature was 800°C.

MOSCAPs were fabricated by deposition of TaN and poly-Si stacked electrode and activation anneal at 1000°C. The capacitor size is 100 μm x 100 μm.

III. Results and Discussion

A. Physical Analyses of HfO₂ Crystalline Films

Figure 3(a) shows the transmission electron microscope image of epitaxial HfO_2 film on Si (100) substrate. The lattice image of HfO_2 crystal is directly connected to the Si substrate, and there is no SiO_2 interfacial layer. Analysis by x-ray photoelectron spectroscopy indicated that Hf-O-Si bonding is formed at the interface. Thus epitaxial growth of HfO_2 film on Si substrate is confirmed.

Figure 3(b) shows an x-ray reflectometry measurement of 5 nm-thick epitaxial HfO_2 film on Si substrate. A clear oscillation demonstrates the uniform thickness of HfO_2 film and the small roughness of both surface and interface, respectively. This result proves that RTC promotes regulated crystallization and suppresses protrusion of random crystal grains in HfO_2 films all over the substrate.

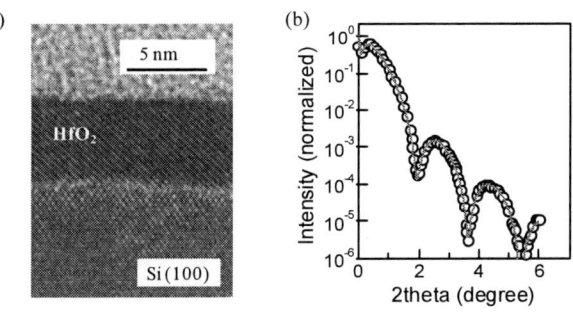

Fig. 3. (a) Transmission electron microscope image of an epitaxial HfO_2 film prepared on Si (100) substrate using ALD and RTC. Direct bonding between HfO_2 crystal and Si substrate is observable, although the image at interface has somewhat disorder which is caused by the lattice mismatch as large as 9%. Absence of interfacial SiO_2 layer is cross-checked by x-ray photoelectron spectroscopy. (b) X-ray reflectometry measurement (open circles) of an epitaxial HfO_2 film (5 nm thickness) on Si substrate. A clear oscillation of reflection intensity manifests the small roughness of both HfO_2 film surface and interface with Si substrate. The roughness is calculated by the simulation (solid line) to be 0.37 nm and 0.25 nm, respectively.

B. Electrical properties of Epitaxial HfO₂ MOSCAPs

C-V characteristics of epitaxial HfO_2 MOSCAPs are shown in Fig. 4(a). The EOT could be scaled to as thin as 0.5 nm. Hysteresis is reasonably small. Comparison of the experimental data with the simulation suggests that the number of interfacial state is sufficiently small. The EOTs are plotted against the physical thicknesses of HfO_2 films in Fig. 4(b). Using the fitting lines in the figure, the k-value of epitaxial HfO_2 films is evaluated to be between 16 and 20. These values are comparable with monoclinic and cubic structures of HfO_2 films.

978-1-4244-9019-6/11 $26.00 © 2011 IEEE

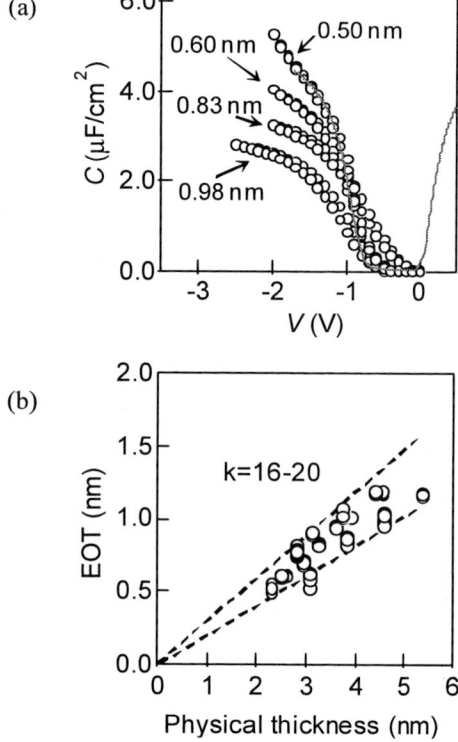

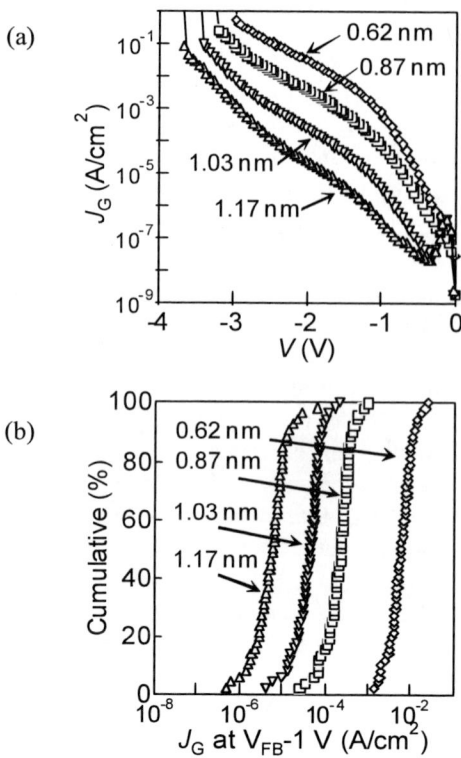

Fig. 4. (a) C-V plots (100 kHz) of epitaxial HfO_2 MOSCAPs on p-Si (100) substrates. Activation anneal was performed at 1000°C. Numbers indicate EOTs calculated by simulation. The solid line is an example of simulation with 0.5 nm EOT where flat band voltage is adjusted to the experiment. (b) .EOT-physical thickness plots of epitaxial HfO_2 MOSCAPs on Si (100) substrates. The slopes of dashed lines show that k-value of epitaxial HfO_2 is between 16 and 20. The intercept of origin demonstrate the absence of SiO_2 interfacial layer.

It is noted that the fitting lines intercept the origin of the plot. This result proves that epitaxial HfO_2 MOSCAPs have no interfacial SiO_2 layer.

I-V characteristics of epitaxial HfO_2 MOSCAPs are shown in Fig. 5(a). The leakage currents are satisfactory small even in the case of extremely thin EOT. Dependence of the leakage current with EOT suggests that the currents are flowing through the bulk of epitaxial HfO_2 films and are not dominated by specific current paths such as grain boundaries and defects. Cumulative plots of leakage currents are summarized in Fig. 5(b). Leakage currents are settled with the EOTs. It demonstrates that the thicknesses of epitaxial HfO_2 films are uniform all over the wafer and variation of thickness in local area scarcely exists.

The relationship between leakage current and EOT is shown in Fig. 6 where a trend of SiO_2 gate dielectric, studies of high-k films with SiO_2 interfacial layers reported between 2002 and 2004, and recent reports of direct bonding high-k films [11-15] are plotted in addition to our results of epitaxial HfO_2 MOSCAPs. Gate stack structures of high-k films with SiO_2 interfacial layers are effective to reduce the leakage current by 3 orders and more. However, existence of SiO_2 interfacial layer is an obstacle of EOT scaling less than 1.0 nm. Introduction of high-k films without SiO_2 interfacial layers has successfully achieved further EOT scaling in which leakage currents are reduced by 6 orders when compared with SiO_2 cases. The trend

Fig. 5. (a) I-V curves of epitaxial HfO_2 MOSCAPs on Si (100) substrates. Numbers indicate EOTs. (b) Cumulative plots of leakage currents at V_{FB}-1 V.

of leakage current with EOT lies on a line until 0.5 nm EOT. These results indicate that SiO_2-less high-k gate stack structure (Fig. 1(c)) is inevitable for sub-1 nm EOT technologies. Among the reports of sub-1 nm EOT, epitaxial HfO_2 MOSCAPs show excellent performance. Furthermore, the advantage of epitaxial HfO_2 films seems to become apparent as the EOT becomes thinner. It is supposed that the dielectric constant of epitaxial HfO_2 can be maintained even with ultra thin films and it contributes to the scaling of EOT and reduction of leakage current.

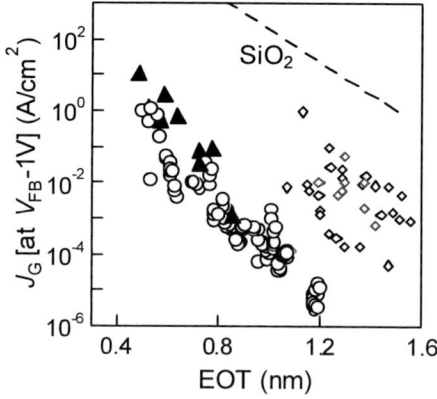

Fig. 6. The relationship between the leakage current at V_{FB}-1 V and EOT. Dashed line shows a trend of SiO_2 gate dielectric films. Open diamonds show the results of high-k films with SiO_2 interface layers reported between 2002 and 2004. Solid triangles show recent results of zero-SiO_2 high-k structures [11-15]. Open circles show the results of epitaxial HfO_2 MOSCAPs.

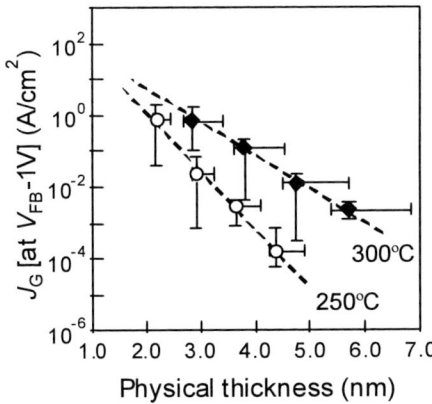

Fig. 7. Impact of ALD temperatures on the leakage currents of epitaxial HfO₂ MOSCAPs. ALD cycles were regulated to 30, 40, 50, and 60. ALD temperatures were set to 250°C and 300°C. RTC anneal at 800°C was the same. Open circles show the results of ALD at 250°C and solid diamonds show the result of ALD at 300°C, respectively.

C. Impact of ALD Temperature for Amorphous HfO₂ films

Importance of temperature management before RTC is further examined by changing the ALD temperatures. HfO₂ films were prepared by ALD at 250°C and 300°C with ALD cycles of 30, 40, 50, and 60. Following processes of RTC at 800°C and MOSCAP fabrication were the same. Figure 7 compares the relationship between the leakage current and physical thickness of HfO₂ films. It is found that growth rate of HfO₂ film differs with temperatures; 0.073 nm/cycle at 250°C and 0.095 nm/cycle at 300°C, respectively. It infers that CVD-mode growth is partially involved in the ALD growth with a small increment of the deposition temperature. The leakage currents increase clearly in 300°C HfO₂ film. It is speculate that 300°C HfO₂ films may include nuclei of crystals in the amorphous films even though they are hard to detect by physical analyses. In other words, transition temperature (Fig. 2(a)) may be extended to 300°C slightly. Thus we must also pay special attention to the ALD temperature.

IV. CONCLUSION

Temperature management is the key parameter for preparation of HfO₂ crystalline films. Unintentional nucleation of crystalline particles in amorphous films at transition temperature region deteriorates the electrical properties of MOSCAPs. Regulated crystallization produces epitaxial HfO₂ films which can scale to 0.5 nm EOT with suppressed leakage current. Although it is hard to detect the beginning of nucleation in amorphous films, insights of crystallization behavior brings improvement of crystalline high-k films and electrical properties.

ACKNOWLEDGMENT

The authors express sincere gratitude to Prof. Masataka Hirose, Prof. Akira Toriumi, Dr. Seiichiro Kawamura, and Prof. Toshihiko Kanayama for their continuous encouragements during the research in MIRAI Project. The authors also thank Drs. Yukinori Morita and Wataru Mizubayashi for fruitful discussions.

REFERENCES

[1] International Technology Roadmap for Semiconductors, Semiconductor Industry Association (http://public.itrs.net).

[2] X. Zhao and D. Vanderbilt, "First-principles study of structural, vibrational, and lattice dielectric properties of hafnium oxide," *Phys. Rev. B*, vol. 65, p.233106, 2002.

[3] K. Kita, K. Kyuno, and A. Toriumi, "Permittivity increase of yttrium-doped HfO₂ through structural phase transformation," *Appl. Phys. Lett.*, vol. 86, p.102906, 2005.

[4] S. Migita, Y. Watanabe, H. Ota, H. Ito, Y. Kamimuta, T. Nabatame, and A. Toriumi, "Design and Demonstration of Very High-k (k=50) HfO₂ for Ultra-Scaled Si CMOS," *Tech. Dig. Symp. VLSI Technol.*, pp. 152-153, 2008.

[5] R. A. McKee, F. J. Walker, and M. F. Chisholm, "Crystalline Oxide on Silicon: The First Five Monolayers," *Phys. Rev. Lett.*, vol.81, pp. 3014-3017, 1998.

[6] Y. Nishikawa, N. Fukushima, N. Yasuda, K. Nakayama, and S. Ikegawa, "Electrical Properties of Single Crystalline CeO₂ High-k Gate Dielectrics Directly Grown on Si (111)," *Jpn. J. Appl. Phys.*, vol. 41, pp. 2480-2483, 2002.

[7] A. Dimoulas, G. vellianitis, G. Mavrou, G. Apostolopoulos, A. Travlos, C. Wiemer, M. Fanciulli, and Z. M. Rittersma, "La2Hf2O7 high-k gate dielectric grown directly on Si (001) by molecular-beam epitaxy," *Appl. Phys. Lett.*, vol. 85, pp. 3205-3207, 2004.

[8] S. Migita, Y. Morita, W. Mizubayashi, and H. Ota, "Preparation of Epitaxial HfO₂ Film (EOT=0.5 nm) on Si Substrate Using Atomic-Layer deposition of Amorphous Film and Rapid thermal Crystallization (RTC) in an Abrupt Temperature Gradient," *IEDM Tech. Dig.*, pp. 269-270, 2010.

[9] D. A. Neumayer and E. Cartier, "Materials characterization of ZrO₂-SiO₂ and HfO₂-SiO₂ binary oxides deposited by chemical solution deposition," *J. Appl. Phys.*, vol. 90, pp. 1801-1808, 2001.

[10] S. Fujii, N. Miyata, S. Migita, T. Horikawa, and A. Toriumi, "Nanometer-scale crystallization of thin HfO₂ films studied by HF-chemical etching," *Appl. Phys. Lett.*, vol. 86, p. 212907, 2005.

[11] M. Takahashi, A. Ogawa, A. Hirano, Y. kamimuta, Y. Watanabe, K. Iwamoto, S. Migita, N. Yasuda, H. Ota, T. Nabatame, and A. Toriumi, "Gate-First Processed FUSI/HfO₂/HfSiOx/Si MOSFETs with EOT=0.5 nm –Interfacial Layer Formation by Cycle-by-Cycle Deposition and Annealing-," *IEDM Tech. Dig.*, pp. 523-526, 2007.

[12] J. Huang, D. Heh, P. Sivasubramani, P. D. Kirsch, G. Bersuker, D. C. gilmer, M. A. Quevedo-Lopez, M. M. Hussain, P. Majhi, P. Lysaght, H. Park, N. Goel, C. Young, C. S. Park, C. Park, M. Cruz, V. Diaz, P. Y. Hung, J. Price, H. –H. Tseng, and R. Jammy, "Gate First High-k/Metal Gate Stacks with Zero SiOx Interface Achieving EOT=0.59 nm for 16 nm Application," *Tech. Dig. Symp. VLSI Technol.*, pp. 34-35, 2009.

[13] L.-A. Ragnarsson, Z. Li, J. Tseng, T. Shram, E. Rohr, M. J. Cho, T. Kauerauf, T. Conard, Y. Okuno, B. Parvais, P.Absil, S. Biesemans, and T. Y. Hoffmann, "Ultra Low-EOT (5 Å) gate-First and Gate-last High Performance CMOS Achieved by Gate-Electrode Optimization," *IEDM Tech. Dig.*, pp. 663-666, 2009.

[14] T. Ando,M. M. Frank, K. Choi, C. Choi, J. Burley, M. Hopstaken, M. Copel, E. Cartier, A. Kreber, A. Callegari, D. Lacey, S. Brown, Q. Yang, and V. Narayanan, "Understanding Mobility Mechanisms in Extremely Scaled HfO₂ (EOT 0.42 nm) Using remote Interfacial Layer Scavenging Technique and Vt-tuning Dipoles with Gate-First Process," *IEDM Tech. Dig.*, pp. 423-426, 2009.

[15] K. Kakushima, T. Koyanagi, D. Kitayama, M. Kouda, J. Song, T. Kawanago, M. Mamatrishat, K. Tachi, M. K. Bera, P. Ahmet, H. Nohira, K. Tsutsui, A. Nishiyama, N. Sugii, K. Natori, T. Hattori, and H. Iwai, "Direct Contact of High-k/Si Gate Stack for EOT below 0.7 nm using LaCe-silicicate layer with Vfb controllability," *Tech. Dig. Symp. VLSI Technol.*, pp. 69-70, 2010.

A New Prediction Model for Effects of Plasma-Induced Damage on Parameter Variations in Advanced LSIs

Koji Eriguchi, Yoshinori Takao, and Kouichi Ono

Abstract—This paper proposes a physics-based variability prediction model integrating the effects of plasma-induced damage (PID) in advanced LSIs. We focus on charging damage to high-k gate dielectrics and physical damage (Si recess by ion bombardment). In addition to gate length-variation which has been discussed so far as a dominant factor for (static) variability, we demonstrate how PID impacts on – increases – the parameter variation (e.g., σ_{Vth}), by employing both experimental PID data for high-k and Si substrate damage and a Monte Carlo method. The model prediction suggests a considerable increase in parameter variations by PID such as threshold voltage and off-state leakage.

Index Terms—gate length, high-k, parameter variation, plasma-induced damage, recess structure, threshold voltage

I. INTRODUCTION

Regarding development of future MOSFETs, the parameter variation has become a key concern in designing an ultra-large-scale integrated (ULSI) circuit where billions of MOSFETs are built-in.[1, 2] From the viewpoint of device technology, threshold voltage (V_{th}) control has been crucial for transistor design, and the short-channel effect (SCE) induced by the shrinkage of gate length (L_g) is regarded as one of the critical phenomena.[3] Suppressing the fluctuation in V_{th} – V_{th}-variation – is believed to be a major challenge for future MOSFETs beyond the 65-nm-technology node.[1, 2] In addition to dopant fluctuation in/near a channel region, L_g-related variations (σ_{Lg}) such as LER (line-edge roughness) and LWR (line-width roughness)[3] is believed to enhance V_{th}-variation, leading to a wider statistical distribution of subthreshold leakage current (I_{off}).[4] From the viewpoint of process technology on the other, critical dimension control of L_g during gate patterning has been crucial challenge in developing plasma etch processes. In suppressing the σ_{Lg}, emphases are placed on controlling the reactions on material surface governed by plasma chemistry. There have been many reports on unexpected negative impact by the interaction between device and plasma; plasma-induced damage (PID).[5] Regarding V_{th}, there are two major phenomena in PID: One is the ion-bombardment damage usually referred to as physical damage which induces "Si recess structure" in a source/drain extension region,[6] and the other is the charging damage such as the antenna effect.[5, 7] Both are found to induce the shift of V_{th} (ΔV_{th}).[8] Although PID is also considered to impact on the V_{th}-variation [Var(V_{th})] from the above discussion, there have been few studies on the prediction of Var(V_{th}) by integrating the PID effect quantitatively. In this paper, we present a comprehensive framework with PID models to predict the V_{th}-variation in advanced LSIs.

II. PARAMETER PREDICTION FRAMEWORK

Fig. 1 summarizes a conceptual flowchart based on (1) Damage Creation Model, (2) Device Degradation Model, and (3) Variability Prediction Model. From the semi-empirical and theoretical equations for the models (1) and (2), one can structure a framework as shown.

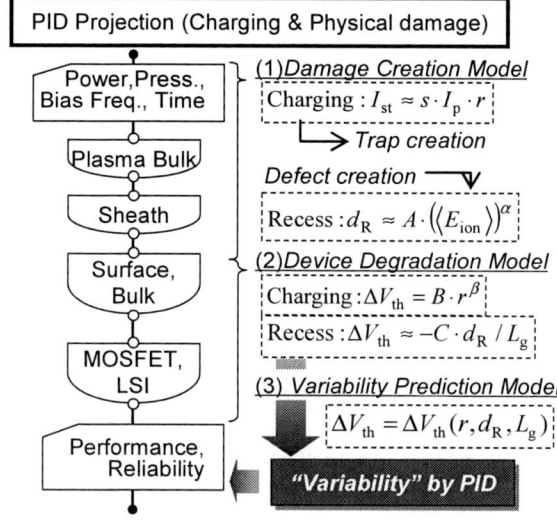

Fig. 1. Flowchart of predicting device parameter variations by plasma-induced damage (PID). Analytical expressions for the prediction are shown. All the parameters are process- and device-structure-dependent, and determined from experiments and simulations. For details, see the text.

Figure 2 illustrates detailed mechanisms to understand

This work was financially supported in part by STARC (Semiconductor Technology Academic Research Center) and Grant-in-Aid for Scientific Research (B) 20360329 from the JSPS.

K. Eriguchi, Y. Takao, and K. Ono are with Kyoto University, Yoshida-Honmachi, Sakyo-ku, Kyoto 606-8501, JAPAN (corresponding author to provide phone: +81-75-753-5983; fax: +81-75-753-5980; e-mail: eriguchi@kuaero.kyoto-u.ac.jp)

978-1-4244-9019-6/11 $26.00 © 2011 IEEE

overall PID phenomena. Charging damage creates trap-site in high-k gate dielectrics, resulting in ΔV_{th}, amplified by the antenna ratio r (= metal area exposed to plasma / gate area). Physical damage (surface-damaged layer and underneath defect sites) is formed during gate and offset spacer etch processes. The depth of Si loss (recess) d_R induces ΔV_{th}[6] and the defect site density (n_{dam}) degrades drain current I_{on}.[9] Validity of this mechanism is confirmed by molecular dynamics simulations as reported previously.[10]

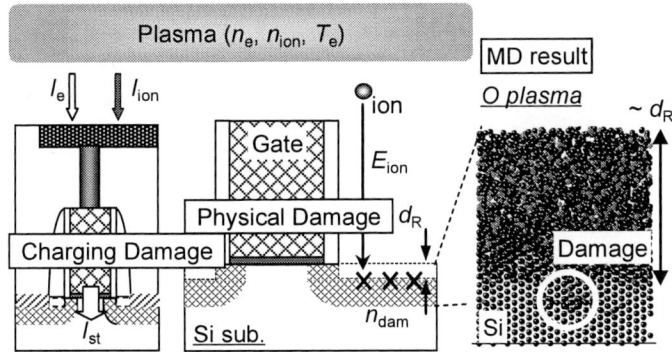

Fig. 2. Illustration of two PID mechanisms, charging damage (left) and physical damage (middle/right). As for physical damage (Si recess and latent defect site), we conducted molecular dynamics (MD) simulations to investigate the mechanisms (right). As seen on the right, displaced Si atoms are identified underneath the interface between surface-damaged layer with Si and Oxygen and Si substrate.

III. EXPERIMENTAL PROCEDURE

Figure 3 shows a schematic view of capacitively coupled plasma (CCP) reactor used in this study. Processed high-k MOSFETs were again exposed to Ar-CCP for 30 and 120 s. In Fig. 3, the schematic view of high-k MOSFET is also illustrated. Al-probing pads with a constant area of 42000 μm^2 served as antennas of the box-type structure, i.e., r is in the range from 4.2 to 4667. Drain current–gate voltage ($I_d – V_g$) characteristic was obtained for at least 12 different devices to determine V_{th}. Si substrate was exposed to inductively coupled plasma (ICP) reactor as reported previously[10] and the damaged-layer (DL) thickness was assigned by spectroscopic ellipsometry. To determine d_R, a modified range theory[11] and the wet-etch stripping model of DL were employed. Details are published elsewhere.[6, 10-12]

IV. RESULTS AND DISCUSSION

A. Device Degradation Models for PID

Figure 4 shows the r-dependence of $|\Delta V_{\text{th}}|$. Including the d_R-dependence of ΔV_{th} previously reported,[6] one can write the following equations for ΔV_{th} by PID as

$$\Delta V_{\text{th}} = -A \cdot d_R , \qquad (1)$$

$$\Delta V_{\text{th}} = B \cdot r^{\beta} (= g(r)) , \qquad (2)$$

where A, B and β are the process- and device-dependent parameters. A is dependent on L_g and we set $A = 0.30 / L_g$ (V/nm)

based on the model[6] and device simulations.[10] B is the

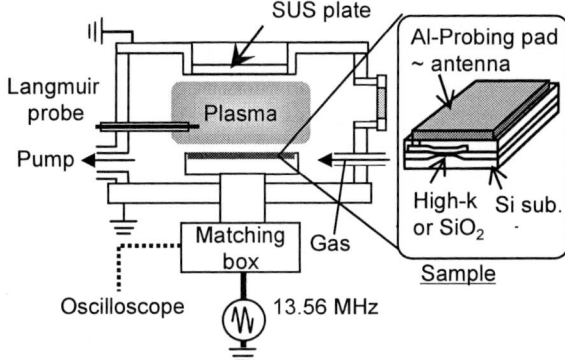

Fig. 3. Illustration of plasma reactor and device structure used in this study.

constant determining the amount of charging damage, hereafter we denote "a charging parameter". By taking into account the result in Fig. 4, we define B and β as -0.01 (V) and 0.25, respectively. Note that $\Delta V_{\text{th}} < 0$ indicates an increase in I_{off} in n-ch MOSFET. Unless otherwise stated, we use these values in the following calculations.

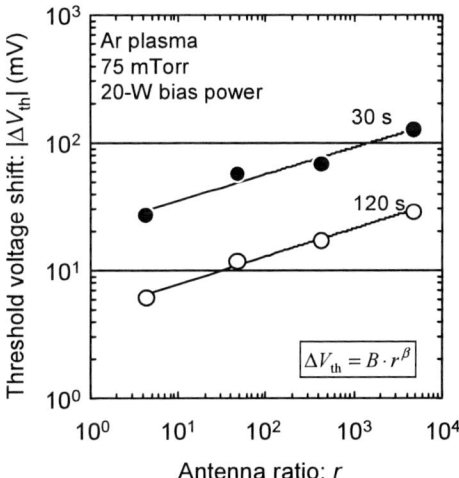

Fig. 4. Threshold voltage shift (< 0) as a function of antenna ratio r.

B. V_{th}-Variability Prediction Models for PID

As reported previously,[6] since d_R affects V_{th} and I_{off}, the d_R-variation induces ΔV_{th}- and I_{off}-variation. To estimate ΔV_{th}- and I_{off}-variation, we define the probability density functions (p.d.f.)[13] for V_{th} and I_{off} as $f_{\Delta V\text{th}}(\Delta V_{\text{th}})$ and $f_{\text{Ioff}}(I_{\text{off}})$, respectively. $f_{\text{dR}}(d_R)$ and $f_r(r)$ are also defined. As for $f_{\text{dR}}(d_R)$, we assume a Gaussian distribution (the mean value μ_{dR} and the standard deviation σ_{dR}). The range of σ_{dR} is assumed to be equivalent to the distance between Si atomic planes in the (100)-direction. Regarding $f_r(r)$, we assume two different distributions: one defined by the Rent's rule[14] and the other, by an exponential distribution (Poisson-type). The Rent's rule corresponds to an interconnect length distribution widely used in the present-day LSIs. The number of nets approximately obeys a power-law dependence on the interconnect length, which is characterized by the Rent's exponent p. The exponential distribution assumed here is expressed as,

$$f_r(r) \propto \exp(-\lambda \cdot r), \tag{3}$$

where λ is a constant. Note that $f_r(r)$ is normalized with respect to r. Although this distribution is artificial, it can be presumably useful from antenna ratio distribution data previously discussed so far.[7, 15] Figure 5 compares two distributions discussed in the above. We reproduced each distribution for more than 10^6 transistors in both cases. Once $f_r(r)$ is determined, one can derive ΔV_{th}-variation induced by the charging damage as,

$$f_{\Delta Vth}(\Delta V_{th}) = \frac{f_r(g^{-1}(\Delta V_{th}))}{|g'(r)|}, \tag{4}$$

where $g(r)$ is in Eq. (2) and $g^{-1}(r)$ is its inverse function. $g'(r)$ is the derivative of $g(r)$.

charging parameter strongly affects on ΔV_{th}^{PID}, one should optimize an ion flux from plasma (Γ_{ion}) and the average energy of ion incident on etching materials (E_{ion}) in terms of physical damage. Regarding charging damage, the protection diodes[5, 16] have been widely introduced in an LSI. The present discussion focuses on the antenna ratio smaller than approximately 10^2.[7, 15] The antenna ratio range evaluated in the present simulation usually satisfies the antenna design rules.[7, 15] Since the antenna ratio range may be difficult to suppression, one should control the plasma charging to high-k devices. PID data with respect to high-k MOSFETs should be carefully evaluated, because high-k is subject to charge trapping mechanisms compared to conventional SiO_2.[8]

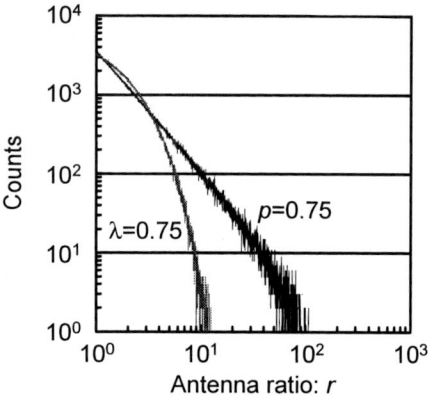

Fig. 5.   Antenna ratio distributions employed in the present simulation. p=0.75 (the Rent's exponent) and λ=0.75 (Eq. (3)) are assumed.

Figures 6(a) and 6(b) show calculated ΔV_{th}-distribution $f_{\Delta Vth}(\Delta V_{th})$ induced by (a) d_R- and (b) r-variations, respectively. In this figure, Eq. (2), the r-dependence of ΔV_{th} is rewritten as $A(r-r_0)^{\alpha}$ to make $\Delta V_{th} \rightarrow 0$ when r approaches the minimum antenna ratio r_0. In the present case, $r_0 = 1$ on the basis of Rent's rule for simplicity. As seen, $f_{\Delta Vth}(\Delta V_{th})$ is strongly dependent on $f_r(r)$ implying that the antenna design rules should be carefully optimized in accordance with charging damage data obtained for the corresponding process. Moreover, from those simulated data, the variances of ΔV_{th}, $Var(\Delta V_{th})$ ($= \sigma_{\Delta Vth}^2$) are determined. The deviation of ΔV_{th} ($3\sigma_{\Delta Vth}$) – more useful in practice – are 3.0 mV (σ_{dR}=0.1 nm), 5.7 mV (σ_{dR}=0.2 nm), 10.5 mV (p=0.75), 7.8 mV (λ=0.75), respectively, as displayed in Fig. 7.

Once obtaining PID-contribution to V_{th}-variation in Fig. 7, one can estimate the total ΔV_{th}-variation,

$$\sigma_{Vth} = \sqrt{(\sigma_{Vth}^0)^2 + (\sigma_{\Delta Vth}^{PID})^2}, \tag{5}$$

where ΔV_{th}^{PID} means the threshold voltage shift by PID as outlined in the above, and V_{th}^0 is the resultant threshold voltage determined by other device and process parameters such as channel doping, L_g, junction depth, LER, and etc. As seen in Fig. 7, PID induces an increase in $Var(\Delta V_{th})$ of MOSFET. These estimated values are not negligible compared to those widely discussed in many literatures.[2, 3] Since d_R and the

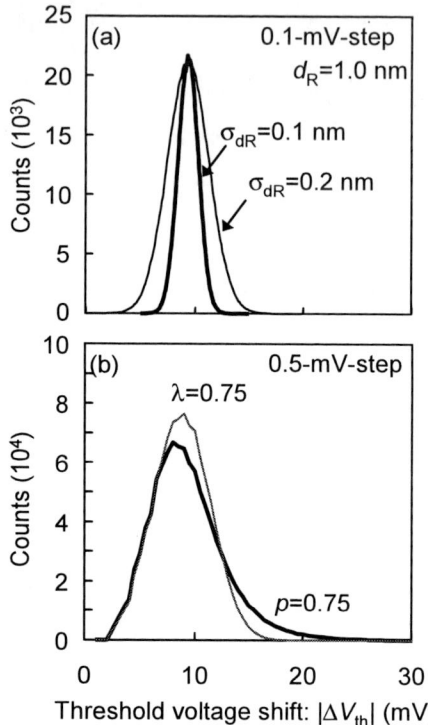

Fig. 6.   Calculated probability density functions of ΔV_{th} for various plasma-induced damage cases. (a) Physical damage by Si recess, (b) Charging damage to high-k MOSFET by the antenna effect.

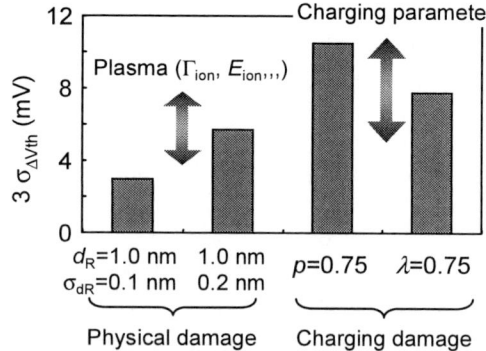

Fig. 7.   Calculated ΔV_{th}-variation for various plasma-induced damage cases. (The charging parameter B in Eq. (2) is assumed to be -0.01 V.)

C. I_{off}-Variability Prediction Model for PID

As discussed previously,[17] based on the above

analytical models, one can predict the I_{off}-variation by PID in addition to L_g. Here we integrate the L_g-variation as the representative fluctuation component with $f_r(r)$ and $f_{dR}(d_R)$ by PID. The joint density function [13] $f_{Ioff}(I_{off})$ is estimated from the derivative of the integral of $f_{Lg}(L_g)$, $f_r(r)$, and $f_{dR}(d_R)$ in the region where L_g, r, and d_R are defined. Figures 8(a) and 8(b) show the calculated results for $> 10^6$ transistors. In this simulation, the variable L_g is assumed to obey a Gaussian distribution. The mean value and the standard deviation are 32 and 1 nm. I_{off} is estimated as the subthreshold leakage current. Since I_{off} does not obey the normal distribution, in Fig. 8(b), the variance is shown as the deviation.

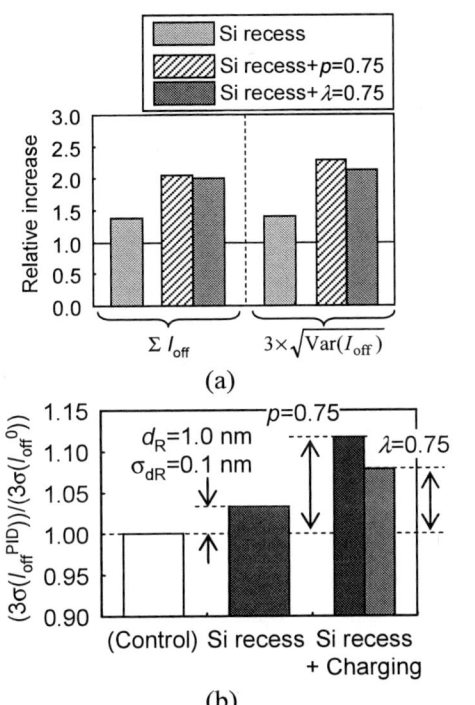

Fig. 8. (a) Calculated total I_{off} increase and the increase in variance by PID for all devices ($> 10^6$) in a chip. (b) Calculated I_{off}-variation increase by plasma induced physical and charging damage for two different $f_r(r)$. Note that the increases in Fig. 8(b) are applicable regardless of process technology.

As seen, both PID mechanisms increase the I_{off} and the variance up to $\sim 100\%$ in for "fully damaged" devices as in Fig. 8(a). These findings imply that, even if d_R-induced ΔV_{th} is considered in designing LSI in advance, the impacts of charging damage ($\sim$ antenna design rule) and σ_{dR} (related to Si lattice constant) on I_{off}-variance are difficult to eliminate, i.e., they enlarge the variability for all L_g-generations. Even for less recessed devices (replacement gate process relatively easy to control d_R or high-k difficult to etch), they may suffer from $> 50\%$ total I_{off}-increase by charging damage as deduced from Figs. 7 and 8. In other words, a variability component by PID already exists in present technology data.

V. CONCLUSION

We have proposed a PID-integrated variability prediction model and statistically clarified impacts of PID on V_{th} and I_{off}, and their variances. Combined with experimental data and a

Monte-Carlo method, the stochastic PID-effects on parameter distributions are identified. The present findings imply that, for accurate variability estimate, the proposed prediction model is useful and should be integrated into conventional device/circuit simulation frameworks (e.g., BSIM + SPICE).

ACKNOWLEDGMENT

We thank Drs. M. Yoshimaru, H. Hayashi, S. Hayashi, H. Kokura, and T. Tatsumi at STARC for their helpful discussion.

REFERENCES

[1] K. A. Bowman, A. R. Alameldeen, S. T. Srinivasan, and C. B. Wilkerson, "Impact of Die-to-Die and Within-Die Parameter Variations on the Clock Frequency and Throughput of Multi-Core Processors," *IEEE Trans. Very Large Scale Integration (VLSI) Systems*, vol. 17, pp. 1679-1690, 2009.

[2] P. Liang-Teck, Q. Kun, C. J. Spanos, and B. Nikolic, "Measurement and Analysis of Variability in 45 nm Strained-Si CMOS Technology," *IEEE J. Solid-State Circuits*, vol. 44, pp. 2233-2243, 2009.

[3] H. Fukutome et al., "Effects of Gate Line Width Roughness on Threshold-Voltage Fluctuation Among Short-Channel Transistors at High Drain Voltage," *IEEE Electron Device Lett.*, vol. 31, pp. 240 - 242, 2010.

[4] C. D'Agostino, P. Flatresse, E. Beigne, and M. Belleville, "Statistical Leakage Modeling in CMOS Logic Gates Considering Process Variations," *Proc. Int. Conf. on Integrated Circuit Design & Technol.*, pp. 301-304, 2008.

[5] K. P. Cheung, *Plasma Charging Damage*: Springer, 2001.

[6] K. Eriguchi et al., "Effects of Plasma-Induced Si Recess Structure on n-MOSFET Performance Degradation," *IEEE Electron Device Lett.*, vol. 30, pp. 712-714, 2009.

[7] C. T. Gabriel and E. d. Muizon, "Quantifying a Simple Antenna Design Rule," *Proc. Int. Symp. Plasma Process-Induced Damage*, pp. 153-156, 2000.

[8] K. Eriguchi et al., "Comprehensive Modeling of Threshold Voltage Variability Induced by Plasma Damage in Advanced Metal–Oxide–Semiconductor Field-Effect Transistors," *Jpn. J. Appl. Phys.*, vol. 49, p. 04DA18, 2010.

[9] K. Eriguchi, Y. Nakakubo, A. Matsuda, Y. Takao, and K. Ono, "Plasma-Induced Defect-Site Generation in Si Substrate and Its Impact on Performance Degradation in Scaled MOSFETs," *IEEE Electron Device Lett.*, vol. 30, pp. 1275-1277, 2009.

[10] K. Eriguchi et al., "A New Framework for Performance Prediction of Advanced MOSFETs with Plasma-Induced Recess Structure and Latent Defect Site," *IEDM Tech. Dig.*, pp. 443-446, 2008.

[11] K. Eriguchi, Y. Nakakubo, A. Matsuda, Y. Takao, and K. Ono, "Model for Bias Frequency Effects on Plasma-Damaged Layer Formation in Si Substrates," *Jpn. J. Appl. Phys.*, vol. 49, p. 056203, 2010.

[12] K. Eriguchi and K. Ono, "Quantitative and comparative characterizations of plasma process-induced damage in advanced metal-oxide-semiconductor devices," *J. Phys. D: Appl. Phys.*, vol. 41, p. 024002, Jan. 2008.

[13] A. Papoulis and S. Pillai, *Probability, Random Variables, and Stochastic Processes*. New York: McGraw-Hill, 2002.

[14] P. Christie and D. Stroobandt, "The interpretation and application of Rent's rule," *IEEE Trans. Very Large Scale Integration (VLSI) Systems*, vol. 8, pp. 639-648, 2000.

[15] W. Maly, C. Ouyang, S. Ghosh, and S. Maturi, "Detection of an antenna effect in VLSI designs," in *Proc. Int. Symp. Defect and Fault Tolerance in VLSI Systems*, 1996, pp. 86-94.

[16] V. Shukla, V. Gupta, C. Guruprasad, and G. Kadamati, "Automated antenna detection and correction methodology in VLSI designs " *Proc. Int. Symp. Plasma Process-Induced Damage* pp. 158-161, 2003.

[17] K. Eriguchi, M. Kamei, Y. Takao, and K. Ono, "Modeling the Effects of Plasma-Induced Physical Damage on Subthreshold Leakage Current in Scaled MOSFETs," *Proc. Int. Conf. on Integrated Circuit Design & Technol.*, pp. 94-97, 2010.

Impact of La on the Bias-Temperature Instability of the HfSiO High-κ N-MOSFET

D. S. Ang and G. A. Du

Abstract—Lanthanum (La), which has been used in recent works to tune the threshold voltage of HfSiO high-κ n-MOSFETs, is shown to introduce a new *bulk* degradation mechanism. Unlike the conventional charge trapping mechanism which exhibits low activation energy (~0.05 eV) and fast post-stress recovery, the La induced degradation mechanism is found to be relatively permanent and has higher activation energy (~0.26 eV). The latter is expected to have a significant impact on the positive-bias temperature instability of n-MOSFETs employing La-doped high-κ gate dielectrics.

Index Terms— Metal/high-κ gate stack, rare-earth oxides, bias-temperature instability, oxide traps, interface states

I. INTRODUCTION

REALIZING a viable high-κ gate stack for future CMOS technologies has been challenging. Besides the thermal stability issue, achieving the "correct" threshold voltage (V_t) is a major obstacle. Fermi-level "pinning" at the polysilicon/high-κ interface [1] had rendered low symmetric V_t's for the n- and p-MOSFET practically impossible. While the use of metal gates has alleviated this problem, the influence of high-κ/SiO$_x$ interface dipoles remains [2], resulting often in too large a V_t. In recent studies [3]-[5], La was introduced to counteract the interface dipoles to achieve a low V_t for the n-MOSFET. To-date, however, the impact of La doping on the gate stack reliability, especially in regards to the bias-temperature instability, is yet to be examined thoroughly [6]-[10].

In this paper, we review our recent finding of an abnormally slow recovery of n-MOSFETs employing the La-doped HfSiON gate dielectric, after subjecting the devices to bias-temperature stressing [11]. Electrical measurement results show that the slow recovery is a consequence of a new and permanent bulk degradation mechanism introduced by La. The activation energy of this degradation mechanism is shown to be relatively large, implying that the defects involved are induced by the bias-temperature stress itself and it is unlike the conventional charge trapping mechanism which involves the

Manuscript received March 22, 2011. This work was supported in part by a Singapore Ministry of Education Tier-2 research grant MOE2009-T2-1-050 and in part by an NTU Ph.D. scholarship grant (G. A. Du).
D. S. Ang is with the School of Electrical and Electronic Engineering, Nanyang Technological University, Singapore 639798 (e-mail: edsang@ntu.edu.sg).
G. A. Du was with the School of Electrical and Electronic Engineering, Nanyang Technological University. He is now with GLOBALFOUNDRIES Singapore Pte. Ltd., Singapore 738406.

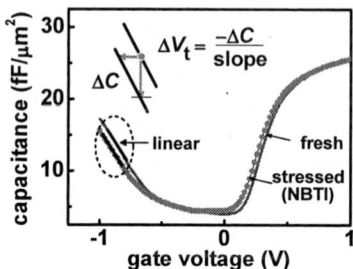

Fig. 1. Capacitance-voltage curves of a La-doped n-MOSFET subjected to negative-bias stress. The thick lines (enclosed by the dashed circle) show linear segments of the *C-V* curves with no change in the slope after stress. This feature was exploited for fast extraction of the voltage shift (~ΔV_t) between the 2 curves without a change in the gate polarity, i.e. a single point capacitance measurement was made at a negative gate voltage (dotted circle) and ΔV_t deduced from the change in capacitance assuming a constant *C-V* slope (inset). The same approach was used to obtain the ΔV_t (at a positive gate measurement voltage) of a p-MOSFET subjected to positive-bias stress.

rapid "filling" or "emptying" of oxide traps [12]-[14]. Due to the relatively permanent nature of the La induced degradation mechanism and its strong temperature dependence, it is expected to have a serious impact on the positive-bias temperature instability (PBTI) of the n-MOSFET.

II. EXPERIMENTAL DETAILS

Test devices were n-MOSFETs with TiN gate and La-doped HfSiON gate dielectric, fabricated by a gate-first CMOS process. A 2-nm HfSiO was formed via atomic layer deposition on an ultrathin SiON (< 1 nm) interfacial layer. A thin layer of LaO$_x$ (~0.3 nm) was then deposited and diffused into the HfSiO via a high temperature rapid annealing step (~1000 °C). Undoped p-MOSFETs fabricated on the same wafer were also tested for comparison. To confirm the new La induced degradation mechanism, undoped n- and p-MOSFETs from another process technology were studied. To make our study more comprehensive, MOSFETs with the polysilicon/ HfO$_2$ as well as the polysilicon/SiON gate stacks were also tested.

Measurement of bias-temperature stressing induced degradation of the linear drain current (I_d) was made at a gate voltage magnitude of 1 V and at a delay of 1 μs (minimum) or longer after stress termination using the Agilent B1500A semiconductor device analyzer. All stresses were performed at 125 °C unless stated otherwise. Measurement of the charge pumping (CP) current (I_{cp}) was also carried out for the evaluation of interface state generation. A 50 % duty cycle, 1

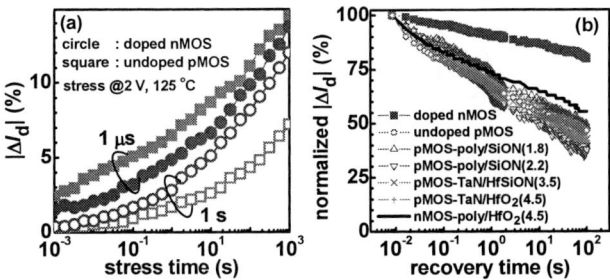
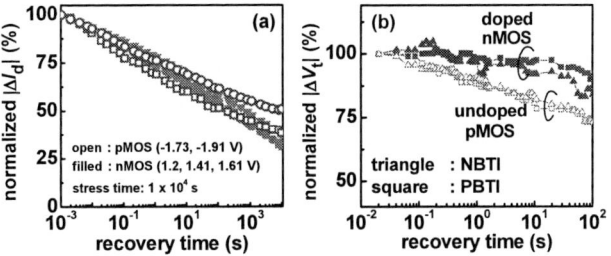

Fig. 2. (a) Evolution of linear drain current degradation (ΔI_d), for two measurement delays (1 μs and 1 s). The undoped p-MOSFET (doped n-MOSFET) was stressed at -2 ($+2$) V. (b) Fractional recovery of ΔI_d (@ 0 V gate voltage) of the doped n-MOSFET and undoped p-MOSFET, as well as a variety of p- and n-MOSFETs having different gate stacks. The brackets indicate the gate dielectric thickness in (nm), inclusive of the interfacial layer for the case of the high-κ gate stack. Stress time was 1×10^2 s.

Fig. 3. (a) Fractional recovery of ΔI_d (@ 0 V gate voltage) of the undoped n-and p-MOSFETs from a different process technology, showing similar recovery speeds. The brackets indicate the gate stress voltages. (b) Similarly, both NBTI and PBTI recover at similar speeds but those of the doped n-MOSFET are consistently slower than those of the undoped p-MOSFET, unlike in (a). Stress time was 1×10^2 s.

MHz gate pulse switching between ± 1 V was used. The measurement delay was in this case was ~20 ms.

A *single-point* capacitance-voltage (*C-V*) measurement was also made within ~20 ms after stress termination, at the same gate polarity as that of the gate stress voltage. The purpose was to obtain the voltage shift ($\sim \Delta V_t$) between the pre- and post-stress curves by assuming that the slope of the *C-V* curve in the vicinity of the measurement point was unaffected by the stress. This was verified in Fig. 1. In this way, ΔV_t shift of n-MOSFET subjected to negative-bias stress or p-MOSFET subjected to positive-bias stress could be deduced without having to change the gate polarity, i.e. any impact on recovery arising from a change in gate polarity as in the case of I_d measurement was entirely avoided.

III. RESULTS AND DISCUSSION

Our observation of a new La degradation mechanism began with the comparison of stress induced I_d shifts at different measurement delays (Fig. 2(a)). For the 1 μs delay, ΔI_d of the undoped p-MOSFET is larger than that of the doped n-MOSFET. It is interesting to note that the reverse happens for the 1 s delay. The apparent influence of measurement delay on the relative difference in the ΔI_d's implies that they are recovering at different rates after stress. The result suggests that ΔI_d of the doped n-MOSFET is recovering more slowly compared to that of the undoped p-MOSFET. Fig. 2(b), which compares the logarithmic recovery behavior of the normalized ΔI_d's, agrees with this inference. Also shown are the recovery characteristics of p-MOSFETs with a variety of gate stacks showing comparable recovery rates as that of the undoped p-MOSFET. The result confirms that it is the doped n-MOSFET which recovers much more slowly.

Studies [15], [16] have suggested that the degradation mechanism of PBTI is different from that of the negative-bias temperature instability (NBTI). The former is related to electron trapping in the high-κ layer while the latter is driven by interface state generation. Hence, it is important for one to ascertain whether this is the reason for the difference in recovery speed of the doped n-MOSFET and undoped

p-MOSFET (Fig. 2). Undoped n- and p-MOSFETs from another process technology were subjected to PBTI and NBTI test, respectively. Fig. 3(a) shows comparable recovery speeds, suggesting that the possibly different PBTI and NBTI mechanisms do not play a major role. It should be emphasized that the undoped n- and p-MOSFETs do not exhibit identical recovery, as different slopes of the logarithmic time dependent characteristics could be observed. This difference, however, pales in comparison to that between the doped n-MOSFET and undoped p-MOSFET in Fig. 2(b). PBTI test of n-MOSFET with the polysilicon/HfO$_2$ gate stack also exhibits similar recovery as the rest of the undoped devices (Fig. 2(b)). The result implies that the much slower recovery of the doped n-MOSFET is related to the presence of La. To further check this inference, the doped n-MOSFET was subjected to both NBTI and PBTI tests and the recovery characteristic of each case was compared in Fig. 3(b). To avoid complications arising from having to change the gate polarity in order to measure the ΔI_d of the n-MOSFET, ΔV_t was derived from capacitance measurement by the manner described above using the same gate polarity as that of the applied stress. As can be observed from Fig. 3(b), the ΔV_t of both NBTI and PBTI recover at comparable speeds. The same observation applies to the undoped p-MOSFET but clearly in this case, NBTI and PBTI both recover much faster than those of the doped n-MOSFET. The results therefore show unambiguously that the slow recovery speed of the doped n-MOSFET is because of the presence of La.

Interface state generation and charge trapping in the high-κ layer are the two main degradation mechanisms of BTI [12]–[16]. To probe the effect of La on these two mechanisms, CP measurement was made to examine the relaxation behavior of interfacial oxide defects. Contrary to the result on the normalized ΔI_d and ΔV_t recovery (Figs. 2(b) and 3(b), respectively), the normalized ΔI_{cp} of the doped n-MOS recovers more than that of the undoped p-MOS (Fig. 4(a)). This observation rules out the role of La on the interfacial oxide defects and suggests that La may have changed the properties of charge trapping in the high-κ layer. Fig. 4(b) shows the

978-1-4244-9019-6/11 $26.00 © 2011 IEEE

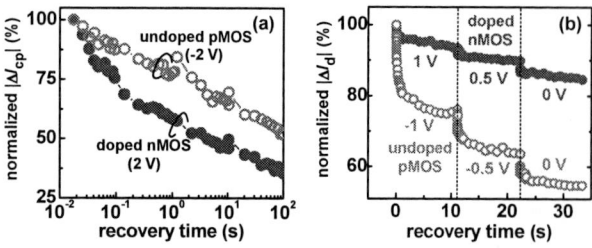

Fig. 4. (a) Fractional recovery of the increase in the charge pumping current (ΔI_{cp}). The brackets indicate the gate stress voltages. (b) Effect of different gate recovery voltages (as indicated) on the relaxation of the linear drain current degradation (ΔI_d).

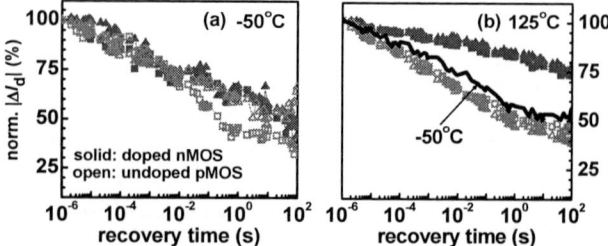

Fig. 6. Recovery of normalized ΔI_d versus time for different temperatures. In (b), the line denotes averaged p-MOS recovery at –50 °C.

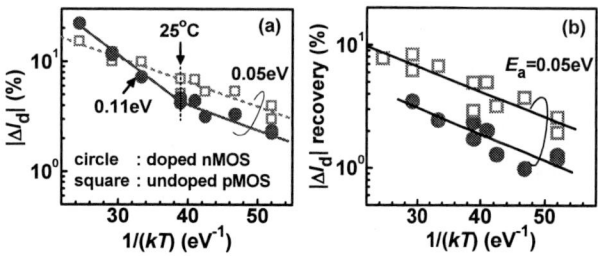

Fig. 5 (a) Thermal activation of ΔI_d degradation for La-doped n-MOS PBTI and undoped p-MOS NBTI (magnitude of gate stress voltage = 2 V; stress time = 100 s). (b) shows the thermal activation of ΔI_d recovery 200 s after stress termination.

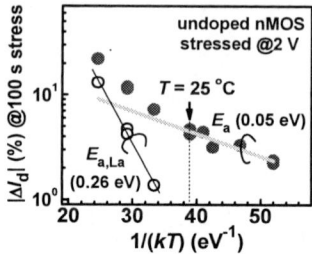

Fig. 7. Extraction of the activation energy $E_{a,La}$ of the La induced degradation mechanism. Contribution of the conventional BTI mechanism in the high temperature regime is estimated through extrapolating the data obtained in the low temperature regime (< 25 °C), on the basis that it is similar to that of the p-MOSFET (cf. Fig. 5(a)) and it does not interact with the La induced mechanism (cf. Fig. 5(b)). This contribution of the usual BTI mechanism in the high temperature regime was then subtracted from the as-measured $|\Delta I_d|$.

effect of a gate-voltage change on the recovery of the normalized ΔI_d. Upon a decrease of the gate voltage, a substantial increase in the rate of the undoped p-MOSFET recovery is evident. This behavior has been ascribed to a lowering of relatively deep hole-trap states below the interface Fermi level when the gate voltage is reduced, thus resulting in spontaneous compensation of the positive charge [13]. For the doped n-MOSFET, however, the gate voltage change has a very small effect on its recovery. This result corroborates the inference that La has changed the nature of charge trapping in the high-κ layer. It further implies that the stress induced La-related defects in the high-κ are deeper and/or more permanent.

The charge trapping mechanism of NBTI is typically characterized by a small activation energy (~0.05 eV) [12]-[14]. This is indeed observed for the undoped p-MOSFET (Fig. 5(a)). We also examined the thermal activation of NBTI recovery, which exhibits an E_a of 0.05 eV as well (Fig. 5(b)), in general agreement with the notion that NBTI recovery is mainly determined by charge (hole) detrapping [12]-[14]. In the relatively low temperature regime, the E_a of the doped n-MOSFET PBTI is identical to that of the undoped p-MOSFET NBTI. This result implies that both devices share a common charge trapping mechanism, which may not be surprising for a given HfSiON gate stack processed under the same conditions (except for the La doping). Oxygen vacancy defects in the high-κ gate stack are known to function both as electron and hole traps. Further support for a similar charge

trapping mechanism stems from the *identical* recovery characteristics of both devices at a low temperature (Fig. 6(a)). The result further implies that the slow PBTI recovery of the doped n-MOSFET is not due to preexisting La defects. It also implies that the possibly larger tunneling barrier, due to La dipoles at the HfSiO/SiON interface [9], is not the main reason for the reduced recovery speed of the doped n-MOSFET (Figs. 2(b) and 3(c)).

In the high temperature regime, however, PBTI of the doped n-MOSFET exhibits a more rapid increase as compared to the NBTI of the undoped p-MOSFET, signifying the onset of another degradation mechanism having stronger thermal activation. This faster rise in degradation corresponds to a slowing down of the doped n-MOSFET recovery (Fig. 4(b)). It should be remarked that such a non-Arrhenius behavior is also observed for the NBTI of the doped n-MOSFET (not shown), corresponding to the slower recovery in Fig. 3(c). In an undoped n-MOSFET, such a non-Arrhenius behavior is absent (not shown) and a slowing down in recovery is not observed (Fig. 3(a)). On the other hand, the temperature dependence of the undoped p-MOSFET NBTI follows the Arrhenius law. Moreover, the NBTI recovery speed at high temperature remains similar to that at low temperature (line in Fig. 6(b)). These results indicate that the reduced BTI recovery of the doped n-MOSFET is a consequence of thermally induced generation of new La defects. Fig. 7 illustrates the separation of this La induced degradation mechanism, of a much higher E_a (~0.26 eV), from the measured data.

No such non-Arrhenius behavior is, however, seen in the thermal activation study of the doped n-MOSFET recovery. The E_a (~0.05 eV) is similar to that of the undoped p-MOSFET, indicating that recovery of both devices is determined by the common charge trapping mechanism. This strongly implies that the La degradation mechanism activated at high temperature is relatively permanent and this is the mechanism which slows down the recovery of the doped n-MOS (Figs. 2(b), 3(c) and 6(b)).

The atomic nature of the La induced degradation mechanism is not clear at this stage. As proposed in a recent simulation study [7], although La tends to suppress the formation of oxygen vacancies via an increase of the formation energy by ~0.7 eV, the greater ionicity of the La–O bond may increase the susceptibility of oxygen interstitial formation during bias-temperature stressing. These oxygen interstitials introduce deep electron traps (π^* states close to the high-κ valence band edge) [17]. However, their proximity to the high-κ valence band does not make them good candidates for deep hole traps. Recall that the NBTI recovery of the doped n-MOSFET is also retarded (cf. Fig. 3(b)). Simulation study [17] have also shown that charged oxygen vacancy defects in the high-κ generally undergo greater structural relaxation, owing to the inherently stronger electron-lattice interaction, leading to permanent defect formation. The introduction of La to HfSiO may have enhanced the electron-lattice interaction in the gate dielectric since its ionicity would be increased. More studies are needed to clarify the exact atomic nature.

IV. CONCLUSION

Results from bias-temperature stressing of the La-doped HfSiO gate dielectric unambiguously show that La introduces a new oxide degradation mechanism which is fundamentally different from the conventional BTI mechanism. The La-related degradation mechanism has a stronger thermal activation and is more permanent. It is expected to have a significant impact on PBTI reliability of the n-MOSFET.

REFERENCES

[1] C. Hobbs, L. Fonseca, V. Dhandapani, S. Samavedam, B. Taylor, J. Grant, L. Dip, D. Triyoso, R. Hegde, D. Gilmer, R. Garcia, D. Roan, L. Lovejoy, R. Rai, L. Hebert, H. Tseng, B. White, and P. Tobin, "Fermi level pinning at the poly-Si/metal oxide interface," in *Symp. VLSI Tech. Tech. Dig.*, 2003, pp. 9-10.

[2] Y. Kamimuta, K. Iwamoto, Y. Nunoshige, A. Hirano, W. Mizubayashi, Y. Watanabe, S. Migita, A. Ogawa, H. Ota, T. Nabatame, and A. Toriumi, "Comprehensive study of V_{FB} shift in high-k CMOS – diople formation, Fermi-level pinning and oxygen vacancy effect –" in *IEDM Tech. Dig.*, 2007, pp. 341-344.

[3] H. N. Alshareef, H. R. Harris, H. C. Wen, C. S. Park, C. Huffman, K. Choi, H. F. Luan, P. Majhi, B. H. Lee, R. Jammy, D. J. Lichtenwalner, J. S. Jur, and A. I. Kingon, "Thermally stable N-metal gate MOSFETs using La-incorporated HfSiO dielectric," in *Symp. VLSI Technol. Tech. Dig.*, 2006, pp. 7-8.

[4] V. Narayanan, V. K. Paruchuri, N. A. Bojarczuk, B. P. Linder, B. Doris, Y. H. Kim, S. Zafar, J. Sathis, S. Brown, J. Arnold, M. Copel, M. Steen, E. Cartier, A. Callegari, P. Jamison, J.-P. Locquet, D. L. Lacey, Y. Wang, P.

E. Batson, P. Ronsheim, R. Jammy, M. P. Chudzik, M. Ieong, S. Guha, G. Shahidi, and T. C. Chen, "Band-edge high-performance high-κ/metal gate n-MOSFETs using cap layers containing group IIA and IIIB elements with gate-first processing for 45 nm and beyond," in *Symp. VLSI Technol. Tech. Dig.*, 2006, pp. 178-179.

[5] P. D. Kirsch, M. A. Quevedo-Lopez, S. A. Krishnan, C. Krug, H. AlShareef, C. S. Park, R. Harris, N. Moumen, A. Neugroshel, G. Bersuker, B. H. Lee, J. G. Wang, G. Pant, B. E. Gnade, M. J. Kim, R. M. Wallace, J. S. Jur, D. J. Lichtenwalner, A. I. Kingon, and R. Jammy, "Band edge n-MOSFETs with high-k/metal gate stacks scaled to EOT=0.9nm with excellent carrier mobility and high temperature stability," in *IEDM Tech. Dig.*, 2006, pp. 370-373.

[6] S. Z. Chang, T. Y. Hoffmann, H. Y. Yu, M. Aoulaiche, M. Rohr, C. Adelmann, B. Kaczer, A. Delabie, P. Favia, S. Van Elshocht, S. Kubicek, T. Scharm, T. Witters, L.-A. Ragnarsson, X. P. Wang, H.-J. Cho, M. Mueller, T. Chiarella, P. Absil, and S. Biesemans, "Low V_T metal-gate/high-k nMOSFETs – PBTI dependence and V_T tune-ability on La/Dy-capping layer locations and laser annealing conditions," in *Symp. VLSI Technol. Tech. Dig.*, 2008, pp. 62-63.

[7] M Sato, N. Umezawa, J. Shimokawa, H. Arimura, S. Sugino, A. Tachibana, M. Nakamura, N. Mise, S. Kamiyama, T. Morooka, T. Eimori, K. Shiraishi, K. Yamabe, H. Watanabe, K. Yamada, T. Aoyama, T. Nabatame, Y. Nara, and Y. Ohji, "Physical model of the PBTI and TDDB of La incorporated HfSiON gate dielectrics with pre-existing and stress-induced defects," in *IEDM Tech. Dig.*, 2008, pp. 119-122.

[8] C. Y. Kang, C. D. Young, J. Huang, P. Kirsch, D. Heh, P. Sivasubramani, H. K. Park, G. Bersuker, B. H. Lee, H. S. Choi, K. T. Lee, Y.-H. Jeong, J. Lichtenwalner, A. I. Kingon, H.-H. Tseng, and R. Jammy, "The impact of La-doping on the reliability of low V_{th} high-k/metal gate nMOSFETs under various gate stress conditions," in *IEDM Tech. Dig.*, 2008, pp. 115-118.

[9] W.-H. Choi, H.-M. Kwon, I.-S. Han, T.-G. Goo, M.-K. Na, C. Y. Kang, G. Bersuker, B. H. Lee, Y.-H. Jeong, H.-D. Lee, and R. Jammy, "A comprehensive and comparative study of interface and bulk characteristics of nMOSFETs with La-incorporated high-k dielectrics," in *IEDM Tech. Dig.*, 2008, pp. 111-114.

[10] M. Sato, S. Kamiyama, Y. Sugita, T. Matsuki, T. Morooka, T. Suzuki, K. Shiraishi, K. Yamabe, K. Ohmori, K. Yamada, J. Yugami, K. Ikeda, and Y. Ohji, "Negatively charged deep level defects generated by yttrium and lanthanum incorporation into HfO$_2$ for V_{th} adjustment, and the impact on TDDB, PBTI, and $1/f$ noise," in *IEDM Tech. Dig.*, 2009, pp. 123-126.

[11] G. A. Du, D. S. Ang, and C. J. Gu, "Abnormal slow recovery characteristic of La-doped HfSiO$_x$ N-MOSFET bias-temperature instability," in *IEDM Tech. Dig.*, 2010, pp. 94-97.

[12] A. Neugroschel, G. Bersuker, R. Choi, and B. H. Lee, "Effect of the interfacial SiO$_2$ layer in high-k HfO$_2$ gate stacks on NBTI," *IEEE Trans. Dev. Mat. Reliab.*, vol. 8, pp. 47-61, Mar. 2008.

[13] D. S. Ang, S. Wang, G. A. Du, and Y. Z. Hu, "A consistent deep-level hole trapping model for negative-bias temperature instability," *IEEE Trans. Dev. Mat. Reliab.*, vol. 8, no. 1, pp. 22-34, Mar. 2008.

[14] Z. Q. Teo, D. S. Ang, and K. S. See, "Can the reaction-diffusion model explain generation and recovery of interface states contributing to NBTI?" in *IEDM Tech. Dig.*, 2009, pp. 737-740.

[15] S. Pae, M. Agostinelli, M. Brazier, R. Chau, G. Dewey, T. Ghani, M. Hattendorf, J. Hicks, J. Kavalieros, K. Kuhn, M. Kuhn, J. Maiz, M. Metz, K. Mistry, C. Prasad, S. Ramey, A. Roskowski, J. Sandford, C. Thomas, J. Thomas, C. Wiegand, and J. Wiedemer, "BTI reliability of 45 nm high-K + metal-gate process technology," in *Proc. IRPS*, 2008, pp. 352-357.

[16] A. Kerber and Eduard Cartier, "A fast four-point sense methodology for extraction of circuit-relevant degradation parameters," *IEEE Electron Dev. Lett.*, vol. 31, no. 9, pp. 912-914, Sep. 2010.

[17] K. Xiong, J. Robertson, M. C. Gibson, and S. J. Clark, "Defect energy levels in HfO$_2$ high-dielectric-constant gate oxide," *Appl. Phys. Lett.*, vol. 87, art. no. 183505, Oct. 2005.

Separation of NBTI Component from Channel Hot Carrier Degradation in pMOSFETs Focusing on Recovery Phenomenon

Y. Mitani, S. Fukatsu, D. Hagishima, and K. Matsuzawa

Abstract— Channel hot-carrier (CHC) degradation becomes more critical as the channel length is reduced. In general, CHC degradation is evaluated using DC stress applying both gate and drain bias. However, in the case of p-channel MOSFETs, negative bias temperature instabilities (NBTI) also degrades threshold voltage (V_{TH}) and saturation drain current (I_{sat}) under DC stress applying gate bias. Therefore, CHC degradation might include the NBTI component, which would lead to over-estimate the CHC degradation. Therefore, a separation of the BTI component from CHC degradation is necessary to predict device lifetime more accurately. In this study, a simple separation method of NBTI and CHC component from CHC test data is proposed, focusing on the recovery phenomenon, which is a distinctive behavior of NBTI.

Index Terms—Reliability, Channel hot carrier, NBTI, pMOSFET, Threshold voltage, Recovery

I. INTRODUCTION

Issues concerning the reliability of ultra-thin gate dielectrics constitute one of the most serious challenges in the scaling of ULSI devices. In particular, CHC (Channel Hot Carrier) degradation has become one of the most serious concerns with the downscaling of the FET devices.

When both gate bias (V_G) and drain bias (V_D) are applied to MOSFETs under CHC test, the threshold voltage shift (ΔV_{TH}) and the degradation of transconductance (Δg_m) cause consequent to the hot carrier generation at drain edge and injection into the gate insulator. On the other hand, the threshold voltage shift in pMOSFETs caused by NBTI (Negative Bias Temperature Instability) has emerged as one of the more interesting and potentially serious reliability limiters for state-of-the art ULSI technology. [1-6]. When the gate is biased so as to invert the state of the channel region in the transistor under higher temperature, NBTI can be observed. That is, the NBTI occurs when the vertical bias to the channel is applied to the gate dielectrics. When applying CHC stress to devices, BTI is also unavoidable because of the vertical electric field. As a result, CHC degradation under DC stress

Y. Mitani, S. Fukatsu, D. Hagishima and K. Matsuzawa are with Advanced LSI Technology Laboratory, Corporate R&D Center, Toshiba Corporation, Yokohama 235-8522 Japan (phone: +81-45-776-5943; fax: +81-45-776-4113; e-mail: yuuichiro.mitani@toshiba.co.jp). S. Fukatsu is now with Device Process Development Center, Toshiba Corporation, and D. Hagishima is now with Toshiba Semiconductor Company.

measurement involves of both the NBTI and the CHC components. Therefore, separation of these components is necessary not only to understand the mechanism of CHC degradation, but it will also enable the improvement of TCAD simulation models, and more accurate lifetime prediction will be possible.

Some separation methods of these components from CHC degradation have been reported previously [7-9]. In one of them, it was reported that the CHC component (ΔV_{TH}^{CHC}) can be separated by subtracting DC-NBTI [$\Delta V_{TH}^{NBTI\_TEST}$] from the HC degradation ($\Delta V_{TH}^{CHC\_TEST}$) [7]. Here, DC-NBTI means the degradation under conventional NBTI test, in which constant negative bias is applied to the gate electrode with source, drain and substrate grounded.

$$\Delta V_{TH}^{CHC} = \Delta V_{TH}^{CHC\_TEST} - \Delta V_{TH}^{NBTI\_TEST} \qquad \cdots (1)$$

In this method, the NBTI component under CHC stress is regarded to coincide with the NBTI obtained under conventional DC stress of the same gate bias. However, the CHC component by using this method might be underestimated, because the BTI is sensitive to the vertical electric field, and the field under CHC stress is different from that under BT stress [9]. Actually, as shown in Ref. [7], the CHC component was estimated to be negative ($\Delta V_{TH}^{CHC}<0$) in the longer gate length pMOSFETs, because NBTI test data ($\Delta V_{TH}^{NBTI\_TEST}$) is more than CHC test data ($\Delta V_{TH}^{CHC\_TEST}$).

We have already proposed the separation method of BTI and CHC components in the devices having High-k gate dielectrics [9]. In this method, the separation of those components performs by using gate leakage current at BT and CHC stress condition. However, this method would be difficult to apply to the case of the devices having thicker gate dielectrics and to the case of the CHC stress condition in which the gate current due to hot carrier injection cannot be negligible, because this method requires the sensing of stationary gate leakage current.

In this paper, a simple separation method of NBTI and CHC components is proposed, focusing on the recovery phenomenon, which is the distinctive behavior of NBTI.

II. PROCEDURE FOR SEPARATION OF NBTI COMPONENT

Fig. 1 shows the degradation of ΔV_{TH} under DC-NBTI stress and its recovery under $V_G=0V$ in pMOSFETs having 1.8nm SiON films as gate dielectrics. As previously reported in a lot of literatures, the recovery phenomenon of NBTI is clearly observed. This recovery phenomenon has been explained as emission (de-trapping) of captured holes and re-passivation of interface states by hydrogen. On the other hand, as shown in the inset, no recovery phenomenon is observed in the CHC degradation in the case of nMOSFETs. This result agrees with the previous reports [10,11]. It has been reported that the recovery phenomenon in CHC degradation can be observed even in nMOSFETs, only when the back bias (body bias) is applied during CHC stress [10]. However, because the back bias was not applied, the recovery is not observed in our experiments. Therefore, the recovery phenomenon can be regarded as a distinctive behavior of NBTI. The proposed separation method in this paper uses this difference of the recovery phenomenon between NBTI and CHC degradation.

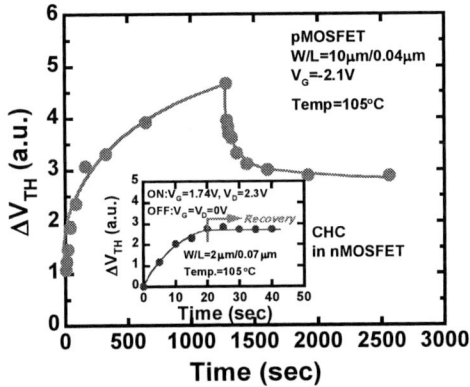

Fig. 1. Threshold voltage shift (ΔV_{TH}) under NBT stress and its recovery. NBTI (V_G=-2.1V) and the recovery (V_G=0V) are measured for 1280 sec. The recovery of ΔV_{TH} can be clearly observed in the pMOSFET. On the other hand, no recovery phenomenon is observed in CHC degradation in nMOSFET. From these results, the recovery phenomenon can be regarded as a distinctive behavior of NBTI.

The separation procedure is schematically shown in Fig. 2. First (STEP 1), the recoveries of NBTI are measured at the same gate voltage (V_G) of CHC test during arbitrary stress time. From these experimental data, the threshold voltage shift just before the recovery begins ($\Delta V_{TH}^{NBTI\_TEST}$) and the portions of the recovery (ΔV_{TH}^{R}) are estimated. Next (STEP 2), the recovery ratio (R) is estimated using the equation given by

$$R = \frac{\Delta V_{TH}^{R}}{\Delta V_{TH}^{NBTI\_TEST}}. \qquad \cdots(2)$$

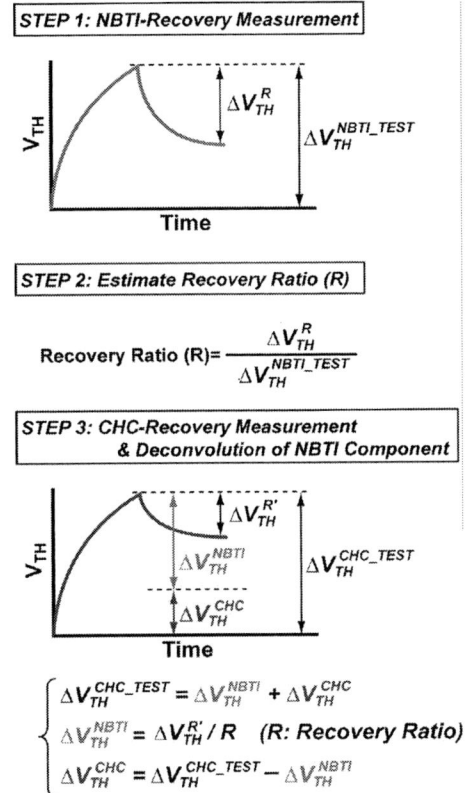

Fig. 2. Characterization procedure of separation method. The experimental data for the recovery of NBTI and HC degradation is used.

For example, the recovery ratio can be estimated to be 0.51 as shown in Fig. 3(a). Then (STEP 3), CHC test and its recovery are measured as shown in Fig. 3(b), and NBTI and CHC components (ΔV_{TH}^{NBTI} and ΔV_{TH}^{CHC}, respectively) are estimated using the equations given by

$$\Delta V_{TH}^{NBTI} = \frac{\Delta V_{TH}^{R'}}{R} \qquad \qquad ...(3)$$

$$\Delta V_{TH}^{CHC} = \Delta V_{TH}^{CHC\_TEST} - \Delta V_{TH}^{NBT} \qquad ...(4)$$

Here, $\Delta V_{TH}^{R'}$ corresponds to the recovery portion of CHC test data as shown in Fig.2 (STEP 3). In the case of Fig.3(b), ΔV_{TH}^{NBTI} and ΔV_{TH}^{CHC} can be estimated as 24% and 76% of $\Delta V_{TH}^{CHC\_TEST}$, respectively.

Here, NBTI component is regarded to be independent to the CHC degradation in this method. However, as previously reported in the literature [12], CHC degradation accelerates NBTI when CHC stress and NBT stress are applied alternately. In our experiments, this influence can be negligible because CHC stress and NBT stress are applied individually.

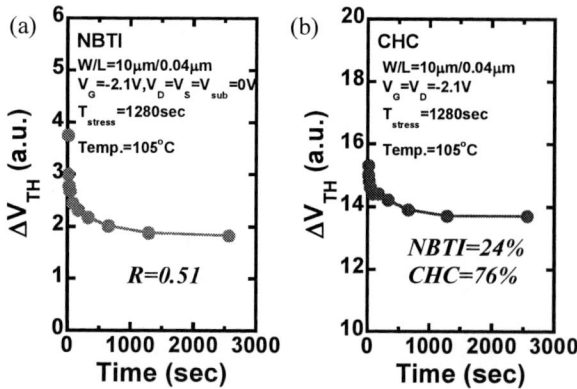

Fig. 3. Recovery in NBTI (a) and CHC degradation (b) in pMOSFETs. The recovery phenomenon is also observed clearly in the case of pMOSFETs. The recovery ratio (R) is estimated to be 0.51 from Fig. 3(a). Using proposed separation method, NBTI and CHC components can be estimated as 24% and 76% of $\Delta V_{TH}^{HC\_TEST}$, respectively.

III. EXPERIMENTAL RESULTS AND DISCUSSION

Fig. 4 shows the drain bias (V_D) dependence of the separation results of CHC degradation in pMOSFET. It should be noted that CHC components increases with increasing V_D, while NBTI component decreases slightly with increasing V_D. Since the effective vertical gate bias decreases toward drain along the channel by applying V_D, the behavior of NBTI component seems to be reasonable.

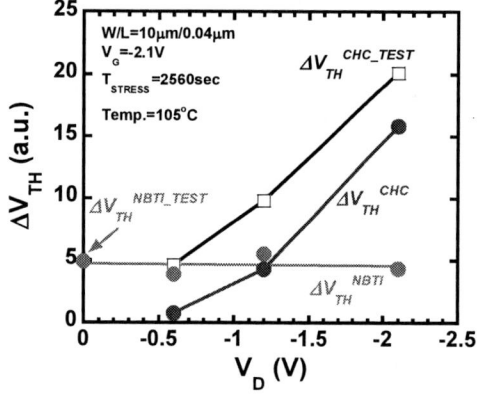

Fig. 4. Drain bias (V_D) dependence of separated NBTI and CHC components. CHC component depend strongly on V_D, while NBTI component slightly decrease with increasing V_D.

Fig. 5 shows the channel length dependence of the separation results of CHC degradation. In this experiment, the recovery ratio estimated from DC-NBTI using respective devices (L=0.04, 0.1 and 0.5μm) was applied. As previously reported in ref. [7], CHC component increases with decreasing channel length, while NBTI component is almost constant. Furthermore, separated NBTI components (ΔV_{TH}^{NBTI}) are smaller than the DC-NBTI data ($\Delta V_{TH}^{NBTI\_TEST}$), which exceeds CHC degradation ($\Delta V_{TH}^{CHC\_TEST}$) at longer gate length. This result

seems to be reasonable because the vertical oxide field during NBT stress is almost independent of the channel length in these experimental conditions. From these results, NBTI and CHC components can be separated from CHC test data in pMOSFETs by simple method by focusing the recovery phenomena.

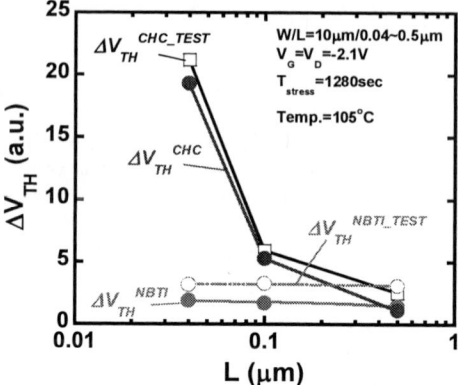

Fig. 5. Separated NBTI and CHC components as a function of channel length (L). CHC component monotonously increases with decreasing gate length. On the contrary, NBTI component is almost constant.

Fig. 6 shows the temperature dependence of CHC degradation, NBTI and CHC components. In this case, the recovery ratio are used the values which were obtained at respective temperature. In general, CHC degradation becomes marked with decreasing temperature. However, ΔV_{TH} value ($\Delta V_{TH}^{CHC\_TEST}$) increases gradually with increasing temperature before the respective components are separated. This is because NBTI component is included, which degrades with increasing temperature. On the other hand, after the separation, it can be seen a trend that extracted CHC component increases and NBTI component decreases at low temperature.

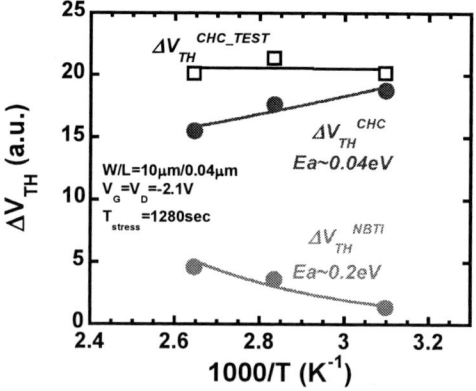

Fig. 6. Temperature dependence of separated NBTI and CHC components. Extracted CHC component monotonously increases with decreasing temperature, which agrees with the previous report [13]. Furthermore, estimated respective activation energies (Ea) also almost correspond to the previously reported values.

From this result, the activation energy can be estimated to be 0.2eV for NBTI component and 0.04eV for CHC component, respectively. The activation energy of NBTI has been reported as 0.1~0.2eV [6], and that of CHC degradation has reported as 0.01~0.03 after ref. [13]. These values estimated using experimental data support the validity of our separation method.

According to the previous report [14], CHC degradation depends strongly on substrate current (I_{SUB}). That is, the degradation is given by

$$Degradation \propto \left(\frac{I_{SUB}}{I_D} \right)^m \qquad \cdots(5)$$

Fig. 6 shows the drain current degradation ($\Delta I_D/I_D^0$) under CHC test as a function of substrate current (I_{SUB}/I_D). Here, I_D^0 corresponds to the drain current of the fresh device. The separated CHC component increases in accordance with the power law of the substrate current. On the other hand, CHC test data which includes NBTI component deviates from the power law of I_{SUB}. This result implies that the simple separation method would be enable TCAD simulation model and the prediction of lifetime for CHC degradation in pMOSFETs.

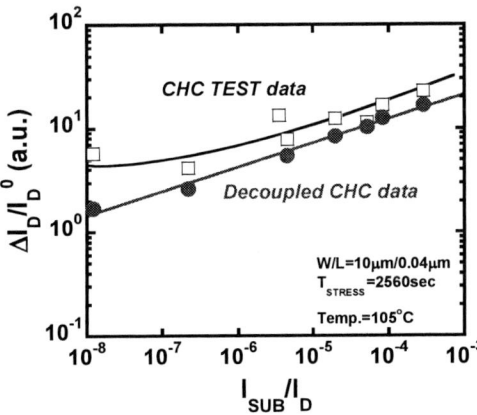

Fig. 6. CHC degradation as a function of substrate current. Separated CHC components are well-fitted to the power law of the substrate current (I_{SUB}/I_D), while CHC test data which includes NBTI component deviates from the power law.

IV. CONCLUSION

In this paper, simple separation method of NBTI and CHC components involved CHC test data in pMOSFETs is proposed. These components are separated successfully by focusing on the recovery phenomenon. The recovery ratio is estimated from NBTI recovery and then, NBTI component in the CHC test data can be extracted by using the recovery ratio. The drain bias dependence, the gate length dependence, temperature dependence of separated NBTI and CHC components suggest the validity of this method. Furthermore, extracted CHC component can be represented by the power law of substrate current.

It can be concluded that this simple separation method is powerful tool for the static CHC degradation. On the other hand, CHC degradation accelerated by NBTI when CHC stress and NBT stress are applied alternately has been discussed in the literature [11]. Therefore, the CHC degradation depending on NBTI should be also taken into account for more accurate TCAD simulation model under AC (dynamic) operation.

ACKNOWLEDGMENT

I would like to thank Drs. K. Inoue and T. Yamamoto for their thoughtful discussions and comments.

REFERENCES

[1] N. Kimizuka, T. Yamamoto, T. Mogami, K. Yamaguchi, K.Imai, T.Horiuchi, "The impact of bias temperature instability for direct-tunneling ultra-thin gate oxide on MOSFET scaling," in Symposium on VLSI Technology, p. 73 (1999).

[2] N. Kimizuka, K. Yamaguchi, K. Imai, T. Iizuka, C. T. Liu, R. C. Keller, T. Horiuchi, "NBTI enhancement by nitrogen incorporation into ultrathin gate oxide for 0.10-µm gate CMOS generation," in Symposium on VLSI Technology, p. 92 (2000).

[3] K. Uwasawa, T. Yamamoto, T. Mogami, "A new degradation mode of scaled p⁺ polysilicon gate pMOSFETs induced by bias temperature (BT) instability," in IEEE International Electron Devices Meeting Technical Digest, p.871 (1995).

[4] T. Yamamoto, K. Uwasawa, T. Mogami, "Bias Temperature Instability in Scaled p⁺Polysilicon Gate p-MOSFET's," IEEE Trans. Electron Devices, p. 921 (1999).

[5] Y. Mitani, M. Nagamine, H. Satake, A. Toriumi, "NBTI Mechanism in ultra-thin gate dielectric- nitrogen-originated mechanism in SiON-," in IEEE International Electron Devices Meeting Technical Digest, p. 509 (2002).

[6] M. A. Alam, "A critical examination of the mechanics of dynamic NBTI for p-MOSFETs," in IEEE International Electron Devices Meeting Technical Digest, p. 346 (2003).

[7] G. La Rosa, F. Guarin, S. Rauch, A. Acovic, J. Lukaitis, E. Crabbe, "NBTI - Channel Hot Carrier Effects in PMOSFETs in Advanced CMOS Technologies," in Proc. IEEE International Reliability Physics Symposium, p. 282 (1997).

[8] B. S. Doyle, B. J. Fishbein, K. R. Mistry, "NBTI-Enhanced hot carrier damage in p-channel MOSFET's," in IEEE International Electron Devices Meeting Technical Digest, p. 529 (1991).

[9] A. Masada, I. Hirano, S. Fukatsu, Y. Mitani, "Method of Decoupling the Bias Temperature Instability Component from Hot Carrier Degradation in Ultrathin High-k Metal–Oxide–Semiconductor Field-Effect Transistors," Japanese Journal of Applied Physics, 49, p. 0711021 (2010).

[10] S. Mahapatra, D. Saha, D. Varghese, "On the Generation and Recovery of Interface Traps in MOSFETs Subjected to NBTI, FN, and HCI Stress," IEEE Trans. Electron Devices, p. 1583 (2006).

[11] C.R. Parthasarathy, M. Denais, V. Huard, C. Guerin, G. Ribes, E. Vincent, A. Bravaix, "Unified perspective of NBTI and hot-carrier degradation in CMOS using on-the-fly bias patterns," in Proc. IEEE International Reliability Physics Symposium, p. 696 (2007).

[12] H. Aono, E. Murakami, K. Ohyama, *K. Makabe, K. Kuroda, K. Watanabe, H. Ozaki, K.Yanagisawa, K. Kuhota, and Y.Ohji, "NBT-induced Hot Carrier (HC) Effect: Positive Feedback Mechanism in p-MOSFET's Degradation," in Proc. IEEE International Reliability Physics Symposium , p.79 (2002).

[13] D. Esseni, L. Selmi, R. Bez, E. Sangiorgi, B. Ricci, "Bias and temperature dependence of gate and substrate currents in n-MOSFETs at low drain voltage," in IEEE International Electron Devices Meeting, p.307 (1994).

[14] P. M. Lee, M. M. Kuo, K. Seki, P. K. Ko, and C. Hu, "Circuit aging simulator (CAS) ," in IEEE International Electron Devices Meeting Technical Digest, p. 134 (1988).

On the impact of the edge profile of interconnects on the occurrence of passivation cracks of plastic-encapsulated electronic power devices

Jan Ackaert[1], Daniel Vanderstraeten[1], Bart Vandevelde [2]

(1) Corporate R&D, ON Semiconductors Westerring 15, 9700 Oudenaarde, Belgium
(2) Imec, Kapeldreef 75, B-3001 Leuven, Belgium.. Tel: (32)-55-332239
Email : jan.ackaert@onsemi.com

Abstract— **Deformations of metal interconnects, cracks in interlayer dielectrics and passivation layers in combination with plastic-packaging are still a major reliability concern for integrated circuit power semiconductors. In order to describe and understand the failure mechanism and its root cause, already a lot of work has been done in the past. However for the first time the impact of the edge profile of the power metal design on the amount of passivation cracks was investigated in detail. It was found that with a sloped edge profile of the power metal, as it is achieved with a combination of an isotropic wet etch followed by a dry etch, the number of passivation cracks is reduced significantly. The observation is confirmed by a 3-D FEM simulation. The simulation enabled to quantify the stress level and to forecast corresponding levels of cracks observed after temperature cycling. As a result, a robust metal edge profile design could be deduced, which led to a distinct reduction of the principal stress at the most critical positions and, consequently, to a reduction of passivation damage.**

I. INTRODUCTION

Deformations of metal interconnects , cracks in interlayer dielectrics and passivation layers in combination with plastic-packaging are still a major reliability concern for integrated circuit power semiconductors In order to describe and understand the failure mechanism and its root cause, already a lot of work has been done in the past [1]–[7].

The main root cause of the issues is the different nature of the silicon chip and the molding compound (MC) embedding this silicon. In general the plastic is softer (low young's modulus) and has a high coefficient of thermal expansion (CTE). In contradiction, the silicon is hard and has a much lower CTE. This mismatch of the material properties leads to a thermo-mechanically induced shear stress in operational conditions with varying thermal conditions. Worst case conditions are observed at the outer edges and in the corner regions of the silicon devices; there the thermo-mechancial stress reaches its maximum value. Interconnects of power devices are often made of aluminum. To protect the aluminum during mechanical handling, from corrosion and to limit moisture penetration in the underlaying interlayer dielectrics (ILD) integrated circuits are coated with plasma silicon nitride (SiN). Again there is a significant mismatch between these two

materials. The aluminum is soft and has a high CTE, while the SiN is hard and brittle and has a low CTE.

Together with the wide and thick metal layers as used in general on power devices, this combination leads to cracks in the SiN. After temperature cycling stressing, cracks mainly occur at the corners and edges of the device where the thermo-mechanical stress conditions are the most extreme. (Fig. 1)

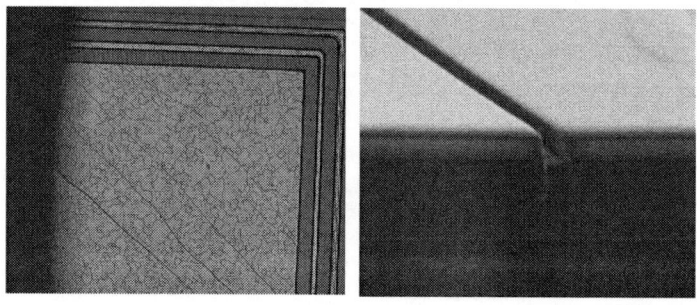

Fig. 1 Left: Typical passivation cracks on large Al areas near the corner of the die after 1000 TCs between −65 and +175 ∘C. Right: FIB Cross section through a typical power IC metalisation layer coated with SiN passivation.

In a number of cases, cracks propagate into the ILD layers and even into the Si underneath, are causing electrical shorts. On top of that, moisture can penetrate the IC from outside causing corrosion or other moisture related defects.. Since the failure mechanism needs time to develop, these failures must be considered as a sever reliability hazard.

To improve the quality and minimize the risk of reliability issues, all components, MC, passivation and interconnect including all possible interactions must be taken in account .

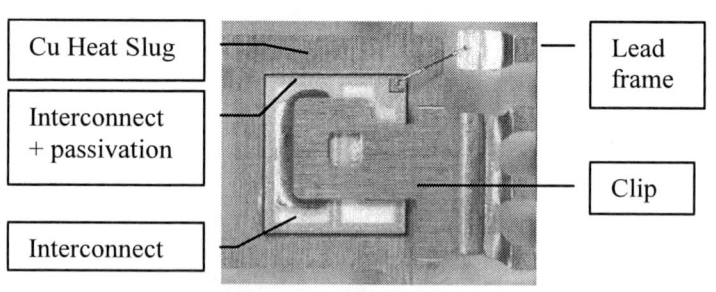

Fig. 2 Typical electronic power device.

In the 1970s a protective coating of the Al metal lines was introduced to prevent corrosion [1] and mechanical handling damage. Thicker layers were shown to be more stable concerning cracking [5]. Also the impact of layout has been investigated [8]–[14]. This resulted in the introduction of slotted broad metal lines and of 45° lines in corner regions. However, in the case with thick metal stacks still a considerable number of passivation cracks were observed after temperature cycling (TC). Within a power device, the stress distribution is governed by two construction features: while the top side of the chip is conventionally covered by the MC, the back side is attached to a solid Cu heat slug. Fig. 2 shows an illustration of a typical power device as, for example, in a clip based S08FL package.

The main features of power ICs are thick and broad metal plates for the wiring of power transistors, which cover a major part of the chip area. The thickness of the top metallization, which is also called power metallization, lies in the range between 3.5 and 5 µm.

Finite element simulation is a cost and time effective approach and is often performed to address these reliability issues. Tremendous growth in computer power has made it possible to perform large scale simulations, and hence, plenty of work has been reported in this field [15,16].

Earlier, the impact of the metal stack, passivation layer, layout and MC has been studied in detail with the help of a finite element model (FEM). The conclusion was that the whole counterforce against the shear stress during TC, is built up by the edges of the passivation layer only [15]. In this study here, for the first time the impact of the shape of this metal edge profile on the resistance to crack formation is investigated in detail. The study is based on both experimental data and FEM simulations.

I. THE FINITE ELEMENT MODEL

A finite element model of a test chip was developed to validate the material models and to explain the experimentally found cracking issues in the passivation. This is done by calculating the stresses in the passivation structure as part of the complete package.

Msc.Marc, a generic commercial finite element software, was used for this study. Simplifications were carried out to reduce the solver time. Therefore, mold compound, lead frame and die attach were included in the model, however, only silicon and power metal were taken into account for the chip. Advanced global-local meshing features were needed to cope with the large package dimension and the small dimensions of the passivation structure. A view of the mesh model is shown in Fig. 3 (the passivation is partially removed to show its structure in the cross-sectional).

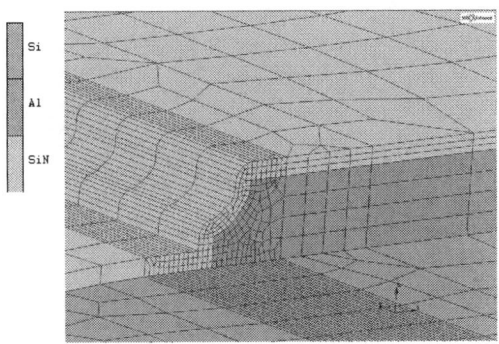

Fig. 3 Finite Element model simulating the thermo-mechanical stresses in the passivation layer.

In the Finite Element model consisting of about 100000 elements, the complete temperature cycling load (operational + environmental temperature changes) is applied and the mechanically induced stresses are calculated over the whole temperature profile. These stresses are induced by mismatch in CTE between the different materials (see Table 1). In addition to the linear properties, the aluminum is simulated as elasto-plastic material with a yield stress of 250 MPa.

Table 1: Material properties used in the FEM

Material	CTE (ppm/°C)	E-modulus (GPa)
Al	24	70
SiN (passiv.)	2.8	143
Overmould	10	24
Leadframe	16	120
Silicon	2.6	169

II. THE ALUMINUM METALLIZATION

Suo [5] and Huang et al. [6], [7] have shown that below the glass transition temperature Tg, the MC gains a high Young's modulus. Due to shrinking relative to the silicon, the MC causes a shear stress in the passivation film covering the metal plates. This shear stress τ caused by the CTE mismatch between the MC and Si is maximum near the chip edges and decreases in the direction of the chip center. As shown in Fig 5, at the chip center, the shear stress becomes zero. It was shown that, after typically 250 temperature cycles (TCs), the yielded metal layer gets properties very similar to a liquid and does not transfer the shear stress to the silicon chip any longer. The only balancing force is now established by the thin brittle passivation layer anchored at the edges of the metallization plate[15]. In order to verify this, samples with and without passivation have been exposed to TC. Fig. 4 shows the pictures of samples with and without passivation after TC. Without the passivationlayer the metal lines are displaced towards the center of the device over a distance of several microns. While on the sample with the passivation, the metal lines are still intact and in their original location.

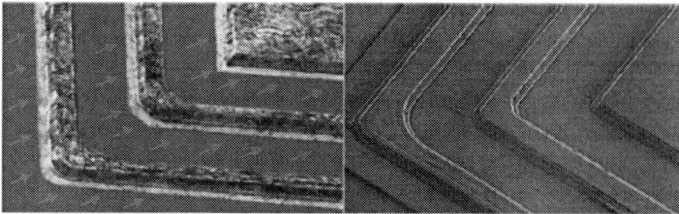

Fig. 4 After 1000 TC's on the sample without passivation (left), metal lines are shifted towards the center of the device. With a SiN passivation (right) metal lines remain intact.

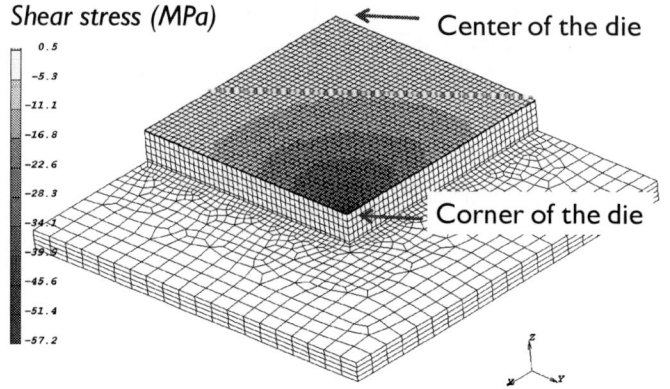

Fig. 5 Typical distribution of shear stress in a packaged die (highest value found in the corner of the die)

In order to investigate the impact of the metal thickness on the occurrence of the passivation cracks, samples with a metal thickness of 2, 3 and 4 μm samples have been exposed to TC. The 2μm metal was patterned with a wet etching resulting in a steep sidewall. The samples of 3 and 4μm were patterend with a combination of wet etching and dry etching resulting in a rounded sloped edge with a small vertical foot. The metal profiles as produced by the patterning are shown in Fig. 6.

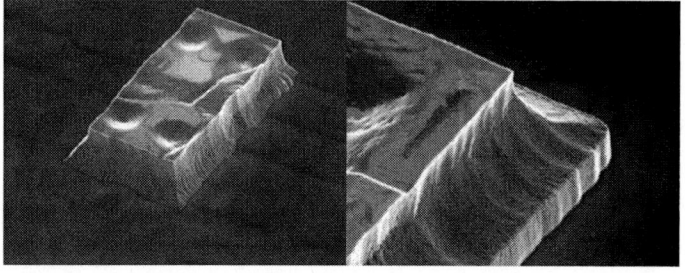

Fig. 6 Left: metal patterned with a wet etching resulting in a steep sidewall. Right: samples patterend with a combination of wet etching and dry etching resulting in a rounded sloped edge with a small vertical foot.

TC was executed between -65C and 175C. After 500, 1000 and 2000 TC.s 3 samples were taken out, decapsulated and inspected for passivation cracks. The results of the inspection as the sum of the cracks for 3 samples as a function of metal thickness and number of TC's are shown on Fig. 7.

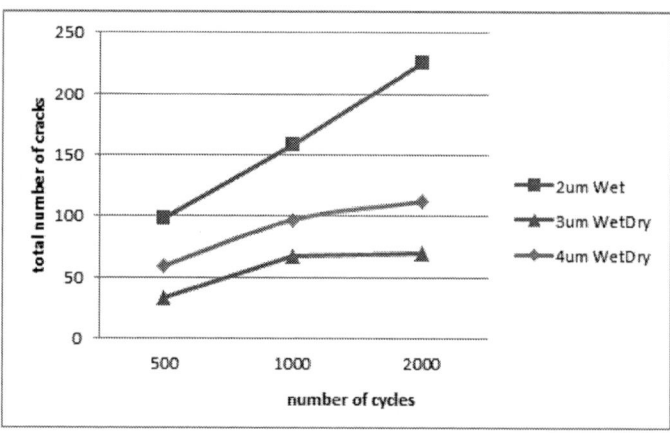

Fig. 7 The number of passivation cracks as the sum of the cracks for 3 samples as a function of metal thickness and number of TC's.

As one can observe in Fig 7, the 2um metal layer yields the highest number of cracks. The 3 and 4 um layers yield significantly less cracks. The number of cracks for the 2um samples keeps increasing with the number of cycles, while for the 3 and 4 um samples there is a stabilizing trend in the number of cracks with increasing numbers of TC's.

III. RESULTS AND DISCUSSION

Earlier in this paper it was discussed that during TC, the shear of the MC has to be absorbed by the edges of the passivation. The shear stress is reaching a maximum with the cold cycles of the TC [6]. With increasing metal thickness one would expect that the impact of the shear force will be increasing causing more cracks on the passivation. What is now observed is that the thicker metal samples perform in fact significantly better than the thin 2um sample. This is exactly the opposite of what one should expect based on [6]. However, the thickness of the metal is not the only difference between the 2μm samples and the 3 and 4μm samples. As shown in fig. 6, the edge profile of sample of the 3 and 4μm samples are significantly different compare to the 2μm sample. Apparently the shape of the edge of the interconnect has a significant impact on the sensitivity for passivation crack formation.

To verify this hypothesis, FEM simulations were used to compare the maximal stress in the passivation. Fig. 8 is showing the results of these simulations both for a vertical edge as for the 2 step sloped edge as it would be the result of the combined wet/dry metal profiling.

As conditions for the FEM, the following parameters were defined: Die size: 2.2 x 2.2 x 0.4 mm^3, leadframe thickness: 0.5 mm, MC thickness: 1.7 mm, metal thickness 4μm.

For both samples maximal principal stress was calculated at -65C, which is the temperature where the highest stresses are found (= the highest temperature difference with the molding curing temperature).

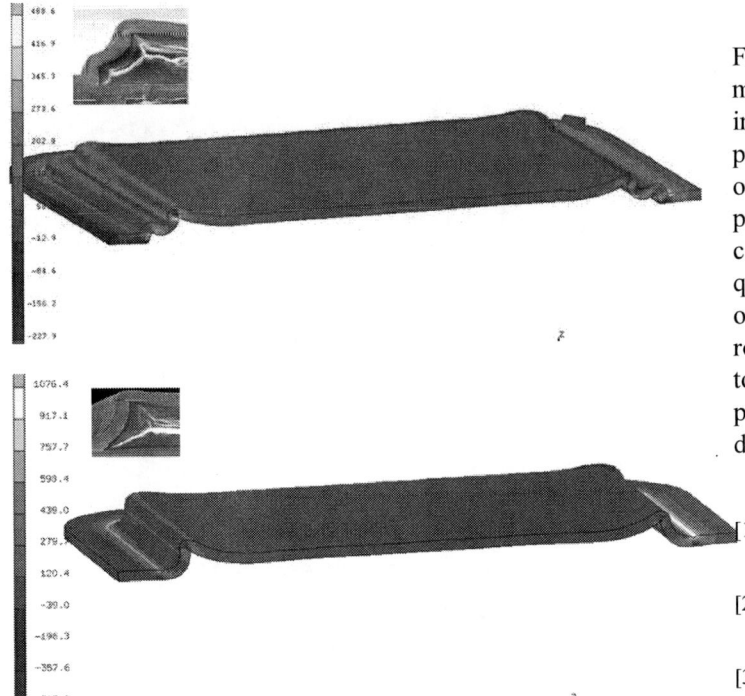

Fig. 8 FEM simulations was used to compare the maximal stress in the passivation. On top the wet/dry etch profile is shown with the maximal principal stress yields 488.6 MPa. at -65C. At the bottom the wet etched sample is shown under the same conditions yields the maximal principal stress yields 1076.4 MPa.

With FEM it was possible to quantify the stress that is occurring during the TC. Maximal stress conditions are found at -65C. Under these conditions, the maximal principal stress for the sample with the wet dry pattering shows significantly lower stress values compared to the sample with the wet etch patterning. This observation corresponds with very well with the corresponding numbers of cracks in the passivation as shown in fig. 8.

One way to reduce cracking is to introduce slots inside the aluminum plates [11], [12], [16] limiting, therefore, the aluminum width L to values that are far below the stress limit for cracks in [15]. In a practical layout, this would mean very narrow metal stripes near the chip corners and edges. For power technologies with the need of a very low wiring resistance of large transistors, however, the increase in resistivity would be too high. Therefore, one should primarily look for other possibilities to reduce the stress in the passivation layer.

Earlier a layered Al/TiN metallization with a distinctly increased yield stress was proposed in order to reduce the number of passivation cracks [15]. Unfortunately, with this approach cracks were now observed in the dielectric layers between the underlying Al metallization leading to metal shorts.

IV. CONCLUSIONS

For the first time the impact of the edge profile of the power metal design on the amount of passivation cracks was investigated in detail. It was found that with a sloped edge profile of the power metal, as is achieved with a combination of an isotropic wet etch followed by a dry etch, the number of passivation cracks is reduced significantly. The observation is confirmed by 3-D FEM simulation. The simulation enabled to quantify the stress level and to forecast corresponding levels of cracks observed after temperature cycling. As a result, a robust metal edge profile design could be deduced, which led to a distinct reduction of the principal stress at the most critical positions and, consequently, to a reduction of passivation damage.

V. REFERENCES

[1] H. Inayoshi, K. Nishi, S. Okikawa, and N. Nakashima, "Moisture induced aluminum corrosion and stress on the chip in plastic encapsulated LSIs," in Proc. IEEE 17th IRPS, 1979, pp. 113–117.

[2] M. Isagawa, Y. Iwasaki, and T. Sutoh, "Deformation of Al metallization in plastic encapsulated semiconductor devices caused by thermal shock," in Proc. IEEE 18th IRPS, 1980, pp. 171–177.

[3] R. J. Usell, Jr. and S. A. Smiley, "Experimental and mathematical determination of mechanical strains within plastic IC packages and their effect on devices during environmental tests," in Proc. IEEE 19th IRPS, 1981, pp. 65–73.

[4] S. Okikawa, M. Sakimoto, M. Tanaka, T. Sato, T. Toya, and Y. Hara, "Stress analysis of passivation film crack for plastic molded LSI caused by thermal stress," in Proc. IEEE ISTFA, 1983, pp. 275–280.

[5] Z. Suo, "Reliability of interconnect structures," in Comprehensive Structural Integrity, vol. 8,W. Gerberich,W. Yang, I. Milne, R. O. Ritchie, and B. Karihaloo, Eds. Amsterdam, The Netherlands: Elsevier, 2002, pp. 1–61.

[6] M. Huang, Z. Suo, and Q. Ma, "Plastic ratcheting induced cracks in thin film structures," J. Mech. Phys. Solids, vol. 50, no. 5, pp. 1079–1098,May 2002.

[7] M. Huang, Z. Suo, and Q. Ma, "Metal film crawling in interconnect structures caused by cyclic temperatures," Acta Mater., vol. 49, no. 15, pp. 3039–3049, Sep. 2001.

[8] C. G. Shirley and R. C. Blish, II, "Thin film cracking and wire ball shear in plastic dips due to temperature cycle and thermal shock," in Proc. IEEE 25th IRPS, 1987, pp. 238–249.

[9] R. C. Blish, II and P. R. Vaney, "Failure rate model for thin film cracking in plastic ICs," in Proc. IEEE 29th IRPS, 1991, pp. 22–29.

[10] S. Okikawa, T. Toida, M. Inatsu, and M. Tanimoto, "Stress analysis for passivation and interlevel-insulation film cracks in multilayer aluminum structures for plastic packaged LSI," in Proc. IEEE ISTFA, 1988, pp. 75–81.

[11] C. F. Dunn and J. W. McPherson, "Temperature cycling acceleration factors for aluminum metallization failure in VLSI applications," in Proc. IEEE 28th IRPS, 1990, pp. 252–258.

[12] P. Alpern, V.Wicher, and R. Tilgner, "A simple test chip to assess chip and package design in the case of plastic assembling," IEEE Trans. Compon.,Packag.,Manuf. Technol. A, vol 17, no. 4, pp. 583-589, Dec. 1994.

[13] P. Alpern, V.Wicher, and R. Tilgner, "Correction to 'A simple test chip to assess chip and package design in the case of plastic assembling'," IEEE Trans. Compon.,Packag.,Manuf. Technol. A, vol 18, no. 4, pp. 862, Dec. 1995.

[14] Zhang GQ, van Driel WD, Fan XJ.Mechanics in microelectronics. Springer; 2006.

[15] Alpern P, Nelle P, Barti E, Gunther H, Kessler A, Tilgner R, et al. On the way to zero defect of plastic-encapsulated electronic power devices—part I to part III, vol. 9; 2009. p. 269–95.

[16] O. Selig, P. Alpern, K. Müller, and R. Tilgner, "Thermomechanical assessment of molding compounds," IEEE Trans. Compon., Hybrids, Manuf. Technol., vol. 15, no. 4, pp. 519–523, Aug. 1992.

Domestic Indirect Feedback Compensation of Multiple-Stage Amplifiers for Multiple-Voltage Level-Converting Amplification

Shang-Hsien Yang and Chua-Chin Wang

Department of Electrical Engineering, National Sun Yat-Sen University

No. 70, Lienhai Rd., Kaohsiung, Taiwan

ccwang@ee.nsysu.edu.tw

Abstract—This investigation presents a Domestic Indirect Feedback Compensation (DIFC) operational amplifier for systems with both low voltage and high voltage circuits. The DIFC operational amplifier is capable of converting a voltage signal from a low voltage circuit and amplifying it into a large voltage signal to drive high voltage load. Since the Metal-Insulator-Metal (MIM) capacitors are not designed for high voltage applications and the feed-forward compensation causes pole-zero doublet as a result from the deviation of high voltage transistor characteristics from shuttle to shuttle, the DIFC is performed only using low voltage circuits. In other words, the feedback by connecting the output node of the operational amplifier is avoided. The proposed design is carried out using the TSMC 0.25 μm 1-poly 3-metal BCD process.

Index Terms—Analog circuits, operational amplifiers, multi-stage amplifiers, frequency compensation, high voltage integrated circuits.

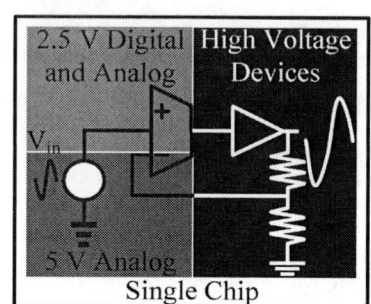

Fig. 1. An operational amplifier capable of converting an analog voltage signal from the circuits in the 5 V domain and amplifying it into a signal of large voltage swing in the high voltage domain, suitable for systems with 2.5 V digital circuits, low power analog circuits, 5 V analog circuits, and high voltage devices all in one single chip.

I. INTRODUCTION

RECENT advance in semiconductor technology allows a digital process to fabricate an integrated circuit with high voltage devices. It is achieved by adding a certain number of extra masks to existing masks needed for the fabrication of digital circuits. Such a semiconductor process, e.g., TSMC 0.25 μm 1-poly 3-metal BCD process, used for this design has a price advantage over conventional high voltage analog process. With the aid of this process and the selection over several high voltage tolerant devices, System-on-chip (SOC) of digital cell-based, low voltage analog, and high voltage driving devices is feasible.

The 0.25 μm 1-poly 3-metal BCD process used to carry out this design offers 0.25 μm transistors for digital cell-based design and low voltage analog circuits operating in 2.5 V, 0.5 μm transistors for analog circuits operating in 5 V, and power devices for high voltage applications supporting up to 60 V. Converting an analog voltage signal from the 2.5 V circuit domain and amplifying it into the 5 V circuit domain, which can be achieved with operational amplifiers designed

This research is partially supported by National Science Council under grant NSC 99-2221-E-110-082-MY3, NSC 99-2220-E-110-001, NSC 99-2923-E-110-002-MY2, and EZ-10-09-44-98.. It is also partially supported by Ministry of Economic Affairs, Taiwan, under grant 99-EC-17-A-01-S1-104, and 99-EC-17-A-19-S1-133.

Prof. C.-C. Wang is with the Department of Electrical Engineering, National Sun Yat-Sen Univers ity, Kaohsiung, Taiwan 80424. (e-mail: ccwang@ee.nsysu.edu.tw)

using 0.5 μm transistors probably would not be too much a problem. However, problems arise when attempting to convert an analog voltage signal from the circuits in the 5 V domain and amplify it into a signal of large voltage swing in the high voltage domain. One major reason of the difficulty lies in the design of these operational amplifiers. First of all, the high voltage transistors offered in this process are bilateral devices. The gate to source voltage is limited to a low voltage of 5 V even when their drain to source voltage can sustain up to 60 V. For this reason, it would be meaningless to design the input stage of an operational amplifier with high voltage transistors. In fact, each high voltage transistor requires an independent isolation rings, which can be several times larger than the actual dimensions of high voltage transistors. On the other hand, all the 5 V transistors can all be fitted into a single isolation ring. Apparently, we should avoid using an excessive number of high voltage transistors, especially if their functionalities can be achieved using 5 V transistors. Thus, the stages of an operational amplifier should be divided into a 5 V voltage domain and a high voltage domain. Furthermore, the Metal-Insulator-Metal (MIM) capacitors offered in this process are not meant to sustain high voltage. This fact prevents the use of Miller capacitors which connect one of its nodes to the output of the operational amplifier working in the high voltage domain, while the other is connected to the node in the 5 V voltage domain.

To resolve this problem, the feed-forward compensation

978-1-4244-9019-6/11 $26.00 © 2011 IEEE

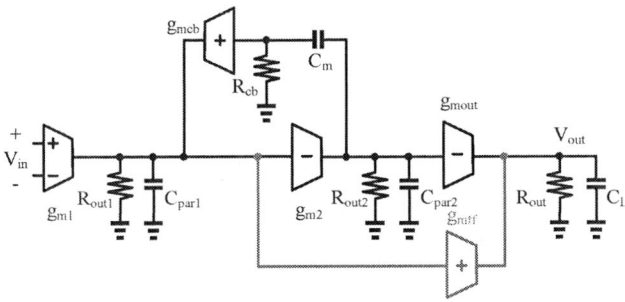

Fig. 2. The proposed DIFC topology.

$$\omega_3 \approx \frac{1}{2C_m R_{out2} + C_L R_{out} + C_m g_{m2} g_{mcb} R_{out1} R_{out2} R_{cb}} \tag{5}$$

$$\omega_4 \approx \frac{1}{g_{m2} g_{mcb} R_{out1} R_{out2} R_{cb} C_m (C_L R_{out} + C_m R_{out2}^2)} \tag{6}$$

$$\omega_5 \approx \frac{1}{C_L C_m^2 g_{m2} g_{mcb} R_{out1} R_{out2}^2 R_{out} R_{cb}} \tag{7}$$

$$\omega_6 \approx \frac{1}{C_L C_m^2 R_{out2} R_{cb} R_{out} (C_{par2} R_{out2} + 2C_{par1} R_{out1})} \tag{8}$$

$$\omega_7 = \frac{1}{C_{par1} C_{par2} C_L C_m^2 R_{out1} R_{out2}^2 R_{cb} R_{out}} \tag{9}$$

and the dominant pole can be approximated to:

$$|p_{-3dB}| \approx \frac{1}{C_m g_{m2} g_{mcb} R_{out1} R_{out2} R_{cb}} \tag{10}$$

Hence, the gain-bandwidth product can be approximated by

$$\omega_t = 2\pi GBW = A_{DC}|p_{-dB}| \approx \frac{g_{m1} g_{mout} R_{out}}{C_m g_{mcb} R_{cb}} \tag{11}$$

If the grey feed-forward amplifier is taken into account, (1) becomes:

$$A_v(s) = \frac{A_{DC} \cdot (1 + \frac{s}{\omega_1} + \frac{s^2}{\omega_2} + \frac{s}{\omega_{f1}} + \frac{s^2}{\omega_{f2}}) + A_{FF}(1 + \frac{s}{\omega_{f3}})}{(1 + \frac{s}{\omega_3} + \frac{s^2}{\omega_4} + \frac{s^3}{\omega_5} + \frac{s^4}{\omega_6} + \frac{s^5}{\omega_7})} \tag{12}$$

and

$$A_{FF} = g_{m1} g_{mff} R_{out1} R_{out} \tag{13}$$

$$\omega_{f1} \approx \frac{1}{C_m R_{out2} + R_{cb} C_m} \tag{14}$$

$$\omega_{f2} \approx \frac{1}{R_{cb} C_m^2 R_{out2}} \tag{15}$$

$$\omega_{f3} \approx \frac{1}{C_m (R_{out2} + R_{cb})} \tag{16}$$

The denominators of (1) and (12) are the same, except the fact that C_{par1} will be slightly increased. The added terms of the numerator in (12) is composed of the additional DC gain contributed by (13) and the zeros contributed by (14), (15), and (16). As will be revealed later, the addition of the feed-forward stage does not affect R_{out}. The gain-bandwidth product can be doubled, while still maintaining the same phase margin of 85 degrees. However, the addition of the g_{mff} is not always beneficial. This will be addressed later. A complete list of system parameters are tabulated below.

As shown in Fig. 3, the proposed DIFC amplifier consists of two 5 V stages and a high voltage stage. The first stage, g_{m1}, is

schemes that do not require any Miller capacitor proposed in [1] may seem promising at first. Unfortunately, such frequency compensation schemes would require additional high voltage transistors. Furthermore, according to many prior empirical experiments, the measurement results of these high voltage transistors often deviate from the characteristic obtained through simulation. It will cause a problem to carry out pole-zero cancelation, which is the main idea of feed-forward compensation. The doublet caused by pole-zero mismatches further degrade the settling time of the operational amplifier. For the sake of robustness, such a compensation scheme should be avoided. In comparison with high voltage transistors, the characteristics of 0.5 μm transistors are found to be more consistent. Hence, reliability of the frequency compensation can be ensured if the compensation is merely carried out in the 5 V voltage domain.

In this paper, a novel frequency compensation scheme is proposed. This scheme solely performs compensation in 5 V voltage domain. The Miller capacitor, C_m, is entirely positioned in the 5 V voltage domain. Additional discussions on applying feed-forward compensation to the proposed DIFC amplifier are also taken into account. Simulation results of gain, frequency response, distortion and slew rate are reported to demonstrate the performance of this operational amplifier. Device dimensions of both 0.5 μm transistors and high voltage transistors are provided for reference.

II. CIRCUIT ARCHITECTURE

The small signal model of the proposed DIFC amplifier without the grey feed-forward amplifier is shown in Fig. 2, and the transfer function can be obtained using KCL.

$$A_v(s) = \frac{A_{DC} \cdot (1 + \frac{s}{\omega_1} + \frac{s^2}{\omega_2})}{(1 + \frac{s}{\omega_3} + \frac{s^2}{\omega_4} + \frac{s^3}{\omega_5} + \frac{s^4}{\omega_6} + \frac{s^5}{\omega_7})} \tag{1}$$

The symbols shown in (1) are defined by (2)-(9), and some of the expressions are approximated by truncating terms that are relatively much smaller than the others.

$$A_{DC} = g_{m1} g_{m2} g_{mout} R_{out1} R_{out1} R_{out} \tag{2}$$

$$\omega_1 = \frac{1}{C_m R_{out2} + R_{cb} C_m} \tag{3}$$

$$\frac{1}{} \tag{4}$$

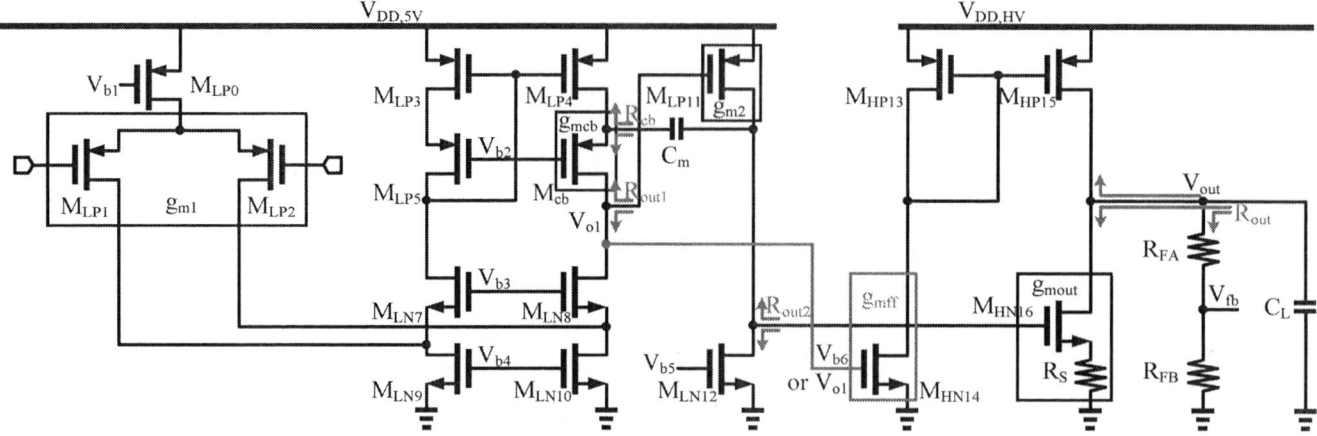

Fig. 3. The proposed DIFC operational amplifier.

TABLE I
SYSTEM PARAMETERS

g_{m1}	2.15 μA/V	R_{out1}	565 MΩ
g_{m2}	19.03 μA/V	R_{out2}	2.25 MΩ
g_{mout}	457.7 μA/V	R_{out}	240 kΩ
g_{mcb}	2.56 μA/V	R_{cb}	1.09 MΩ
g_{mff}	501 μA/V	C_m	577 fF

realized using a classical folded cascode differential amplifier, consisting of 0.5 μm transistors M_{LP1} to M_{LN10}. The output resistance, R_{out1} is made large to produce a high RC time constant. 0.5 μm transistors M_{LP11} and M_{LN12} form g_{m2}, i. e., a common source amplifier used to provide large DC gain and multiply the Miller capacitor, C_m. The capacitor connects to the output of the common source amplifier the source of M_{cb} to avoid the RHP zero. This forms the domestic indirect feedback path. M_{cb} serves as a current buffer, with a transconductance of g_{mcb}, which also prevents the high frequency shorting effect of the output terminals of the folded cascode differential amplifier and the common source amplifier.

The high voltage part of this operational amplifier has 2 high voltage N-type transistors, M_{HP14}, M_{HP16} (NLD60G5 1 μm transistor in 0.25 μm BCD process), and 2 P-type transistors, M_{HP13}, M_{HP15} (PA60G5 0.8 μm transistor in TSMC 0.25 μm BCD process). Depending on configuration, the gate of M_{HN14} is either connected to the output terminal of the folded cascode differential amplifier, V_{o1}, or to a bias voltage V_{b6}. If it is connected to V_{b6}, it will provide a DC bias current to M_{HP13}, and mirrored to M_{HP15}. M_{HN16} serves as g_{mout}, which is the third stage of the operational amplifier, and uses M_{HP15} as an active load. The output resistance of the third common source stage is kept small to prevent loading effect of the operational amplifier when driving resistors. Furthermore, an external source degeneration is added to M_{HN16} to adjust g_{mout} corresponding to any possible process variation. Feedback resistors R_{FA} and R_{FB} with values of 220.1 kΩ and 19.9 kΩ are selected to reduce the quiescent current consumption caused by the DC path these resistors introduced. The output resistance of the DIFC operational amplifier, R_{out} is the shunt resistance of $R_{FA} + R_{FB}$ with r_{oHN16} and r_{oHP15}. By connecting the gate of

M_{HN14} to V_{o1}, the feed-forward stage, g_{mff}, is realized. Since its transconductance is mirrored to the output through current mirror formed by M_{HP13} and M_{HP15}, R_{out} will not be affected. However, due to the uncertainty in transconductance of high voltage transistors, implementing the feed-forward stage might result in undesired pole-zero doublets, prolonging settling time of the operational amplifier. It is clearly evident by inspecting (12), (13), and (16). If the zero dominated by g_{mff} if shifted to high frequency, the stability of the operational amplifier may possibly be jeopardized. Similarly to other multi-stage operational amplifiers [2][3], the slew rate of this amplifier is constrained by the slowest stage. Generally, the slew rate can be estimated with:

$$ SR = min(\frac{I_{D1}}{C_m}, \frac{I_{Dout}}{C_L}) \quad (17) $$

where I_{D1} and I_{Dout} represent the currents of the folded cascode differential amplifier and the output stage, respectively. Apparently, (17) is dependent of the capacitive load driven by the operational amplifier.

The device parameters used in this operational amplifier are shown in Table II.

TABLE II
DEVICE PARAMETERS

M_{LP0}	0.7 μm / 1 μm	M_{LP11}	1.65 μm / 0.5 μm
M_{LP1}	1 μm / 1 μm	M_{LN12}	0.6 μm / 0.5 μm
M_{LP2}	1 μm/ 1 μm	M_{HP13}	70 μm / 0.8 μm
M_{LP3}	0.6 μm / 2.4 μm	M_{HN14}	8 μm / 3 μm
M_{LP4}	0.6 μm / 2.4 μm	M_{HP15}	70 μm / 0.8 μm
M_{LP5}	8 μm / 0.5 μm	M_{HN16}	4.4 μm / 1 μm
M_{cb}	8 μm / 0.5 μm	C_m	577 fF
M_{LN7}	0.6 μm / 1.5 μm	R_S	900 Ω
M_{LN8}	0.6 μm / 1.5 μm	R_{FA}	220.1 kΩ
M_{LN9}	0.73 μm / 1.5 μm	R_{FB}	19.9 kΩ
M_{LN10}	0.73 μm / 1.5 μm	C_L	10 pF

III. SIMULATION RESULTS

This design is carried out using TSMC 0.25 μm 1-poly 3-metal BCD process. It consumes 15 μW from the 5 V supply and 127.7 mW from the 60 V supply. The simulation result

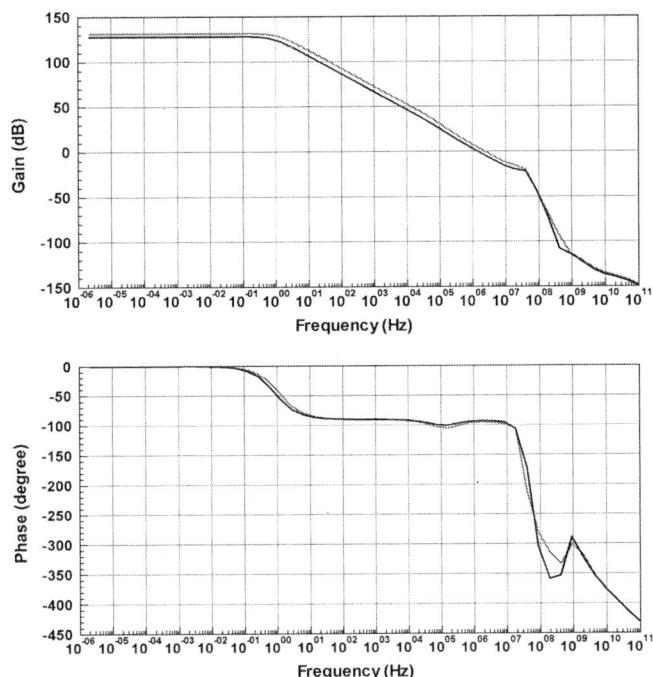

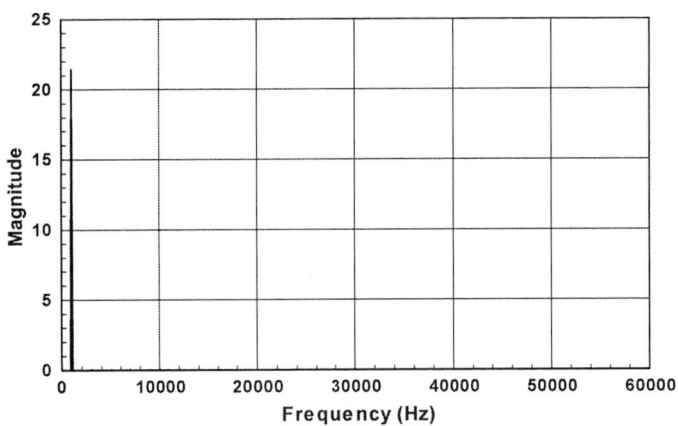

Fig. 6. The Discrete Fourier Transform of a 10 mHz sine wave output with a 50 V voltage swing.

TABLE III
SYSTEM PARAMETERS

C_L	10 pF	A_{DC}	133 / 128 dB
GBW	2.6 / 1.3 MHz	PM	85 °
Power	15 μW + 127.7 mW	GM	24 dB
V_{DD}	5V, 60V	SR	3.6 V/μs

IV. CONCLUSION

In this work, we have presented a novel multiple-stage operational amplifier dedicated to multiple-level level-converting amplification. It has been shown that the domestic indirect feedback compensation can avoid connecting a MIM capacitor terminal to couple a high voltage node, while still maintaining acceptable performance in gain, phase margin, slew rate, and power consumption.

ACKNOWLEDGEMENT

This research was partially supported by National Science Council under grant NSC 99-2221-E-110-082-MY3, NSC 99-2220-E-110-001, NSC 99-2923-E-110-002-MY2, and EZ-10-09-44-98.. It was also partially supported by Ministry of Economic Affairs, Taiwan, under grant 99-EC-17-A-01-S1-104, and 99-EC-17-A-19-S1-133. The authors would like to express their gratitude toward Chip Implementation Center, Taiwan, for EDA Tools and TSMC 0.25 μm 1-poly 3-metal BCD process information.

REFERENCES

[1] B. K. Thandri and J. Silva-Martinez, "A robust feedforward compensation scheme for multistage operational transconductance amplifiers with no Miller capacitors," *IEEE J. Solid-State Circuits*, vol. 38, no. 22, pp. 237-243, Feb. 2003.

[2] X. Peng and W. Sansen, "Transconductance with capacitances feedback compensation for multistage amplifiers," *IEEE J. Solid-State Circuits*, vol. 40, no. 7, pp. 1514-1520, Jul. 2005.

[3] H. Lee and P. K. T. Mok, "Advances in active-feedback frequency compensation with power optimization and transient improvement," *IEEE Trans. Circuits Syst. I, Regul. Papers*, vol. 51, no. 9, pp. 1690-1696, Sep. 2004.

[4] A. Pugliese, F. A. Amoroso, G. Cappuccino, and G. Cocorullo, "Settling time optimization for three-stage CMOS amplifier topologies," *IEEE Trans. Circuits Syst. I, Regul. Papers*, vol. 56, no. 12, pp. 2569-2581, Sep. 2009.

Fig. 4. The Bode plot of proposed operational amplifier with feed-forward stage (gray) and without (black).

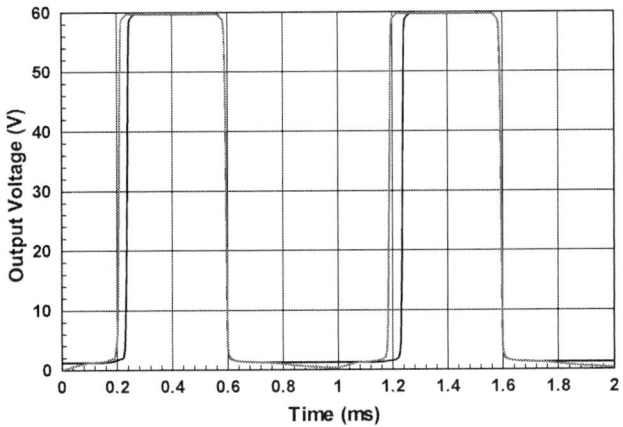

Fig. 5. The slew rate of proposed operational amplifier with feed-forward stage (gray) and without (black).

in Fig. 4 suggests that the gain-bandwidth of this operational amplifier is 2.6 MHz and 1.3 MHz with or without the feed-forward stage, respectively. Both configurations have a phase margin of 85 degrees. DC gains are 133 dB with the feed-forward stage and 128 dB without the feed-forward stage, respectively. As shown in Fig. 5, the slew rate is 3.6 V/μs, but it will require an additional delay of 48 μs for the output stage to charge the capacitive load. The feed-forward stage has minor significance discharging the capacitive load. By performing Discrete Fourier Transform on the output waveform of the operational amplifier, we can assume that the distortion caused by this operational amplifier is negligible.

978-1-4244-9019-6/11 $26.00 © 2011 IEEE

Microwatt Low-noise Variable-Gain Amplifier

Chun-Yi Li, Yu-Bin Lin, Robert Rieger

Electrical Engineering Department, National Sun Yat-Sen University
804 Kaohsiung, Taiwan
rrieger@mail.nsysu.edu.tw

Abstract—**A fully integrated variable gain amplifier circuit is reported in this paper. The amplifier is based on an integrating topology allowing the gain to be controlled by the timing of a clock signal. The recording of physiological signals such as the electroneurogram (ENG) or electromyogram (EMG) is a targeted application. Therefore, low-noise performance and low power consumption are important. Simulated and measured results for a chip fabricated in 0.35µm CMOS technology show a gain range from 10-133 V/V, 169 nV/√Hz input spot noise, a NEF of 10.1 and an active area of 0.017 mm^2 with a power consumption of 1.44 µW using ±0.9 V supplies.**

I. INTRODUCTION

The acquisition of bio-signals has become an important feature of advanced medical applications. Examples include the use of nerve signals to control functional electrical stimulation (FES) prostheses [1]-[3] the detection and localization of brain activity [3][4] or the recording of the electrocardiogram (ECG) [5] or electromyogram (EMG) [6] as part of a wearable monitoring system. The signals thus obtained are small, on the order of millivolts or less. With the advance in technology the recording circuit can be integrated on a chip, which due to its small size can be placed near to the recording site and often even be implanted [7]. A relatively high gain is required to ensure that the SNR can be maintained throughout the signal processing cascade. At the same time the bandwidth of the signals are low, typically in the kilo-hertz range. Amplifiers with controllable gain allow adjusting the gain to the optimum value during recording, providing maximum amplification without saturating the channel. Moreover, amplifiers with a wide gain range and gain controllable in fine steps are essential for the implementation of interference rejection circuits in ENG and EMG recording arrangements [8], [9]. A double-differential amplifier arrangement is shown in Fig. 1 consisting of first rank amplifiers (*G1* and *G2*) and a second rank amplifier providing output signal V_o. In this configuration the input signal is detected between two outer electrodes and a centre electrode used as the reference. Interference which is common to both differential inputs is thus rejected. However, to work effectively, the common signal must appear with the same amplitude at both first rank amplifier outputs. This is difficult to achieve in practice mainly due to electrode impedance imbalance (which may also change over time). Variable gain amplifiers can restore the signal balance. A gain control system to automatically restore balance in ENG tripolar recording was proposed in [9]. The principle of this system is also shown in Fig. 1. The signal obtained at the first rank amplifiers *G1* and *G2* is rectified, integrated and used as a feedback signal to adjust the gain of these pre-amplifiers. Cuff impedance mismatch of up to 30% may be compensated with this approach requiring a gain tuning range from around 1/3 to 3 times the nominal gain [10].

Switched-capacitor (SC) techniques can be used to implement a gain stage with very tightly controlled gain and excellent noise performance. However, as the gain is determined by the ratio of capacitors, in-circuit gain adjustment is difficult to achieve and large capacitor ratios are required for high gain settings.

In this paper we discuss the ASIC realization of two variable gain amplifiers, where the first implementation provided minimum noise and a second implementation is operates with reduced power consumption and achieves a smaller active circuit area. The chips are combined with a microcontroller which provides the control signals and

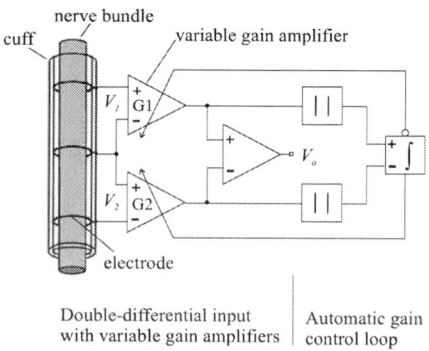

Fig. 1: Variable gain amplifiers connected to a cuff for nerve recording. The gain may be adjusted using a feedback control loop consisting of rectifiers and integrator. Interference suppression is thus achieved.

This work is supported in parts by the Taiwan Chip Implementation Center (CIC) and National Science Council Grant NSC-99-2221-E-110-023.

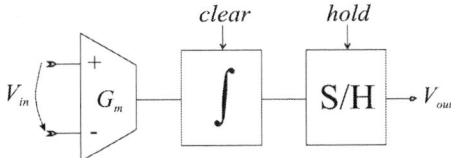

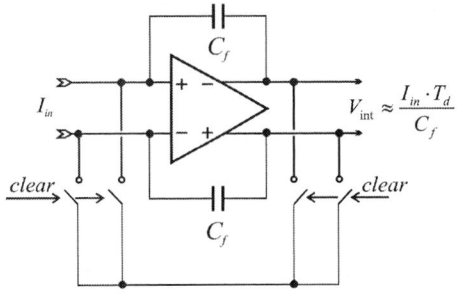

Fig. 2: Principle diagram of the variable gain amplifier consisting of a transconductance input stage, a current integrator and sample-and-hold stage.

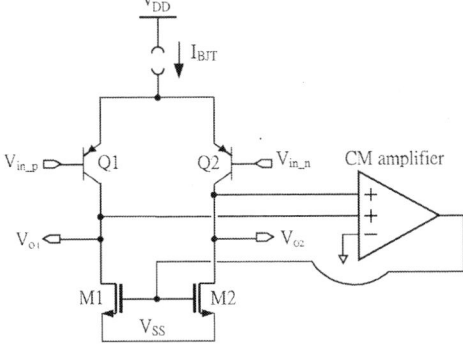

Fig. 3: Schematic diagram of the transconductance stage. Bipolar transistors Q1 and Q2 are realized as lateral devices suitable for CMOS technology.

Fig. 4: Diagram of the current integrator stage with reset switches.

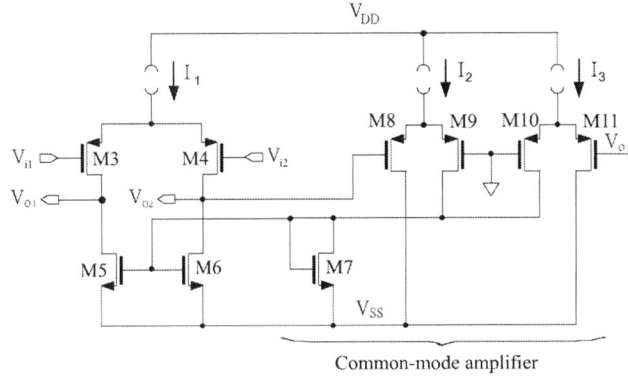

Fig. 5: Schematic diagram of the fully differential integrator amplifier. Setting a low common-mode level enables operation with low supply voltage.

determines the channel gain. The circuit design is presented in Section II followed by simulated and measured results in Section III. Conclusions are given in Section IV.

II. CIRCUIT DESIGN

The principle diagram of the amplifier is shown in Fig. 2. The circuit consists of an input operational transconductance stage (OTA) whose output current is integrated on a capacitor providing a voltage output. A sample and hold stage captures this signal periodically and provides it as the output signal V_{out}. The integrator is cleared (the output is reset to zero) by the clear signal. Thus, the effective integration time T_d of the transconductor current is the period between integrator reset and sampling of the output voltage. If the integrator is reset frequently the input signal does not change much during this interval and can be approximated by a constant. Analysis of the circuit then yields the overall voltage gain:

$$A_V = \frac{V_o}{V_{in}} = G_m T_d \qquad (1)$$

Thus, the gain of the circuit can be varied by adjusting the integration timing T_d. Notably T_d is independent of the sampling period. The sample frequency is therefore denoted by $f_s=1/T$, and T is always larger than T_d.

The transconductance stage is realized as a fully differential, symmetrical OTA shown in Fig. 3 [11]. Lateral bipolar transistors Q1 and Q2 are used as input devices to reduce the effect of flicker noise compared to a CMOS input stage [12]. It trades off with a reduced input impedance, which is tolerable for electrode resistance on the order of a few kilo-ohm in the targeted application [13]. Also, the lateral structures are expected to be less effective than dedicated structures in BiCMOS technology. Therefore, a reduced

transconductance of 70% of the ideal value is initially assumed for simulation. The common-mode reference level is set slightly below the zero-volt level. This yields a higher collector-emitter voltage and allows for a wider input common-mode range.

The integrator is realized using a differential amplifier with capacitive feedback as shown in Fig. 4. The amplifier employs the same structure as the transconductance stage in Fig. 3, but using conventional MOSFET devices instead of bipolar transistors. The feedback capacitor C_f is 500 fF. The feedback integrator approximates ideal integration however with an additional gain factor $1/C_f$ which multiplies with (1) to yield the overall system gain. The schematic of the amplifier is shown in Fig. 5. Using a low common-mode reference ground enables operation at low supply voltage.

The sample-and-hold circuit (S&H) is conventional and consists of complementary MOS devices operating as sampling switch in series with a sampling capacitor. The sampling capacitance was chosen as 100 fF. The input BJTs are each biased with a DC current of 50 nA resulting in a transconductance G_m of around $0.7 \cdot 1.9\ \mu\text{A/V}=1.33\ \mu\text{A/V}$.

The amplifier circuit is simulated using Cadence design tools and MOSFET models provided by the foundry. A sine wave of 1.6 mV peak-to-peak amplitude and a frequency of 300 Hz is applied differentially as the test signal. The simulated output voltage is shown in Fig. 6 together with the input

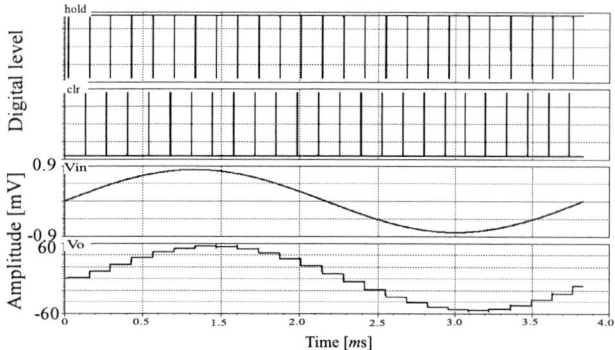

Fig. 6: Simulated clock signals, input and output voltage of the integrating amplifier. The input is a 300 Hz sinewave with 1.6 mV$_{pp}$ amplitude. T_d =25 µs.

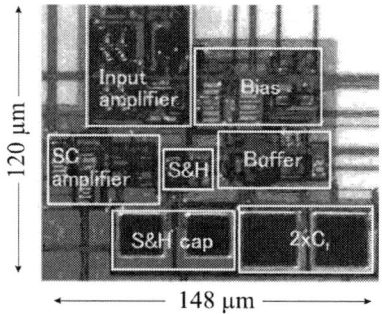

Fig. 7: Microphotograph of the active area of the circuit fabricated in TSMC 0.35 µm CMOS technology.

signal and clock phases. The simulated gain is 66 V/V for a T_d of 25 µs.

III. MEASURED RESULTS

The circuit was laid-out and fabricated in TSMC 0.35 µm CMOS technology occupying an active die area of less than 0.017 mm^2. The chip microphotograph is shown in Fig. 7. The chip is operated with ±0.9V dual supply voltage. For amplifier testing a clock pattern is generated using a microcontroller added to the testing PCB. The clock pattern uses T_d=25µs and T=50µs. The closing period for the clear and sample switches is 5µs. The first test input is a sinusoidal voltage with 1.6 mV peak-to-peak amplitude. An oscilloscope screenshot of the circuit output observed through an on-chip test buffer is shown in Fig. 8. The buffer yields a gain of 3.2 V/V, thus referring the measured amplitude back to the input of the buffer yields the expected gain of 62 V/V. The measured gain (at the buffer input) is plotted in Fig. 9 for different gain settings varying T_d from 5µs to 50µs. It confirms that the gain is controlled by adjusting the clock phase delay with a nearly linear dependency on T_d. The incremental gain variation is approximately 2.7 V/(V·µs). The measured system transfer gain is shown in Fig. 10 for a sampling frequency of 15 kHz (T≈66 µs) and different T_d settings. The practical bandwidth of the system is limited to half the sampling frequency to avoid undersampling.

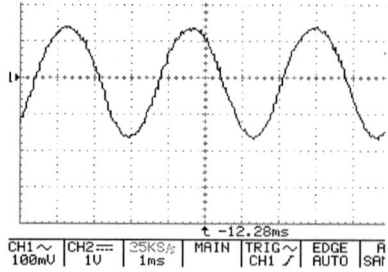

Fig. 8: Oscilloscope screenshot of the buffered amplifier output voltage with T_d=25 µs. The input is a sinewave with 1.6 mV$_{pp}$ amplitude. The gain referred to the buffer input is 62 V/V.

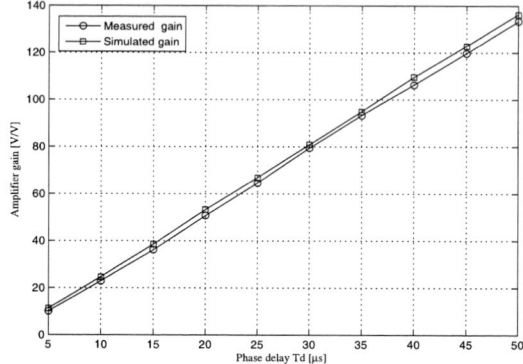

Fig. 9: Measured variation of the amplifier gain w. r. t. delay T_d. Clock period T is 50 µs. Measured and simulated results are shown for comparison.

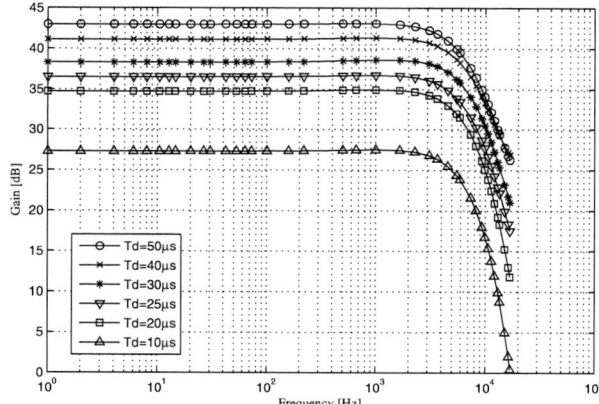

Fig. 10: Measured transfer gain for different integration time T_d. The sample period T is 50 µs in all cases.

The current consumption of the amplifier I_{VDD} is measured as approximately 800 nA excluding the buffer circuit and microcontroller, resulting in 1.44 µW power consumption.

The circuit noise is measured using a spectrum analyzer connected to the buffer output and dividing by the buffer gain. In this way, the analyzer records the continuous-time amplifier output voltage noise spectral density which is shown in Fig. 11. It is observed that the output noise increases with an increase in T_d, i.e. with the circuit gain. This increase is not linear resulting in a minimum input-referred noise level and best signal-to-noise performance for high gain settings. The noise-efficiency factor (NEF) [14] is a measure

978-1-4244-9019-6/11 $26.00 © 2011 IEEE

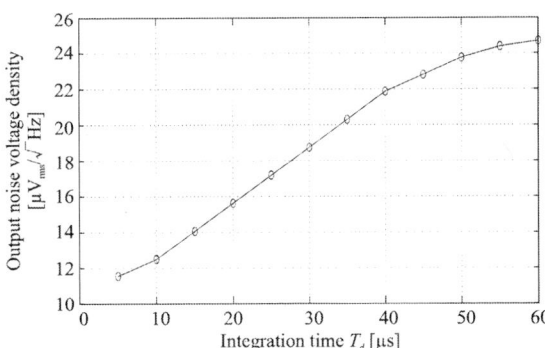

Fig. 11: Output voltage noise density measured with grounded input terminals for different T_d settings.

TABLE I. PERFORMANCE SUMMARY IN COMPARISON WITH PREVIOUS VARIABLE GAIN DESIGN.

Parameter	Measured	Simulated	[16]
Gain range @f_s=7.5 kHz [V/V]	10~133	11~136	10-1260
Bandwidth	DC-f_s/2	DC-f_s/2	DC-f_s/2
Input noise [nV/√Hz]	169	157	16
Power [µW]	1.44	1	280
CMRR [dB]	>97 dB	-	>97 dB
Minimum supply voltage [V]	±0.9 V	±0.9 V	±1.5 V
NEF (T_d=50 µs)	7.3	6.8	7.5
Active area [mm²]	0.017	-	0.064
CMOS Technology	0.35 µm	0.35 µm	0.35 µm

of the amplifier noise performance by comparing its input referred noise to the noise generated by a single bipolar transistor biased to yield the same current consumption as the amplifier under consideration. Expression (2) uses the spot noise voltage density to calculate the *NEF*:

$$NEF = \frac{V_{n\_out}}{A_V} \sqrt{\frac{I_{VDD}}{4qU_{th}^2}} \quad (2)$$

For a T_d of 25 µs a *NEF* of 10.1 results and for T_d of 50 µs the efficiency increases to a *NEF* of 7.3. Finally, the input CM rejection ratio (CMRR) was measured using a 400 mV$_{pp}$ input sinusoidal with a frequency of 1 kHz. The resulting output signal was below the noise floor of the measurement equipment of 2 mV. Therefore, the CMRR is larger than 97 dB.

The measured results are summarized in Table I and compared to the target values expected from simulation or calculation. Also shown is the comparison to a previous variable gain design targeted at higher gain factors and consuming higher power and silicon area [16].

IV. CONCLUSIONS

We have implemented a fully differential low-power and low-noise amplifier with configurable gain which has particular relevance for the recording of biomedical signals. Directly loading a transconductance input amplifier with a switched-capacitor transresistance stage allows setting the gain by variation of the delay between two control signals. A microcontroller was used to generate the control signals in the reported test setup leading to well defined and precisely tunable gain. Measurement shows that the circuit noise level is very low and highest performance is achieved for high gain settings as required for the recording of very small signals. Furthermore, the circuit occupies small silicon area, consumes very little power and operates down to ±0.9 V supply voltage (which is a typical battery end-of-life voltage). These parameters constitute a considerable improvement over a previously reported implementation. The sample rate can be adjusted independent of the gain control and the differential circuit achieves a high CMRR, making it generally attractive for use in a wide range of recording front-ends.

REFERENCES

[1] M. Haugland, J. Hoffer, "Slip information obtained from the cutaneous electroneurogram: Application in closed loop control of functional electrical stimulation." *IEEE Trans. Rehab. Eng.*, vol 2, pp. 29-36, 1994.

[2] M. Haugland, J. Hoffer, and T. Sinkjaer, "Skin contact force information in sensory nerve signals recorded by implanted cuff electrodes." *IEEE Trans. Rehab. Eng.*, vol 2, pp. 18-28, 1994.

[3] B. Popovic, R. B. Stein, et al., "Sensory nerve recording for closed-loop control to restore motor functions," *IEEE Trans. Biomed. Eng.*, vol. 40, no. 10, pp. 1024-1031, 1993.

[4] M. A. Lebedev, M. A. L. Nicolelis, "Brain-machine interfaces: Past, present and future," *Trends in Neuroscience*, vol. 29, no. 9, pp. 536-546, 2006.

[5] M. Shojaei-Baghini, R. K. Lal, D. K. Sharma, "A Low-Power and Compact Analog CMOS Processing Chip for Portable ECG Recorders," in Proc. ASSCC, 2005, pp. 473-476.

[6] R. G. Haahr, S. Duun, E. V. Thomsen, K. Hoppe, J. Branebjerg, "A Wearable "Electronic Patch" for Wireless Continuous Monitoring of Chronically Diseased Patients," in Proc. 5th Int. Workshop on Wearable and Implantable Body Sensor Networks, 2008, pp. 66-70.

[7] R. Rieger, M. Schuettler, D. Pal, C. Clarke, et al. ,"Very low-noise ENG amplifier system using CMOS technology," *IEEE Trans. Neural Systems and Rehab. Eng.*, vol. 14, no. 4, pp. 427-437, 2006.

[8] J.P.P. van Vugt, J.G. van Dijk, "A convenient method to reduce crosstalk in surface EMG," Clinical Neurophysiology, no. 112, pp. 583-592, 2000.

[9] J. Taylor, J, A. Demosthenous, I. Triantis, R. Rieger, N. Donaldson, "Design of an Adaptive Interference Reduction System for Nerve Cuff Electrode Recording," IEEE Trans. Circuits and Systems - I, vol. 51, no. 4, pp. 629-639, 2004.

[10] I. F. Triantis, A. Demosthenous,N. Donaldson, « On Cuff Imbalance and Tripolar ENG Amplifier Configurations," IEEE Trans. Biomed. Eng., vol. 52, no. 2, pp. 314-320, 2005.

[11] D.A Johns, K. Martin, *Analog Integrated Circuit Design*, John Wiley & Sons, 1997.

[12] R. Rieger, J. Taylor, A. Demosthenous, N. Donaldson, and P. Langlois, "Design of a low-noise preamplifier for nerve cuff electrode recording," *IEEE J. Solid-State Circuits*, vol. 38, no. 8, pp. 1373–1379, Aug. 2003.

[13] J. G. Webster, *Medical Instrumentation – Application and Design*, 3rd ed., John Wiley & Sons, 1998.

[14] M. S. J. Steyaert, W. M. C. Sansen, Z. Chang, "A Micropower Low-Noise Monolithic Instrumentation Amplifier For Medical Purposes," *IEEE J. Solid-State Circuits*, vol. SC-22, no. 6, pp. 1163-1168, 1987.

[15] R. Rieger, J. Taylor, "Design Strategies for Multi-Channel Low-Noise Recording Systems," *Analog Integrated Circuits & Signal Processing*, vol. 58, no. 2, pp. 123-133, 2009.

[16] R. Rieger, Y.-R. Huang, "A High-gain, Low-noise CMOS Amplifier for Sampled Bio-potential Recording," in Proc. IEEE ISCAS 2010, May-June 2010, pp. 1220 - 1223.

SRAM Bitcell Design for Low Voltage Operation in Deep Submicron Technologies

Young Hwi Yang, Jisu Kim, Hyunkook Park, Joseph Wang, Geoffrey Yeap, and Seong-Ook Jung

Abstract—**As technology scales down, an increasing number of transistors can be integrated into a single chip but process variation becomes more serious. SRAM is one of the key components in a SoC and it occupies a large portion of the SoC. Thus, the SRAM bitcell is typically designed using very small transistors for high integration, which limits the minimum operating voltage (V_{CCmin}) of the SoC because of the large threshold voltage (V_{th}) mismatch between paired transistors caused by small feature size. As process technology scales down to sub-32nm technology, the 6T SRAM bitcell that is currently used may not achieve proper stability, write-ability, and read-ability at the required operating voltage. In this paper, several approaches are investigated to resolve the issue, such as upsized 6T SRAM bitcell, 8T SRAM bitcell, read- and write-preferred bitcells, and read- and write-assist circuits. HSPICE simulations are performed using PTM 32nm model parameters.**

Index Terms—**SRAM, minimum operating voltage, stability, write-ability, preferred bitcell, assist circuit**

I. INTRODUCTION

THE process variation increases and the supply voltage decreases as technology shrinks. Thus, the stability and write-ability of SRAM are seriously reduced in deep submicron technologies. Many device and circuit solutions to overcome the low stability and write-ability of SRAM have been proposed. However, because technology scaling will continue, the low stability and write-ability issue of SRAM remains problem to be solved.

The use of mobile devices powered by batteries is growing exponentially as ubiquitous technology is advanced. In addition, the requirement for high-performance results in increased power consumption. Thus, the importance of low power SoC has been increased and research has focused on obtaining low power consumption while minimizing performance degradation. Dynamic voltage and frequency scaling (DVFS) is an effective technique to reduce the dynamic power and has been researched intensively. An SoC with DVFS can achieve low power consumption by reducing the supply voltage and operating frequency when a workload does not require the maximum throughput. Because the workload rarely requires the maximum throughput, DVFS can greatly reduce power consumption. SRAM is one of the main components in an SoC,

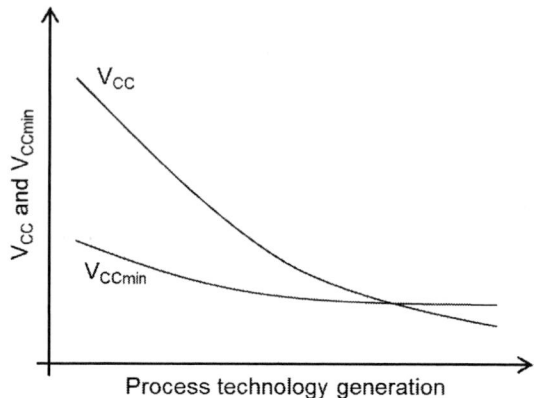

Fig. 1. The trend of V_{CC} and V_{CCmin} with technology scaling.

and it takes up a large portion of the SoC's power consumption. In order to apply DVFS to SRAM, SRAM needs to operate at low supply voltage. However decreasing the supply voltage reduces the stability and write-ability of SRAM. Therefore, a radical improvement in stability and write-ability is required to apply DVFS to SRAM in deep submicron technologies.

Read- and write-preferred bitcells are designed by adjusting the transistor characteristics in the SRAM bitcell to obtain high stability and write-ability, respectively. In general these preferred bitcells are combined with write- and read-assist circuits to simultaneously achieve high stability and write-ability. There are many design options for the combination of preferred bitcells and assist circuits. In this paper, several options for 6T and 8T SRAM are investigated in terms of stability, write-ability, and read-ability with layout area constraint when a preferred bitcell and an assist circuit are simultaneously applied.

II. LOW V_{CCMIN} ISSUE IN DEEP SUBMICRON TECHNOLOGIES

Fig. 1 shows the trend of the nominal operating voltage (V_{CC}) and the minimum operating voltage (V_{CCmin}) in the conventional 6T SRAM bitcell with technology scaling [1]. As technology scales downs, V_{CC} is scaled to guarantee the device reliability and to reduce the power consumption, and V_{th} is also scaled to retain device performance. However in deep submicron technologies, V_{th} scaling is limited because of increased leakage current. Thus, V_{CC} scaling becomes slower as technology scales down, as shown in Fig. 1.

The V_{CCmin} is lowered because of process optimization efforts that minimize channel length and width variation as well

---

Y. H. Yang, J. Kim, H. Park, and S. O. Jung are with the School of Electrical and Electronic Engineering, Yonsei University, Seoul 120-749, Korea (e-mail: sjung@yonsei.ac.kr).
J. Wang and G. Yeap are with Qualcomm Inc., San Diego, CA 92121 USA.

978-1-4244-9019-6/11 $26.00 © 2011 IEEE

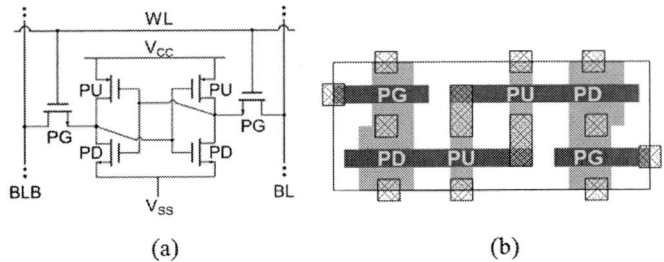

Fig. 2. 6T SRAM bitcell (a) schematic and (b) layout.

as thermal gradient. However, the V_{CCmin} has a lower decreasing rate than V_{CC} as a result of random dopant fluctuation and small feature size [1]. Thus, in deep submicron technologies, the difference between V_{CC} and V_{CCmin} nearly disappears and V_{CCmin} can even be increased greater than V_{CC}. Thus, this problem should be resolved to keep the V_{CC} scaling trend with technology scaling. In addition, V_{CCmin} needs to be further lowered to make DVFS applicable to SRAM to achieve low power consumption in deep submicron technologies

III. PREFERRED 6T SRAM WITH ASSIST CIRCUIT

Fig. 2 shows the schematic and layout of a 6T SRAM bitcell. SRAM stability and write-ability depend on the strength ratio of bitcell transistors. Stability is measured by SNM, which is determined by the strength ratio of pull down transistor (PD) to pass gate transistor PG (β ratio) and that of PD to pull up transistor (PU) (γ ratio), whereas write-ability is measured by WAM, which is determined by the strength ratio of PG to PU (α ratio), where SNM (WAM) is the maximum value of noise at the storage node that can be tolerated in read (write) mode [2]. Thus, SRAM stability and write-ability can be improved by controlling strengths of bitcell transistors.

Preferred bitcell is designed to improve SNM (or WAM) by adjusting the strength of bitcell transistors by transistor size (width and length) and V_{th0}, which necessarily degrades WAM (or SNM). Then, whichever is worse (SNM or WAM) as a result of the preferred bitcell design is improved by using an assist circuit that controls the source and gate voltages of the bitcell transistors. Thus, both SNM and WAM can be improved with a preferred bitcell combined with an assist circuit. In this section, bitcell design methods using the preferred bitcell and assist circuit are described. PTM 32nm model parameters are used for HSPICE simulation. In deep submicron technologies, the effect of systematic variation (width, length, and T_{ox} variation) is relatively small and becomes smaller in SRAM bitcells because of small feature size, whereas random variation is more significant. Thus, device mismatch caused by random threshold voltage variation is considered to be a major source of the variation in the Monte Carlo simulation [3].

A. Write-Preferred Bitcell and Read-Assist Circuit

A large α ratio makes WAM higher. Thus, WAM can be improved by increasing the strength of PG or decreasing the strength of PU. There are three options for achieving a write-preferred bitcell:

1) Increasing V_{th0} of PU: Because V_{th0} of PU (V_{th0_PU}) is

typically high enough for low leakage current and increasing V_{th0_PU} reduces the performance, increasing V_{th0_PU} has limitations.

2) Increasing the width of PG: This method increases the read current (I_{cell}) as well as WAM. However, the bitcell size is increased.

3) Decreasing V_{th0} of PG and PD: This method is advantageous in that it has no area overhead and I_{cell} is increased. However, leakage current is exponentially increased with decreasing V_{th0} of PG and PD (V_{th0_nmos}). Thus, there is a limitation in WAM improvement by decreasing of V_{th0_nmos}.

Because the SNM of the write-preferred bitcell is degraded, it should be recovered and improved to achieve low V_{CCmin}. This can be accomplished by the read-assist circuit. There are several techniques for achieving read-assist circuit:

1) Increasing cell V_{CC}: This method directly increases the strength of PD and thus sufficiently increases SNM, but it requires column by column control of cell V_{CC} (V_{CC_cell}) lines to improve SNM of all bitcells in read mode. Thus, increasing V_{CC_cell} is required for all bitcells in the read operation and almost all bitcells except for bitcells in the selected column in the write operation, leading to large power consumption. Thus, V_{CC_cell} control is not efficient for the read-assist circuit.

2) Decreasing bitline voltage: SNM improvement by decreasing bitline voltage (V_{BL}) is not sufficient. In addition, read access time is increased due to reduced I_{cell} by lowered V_{BL}. Thus, this method is not efficient for the read-assist circuit.

3) Decreasing wordline voltage: Wordline voltage (V_{WL}) control has no WAM degradation problem in the read operation because all bitcells are in read mode. However, as the half-selected cells in which the wordline is selected but the bitline (BL) is not selected are in read mode in the write operation, the wordline needs to be decreased for the half-selected cells, leading to the degradation in WAM for the selected bitcells. Thus, it is difficult to reduce V_{CCmin} by decreasing V_{WL} because of lowered WAM caused by the decreased V_{WL}.

In summary, although increasing the PG width can be used to obtain write preferred bitcells with improved read access time, all read assist circuits are not efficient. Therefore, a write preferred bitcell combined with a read assist circuit is not useful for improving V_{CCmin}.

B. Read-Preferred Bitcell and Write-Assist Circuit

A large β ratio makes SNM higher and a large γ ratio achieves small SNM improvement. Thus, SNM can be improved by increasing the strength of PD or decreasing the strengths of PG and PU. There are six options for achieving a read-preferred bitcell:

1) Increasing V_{th0} of PG and PD: This method is advantageous in that it has no area overhead and achieves SNM improvement. However, I_{cell} is seriously decreased with increasing V_{th0_nmos}.

2) Decreasing the width of PG: This method also has no area overhead. However, I_{cell} is decreased with decreasing PG width. Typically, PG width in the practical SRAM is made to be a minimum size, and thus there is a limitation of decreasing the

978-1-4244-9019-6/11 $26.00 © 2011 IEEE

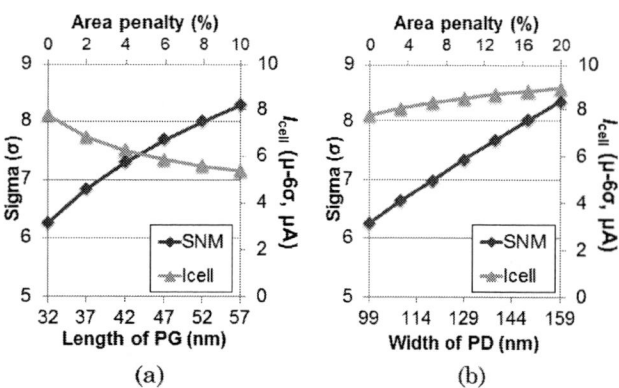

Fig. 3. *SNM*, I_{cell}, and area penalty according to (a) PG length and (b) PD width.

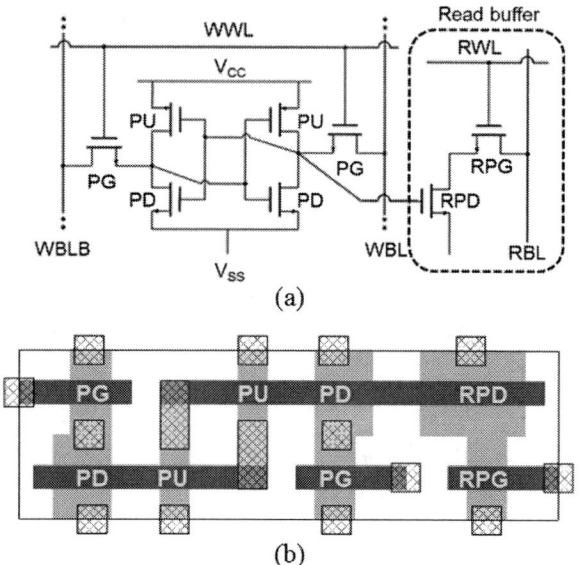

Fig. 4. 8T SRAM bitcell (a) schematic and (b) layout.

width of PG.

3) Increasing the width of PU: This method improves *SNM* by changing the γ ratio. When considering the area overhead, the increase in *SNM* may not be sufficient compared with other options.

4) Decreasing V_{th0} of PU: This method also increases the γ ratio, but *SNM* improvement is not sufficient. In addition, the leakage current is exponentially increased by decreasing V_{th0} of PU (V_{th0}_PU).

5) Increasing the length of PG: This method can sufficiently increase *SNM*, but I_{cell} is slightly decreased. Fig. 3(a) shows the *SNM* improvement with increasing PG length. When the area penalty due to the increase in PG length is 10%, *SNM* is improved by 2σ and I_{cell} is decreased by 2uA at the 6σ point.

6) Increasing the width of PD: With the same area constraint, this method achieves smaller *SNM* improvement compared to the method that increases PG length but increases I_{cell} whereas the latter reduces I_{cell}. Fig. 3(b) shows *SNM* improvement with increasing the width of PD. When the area penalty due to the increase in PD width is 20%, *SNM* is improved by 2σ and I_{cell} is increased by 1uA at the 6σ point.

TABLE I
SNM, *WAM*, I_{CELL}, AND V_{CCMIN} OF READ-PREFERRED BITCELLS

Characteristics	Spec1	Spec2	Spec3
SNM	9.17 σ	8.33 σ	9.24 σ
WAM	5.78 σ	7.51 σ	6.61 σ
I_{cell} (μ-6σ)	4.80 μA	8.9 μA	5.74 μA
$I_{leakage}$ (μ+6σ)	212.64 pA	268.56 pA	224.71 pA

Spec 1 : 20% area overhead by increasing only PG length
Spec 2 : 20% area overhead by increasing only PD length
Spec 3: 20% area overhead by increasing both PG length and PD width

In summary, increasing PG length is a useful method in improving *SNM*, but it slightly decreases I_{cell}. Thus, increasing both PG length and PD width is more efficient, because increasing PD width increases I_{cell}. Table I presents the characteristics of three bitcells: increasing only PG length, only increasing PD width, and increasing both PG length and PD width.

To improve *WAM*, the write-assist circuit needs to be applied to the read-preferred bitcell. There are several techniques for achieving the write-assist circuit:

1) Increasing V_{WL} (BV_{WL}): This method cannot be applied because *SNM* for half-selected cells in the same row decreases.

2) Negative V_{BL} (NV_{BL}): This method can be used as the write-assist circuit, because *SNM* for half-selected cells is not affected. However, an additional circuit for the negative charge pump is required [4].

3) Lowering V_{CC_cell} (LV_{CC_cell}): This method can be used for the write-assist circuit without decreasing *SNM* and it requires column by column control of V_{CC_cell} lines, dual supply voltage lines, to improve the write-ability for the selected bitcells in the write operation [5]. Even though it requires column by column control of V_{CC_cell}, decreasing V_{CC_cell} is required only for the cells in the selected column, leading to insignificant increase in power consumption.

In summary, Increasing PG length and PD width can be used to obtain a read-preferred bitcell, and NV_{BL} and LV_{CC_cell} can be used as a write-assist circuit. Therefore, a read-preferred bitcell combined with a write-assist circuit can efficiently improve V_{CCmin}.

IV. 8T SRAM WITH ASSIST CIRCUIT

An 8T SRAM bitcell, which adds read buffer (RPG and RPD) to a 6T SRAM bitcell, does not need to consider *SNM* because the BL and storage node are disconnected by turning off the PG transistor even in read mode, as shown in Fig. 4. By virtue of removing *SNM*, the 8T SRAM bitcell only needs to consider hold stability (*HSNM*) and *WAM*. Because *HSNM* is typically sufficiently high, the methods to increase *WAM* are focused. One of the methods to obtain a write-preferred bitcell of the 8T SRAM bitcell is to increase PG width. However, because the 8T SRAM bitcell typically has 30% area overhead compared to 6T SRAM, the write-preferred bitcell realized by increasing the transistor size worsens the area penalty. As the write-assist circuit of the 8T SRAM bitcell, the control of V_{CC_cell} and V_{BL} can be used similar to the write-assist circuit of the 6T SRAM bitcell. The V_{WL} control can also be used in the

978-1-4244-9019-6/11 $26.00 © 2011 IEEE

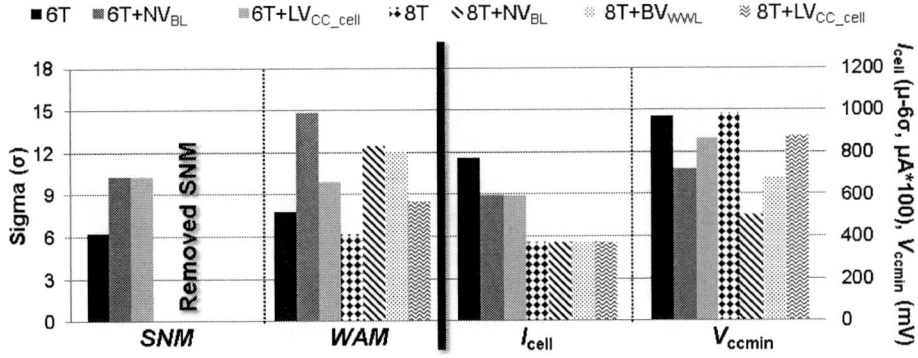

Fig. 5. *SNM*, *WAM*, I_{cell}, and V_{CCmin} in 32nm technology process.

8T SRAM bitcell because the read wordline (RWL) and the write wordline (WWL) are separated.

However, 8T SRAM still suffers from the bit-interleaving issue for unselected bitcells in the write operation. To solve this issue, the write back scheme with 8T SRAM [6] and different bitcell structures have been proposed [7].

V. COMPARISONS

To compare the 6T and 8T SRAM bitcells with the same area penalty, transistors in the 6T SARM bitcell with preferred bitcell scheme are upsized by 30% (10% for PG length and 20% for PD width). The 8T SRAM bitcell is designed using solely the write assist circuit without the write-preferred bitcell scheme.

Fig. 5 shows the Monte Carlo simulation result for *SNM*, *WAM*, I_{cell} and V_{CCmin} of 6T, U6T+NV$_{BL}$, U6T+LV$_{CC\_cell}$, 8T, 8T+NV$_{BL}$, 8T+BV$_{WWL}$ and 8T+LV$_{CC\_cell}$ in 32nm technology, where 6T denotes basic 6T SRAM bitcell, U6T+NV$_{BL}$ denotes the 30% upsized 6T SRAM bitcell with negative V_{BL}, U6T+LV$_{CC\_cell}$ denotes 30% upsized 6T SRAM bitcell with lowered V_{CC_cell}, 8T denotes the basic 8T SRAM bitcell, 8T+NV$_{BL}$ denotes the 8T SRAM bitcell with negative V_{BL}, 8T+BV$_{WWL}$ denotes the 8T SRAM bitcell with boosted V_{WWL} and 8T+LV$_{CC\_cell}$ denotes the 8T SRAM bitcell with lowered V_{CC_cell}. V_{BL} in NV$_{BL}$ and V_{CC_cell} in LV$_{CC\_cell}$ is reduced by 250mV. V_{WL} in BV$_{WWL}$ is boosted by 250mV. Compared with basic 6T and 8T SRAM bitcells, bitcells using a preferred bitcell combined with an assist circuit achieve higher *SNM* and *WAM*, and thus reduce V_{CCmin}. U6T+NV$_{BL}$ and U6T+LV$_{CC\_cell}$ decrease the V_{CCmin} by 252mV and 105mV, respectively. 8T+NV$_{BL}$, 8T+BV$_{WWL}$ and 8T+LV$_{CC\_cell}$ decrease V_{CCmin} by 483mV, 306mV and 112mV, respectively. Thus, 8T+NV$_{BL}$ can achieve the lowest V_{CCmin} but it has relatively small I_{cell}. Therefore, U6T+NV$_{BL}$ is an alternative candidate because of its large I_{cell} with relatively low V_{CCmin}.

VI. CONCLUSION

To reduce V_{CCmin}, the stability and write ability should be increased. The preferred bitcells combined with the assist circuits are investigated to increase the stability and write-ability. Compared with basic 6T and 8T SRAM bitcells,

bitcells using a preferred bitcell combined with an assist circuit achieves higher *SNM* and *WAM*, and thus reduce V_{CCmin}. In 32nm technology, basic 6T and 8T SRAM bitcells have negligibly low V_{CCmin} compared to V_{CC} (1.0V). U6T+NV$_{BL}$ and U6T+LV$_{CC\_cell}$ lower V_{CCmin} to 719mV and 866mV, respectively. 8T+NV$_{BL}$, 8T+BV$_{WWL}$ and 8T+LV$_{CC\_cell}$ lower V_{CCmin} to 506mV, 683mV and 877mV, respectively. This result will contribute to the implementation of the DVFS system for low power consumption.

REFERENCES

[1] S.C. Song , M. Abu-Rahma, and G. Yeap, "FinFET based SRAM bitcell design for 32 nm node and below", *Microelectronic Journal*, 2011, pp. 520-526

[2] J. Lohstroh, E. Seevinck, and J. de. Groot, "Worst-Case Static Noise Margin Criteria for Logic Circuits and Their Mathematical Equivalence", *IEEE J. Solid-State Circuits*, 1983, 18, (6), pp. 803-807

[3] S.H. Woo, H. Kang, K. Park, and S.O. Jung, "Offset voltage estimation model for latch-type sense amplifiers", *IET Circuits Devices Systems*. 2010, 4, (6), pp. 503-513

[4] Y. Fujimura, et al., "A Configurable SRAM with Constant-Negative-Level Write Buffer for Low-Voltage Operation with 0.149μm2 Cell in 32nm High-K Metal-Gate CMOS," *Int. Solid-State Circuits Conf.(ISSCC)*, 2010, pp. 348-349

[5] K. Zhang et al., "A 3-GHz 70 Mb SRAM in 65 nm CMOS technology with integrated column- based dynamic power supply," *IEEE J. Solid-State Circuits*, 2006, 42, (1), pp. 146-151

[6] Y. Morita, H. Fujiwara, H. Noguchi, Y. Iguchi, K. Nii, H. Kawaguchi, and M. Yoshimoto, "An area-conscious low-voltage-oriented 8T-SRAM design under DVS environment," *in Symp. VLSI Circuits Dig.*, 2007, pp. 256–257.

[7] M. Yabuuchi, K. Nii, Y. Tsukamoto, S. Ohbayashi, Y. Nakase, and H. Shinohara, "A 45nm 0.6V Cross-Point 8T SRAM with Negative Biased Read/Write Assist", *in Symp. VLSI Circuits Dig.*, 2009, pp. 158-159

An Ultra-Low Power K-Band Low-Noise Amplifier Co-Designed With ESD Protection in 40-nm CMOS

Ming-Hsien Tsai[1,2], Shawn S. H. Hsu[1] Fu-Lung Hsueh[2], Chewn-Pu Jou[2], Tzu-Jin Yeh[2] Ming-Hsiang Song[2], and Jen-Chou Tseng[2]

[1]Dept. of Electrical Engineering and Institute of Electronics Engineering,
National Tsing Hua University, Hsinchu, Taiwan
[2]Design Technology Division,
Taiwan Semiconductor Manufacturing Company, Hsinchu, Taiwan

Abstract—This paper presents a K-band low noise amplifier (LNA) co-designed with ESD protection circuit in 40-nm CMOS technology. By treating ESD devices as a part of the input matching network, an ESD protected 24-GHz LNA is demonstrated with a NF of 3.2 dB under a power consumption of only 4.1 mW. The ESD protection network is composed of dual-diode and a gate-driven power clamp achieving an ESD level of 2.8 kV human body model (HBM). Owing to the co-design approach, the NF only degrades by 0.2 dB compared with the reference LNA without the ESD network. The ESD-LNA presents a power gain of 13.0 dB with the input and output return losses both greater than 10 dB. To the best of our knowledge, this is the first report on a 24-GHz ESD-protected LNA in 40-nm CMOS.

Index Terms— electrostatic discharge (ESD), low-noise amplifier (LNA), radio frequency (RF), shallow-trench isolation (STI).

I. INTRODUCTION

RAPID scaling of the feature size has made CMOS technology the most attractive candidate for system-on-chip (SOC) applications due to its high integration level, high operation frequency, and low cost [1]-[3]. Recently, the applications such as car radars and wireless communications with SOC using K-band (18 to 26.5 GHz) have drawn significant attentions from both academia and industries [4]-[6]. In such a system, the critical low noise amplifier (LNA) is often exposed to the risk of electrostatic charge directly, and hence the on-chip electrostatic discharge (ESD) devices are essential. However, the parasitic effect introduced by ESD protection device could seriously degrade the RF performance of the LNA when operating at such a high frequency. Therefore, there exists a strong motivation to optimize RF and ESD design simultaneously. For achieving an overall good system performance with adequate reliability, the ESD protection is a major concern and should be taken into consideration at the early design stage together with the RF core circuit, especially for the design using advanced CMOS technology with a very thin gate oxide [7], [8].

In this study, we propose a LNA co-designed with the ESD protection circuits in advanced 40-nm CMOS technology. Differing from most of previous works [9]-[11] treating the

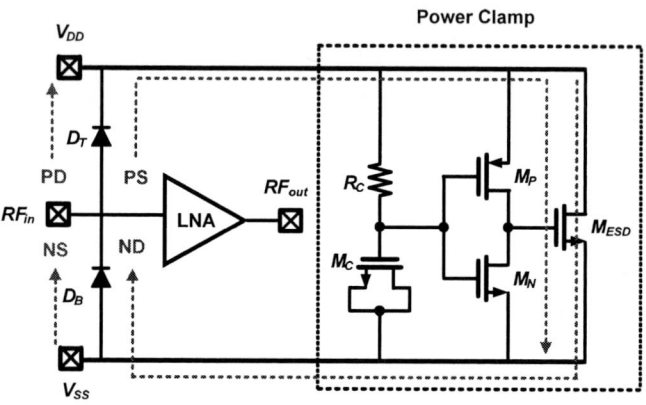

Fig. 1. Circuit blocks of the LNA with ESD protection network, consisting dual-diode and gate-driven power clamp. Four ESD testing modes with ESD current path are also indicated.

ESD blocks separately with the RF core circuit, the ESD blocks are utilized as a part of the input matching network directly. To achieve good RF performance and meet the ESD specification simultaneously, the dual-diode together with a power clamp is employed. In addition, the linear gate-driven mode power clamp is employed to provide an efficient detection on ESD events, which is beneficial for the core transistors with a low gate-oxide breakdown. Under a supply voltage of 1.1V and an associated drain current of 3.7 mA, the ESD-protected LNA presents a peak power gain of 13.0 dB and a NF of only 3.2 dB at 24 GHz. The ESD-protected LNA also demonstrates a 1.83-A TLP failure level, corresponding to an over 2.8 kV HBM ESD protection. To the best of our knowledge, this is the first reported 24-GHz LNA with ESD protection in 40-nm CMOS technology.

II. CIRCUIT TOPOLOGY

A. ESD Protection Network

Fig. 1 shows the ESD protection scheme employed in this design, which consists three components, namely the diodes D_T and D_B, and also the power clamp. The figure also indicates the

978-1-4244-9019-6/11 $26.00 © 2011 IEEE

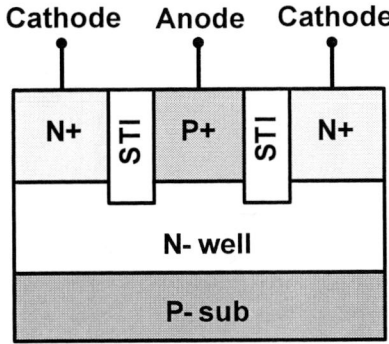

Fig. 2. Cross section view of the P+/N-well STI diode.

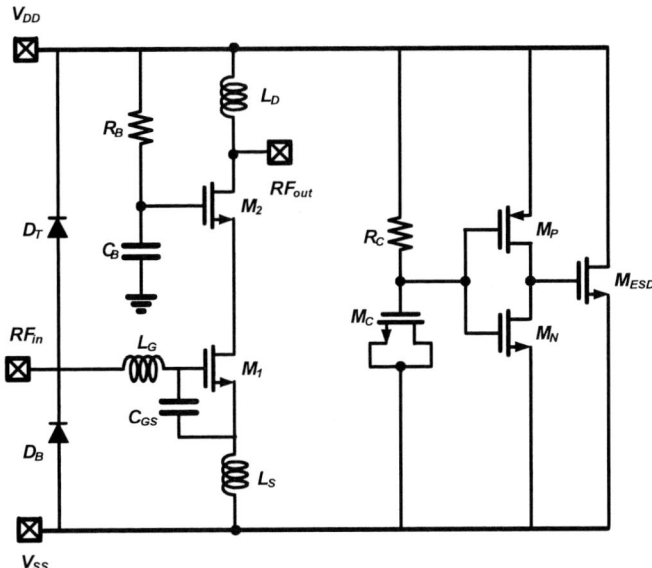

Fig.3. Complete circuit topology of the proposed ESD-protected LNA

discharge paths for the four different ESD testing modes, i.e., positive (PD mode) and negative (ND mode) to V_{DD}, and positive (PS mode) and negative (NS mode) to V_{SS}. When an ESD zapping occurs to the circuit, the diodes function together with the power clamp to provide a low-impedance path for discharging the electrostatic current. Fig. 1 also shows the circuit configuration of the gate-driven power clamp. The MOS capacitor M_C and P-type poly resistor R_C produce a RC time delay to ensure M_{ESD} functions correctly during an ESD event. The transistor M_{ESD} with a multi-finger topology has a total gate width up to ~ 2000 μm to sustain a high ESD current level. The large size of low V_t device M_{ESD} with a low on-resistance and leakage current also allows the power clamp to consume less voltage budget under a certain current level, and thus relaxes the ESD design requirement for other ESD elements. Note the parasitic resistances and capacitances introduced by the power clamp are not that crucial for the RF characteristics since this block is connected between the power and ground rails only.

The diodes at the input of LNA play an important role for the ESD discharge paths, which are also critical for the input matching network. The diodes with the P+/N-well and N+/P-well structures surrounded by STI are utilized for D_T and D_B, respectively. Fig. 2 shows the cross section view of the P+/N-well STI diode, respectively. For an area efficient design, the geometry of diode has a large L/W ratio. Under a fixed chip area, as the total perimeter increases, the overall current handling capability can be enhanced associated with a reduced parasitic resistance. Moreover, the power clamp is placed as close as possible to the RF input pins to reduce the interconnect resistance and the associated voltage drop.

B. LNA Configuration

Fig. 3 shows the complete proposed ESD-protected LNA using the dual-diode in conjunction with a power clamp as ESD protection network. The LNA is designed as a cascode configuration with inductive degeneration applied in the common-source stage. A single-stage topology is designed to prevent unpredicted parasitic effects from the complicated layout. Besides, metal 1 and metal 2 cross pattern ground

shielding is employed for interconnects to alleviate the signal attenuation and substrate coupling effect [12]. The input matching network includes dual-diode D_T and D_B, gate inductor L_G, source inductor L_S, and gate-source capacitor C_{GS}. The extra gate-source capacitor is used for achieving power-constrained simultaneous noise and input matching (PCSNIM) [13]. The drain inductor L_D is used for inductive peaking and also output matching.

Using the ESD and matching network co-design approach, the size of M_1 and M_2, and the transconductance of the transistors are determined first to meet the requirements of power dissipation, gain, and noise characteristics. The ESD blocks, D_T and D_B) are then designed based on the estimated ESD protection levels and parasitic capacitances. The shunt parasitic capacitances introduced by the ESD blocks are co-designed with L_G, L_S, and C_{GS} to achieve simultaneous noise and power matching. The ESD design using the dual-diode increases the parasitic capacitance by about 87 fF at the RF input, which is absorbed into the input matching network and hence with a small impact on RF performance.

III. RESULTS AND DISCUSSION

The LNA was fabricated in a 40-nm CMOS low-power process. The chip micrograph of the proposed LNA is with a chip area size of ~ 0.22 mm^2. The reference LNA without the ESD protection circuit was also implemented.

A. RF Measurements

The S-parameters and noise figures measurements were performed on-wafer by the PNA network analyzer and the noise figure analyzer, respectively. Under a supply voltage of 1.1 V and an associated drain current of 3.7 mA (P_{diss}= 4.1 mW),

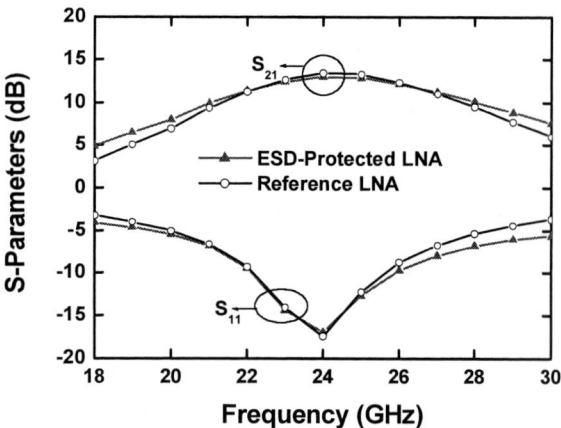

Fig. 4. Measured S_{11} and S_{21} of the LNAs with/without ESD protection.

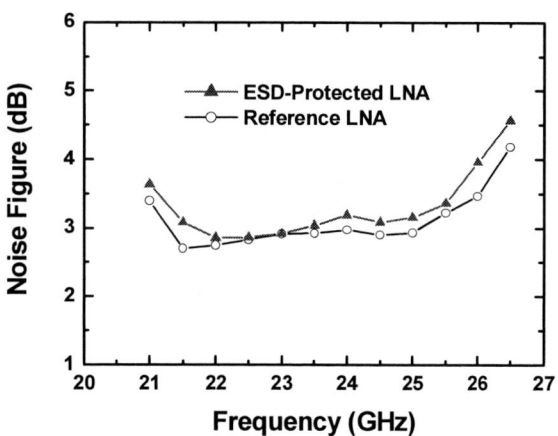

Fig. 5. Measured NFs of the LNAs with/without ESD protection.

the measured S-parameters (S_{11} and S_{21}) of the LNAs with/without ESD protection are shown in Fig. 4. The LNA with ESD protection circuit presents a peak power gain of 13.0 dB, which is only 0.4 dB lower compared with reference LNA at the center frequency of 24 GHz. The input return losses of LNAs are both greater than 10 dB demonstrating the successfully designed input matching network. Fig. 5 shows the measured noise figures of the LNAs with/without ESD protection. At 24 GHz, a noise figure of 3.2 dB is achieved, which is only about 0.2 dB higher compared to the reference LNA.

B. ESD Testing Results

The ESD testing was performed on-wafer by DC probes using the Barth 4002 transmission line pulse (TLP) test system. The pulse of a 10-ns rise time with a 100-ns pulse width was

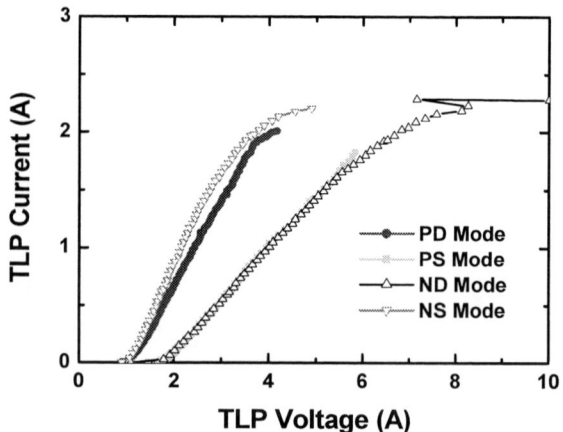

Fig. 6. Measured TLP *I-V* curves including four testing modes. The end point of TLP measurement represents the significant increased leakage current, indicating the TLP failure point *It2*

used to simulate the HBM ESD condition. Fig. 6 shows the transmission line pulse (TLP) test results of different testing modes (PD, PS, ND, and NS) for the ESD-protected LNA [8], [14]. In the PD mode, the *I-V* curve presents a linear characteristic, indicating the ESD bypass current enters the RF input pad and flows through D_T to V_{DD}. Note the *I-V* curve of the NS mode is almost identical to that for the PD mode, indicating the balanced ESD capability of D_T (P+/N-well) and D_B (N+/P-well). In the PS mode, the *I-V* curve also illustrates a linear characteristic, suggesting the ESD bypass current travels through D_T to V_{DD} and flows to power clamp, and then reaches V_{SS}. The *I-V* curve of the ND mode is close to the PS mode representing the same voltage drop of the ESD diode and power clamp. In the four testing modes, a minimum second breakdown current (It_2) up to 1.83 A is obtained, corresponding to an ESD level over 2.8 kV. The ESD test combination is summarized in Table I. As can be seen, the worst HBM ESD case occurs in the PS mode, but still can achieve a 2.8 kV HBM ESD level. Table II compares this work with other published RF LNAs. The proposed LNA achieves a lowest NF of 3.2 dB under a lowest power consumption of only 4.1 mW. The LNA also demonstrates an over 2.8 kV HBM ESD level by 40-nm technology with the thinnest gate oxide compared with other published works.

TABLE I
SUMMARY OF ESD TEST COMBINATION

ESD	It_2 (A)	HBM (kV)
PD mode	2.0	~ 3.0
PS mode	1.83	~ 2.8
ND mode	2.3	~ 3.5
NS mode	2.2	~ 3.4

* ESD level are estimated from the TLP tests ($\sim It_2 * 1.5k\Omega$).

TABLE II
Performance Comparison of the Proposed RF LNAs with Prior Arts

Reference	Tech. (nm)	Freq. (GHz]	Gain (dB)	NF (dB)	S_{11} (dB)	S_{22} (dB)	Power (mW)	IIP3 (dBm)	ESD (kV)
[5]	180 CMOS	22	10.1	4.3	-12	NA	7.2	-1.0	NA
[6]	180 CMOS	22	15.0	6.0	-21	NA	24	NA	NA
[15]	180 CMOS	23.7	12.86	5.6	-11	-22	54	2.04	NA
[16]	180 CMOS	24	13.1	3.9	-15	-20	14	NA	NA
[17]	180 CMOS	24	12.8	3.3	-7.5	-17	8	NA	NA
[18]	130 CMOS	24	14.0	5.0	-7.0	-15	18	-1.7	2.5
[19]	90 CMOS	24	7.5	3.2	-16	-30	10.6	NA	NA
This work	40 CMOS	24	13.0	3.2	-17	-13	4.1	-5.0	2.8

IV. CONCLUSION

In this paper, an ultra-low power 24-GHz LNA with excellent ESD and noise performance by ESD/matching co-design methodology was proposed and demonstrated. The LNA was fabricated in a standard 1P6M 40-nm CMOS process presenting an excellent NF of 3.2 dB under a power consumption of only 4.1 mW. An HBM ESD level over 2.8 kV was also demonstrated. With careful considerations of the ESD blocks during the RF design, the RF performance of the LNA was virtually unaffected by the ESD protection circuit. With the additional ESD blocks, The NF only increased by 0.2 dB and the power gain only degraded by 0.4 dB at the center frequency.

ACKNOWLEDGEMENTS

The authors would like to thank Dr. Ming-Dou Ker, Dr. Chun-Yu Lin, and Mr. Li-Wei Chu for their helpful technical discussion.

REFERENCES

[1] A. Amerasekera and C. Duvvury, *ESD in Silicon Integrated Circuits*, 2nd Edition, John Wiley & Sons, 2002.

[2] A. Wang, On-Chip ESD Protection for Integrated Circuits, Kluwer, 2002.

[3] S. Voldman, *ESD: Circuits and Devices*, John Wiley & Sons, 2006.

[4] Y. Cao, V. Issakov, and M. Tiebout, "A 2kV ESD-protected 18GHz LNA with 4dB NF in 0.13μm CMOS," *ISSCC Dig. Tech. Papers*, pp. 194-195, Feb. 2008.

[5] Y. Wei, S. Hsu, and J. Jin, "A low-power low-noise amplifier for K-band applications," *IEEE Microw. Wireless Compon. Lett.*, vol. 19, no. 2, pp. 116-118, Feb. 2009.

[6] X. Guan and A. Hajimiri, "A 24-GHz CMOS front-end," *IEEE J. Solid-State Circuits*, vol. 9, no. 2, pp. 368-373, Feb. 2004.

[7] J. Borremans, S. Thijs, P. Wambacq, Y. Rolain, D. Linten, and M. Kuijk, "A fully integrated 7.3 kV HBM ESD-protected transformer-based 4.5-6 GHz CMOS LNA," *IEEE J. Solid-State Circuits*, vol. 44, no. 2, pp. 344-353, Feb. 2009.

[8] M. Tsai, S. Hsu, F. Hsueh, and C. Jou, "A multi-ESD-path low-noise amplifier with a 4.3-A TLP current level in 65-nm CMOS," *IEEE Trans. Microwave Theory and Tech.*, vol. 58, no. 12, pp. 4004-4011, Dec. 2010.

[9] M.-D. Ker, et al., "ESD protection design for 1- to 10-GHz distributed amplifier in CMOS technology," *IEEE Trans. Microwave Theory and Techniques*, vol. 53, no. 9, pp. 2672-2681, Sep. 2005

[10] C.-Y. Lin, L.-W. Chu, M.-D. Ker, T.-H. Lu, P.-F. Hung, and H.-C. Li, "Self-matched ESD cell in CMOS technology for 60-GHz broadband RF applications," in *Proc. IEEE Radio Frequency Integrated Circuit Symp.*, 2010, pp. 573-576.

[11] B. Kleveland, T. J. Maloney, I. Morgan, L. Madden, T.H. Lee, and S. S. Wang, "Distributed ESD protection for high-speed integrated circuits," *IEEE Electron Device Lett.*, vol. 21, pp. 390-392, Aug. 2000.

[12] Y. Lin, S. Hsu, J. Jin, and C. Chan, "A 3.1-10.6 GHz ultra-wideband CMOS low noise amplifier with current-reused technique," *IEEE Microw. Wireless Compon. Lett.*, vol. 17, no. 3, pp. 232-234, Mar. 2007

[13] T. Nguyen, C. Kim, G. Ihm, M. Yang, and S. Lee, "CMOS low-noise amplifier design optimization techniques," *IEEE Trans. Microwave Theory and Tech.*, vol. 52, no. 5, 1433-1442, May 2004.

[14] M. Tsai, S. Hsu, F. Hsueh, C. Jou, S. Chen, and M. Song, "A wideband low noise amplifier with 4 kV HBM ESD protection in 65 nm RF CMOS," *IEEE Microw. Wireless Compon. Lett.*, vol. 19, no. 11, pp. 734-736, Nov. 2009.

[15] K. Yu, Y. Lu, D. Chang, V. Liang, and M. Frank Chang, "K-band low-noise amplifier using 0.18 μm CMOS technology," *IEEE Microw. Wireless Compon. Lett.*, vol. 14, no. 3, pp. 106-108, Mar. 2004.

[16] S. Shin, M. Tsai, R. Liu, K. Lin, and H. Wang, "A 24-GHz 3.9-dB NF low-noise amplifier using 0.18 μm CMOS technology" *IEEE Microw. Wireless Compon. Lett.*, vol. 15, no. 7, pp. 448-490, July 2005.

[17] A. Sayag S. Levin, D. Regev, D. Zfira, S. Shapira, D. Goren, and D. Ritter, "A 25 GHz 3.3 dB NF low noise amplifier based upon slow wave transmission lines and the 0.18 μm CMOS technology" in *Proc. IEEE RFIC Symp.*, pp. 373-376, June 2008.

[18] V. Issakov, M. Tiebout, Y. Cao, A. Thiede, and W. Simburger, "A low power 24 GHz LNA in 0.13 μm CMOS," in *Proc. IEEE COMCAS Conf.*, pp. 1-10, May 2008.

[19] O. Dupuis, X. Sun, G. Carchon, P. Soussan, M. Ferndahl, S. Decoutere, and W. De Raedt, "24 GHz LNA in 90nm RF-CMOS with high-Q above-IC inductors," in *Proc. ESSCIRC*, pp. 89-92, Sep. 2005.

Low Power Embedded Memory Design – Process to System Level Considerations

Esin Terzioglu, Sei Seung Yoon, ChangHo Jung, Ritu Chaba, Venu Boynapalli, Mohamed Abu-Rahma, Joseph Wang, Sam Yang, Giri Nallapati, Aaron Thean, Chidi Chidambaram, Michael Han, Geoffrey Yeap, and Mehdi Sani

Qualcomm Inc., 5775 Morehouse Drive, San Diego, CA 92121, USA

Abstract—**Embedded memories are widely used in low power System-on-Chip (SoC) applications. Low power performance can be optimized with process, circuits, architecture and system level co-development. In this paper, low power design considerations are described in advanced technology nodes to address memory leakage and active power dissipation. Memory bit cell design in context of process technology definition, circuit techniques at the macro design level, and chip-level integration considerations for low power are described.**

Index Terms—**Semiconductor memory, random access memory, low power design**

I. INTRODUCTION

Embedded memories are widely used in System-on-Chip (SoC) applications with increasing amounts for highly integrated products. Low power static memory designs present challenges that are distinct from logic blocks. Process, IP design and system integration/usage collaboration is essential to achieve optimized memory technology for SoC designs. This paper describes challenges and opportunities for low power memory in most advanced technology nodes (28nm and beyond) in three general areas: process and bit cell design, peripheral circuit design, and system level integration/usage.

II. LOW POWER PROCESS AND BIT CELL DESIGN

Traditional 6-transistor (6T) SRAM bit cell has been widely used in System-on-Chip applications due to its versatile balance of area-performance-power (Fig. 1). However, 6T SRAM bit cell suffers from read stability and writeability challenges, especially at low voltage operation.

There is a conflicting device strength requirement for the access transistors N1 and N2 – they need to be weak enough to limit the internal node bounce during read and strong enough to overcome pull up device during write operation. Voltage headroom in the presence of local variations becomes an increasing challenge in advanced process nodes, exacerbated by product pressures to reduce supply voltage, and scale down bit cell area. As the supply voltage is reduced, the memory bit cell threshold voltage also has to be reduced to maintain the required effective gate overdrive (including local variations), resulting in substantially increased subthreshold leakage currents. Also note that, as device area decreases, local threshold variation (device mismatch) increases, as dictated by Pelgrom's Law [1], [2]. Since one of the matching requirements is between NMOS access device and PMOS pull up device, global process corners (such as slow NMOS and fast PMOS) as well as local mismatch considerations have to be taken into account. The shift in PMOS threshold voltage during the lifetime of a product also needs to be considered (NBTI shift). The memory bit cell threshold voltage is typically tuned to be as high as possible to allow minimum subthreshold leakage, shifting the challenge to achieve satisfactory minimum operational voltage, known as memory V_{ccmin}. In most advanced technology nodes (e.g. 28nm), end-of-life memory V_{ccmin} is specified to be $0.9V_{dd}$, effectively leaving no room for voltage scaling for memory.

A number of techniques have been proposed to mitigate bit cell leakage and V_{ccmin} challenges in advanced technology nodes [3]-[9]. These techniques include voltage bias techniques to provide "read assist" and "write assist" to the bit cell, new bit cell topologies with larger area to mitigate supply bounce and write conflict, as well as technology proposals to create asymmetric devices providing strong access device for write operation when current flows into bit line and weaker access device when current flows in the opposite direction. Adaptive assist techniques increase design complexity, delay time to market, and add product test challenges.

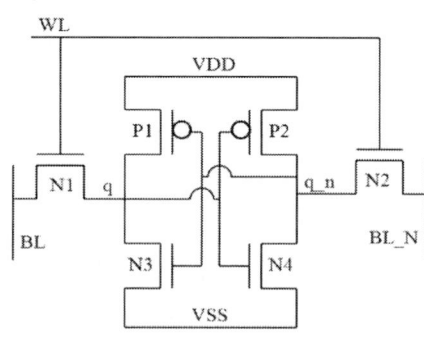

Figure 1: Circuit schematic of 6T SRAM bit cell

Manuscript received March 31, 2011.

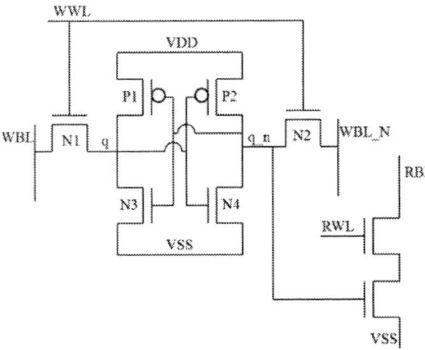

Figure 2: Circuit schematic of 8T register file bit cell

One commonly used approach to achieve lower V_{ccmin} in a low cost process is to separate the write port from the read port and use a single-ended (full-swing) read bitline, known as "8T register file bit cell" and shown in Fig. 2. Read bitline does not disturb the storage node in this bit cell architecture. The write port has the same topology as the 6T bit cell, however the only electrical optimization required is for write operation – the conflicting read/write requirements of the access device has been eliminated. Since larger devices are used in 8T bit cell, local variation is smaller than for 6T. Register file bit cell typically does not allow multiplexing or bitwise write selection operation i.e. once wordline for a given row is driven high, entire row has to be written with new data. The 8T bit cell is typically about twice the size of dense 6T bit cell, however it has wide application in SoC design. Many memory blocks on an SoC have very few bits (e.g., 16kbits or fewer) and are dominated by peripheral circuit area when implemented with compilers based on 6T bit cell. Since 6T bit cell has to be carefully optimized for read stability, read current is small especially when local mismatch of the constituent small devices is taken into account. Large sense amplifiers are used to detect small swing voltage generated by 6T bit cell which results in peripheral circuit area dominance for small macros. On the other hand, 8T bit cell read port is optimized for high speed full voltage swing sensing with a simple logic gate. For small memory instances, 8T register file bit cell results in smaller area and better performance, as well as better V_{ccmin}. 8T bit cell V_{ccmin} can be tuned so that it approximately tracks logic circuits.

Bit cell leakage is becoming a larger portion of overall memory macro leakage in advanced technology nodes. As shown in Fig. 3, in 28nm technology node, bit cell leakage can be as high as ~80% of the total macro leakage. One technique to achieve lower bit cell leakage is separating the memory and logic supply voltages so that logic voltage can be independently reduced for adaptive active logic power savings without corrupting memory operation (memory-logic dual rail). Once the logic and memory supplies are separated, the memory bit cell supply range and threshold voltage can be defined for optimal desired characteristics, with the constraint that logic and bit cell share the same gate dielectric (for low process complexity and cost). The gate dielectric thickness may need to be increased to support higher memory supply

TABLE I
VOLTAGE OVERDRIVE METRIC COMPARISON

Parameter	Thin oxide	Thick oxide (same Vth)	Thick oxide (higher Vth)
σ_{Vth} (mV)	50	55	55
n	5.5	5.5	5.5
V_{th} (V)	0.4	0.4	0.46
V_{dd} (V)	0.9	0.99	0.99
Effective Overdrive(V)	0.225	0.2875	0.2275

Bit cell effective gate overdirve metric (including local variations) for a hypothetical advanced process node

voltage. One critical aspect to consider is device local mismatch. Device threshold mismatch increases as the device area decreases, as described by Pelgrom's Law:

$$\sigma_{\Delta Vt} = \frac{A_{Vt}}{\sqrt{W\,L}}$$

Where $\sigma_{\Delta Vt}$ is the standard deviation of device threshold mismatch, W is device width, L is device gate length, and A_{Vt} is a technology-dependent constant, typically proportional to effective gate oxide thickness in the technology [10], [11].

If we take net effective gate voltage overdrive as a simple metric of circuit robustness at a given supply voltage in presence of local device variation out to n-sigma:

Thin oxide effective overdrive = $V_{dd} - V_{th0} - n\,\sigma_{\Delta Vt}$

Increasing the oxide thickness by a factor α with a corresponding approximately the same increase in supply voltage, the net overdrive metric is given by:

Thick oxide effective overdrive = $\alpha\,V_{dd} - V_{th1} - n\,\alpha\,\sigma_{\Delta Vt}$

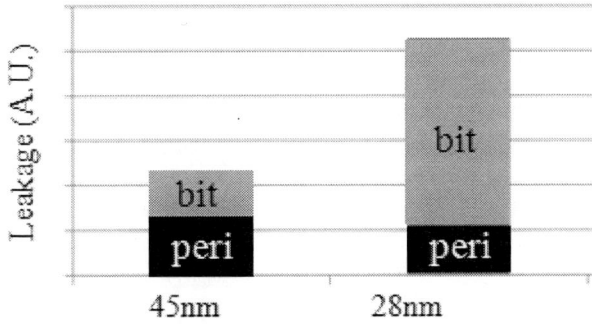

Figure 3: Memory leakage breakdown in 45nm and 28nm low power technology nodes between bit cell (bit) and periphery (peri). The 45nm design used low threshold devices in the periphery for higher speed.

We see that if V_{th0} is equal to V_{th1}, the effective thick oxide overdrive is larger than thin oxide case, when $V_{dd} > n \, \sigma_{\Delta Vt}$. Therefore, once oxide thickness and memory supply voltage are increased, the bit cell threshold voltage can also be increased with the same or better net overdrive metric, resulting in smaller subthreshold and gate leakage currents. This simple analysis is intended to demonstrate the trend in technology – the overall result remains the same when a more rigorous analysis is applied. Clearly, lower memory leakage is traded off for higher memory active power, however in mobile applications memory standby power is a significant portion of power budget. Table I demonstrates improvement in the simple effective overdrive metric for a hypothetical advanced technology node with assumed reasonable parameters. Last column indicates that the thicker oxide option can support ~60mV higher bit cell threshold with approximately the same gate overdrive metric as the thin oxide/low supply option (but with lower leakage). Note that the increase in oxide thickness has to be balanced to limit short channel effects on logic devices, which share the same gate dielectric as memory bit cell and may choose to operate at a lower supply voltage.

III. Low Power Memory Design

Numerous design techniques are necessary to reduce the active (switching) power and leakage power in memory design. In this section, we will briefly summarize a few techniques to reduce leakage at the circuit design level, and then focus switching power reduction considerations.

A. Leakage Power Reduction

One approach to reducing bit cell leakage at the marco level is to apply "source biasing" where the array source node is isolated and biased to an intermediate potential (typically several hundred millivolts) when the memory is not in use. The resulting combination of body effect (on pull down devices) and negative gate-source biasing (access device) will typically lead to 30%-45% bit cell leakage savings. However, the array source node has to be pulled back to ground before the memory can be accessed in order to avoid memory corruption. If source biasing is enabled in active usage mode, it will result in access delay and an increase in active power. It is desirable to explicitly enable source biasing during known idle periods. If data retention is not required for the memory, a local head switch can be used to shut down the entire array saving significant leakage power. For large memory arrays, each constituent sub-block has its own local headswitch control so that system can decide on the amount of available active memory by shutting down the subblocks that are not needed.

B. Active Switching Power Reduction

Majority of memory switching power is dissipated in the data path where many parallel lines are activated, compared with the decode and access path where only one wordline is ultimately enabled. In 6T SRAM designs, a minimum level of multiplexing is typically applied at the local sense amplifier

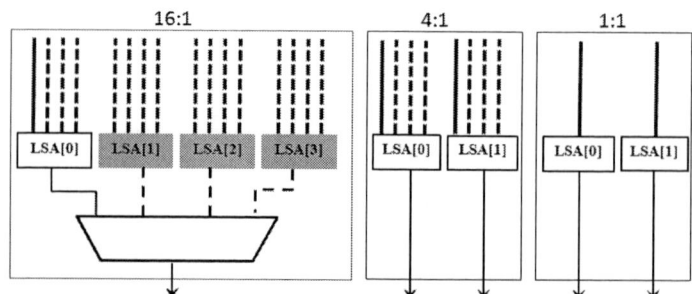

Figure 4: Data line power dissipation increases as larger multiplexing option is used in memory design. Dashed lines represent power wasted in unused local bitline columns connected to local sense amplifiers. Register files based on 8T bit cell typically use 1:1 multiplexing because inactive writes are not allowed.

level to allow compact local sense amplifier physical implementation and minimize noise coupling. While multiplexing results in more compact memory implementation, it leads to higher active power dissipation due to switching inactive columns wasting power, as illustrated in Fig. 4. Register file designs based on 8T bit cell typically only allow 1:1 multiplexing. Also, register files use much shorter bitlines with simpler (and lower capacitance) sense circuits, resulting in lower active power dissipation compared with 6T SRAM designs.

In 6T SRAM design, the local bit line power can be reduced by partitioning the memory to have shorter bit lines (and more internal sub-blocks). Smaller partitioning results in smaller bitline capacitance and lower active power (as well as faster access time). However, more finely paritioned memory blocks are larger due to increased peripheral circuit area. A critical parameter in memory designs with limited swing bit lines and strobed sense amplifiers, is the required bit line voltage split ("sense margin") when the sense amplifier is enabled. If the sense margin is not sufficient, memory read may fail. If the bit line split is too conservative, active power and access time are compromised. Sense margin related yield results from the statistical interaction of sense amplifier offset and bit cell read current variations. When a very slow bit cell happens to be sensed by a sense amplifier with a large offset, a read failure may occur. Rigorous statistical modeling is required to accurately estimate the resulting yield given bit cell read current and sense amplifier offset probability distribution curves. For a 1Mbit memory instance, the 6-sigma/3-sigma rule of thumb can be used to demonstrate the importance of sense amplifier design in advanced technology nodes. The 6-sigma/3-sigma rule assumes that a bit cell that is 6-sigma slower than the average is paired with a sense amplifier that has a 3-sigma offset. Therefore, in the worst case a 6-sigma slow bit cell must produce bitline split voltage equal to 3-sigma sense amplifier offset:

$$\Delta V_{BL}(6\sigma) = 3\sigma_{SA}$$

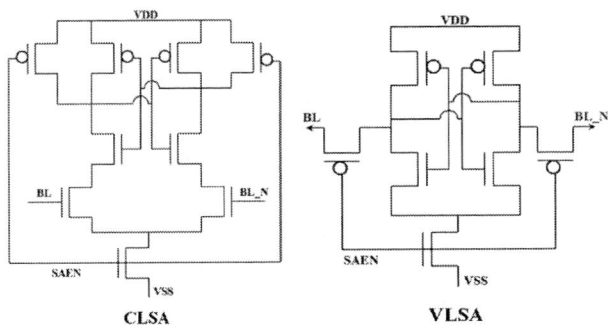

Figure 5: Current Latch Sense Amplifier (CLSA) and Voltage Latch Sense Amplifier (VLSA) topologies characterized in 28nm process node

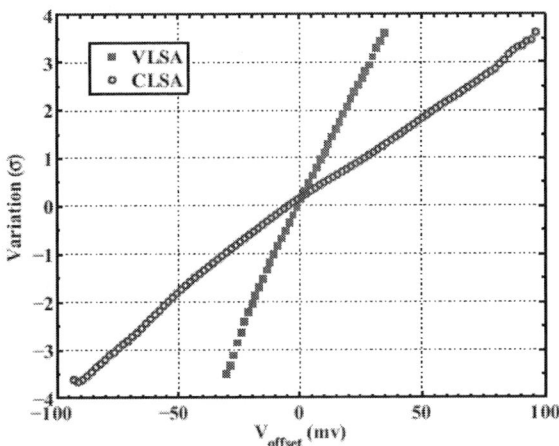

Figure 6: Current Latch Sense Amplifier (CLSA) and Voltage Latch Sense Amplifier (VLSA) input referred offset measurement results in 28nm technology node

A center bit cell (as given in spice models for slow process corner) must produce a bitline split given by:

$$\Delta V_{BL}(\text{center}) = 3 \frac{I_{center}}{I_{6\sigma}} \sigma_{SA}$$

Where I_{center} is the center bit cell read current and $I_{6\sigma}$ is current from a bit cell 6-sigma slower than the average. In the worst case corner (slow process and low supply voltage), due to large local variations, the ratio $I_{center}/I_{6\sigma}$ is in the range of 4 for 28nm process. In other words, the average bit cell must produce 12mV sense margin for every millivolt of σ_{SA}, based on this simple analysis. Using rigorous statistical simulation for the bit cell and the sense amplifiers, it can be shown that required sense margin is about 10mV for every millivolt of σ_{SA} [12].

It is critical to design a compact sense amplifier with relatively small offset voltage in order to enable low power (and high performance) memory design. We have experimentally investigated offset characteristics for two types of sense amplifiers in a low power 28nm technology node described in [13]. These sense amplifier topologies are shown in Fig. 5. The Current Latch Sense Amplifier (CLSA) is usually favored due to its gate-fed topology which isolates the bit lines from the internal nodes of the sense amplifier, minimizing capacitive loading on the bit lines. Voltage Latch Sense Amplifier (VLSA) has fewer transistors, but due to its drain-fed configuration adds more capacitive loading to the bit lines. The two sense amplifier devices were sized so that they fit into the same physical area. Silicon data from 28nm process node shown in Fig. 6 indicate that VLSA has a significant advantage with input-referred offset voltage [14]. VLSA has offset standard deviation of 9.5mV, compared with 26.7mV for CLSA. Using the relation derived above, VLSA would require about 114mV of sense margin, compared with 320mV for CLSA in this particular example. There is a significant advantage to using low offset sense amplifier, especially with long bit lines where the larger input capacitance of VLSA becomes a smaller percentage of the total bit line capacitance.

There is also a need to accurately characterize bit cell current distribution out to 6-sigma tail in order to feed accurate models to statistical analysis tools. As the local variation becomes larger, the need for accurate modeling in the tail region becomes more critical since the ratio that determines sense margin requirement $I_{average}/I_{6\sigma}$ becomes very sensitive as $I_{6\sigma}$ becomes very small. Read current deviation from normal distribution in the tail region has been observed on 28nm silicon as reported in [15].

IV. SYSTEM LEVEL INTEGRATION FOR LOW POWER

Low power memory requirements should be considered at initial planning and throughout system design process. There are many opportunities to trade off memory power for performance and area in terms of choice of memory type, organization and enabling power saving features.

Memory macros that have fewer than approximately 16kbits of memory will benefit from register file implementation using 8T bit cell. Active power savings becomes significant when the memory has a wide input-output (IO) bus. Typically, 6T SRAM compilers use a minimum multiplexing (2:1 or 4:1) in order to enable compact local sense amplifier layout. For wide IO memories, there is a significant amount of wasted active power in the inactive columns that are multiplexed out, as indicated in Fig. 4.

Chip designers use memory compilers to generate a long list of memory instances used on an SoC. Memory organization options affect area, speed and power. For a given memory, multiplexing changes the aspect ratio and also affects active power. As the multiplexing depth is increased, the number of passive columns increase – passive bit lines are accessed, and power is spent developing sense voltage on these bit lines, however the data are not sensed. From an active power

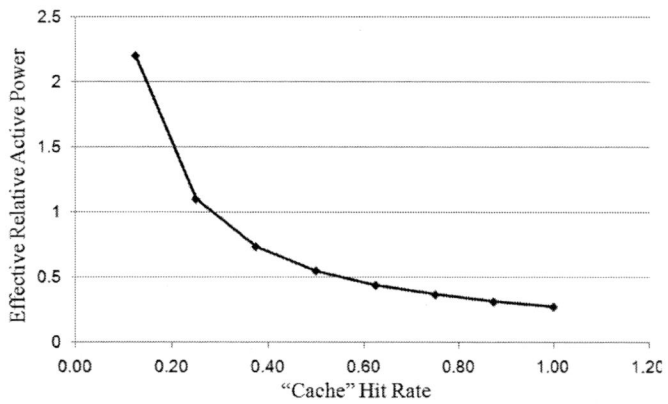

Figure 7: Memory effective active power as a function of "cache hit rate" at the memory output registers. Memory is implemented with 128 bit wide IO in a 16-bit system (i.e. 8 words per cache entry). The active power break-even cache hit rate is about 30%.

efficiency point of view, for a given number of memory bits, optimal instance is achieved when minimum multiplexing is used and the memory is close to square aspect ratio, so that wasted power in inactive bit lines is minimized. In order to achieve such an instance, it is clear that the memory IO width has to be adjusted and in many cases will be quite wide to achieve square aspect ratio with minimum multiplexing. However, if only a small portion of the data on this wide IO is then used by the system, we would still be wasting power since power is required to sense and ship out the additional bits. For cases where memory access pattern is sequential, the benefit of using wide IO embedded memory is clear – all the data in the wide IO registered at the output of the memory will eventually be used by the system before a new access is initiated into the memory (i.e. 100% data utilization rate). The number of system accesses into the actual memory will be reduced by the size of this "output cache" formed by memory output registers. In fact the analogy to a system cache memory, or page-access DRAM can be extended further. In cases where the system access pattern into the memory is not strictly sequential, a reasonable cache hit rate would still result in significant power savings, as indicated in Fig. 7. In this example shown in Fig. 7, a 128kbit memory is physically implemented as a 1k x 128 instance in a 16-bit wide system (i.e. output cache stores 8 words of data). At 50% cache hit rate, approximately 50% power savings is possible compared with 8k x 16 implementation.

ACKNOWLEDGMENT

The authors would like to acknowledge stimulating discussions with many colleagues across all divisions at Qualcomm.

REFERENCES

[1] Pelgrom, M.J.M.; Duinmaijer, A.; Welbers, A.P.G., "Matching properties of MOS transistors," *IEEE Journal of Solid-State Circuits*, vol. 24(5), pp. 1433-1439, 1989

[2] Pelgrom, M.J.M.; Tuinhout, H.P.; Vertregt, M., "Transistor matching in analog CMOS applications," *Proceedings of the International Electron Devices Meeting IEDM*, pp. 915-918, 1998

[3] *Hirabayashi, O.; et al., "A process-variation-tolerant dual-power-supply SRAM with 0.179μm² cell in 40nm CMOS using level-programmable wordline driver," ISSCC, pp. 458-459, 2009*

[4] Abu-Rahma, M.; Anis, M.; Yoon, S., *"A robust single supply voltage SRAM read assist technique using selective precharge,"* Proc. of IEEE European Solid-State Circuits Conference (ESSCIRC), pp. 234-237, 2008

[5] Khellah, M.; et al., *"A 256-Kb dual-Vcc SRAM building block in 65-nm CMOS process with actively clamped sleep transistor,"* IEEE Journal of Solid State Circuits, vol. 42, no. 1, pp. 233-242, 2007

[6] Kawasumi, T.; et al., *"A single-power-supply 0.7V 1GHz 45nm SRAM with an asymmetrical unit-alpha-ratio memory cell,"* ISSCC, pp. 382-622, 2008

[7] Hyunwoo, N.; et al., *"A 32nm high-k metal gate SRAM with adaptive dynamic stability enhancement for low-voltage operation,"* ISSCC, pp. 346-347, 2010

[8] Yabuuchi, M.; et al., *"A 45nm low-standby-power embedded SRAM with improved immunity against process and temperature variations,"* ISSCC, pp. 326-606, 2007

[9] Leland, Y.; et al., *"A 5.3GHz 8T-SRAM with operation down to 0.41V in 65nm CMOS,"* VLSI Circuit Conference, pp. 252-253, 2007

[10] Posch, W.; Rappitsch, G.; Leonardelli, G., *"MOS transistor mismatch modelling for high voltage CMOS processes,"* Advanced Semiconductor Manufacturing Conference and Workshop IEEE/SEMI, pp. 256-260, 2005

[11] Croon, J.A.; Decoutere, S.; Sansen, W.; Maes, H.E., *"Physical modeling and prediction of the matching properties of MOSFETs,"* Solid-State Device Research Conference ESSDERC, pp. 193-196, 2004

[12] Abu-Rahma, M.; Chowdhury, K.; Wang, J.; Chen, Z.; Yoon, S.; Anis, M., *"A methodology for statistical estimation of read access yield in SRAMs,"* Design Automation Conference (DAC), pp. 205-210, 2008

[13] Chidambaram, P.R.; et al., *"Cost effective 28nm LP SoC technology optimized with circuit/device/process co-design smart mobile devices,"* Electron Devices Meeting (IEDM), pp. 27.3-27.3.4, 2010

[14] Abu-Rahma, M.; et al., *"Accurate characterization of sense amplifier input offset for SRAM yield optimization,"* unpublished

[15] Wang, J.; et al., *"Non-gaussian distribution of SRAM read current and design impact to low power memory using voltage acceleration method,"* VLSI Tech.Symposium, 2011

978-1-4244-9019-6/11 $26.00 © 2011 IEEE

65nm PD-SOI Glitch-Free Retention Flip-Flop for MTCMOS Power Switch applications

J. Le-Coz[1], P. Flatresse[1], S. Clerc[1], M.Belleville[2], A. Valentian[2]

[1]STMicroelectronics, Crolles, France
[2]CEA LETI, MINATEC campus, Grenoble, France

Abstract— **This work presents a partially depleted Silicon-on-Insulator (PD-SOI) low-static power consumption Retention Flip-Flop (REFF). This flip-flop is designed in order to avoid wake-up transient glitches. In addition specific leakage reduction techniques are used to compensate the extra leakage currents induced by the SOI floating body effects. This leads to a static power consumption reduced by 2 for only 6% of extra silicon area, compared to a regular floating body implementation.**

I. INTRODUCTION

Low-Power Partially Depleted Silicon On Insulator Technology (LP PD-SOI) is mainly used for Low-Dynamic-Power applications. CMOS 65nm LP PD-SOI technology achieves, compared to Bulk, a 20% higher speed at same nominal Vdd. Whereas at same speed, by reducing PD-SOI Vdd supply, 30% of dynamic power consumption is saved [1]. This is obtained thanks to lower junction capacitances and Floating Body Effect (FBE) which decreases the transistors threshold voltages (Vt).

In Submicron technologies power consumption has been saved by reducing its main component – the dynamic power consumption – thanks to a Vdd and Vt lowering [2]. This has directly increased the static power consumption. because of its exponential dependence on Vt, making it very significant for 65nm and beyond technologies. In PD-SOI technology, due to the Floating Body Effect, this static power is further amplified. Thus, in order to reduce as much as possible this static power consumption, energy-efficient PD-SOI circuits are usually implemented with a low static power design technique: MTCMOS power switch.

MTCMOS power switches are used to switch ON and OFF the supply of the power domains [3]. During OFF mode Retention Flip-Flops are required to store the logical state of a circuit, in order to restore it when power is switched ON again [4]. In this paper a Retention Flip-Flop, optimized for robust functionality with Glitch suppression and Low-Stand-by Power features in PD-SOI technology, is proposed. It is organized as follow: section II presents some key characteristics of PD-SOI technologies, and the usage of MTCMOS techniques combined with retention flip-flops to cope with standby leakage. Next, the proposed PD-SOI retention flip-flop is detailed section III and compared to prior art SOI alternatives. Section IV concludes this paper.

II. PRIOR ART

A. PD-SOI technology

PD-SOI technology is similar to Bulk CMOS in terms of circuit implementation. The only difference is the substrate material which is composed of a thin silicon film, on top of an oxide layer called Buried Oxide or "box" as illustrated in Fig. 1. The thicknesses of the upper silicon film and the "box", respectively t_{si} and t_{box}, define the kind of SOI technology. In PD-SOI, the upper silicon thickness is twice higher than the channel depletion one (thus, the MOS channel is only partially depleted in conduction mode) and the buried oxide is thick enough to disable any influence from the back side.

This leads to the creation of an electrically neutral zone called Body, close to the "box". This Body is said to be floating since it is encased by insulators..

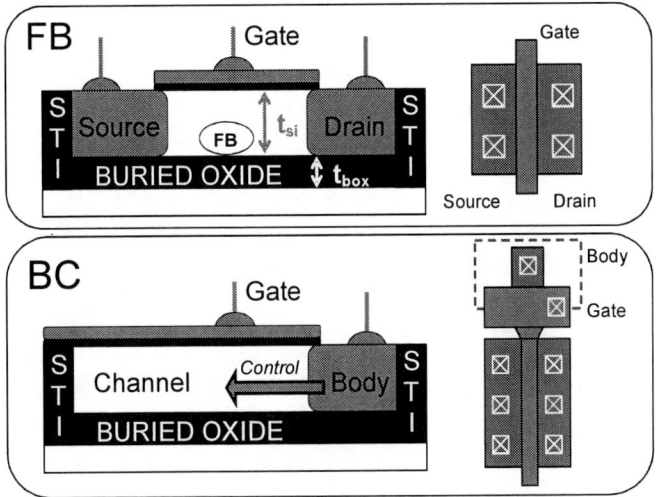

Figure 1: Cross-section and Layout views of Floating Body FB and Body Contacted BC PD-SOI MOSFETs.

When implementing MOSFETs, two design and layout options are available: Floating Body MOS (FB) and Body

---

978-1-4244-9019-6/11 $26.00 © 2011 IEEE   147

Contacted MOS (BC). The first one, FB, is mainly used in the logic and memories because of its higher speed, smaller area and lower dynamic power consumption compared to Bulk. It takes advantages of the Floating Body Effect which dynamically and statically reduces the Vt, increasing the channel current and consequently the speed. Lower dynamic power consumption is obtained, compared to Bulk, thanks to smaller source and drain parasitic capacitances, and also because of the possibility to reduce the supply voltage Vdd while operating at same speed. Additionally, FB MOS does not need any well tap because of its isolated body, reducing also the overall implementation area.

The second MOS, BC, is electrically similar to a Bulk one, because of the floating body suppression thanks to a body access pin acting as a well tap. The drawbacks of BC MOS are its larger layout area and its lower speed due to extra gate capacitance and lower channel current. Thus BC MOS are avoided in Logic and exclusively used for Low-Power management and others features when body control is mandatory.

B. MTCMOS power switch

Multi Threshold CMOS is a low static power consumption design technique. In the "HEADER" case, it consists in adding some PMOS power switches between Vdd supply and the power domain supply line ("virtual supply"). The goal of this technique is to switch OFF power domain supply during stand-by mode in order to reduce current as much as possible. In a PD-SOI implementation, logic connected to the power domain supply is composed of Floating Body, mostly low-Vt, MOS transistors for maximum speed performances when ON; while Power switches are BC and high-Vt ones for minimum leakage when OFF. Drawbacks are a slight voltage drop (Vdrop) introduced by power switches when ON, a wake-up transient time and a loss of data in OFF state.

C. Retention Flip-Flop

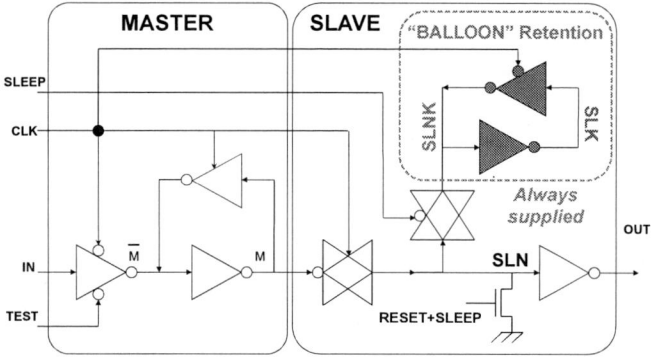

Figure 2: Prior art retention flip-flop schematic.

During wake-up transient phase, Flip-Flops have their data nodes unset. This can cause a random uncontrolled level on any logic node and could be very leaky because of short-circuits [5]. Thus it is important to restore or reset predefined states. To solve this problem, one solution is to adopt Retention Flip-Flops (REFF). In OFF state, when the logic supply becomes grounded, REFF keeps in memory the last value of its output in order to restore it during wake-up phase [6], thus making the logic immediately ready to continue its task. This enables to save time compared to resetting which requires a higher level complexity and moreover time to reload internal data from cache memory.

Classical schematic of such a flip-flop with "balloon" retention [4] is presented Fig. 2. It is made of 3 parts. The MASTER, composed of a latch stage is followed by the SLAVE. Data is transferred from the MASTER to the SLAVE on the clock rising edge. SLAVE Data is memorized in the always powered RETENTION memory (SLK and SLNK nodes), which is isolated from the rest of the REFF in OFF state.

Such retention flip-flop exhibits some potential glitch during wake-up transient phase, due to the pass-gate configuration. Indeed, when SLNK data is high in retention mode and because during this period the SLN node is grounded to avoid any short-circuit current, there is a transient conflict when SLNK restores its high level signal to SLN. Silicon results have shown that this glitch, although with very weak probability, can become high enough to switch the SLK node leading to a failure. To cope with this, a glitch-free retention flip-flop is proposed in this paper, with dedicated PD-SOI low-static power features.

III. GLITCH-FREE RETENTION FLIP-FLOP

A. Proposed schematic

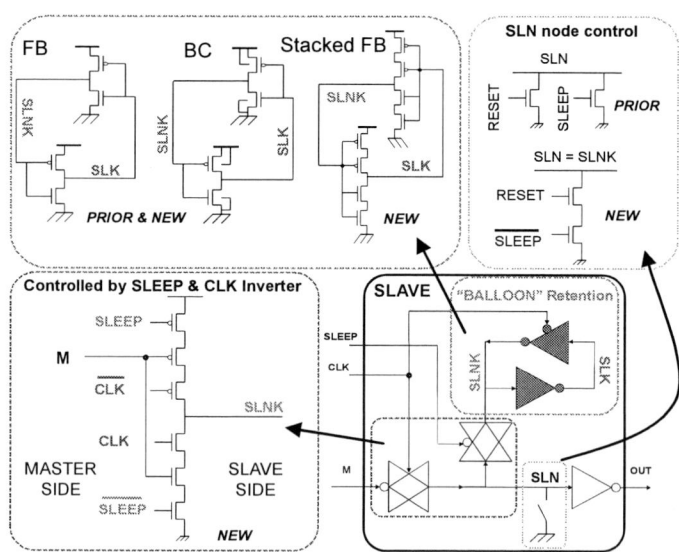

Figure 3:Evolutions of the proposed schematics compared to prior art.

For the sake of comparison, Retention Flip-Flops were initially designed following design guidelines found in prior art, i.e. using pass-gates between the MASTER and the

SLAVE and between SLN and SLNK nodes. The purpose of these pass-gates is to isolate the retention memory, also called "balloon", which is the only supplied part as illustrated in Fig. 3. These reference REFF come in two flavors, called REFF_OLD_FB and REFF_OLD_BC, using respectively Floating Body (FB) and Body Contacted (BC) transistors in the retention memory. Those 2 REFF are also designed with SLEEP and RESET parallel NMOS connected to the SLN node; resetting the retention Flip-Flop at any time in ON mode and enabling to set the SLN node to GND during SLEEP mode in order to avoid floating node. None optimization was done on Master because it is connected to the power domain supply and is not the reason of the glitches apparition.

To get around the potential glitch issue, a new GLITH-FREE retention flip-flop is also proposed, extending this prior solution. This Retention Flip-flop is designed by replacing the two pass-gates by an inverter controlled by SLEEP and CLK signals as described in Fig. 3. In this NEW version SLN and SLNK merge into the same net, with the consequence that SLN is never left floating. Thus the SLEEP signal setting the SLN node at GND in OFF state becomes useless. Because resetting SLN must be done only during ON mode, when SLEEP is low, the 2 parallel MOS, SLEEP and RESET become 2 stacked MOS, driven by RESET and not SLEEP, further reducing the leakage current. As for the reference solution, 2 versions of this new REFF are proposed, namely FB and BC. In addition, a third one is designed using stacked FB MOS. Indeed, a key challenge in PD-SOI is to minimize the leakage. This can be achieved by using BC devices, but stacking devices is also advantageous in terms of leakage reduction [7]. Those three versions are respectively called REFF_NEW_FB, REFF_NEW_BC and REFF_NEW_Stack.

They are compared to the reference ones, considering a 65nm PD-SOI technology. Results in terms of static power consumption, performances and area are given in the following sections.

B. Static power consumption

Two supplies are used in a MTCMOS circuit: the first one, external, called Vddo, is always ON; the second one, called Vddi, depends on the power switch mode. The retention Flip-flops are configured with the retention memory "balloon" connected to Vddo in order to be always supplied to keep the data. The rest is connected to Vddi, in order to be switched OFF to reduce as much as possible the leakage currents. All the virtually supplied parts of the flip-flops were designed using FB transistors, in order to achieve good dynamic performance. On the contrary, to minimize leakage, the retention is preferably based on BC or Stacked FB devices. FB-based retention has been used as a reference in the glitch-free and prior-art solution but is intended to be dismissed because of the high leakage of the floating body transistors. In any case, high-Vt transistors are used for the slave part, except the output inverter.

The following Fig. 4 displays the always ON Retention

memory Static leakage currents of the 5 solutions for a wide range of Vdd supplies at 25°C and 125°C. It clearly demonstrates that glitch-free solutions using BC and Stacked FB devices for retention have static power consumption 50% lower than prior solution with BC retention and 70% lower than prior solution with FB retention at 25°C and 1.4V. For higher temperature (125°C), the REFF_NEW_Stack brings even lower static power consumption (down to 30%). in comparison with REFF_NEW_BC This is mainly due to the stacking effect which divides the Vds of each MOS by 2, reducing the floating body effect even at high temperature.

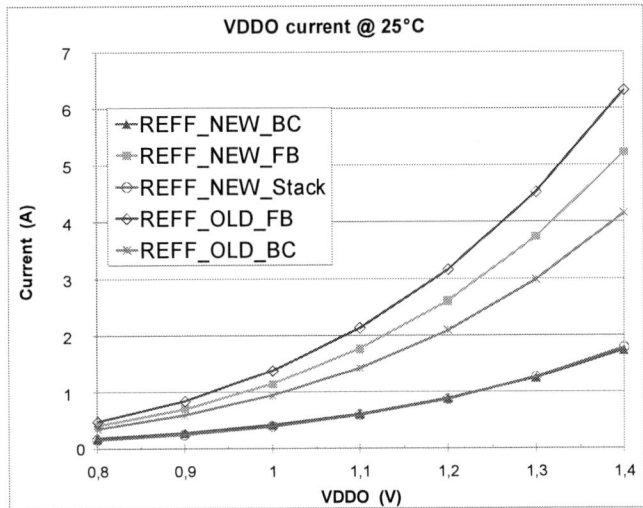

Figure 4.a: Normalized vddo current comparison of the 5 REFF at 25°C.

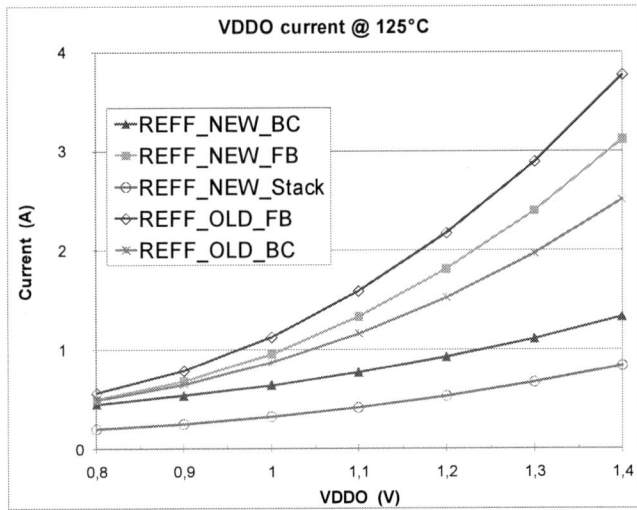

Figure 4.b: Normalized vddo current comparison of the 5 REFF at 125°C.

C. Speed performances Degradation

The five Retention Flip-Flops were simulated. The CLK_to_OUT worst case delay is given by the falling edge and is presented in Fig. 5 for a typical process and temperatures equal to 25°C and 125°C.

The results show performances degradation with the NEW REFF architecture. When the retention is in the stacked MOS configuration (REFF_NEW_Stack), the speed is reduced by 30% compared to REFF_OLD_FB and by 25% compared to

978-1-4244-9019-6/11 $26.00 © 2011 IEEE   149

REFF_OLD_BC at 0.8V and 125C This delay degradation is twice higher with REFF_NEW_BC whereas REFF_NEW_FB behaves quite similarly to REFF_NEW_Stack, as "balloon" equivalent capacitance on SLNK node is lower with stacked than body contacted devices. This degradation is due to the greater Fan-in and Fan-out of the gate controlled by the SLEEP and CLK signals. However, the main objective of such a retention flip-flop is not to reduce the propagation delay but to be more robust, by suppressing the potential wake-up glitches.

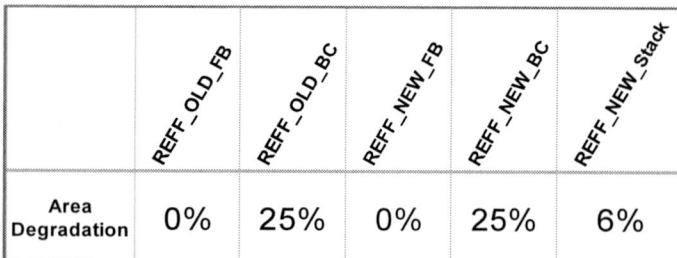

	REFF_OLD_FB	REFF_OLD_BC	REFF_NEW_FB	REFF_NEW_BC	REFF_NEW_Stack
Area Degradation	0%	25%	0%	25%	6%

Table 1: Area degradation compared to BULK solution.

In terms of area, a first conclusion is to notice that the proposed glitch-free schematic has no noticeable. impact (e.g. see REFF_NEW_FB and REFF_OLD_FB). Secondly, Floating Body cells exhibit the same area as the bulk reference: this was expected as the schematics and devices layout are the same. Finally, in the solutions minimizing PD-SOI leakage, REFF_NEW_Stack is the one with the lowest impact on the implementation area. Using Body Contacted devices for retention leads to a 25% higher area and needs 2 metal layers for routing when others needs only one.

By merging all the comparison results in terms of speed, area and static power consumption, the best trade-off is the REFF_NEW_Stack solution. This solution gives to the PD-SOI circuit the robustness needed as well as a good retention leakage current.

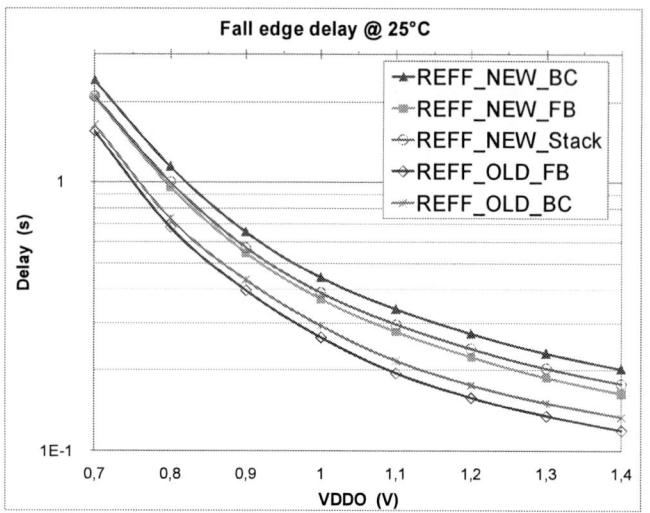

Figure 5.a: Normalized speed comparison of the 5 REFF at 25°C.

IV. CONCLUSION

A new glitch-free retention flip-flop has been proposed. To compensate for PD-SOI floating body devices extra leakage current, the use of body contacted or stacked floating body devices in the retention memory has been explored. The best tradeoff is based on the stacked devices, leading to the best retention leakage gain (50 to 70% reduction), for a small area overhead (6%) and no speed penalty, when compared to a similar floating body implementation. In addition to [1], this result paves the way toward PD-SOI circuits exhibiting a lower operating power for an equivalent stand-by power, than equivalent CMOS bulk circuits.

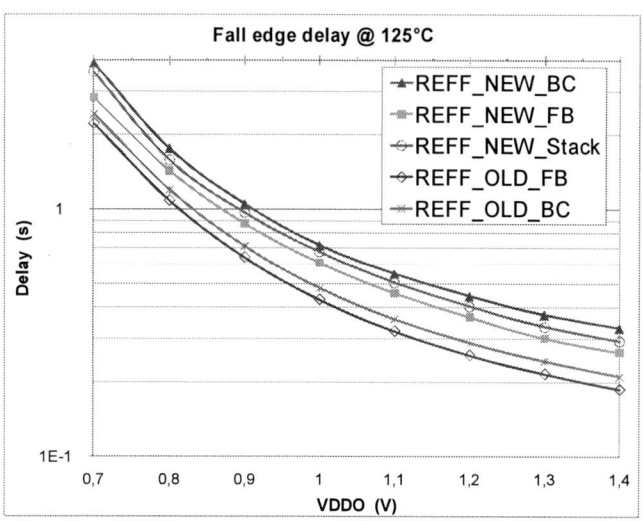

Figure 5.b: Normalized speed comparison of the 5 REFF at 125°C.

D. Layout Area

Concerning the area, the proposed and compared Retention Flip-Flops have been laid-out and compared. Results are presented in Table 1. This table uses a Bulk cell as reference. Retention Flip-Flops have all the same height, like standard cells, to be implemented in-between the VDD and GND rails.

REFERENCES

[1] J. Le-coz , et al. "Comparison of 65nm LP Bulk and LP PD-SOI with Adaptive Power Gate Body Bias for an LDPC Codec" ISSCC 2011.

[2] M. Keating "Low power methodology manual: for system-on-chip design" Springer, 2007.

[3] S. Mutoh, et al. "1-V power supply high-speed digital circuit technology with multithreshold-voltage CMOS" JSSCC 1995.

[4] S. Shigematsu "A 1-V High-Speed MTCMOS Circuit Scheme for Power-Down Application Circuits" JSSCC, 1997.

[5] Z. Liu, et al. "Characterization of Wake-up Delay versus Sleep Mode Power Consumption and Sleep/Active Mode Transition Energy Overhead Tradeoffs in MTCMOS Circuits" NWSCAS, 2008.

[6] Z. Liu, et al. "New MTCMOS Flip-Flops with Simple Control Circuitry and Low Leakage Data Retention Capability" ICECS, 2007.

[7] S. Narendra, et al. "Scaling of Stack Effect and its Application for Leakage Reduction" ISLPED, 2001.

978-1-4244-9019-6/11 $26.00 © 2011 IEEE

An ultra-low energy capacitive DAC array switching Scheme for SAR ADC in biomedical applications

Chao Yuan, Yvonne Y. H. Lam
School of Electrical and Electronics Engineering
VIRTUS, IC Design Centre of Excellence
Nanyang technological University Singapore 639798
Email: yuan0042@e.ntu.edu.sg

Abstract—**This paper presents a novel switching scheme for an ultra-low energy charge-redistribution digital-to-analog converter (DAC) to be used in successive-approximation register (SAR) analog-to-digital converter (ADC). The proposed scheme employs unit capacitors for voltage sampling and charge redistribution. Compared with previously published capacitive DAC which uses the same unit size of capacitor array, the proposed DAC needed only 33% of the total switches. SPICE simulation results show that the average switching energy can be reduced by more than 50%. An 8-bit SAR-ADC using the proposed switch scheme is designed in Global foundries 65nm CMOS process. The power consumption of the capacitive DAC is 160 nW at 1.2V power supply and 100KS/s.**

I. INTRODUCTION

Recently there is a growing trend of acquiring biomedical signals with low-voltage, low-power, miniaturized CMOS electronic devices [1]. In these systems, the sampling speed can be relaxed and is usually less than 200KS/s [2]. The resolution is not very critical, generally 6-10 bits [2]. However, power consumption and area are the most crucial factors in these applications. Successive approximation register (SAR) ADC has been a very popular candidate due to its simplicity, low power consumption, medium speed and resolution [3]. In most of the SAR-ADC designs, charge-redistribution capacitive DACs are preferred because of the elimination of DC current flow.

As shown in Fig. 1(a), a conventional charge-redistribution (CR) SAR-ADC includes a comparator, a binary-weighted capacitor array, sample-and-hold (S/H) circuit and SAR digital logic. A n-bit DAC array shown in Fig. 1(b) consists of $n+1$ binary-weighted capacitors, having a total capacitance of $2^n C_0$, where C_0 is the unit capacitor. As a result, the total area of the capacitor array rises exponentially as the resolution increases. In CR SAR-ADC operations, most of the energy is used to charge the large capacitor array [5]. Several methods focused on saving capacitor switching energy have been published [1-11]. The capacitor-splitting [5][8] technique reduces the switching energy by charge-recycling. However, the MSB evaluation still consumes most of the energy. Junction-split [4][10] can reduce the switching energy by 75% compared with conventional switching scheme. But the number of switches is increased by more than 50% compared with conventional DAC. Furthermore, the charge-injection errors and switch isolation have deteriorated the

linearity. The dual-capacitor array technique [7] can reduce both capacitive array area and switching energy. However, the additional scaled reference voltage requires a resistive voltage divider and incurs DC power consumption. In this paper, a novel capacitive DAC switching scheme designed to reduce both the switching energy and capacitor array area is proposed. Unlike conventional binary-weighted capacitive DAC, the proposed DAC uses only $n+1$ unit capacitors for n-bit resolution. Compared with previously published unit capacitor DAC switching scheme, the proposed switching scheme can reduce DAC area by at least 67.8% and DAC switching energy by more than 50%.

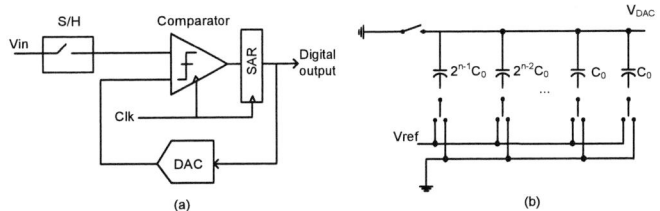

Fig. 1.   (a) Block diagram of a SAR-ADC. (b) n-bit capacitive DAC array

The rest of this paper is organized as follows. Section II briefly explains the proposed method and operations. Section III discusses the non-idealities and limiting factors of the proposed scheme. Section IV compares the proposed technique with previous design techniques. Some simulation results are also included. Section V concludes the paper.

II. THE PROPOSED DAC SWITCHING SCHEME

A SAR-ADC using the proposed DAC scheme is shown in Fig. 2. The proposed DAC array consists of $n+1$ unit capacitors and $n+3$ switches for n-bit resolution. Initially one unit capacitor is charged to the reference voltage, V_{ref}. In subsequent conversion, charge redistribution occurs in 2 unit capacitors in each clock cycle and the corresponding reference voltage levels e.g. $V_{ref}/2$, $3V_{ref}/4$, $V_{ref}/4$ etc., are generated. A comparator compares sampled input voltage and the generated reference voltage. Triggered by the output of the comparator, the SAR logic generates control signals for all the switches to produce the correct reference voltage level for the next bit decision. Similar switching methods are reported in [1] and [2].

978-1-4244-9019-6/11 $26.00 © 2011 IEEE

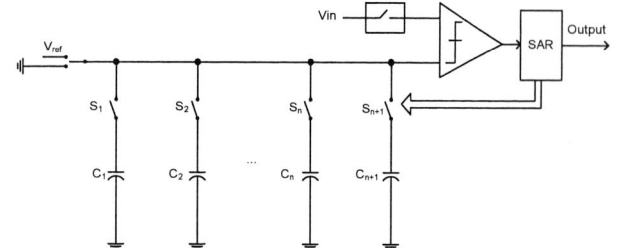

Fig. 2. The proposed charge-redistribution DAC: all $n+1$ capacitors are identical.

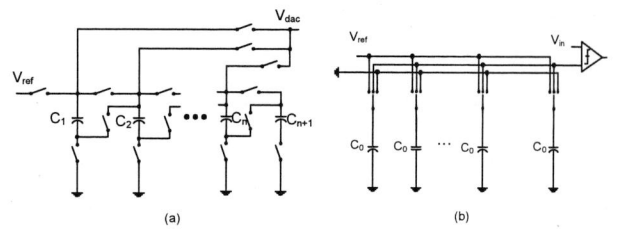

Fig. 3. (a) The DAC array in [1] consisting of $n+1$ unit capacitors and $4n+1$ switches (b) The DAC array presented in [2]. It consists of 28 unit capacitors for 8-bit, 44 unit capacitors for 9-bit and 64 unit capacitors for 10-bit. The number of switches is three times of the number of capacitors.

The schematics of charge-redistribution DAC schemes employing unit capacitors in [1] and [2] are shown in Fig. 3 (a) and (b). For the DAC scheme in [1], it charges only one unit capacitor to V_{ref} for every conversion cycle. It consumes the least amount of switching energy. However, it requires $4n+1$ switches for n-bit DAC. Thus the charge injection errors and leakage current limit the resolution of this scheme to 6-bit only [1]. The DAC scheme in [2] consists of much more capacitors and switches than the proposed DAC scheme does. For DAC scheme in [2], all the unit capacitors have a voltage of $V_{ref}/2$ for the most significant bit (MSB) conversion. If MSB is 1, several unit capacitors are charged to V_{ref} again. If MSB is zero, the same number of capacitors are discharged to ground, resulting in a waste of energy. In the proposed DAC switching scheme only two unit capacitors are $V_{ref}/2$ for MSB conversion. No energy is dumped to ground even if MSB is 0. One of the two unit capacitors will be charged to V_{ref} if MSB is 1. The switching energy is adaptively increased as the input voltage increases. Thus, the proposed DAC scheme uses energy more efficiently while maintaining a small area.

The operation of a 5-bit DAC using the proposed DAC switching scheme is illustrated in Fig. 4. For simplicity, the input voltage is assumed to be 0.34V and V_{ref}=1.2V. In the reset phase, all capacitors are discharged to ground. The conversion begins by charging C_1 to V_{ref} whereas the rest of the capacitors are disconnected. In the next phase, C_1 and C_2 are connected in parallel for charge redistribution to generate $V_{ref}/2$. The comparator outputs 0 since $V_{in} < V_{ref}/2$. After that, C_1 is disconnected from V_{DAC}. C_2 and C_3 are connected for charge-redistribution to generate $V_{ref}/4$. In subsequent clock cycles, the corresponding reference voltage level for each bit conversion is generated in the same manner.

It can be observed from Fig. 4 that phases (g)(h)&(i) are

three intermediate steps to generate $V_{ref}/4$. After generating $5V_{ref}/16$ in phase(f), the DAC needs a voltage level of $V_{ref}/4$ in order to generate the desired voltage level of $9V_{ref}/32$ for the next bit conversion. Thus, the three intermediate phases are required. Generally the number of intermediate steps is unknown for each conversion. It needs to be worked out bit by bit. For an 8-bit DAC with the proposed scheme, the maximum number of clock cycles is 19 and the minimum number is 10. The comparator outputs in the intermediate clock cycles are useless. In order to reduce comparator power consumption, it is possible to suspend the clock to the comparator in these clock cycles and resume operation after that.

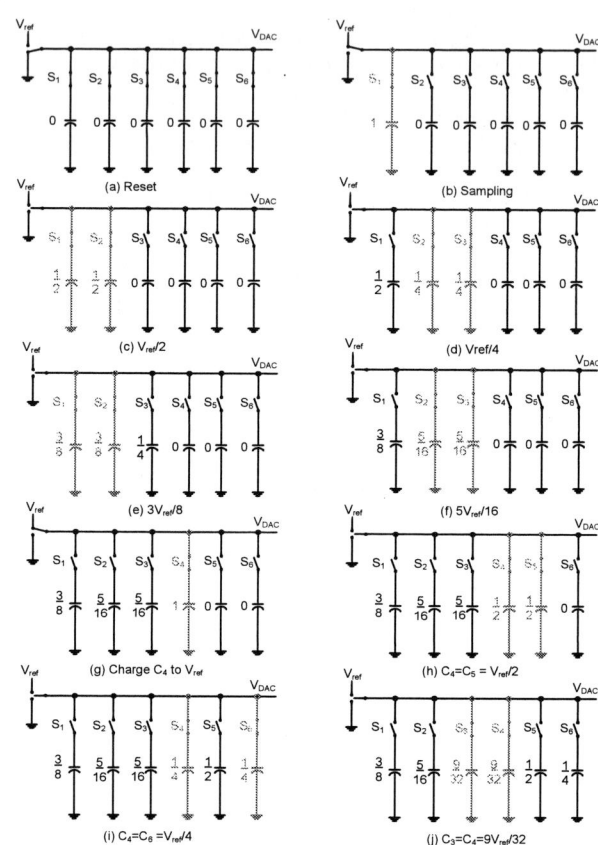

Fig. 4. An example of a 5-bit DAC using the proposed DAC scheme for V_{in}=0.34. The resulting digital output is B[4:0]=01001.

III. DAC OUTPUT ERROR ANALYSIS

The fundamental principle of the proposed switching scheme is connecting two identical capacitors for charge redistribution and generating the desired voltage level. Charge injection and parasitic capacitance [2] are the two primary sources of errors. In this section, these error sources are analyzed and discussed.

To analyze the charge injection errors, it is assumed that a voltage level $V_X(t_n)$ must be generated at the DAC output at time t_n. As shown in Fig.5(a), C_1 and C_2 are connected at t_{n-1} and the voltage on both C_1 and C_2 is $V_X(t_{n-1})$. R_1 represents the on resistance of switch S_1. The voltage stored on capacitor C_3 is $V_X(t_m)$. t_m is some time point prior to

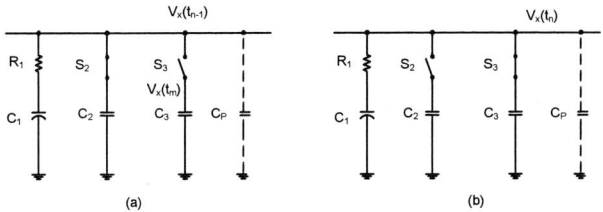

Fig. 5. (a) C_1 and C_2 are connected at t_{n-1}. (b) C_2 is disconnected and C_3 is connected to C_1 and the desired voltage level $V_X(t_n)$ is generated.

time t_n. At time t_n, C_2 is disconnected and C_3 is connected to C_1 to generate the desired voltage level, as in Fig.5(b). The DAC output voltage generated can be expressed as:

$$V_X(t_n) = V_{X0}(t_n) + \varepsilon_X(t_n) \qquad (1)$$

$$V_{X0}(t_n) = \frac{V_{X0}(t_{n-1}) + V_{X0}(t_m)}{2} \qquad (2)$$

Thus the error voltage on the DAC output at t_n can be expressed:

$$
\begin{aligned}
\varepsilon_X(t_n) = &\frac{(C_1 + C_P)\varepsilon_X(t_{n-1}) + C_1\varepsilon_X(t_m)}{2C_1 + C_P} \\
&+ \frac{\frac{C_P}{2}(C_1 + C_P)[V_{X0}(t_{n-1}) - V_{X0}(t_m)]}{2C_1 + C_P} \\
&+ \frac{\Delta Q_{S2}(t_n) + \Delta Q_{S3}(t_n)}{2C_1 + C_P}
\end{aligned}
\qquad (3)
$$

where $V_{X0}(t_n)$, $V_{X0}(t_{n-1})$ and $V_{X0}(t_m)$ are the desired DAC output voltage levels at time t_n, t_{n-1} and t_m, respectively. $\varepsilon_X(t_n)$, $\varepsilon_X(t_{n-1})$ and $\varepsilon_X(t_m)$ are the error voltages at time t_n, t_{n-1} and t_m, respectively.

It can be seen from (3) that the error voltage for a particular DAC voltage level depends on the error voltages of previously generated two voltage levels which are used to produce the present DAC output voltage, the parasitic capacitance and the charge injection due to two switches. Charge injection errors caused by switches S_2 and S_3 are shown in Fig.6.

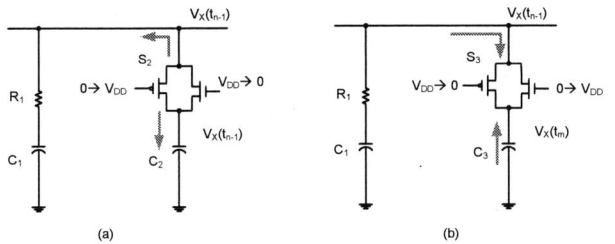

Fig. 6. (a) Channel charges injected into C_1 and C_2 when S_2 is turned off. (b) Charges are attracted into the channel when S_3 is turned on.

The switches are implemented as transmission gates (TG). Thus, the amount of charge injection depends on the biasing voltages of the switches. The biasing voltages, $V_X(t_{n-1})$ and $V_X(t_m)$, determine whether NMOS or PMOS is on. It is estimated that there are 8 combinations for ΔQ_{S2} and ΔQ_{S3}. All the conditions for the 8 combinations are summarized in table I. It is not possible to list down all the equations to

TABLE I
8 COMBINATIONS OF NMOS AND PMOS IN TG FOR DETERMINING CHARGE INJECTION

Conditions			$\Delta Q_{S2}(t_n)$	$\Delta Q_{S3}(t_n)$
$V_X(t_{n-1}) > V_{DD} - V_{tn}$	$V_X(t_m) > V_X(t_{n-1})$		NMOS ON	PMOS ON
	$V_X(t_m) < V_X(t_{n-1})$	$V_X(t_m) < V_{DD} - V_{tn}$		NMOS PMOS ON
		$V_X(t_m) > V_{DD} - V_{tn}$		PMOS ON
$V_X(t_{n-1}) < V_{tP}$	$V_X(t_m) > V_X(t_{n-1})$	$V_X(t_m) < V_{tp}$	PMOS ON	NMOS ON
		$V_X(t_m) > V_{tp}$		NMOS PMOS both ON
	$V_X(t_m) < V_X(t_{n-1})$			NMOS ON
$V_{tp} < V_X(t_{n-1}) < V_{DD} - V_{tn}$	$V_X(t_m) > V_X(t_{n-1})$		NMOS PMOS both ON	NMOS PMOS both ON
	$V_X(t_m) < V_X(t_{n-1})$			NMOS PMOS both ON

determine ΔQ_{S2} and ΔQ_{S3} for all eight combinations. An example of the charge injection error of ΔQ_{S2} and ΔQ_{S3} is shown below for both NMOS and PMOS on in both S_2 and S_3, i.e. $V_{tp} < V_X(t_{n-1}) < V_{DD} - V_{tn}$ and $V_X(t_m) > V_X(t_{n-1})$.

$$
\begin{aligned}
\Delta Q_{S2}(t_n) = &\frac{1}{2}C_{oxp}W_pL_p[V_X(t_{n-1}) - V_{tp}] + C_{ovp}V_{DD} \\
&-\frac{1}{2}C_{oxn}W_nL_n[V_{DD} - V_X(t_{n-1}) - V_{tn}] \\
&-C_{ovn}V_{DD}
\end{aligned}
\qquad (4)
$$

$$
\begin{aligned}
\Delta Q_{S3}(t_n) = &C_{oxn}W_nL_n[V_{DD} - V_X(t_{n-1}) - V_{tn}] \\
&+2C_{ovn}V_{DD} - C_{oxp}W_pL_p[V_X(t_m) - V_{tp}] \\
&-2C_{ovp}V_{DD}
\end{aligned}
\qquad (5)
$$

where C_{oxn}, C_{oxp}, W_n, W_p, L_n, L_p, C_{ovn}, C_{ovp}, V_{tn} and V_{tp} are the NMOS and PMOS gate oxide capacitance per unit area, device width, device length, overlap capacitance and threshold voltage with the consideration of body effect, respectively.

From the error voltage model derived in (3) and the conditions in table I, the error voltages for 3, 4, 5 and 6-bit DACs at various DAC output voltage levels are calculated and shown in Fig. 7. In the calculation, the model parameters are obtained from 65nm CMOS process. The parasitic capacitance is modeled as the sum of parasitic capacitance of the switches and the input capacitance of comparator. The SPICE simulated parasitic capacitance is $3.03fF$. V_{DD} and V_{ref} are set to 1.2V. The unit capacitor is $338fF$ in the calculation.

It can be seen from Fig. 7 that the analytical model agrees with the simulation results. The derived model can be used to calculate the DAC output error voltage. However, there is no compact-form equation to predict the maximum error voltage due to the nature of the proposed switching scheme. The maximum deviation has to be determined bit-by-bit.

IV. SIMULATION RESULTS AND COMPARISONS

An 8-bit SAR-ADC based on the proposed DAC switching scheme is designed and simulated in Global foundries $65nm$ CMOS process. For n-bit converter it only requires $n+1$ identical capacitors. It has been shown in previous section that the DAC output error voltage is inversely proportional to the unit capacitance value. Thus, the unit capacitance value

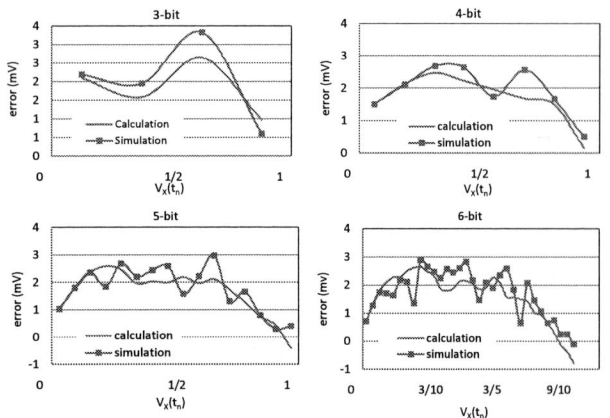

Fig. 7. Comparison between analytical model and spice simulation results for 3, 4, 5 and 6-bit DAC.

is generally large, in the order of hundreds of fF. In a conventional binary-weighted capacitor array, the unit capacitor value is limited by the KT/C noise. The unit capacitance is usually smaller, in the order of tens of fF. Hence, comparison in this paper is mainly performed with other similar schemes which also use identical "large" capacitors for charge redistribution rather than conventional binary-weighted DACs, namely [1] and [2]. Fig. 8 shows the simulated switching energy of 8-bit DAC capacitor array for 17 output codes with the switching schemes in [1], [2] and the proposed scheme, respectively. All simulations are conducted with ramp input signals and sampling rate of 100KS/s. The simulated power consumption for the proposed DAC is 160nW at 1.2V power supply..

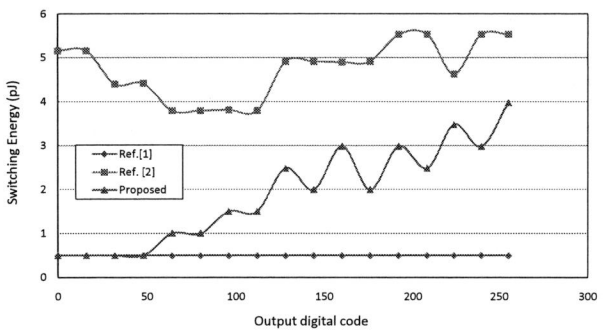

Fig. 8. Switching energy comparison

Table II compares the three switching schemes in terms of the number of capacitors, the number of switches and the number of clock cycles. One third of total capacitance is used in the proposed DAC scheme compared with the DAC scheme in [2]. Compared with DAC scheme in [1], the proposed scheme uses only 33% of analog switches. The proposed switching scheme has the smallest area at the expenses of increased conversion time.

V. CONCLUSION

This paper reports a novel capacitive DAC switching scheme for SAR-ADC designs. An analytical model for the DAC

TABLE II
COMPARISON OF AREA

DAC scheme	No. of capacitors	No. of switches	No. of clock cycles
conv.	2^n	$2*n$	$n+1$
[1]	$n+1$	$4n+1$	$n+3$
[2] 8bit	28	$3*28$	$10-16$
Proposed (8bit)	$n+1$ (9)	$n+3$ (11)	$(10-19)$

output error voltage is also presented and compared with simulation data. Compared with previous publications, the proposed DAC has the least number of capacitors and switches. Thus the area is the smallest. The switching energy is reduced by more than 50% compared with [2]. The max DAC output error voltage has been reduced by two times from [1]. The switching energy of the proposed scheme increases adaptively as the digital output code increases, avoiding dumping charges into ground as the scheme in [2] and conventional binary-weighted DAC do.

REFERENCES

[1] H. V. Gopal and M. S. Abaghini, "An Ultra Low-Energy DAC for Successive Approximation ADCs," in *Proc. IEEE Int. Symp. Circuits and Systems*, 2010, pp. 3349-3352.

[2] F. Chen, A. P. Chandrakasan and V. Stojanovic,"A Low-power Area-efficient Switching scheme for charge-sharing DACs in SAR ADCs," in *Proc. IEEE CICC 2010*

[3] D. B. Stankovic, M. K. Stojcev and G. L. Djordjevic, "Power Reduction Technique for Successive-approximation Analog-to-digital Converters," in *Proc. IEEE TELSINKs 2007*

[4] J. S. Lee and I. C. Park, "Capacitor array Structure and Switch Control for Energy-efficient SAR Analog-to-Digital Converters," in *Proc. IEEE Int. Symp. Circuits and Systems*, 2008, pp. 236-239.

[5] B. P. Ginsburg and A. P. Chandrakasan,"An Energy-efficient Charge Recycling Approach for a SAR Converter with Capacitive DAC," in *Proc. IEEE Int. Symp. Circuits and Systems*, 2005, pp. 184-187.

[6] R. Y. Choi and C. Y. Tsui, "A Low Energy Two-step Successive Approximation Algorithm for ADC design," in *Proc. IEEE Int. Symp. Circuits and Systems*, 2009, pp. 17-20.

[7] S. Chang and E. Yoon, "A Low-Power Area-efficient 8-bit SAR ADC Using Dual Capacitor Arrays for Neural Microsystems," in *Proc. IEEE EMBC*, 2009.

[8] Y. K. chang, C. S. Wang and C. K. Wang, "A 8-bit 500-KS/s Low Power SAR ADC for Biomedical Applications," in *Proc. IEEE Asia solid-state conf.*, 2007.

[9] M. Kandala, R. Sekar, C. Zhang and H. Wang, "A Low Power Charge-redistribution ADC With Reduced Capacitor Array," in *Proc. IEEE ISQED* , 2010.

[10] W. Yu, J. Lin and G. C. Teme, "Two-step split-junction SAR ADC," *Electronics Lett.*, vol. 46, no. 3, Feb. 2010.

[11] B. Kim, L. Yan, J. Yoo, N. Cho and H. J. Yoo, "An Energy-efficient Dual Sampling SAR ADC with Reduced Capacitive DAC," in *Proc. IEEE Int. Symp. Circuits and Systems*, 2009, pp. 17-20.

[12] M. Elzakker, E. V. Tuijl, P. Geraedts, D. Schinkel, E. Klumperink and B. Nauta, "A 10-bit Charge-redistribution ADC Consuming 1.9uW at 1MS/s," *IEEE J.Solid-State Circuits*, vol. 45, no. 5, pp. 1007-1015, May 2010.

[13] R. M. Rangayyan, *Biomedical Signal Analysis: A Case-Study Approach* , Wiley-IEEE Press, 2001

Slew-Rate Controlled Output Stages for Switching DC-DC Converters

Jia-Ming Liu, Yi-Cheng Huang, Yu-Chun Ying, and Tai-Haur Kuo

Department of Electrical Engineering, National Cheng Kung University, Tainan City 70101, Taiwan

TEL: 886-6-2081999, FAX: 886-6-2345482

Abstract—**Large supply bouncing due to the fast switching current and parasitic inductance of the supply rail may cause reliability and electromagnetic interference (EMI) problems, especially for ICs with the pulse-width modulation (PWM) technique, such as switching DC-DC converters. In this paper, a new slew-rate controlled (SRC) output stage is proposed to appropriately increase the rise and fall times of the PWM output by combining a feedback capacitor technique and a distributed-and-weighted design. Therefore, the supply bouncing during PWM switching can be reduced. The SRC output stage is successfully integrated into a DC-DC converter implemented with a 0.35μm 1P4M 3.3V mixed-signal CMOS process for verification. With an input voltage of 3.3V, an output voltage of 1.8V, a switching frequency of 500 kHz, and a load current range of 700mA, the active area of the converter is 2.3mm². With a merely 0.035mm² control circuit for the SRC output stage, the measured supply bouncing of the designed converter can be reduced by 40% and thus the reliability and the EMI can be improved.**

Index Terms—**electromagnetic interference (EMI), output stage, pulse-width modulation (PWM), slew-rate, supply bouncing**

I. INTRODUCTION

SWITCHING DC-DC converters are widely used in portable devices, such as personal digital assistants (PDAs), cellular phones, and eBooks. By using the pulse-width modulation (PWM) technique, the output stage of a switching converter is operated with fast rise and fall times. During PWM switching, the rapid change of the supply current to the power MOSFET and the parasitic inductances of the supply rails will cause large supply bouncing. This large supply bouncing may lead to reliability and electromagnetic interference (EMI) problems for the switching converter ICs [1]. Supply bouncing can be reduced by increasing the rise and fall times of the output stage during PWM switching [2].

To increase the rise and fall times, a distributed-and-weighted (DAW) technique [3]-[4], or a feedback capacitor technique [5]-[7], has been applied to the CMOS output buffer design for capacitive load but not inductive load applications. In this paper, a new slew-rate controlled (SRC) output stage for

Manuscript received Feb. 9, 2011. This work was supported by the National Science Council of Taiwan under Grant NSC 99-2218-E-006-003.

Jia-Ming Liu, Yi-Cheng Huang , Yu-Chun Ying, and Tai-Haur Kuo are with the Department of Electrical Engineering, National Cheng Kung University, Tainan City 70101, Taiwan. (e-mail: jmliu@msic.ee.ncku.edu.tw).

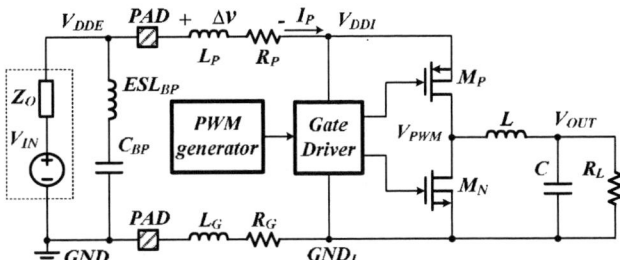

Fig. 1. The typical block diagram for the conventional PWM output stage

inductive load applications is developed. The new SRC output stage combines the DAW and feedback capacitor techniques and is successfully integrated into a switching DC-DC converter to increase both the rise and fall times of the PWM output. The new SRC output stage can significantly reduce the supply bouncing of the switching converter with negligible control circuit overhead.

The organization of this paper is as follows. In section II, the system-level design of the SRC output stage is presented. Section III describes the design of the proposed SRC output stage and its integration with the DC-DC converter. The measurement results are illustrated in section IV, and finally, conclusions are drawn in section V.

II. SYSTEM-LEVEL DESIGN OF THE SRC OUTPUT STAGE

Fig. 1 shows a typical block diagram for a switching converter with a conventional output stage. The PWM code, which is generated by the PWM generator, passes through the gate driver to provide enough driving capability for the power PMOS M_P and NMOS M_N, which have large gate capacitances. The rise and fall times of the PWM output node, V_{PWM}, are fast for the conventional output stage, resulting in a fast switching transient current I_P being induced. The power supply, with the voltage V_{IN} and its output impedance Z_O, is used as the input of the switching converter IC. Though the bypass capacitor C_{BP}, with its equivalent series inductance ESL_{BP}, is added outside the IC for reducing the bouncing at the external supply rail V_{DDE}, large supply bouncing at the supply rail in the chip, i.e, V_{DDI} and GND_I, can still be generated due to the parasitic inductances of the supply rail L_P and L_G and the switching current I_P. This large supply bouncing may lead to reliability and even EMI problems of the ICs. Since the parasitic inductance mainly

978-1-4244-9019-6/11 $26.00 © 2011 IEEE

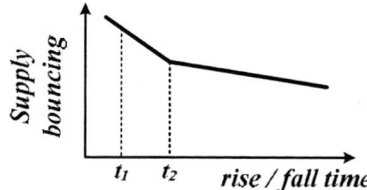

Fig. 2. The relationship between the rise / fall times versus supply bouncing

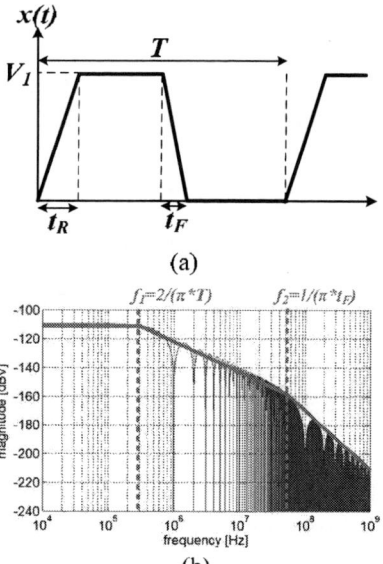

Fig. 3. (a)The time domain waveforms and (b) its FFT plot of the PWM output node V_{PWM}

$$X(s) = \frac{V_1}{s^2} \cdot \left[\frac{1}{t_R} - \frac{1}{t_R} e^{-st_R} - \frac{1}{t_F} e^{-s\frac{T}{2}} + \frac{1}{t_F} e^{-s\left(\frac{T}{2}+t_F\right)} \right], \quad (2)$$

where t_R, t_F and T are the rise time, fall time, and period of V_{PWM}, respectively. As can be seen from (2), the dominant pole f_1 is determined by the period of V_{PWM}, while the second pole f_2 is determined by the shorter time of the rise or the fall time. In this case, f_1 and f_2 are $(2/\pi T)$ and $(1/\pi t_F)$, respectively, as shown in Fig. 3 (b). The spectrum envelope for $X(s)$ rolls off at -20dB/decade until f_2, while above f_2 the spectrum envelope for $X(s)$ falls off at -40dB/decade. Once the rise and fall times are increased, the energy for the EMI regulation frequency range of 30MHz to 1GHz, which is addressed by EN55022 EMI regulation [8], will be decreased and thus the radiated EMI can be reduced.

III. CIRCUIT DESIGN OF THE SRC OUTPUT STAGE

The new SRC output stage is integrated into a current-mode switching DC-DC converter for verification. The block diagram of the designed converter is shown in Fig. 4. The converter consists of a bandgap reference, a current sensing circuit, an error amplifier with compensating circuit for the system, and the SRC output stage parallel-combined with a conventional power stage. The output voltage V_{OUT} is fed back through the resistor ladders, R_1 and R_2, and compared with a constant reference voltage V_{ref} generated by the bandgap reference. The duty cycle of the system is decided by the comparison of Vc, generated by the error amplifier, and a summation voltage V_{sum} of the compensation ramp and sensing voltage from the current sensor. The resultant duty cycle regulates the V_{OUT} to a constant voltage.

The schematic of the new SRC output stage is shown in Fig. 5. The proposed SRC output stage is composed of current-bias with the feedback capacitor C_{FP} and the distributed-and-weighted (DAW) output stage. For the rising edge of V_{PWM}, the switch S_1 will be turned on, so that the rise time of V_{PWM} can be controlled by a current source I_a and a capacitor C_{FP} added between the gate and the drain terminal of the power PMOSFET M_P, as shown in Fig. 5. For the falling edge of V_{PWM}, the switch S_1 will be turned off. The DAW output stage is enabled to increase the fall time of V_{PWM}. The details of the operation of these circuits will be explained in the following paragraph.

results from the IC's packages and the bonding wires, it is not easily reduced for lowering the supply bouncing. A new slew-rate controlled (SRC) output stage is therefore required for increasing the rise and fall times to reduce the effective di/dt so that supply bouncing can be decreased.

The selections of the rise and fall times of V_{PWM} are based on a compromise between the reduced supply bouncing and degraded efficiency. Fig. 2 illustrates the relationship between the rise time and supply bouncing. The supply bouncing Δv can be estimated by

$$\Delta v = L_P \frac{dI_P}{dt} + I_P \cdot R_P \quad (1)$$

where R_P is the parasitic resistance of L_P. In general, the supply bouncing can be notably reduced when the designed rise / fall time is less than a specific time t_2. This is because the first term of (1) will be insignificant when the rise time of V_{PWM} exceeds t_2. Slower rise and fall times may slightly degrade the converter's efficiency due to the V-I overlap loss of the output stage. Therefore, if the rise / fall time can be increased from t_1 to t_2, the supply bouncing can be significantly reduced with acceptable efficiency degradation.

The time domain waveforms $x(t)$ of V_{PWM} is shown in Fig. 3(a). Assuming the duty-cycle is 50%, the Fourier transformation of $x(t)$ can be derived as

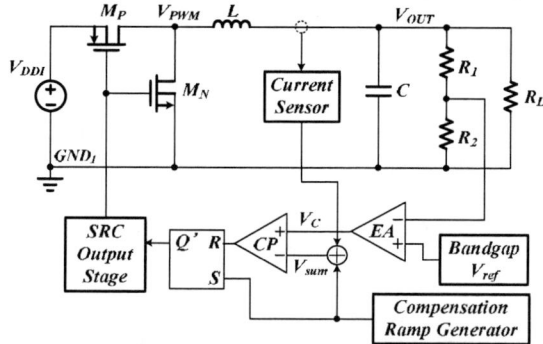

Fig. 4. The block diagram of the current-mode switching DC-DC converter

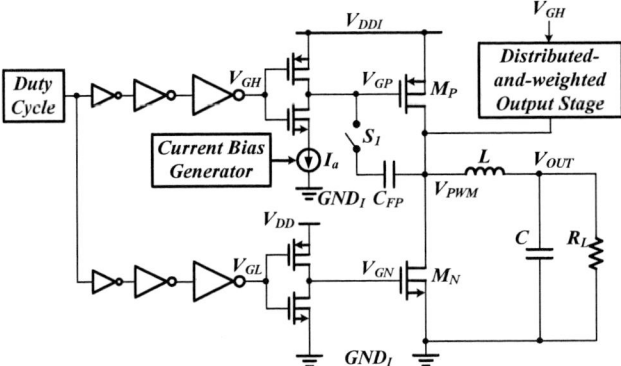

Fig. 5. The schematic of the new slew-rate controlled (SRC) output stage

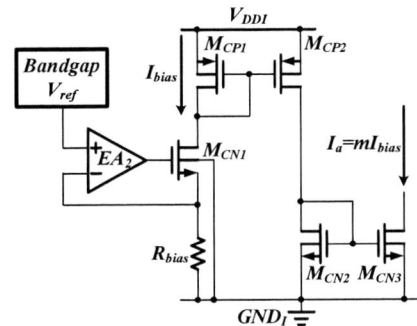

Fig. 6. The current-bias generator

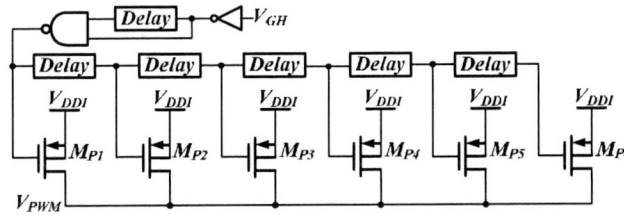

Fig. 7. The distributed-and-weighted power stage

Fig. 6 shows the current-bias circuit composed of an OPAMP EA_2, two current mirrors M_{CP1}-M_{CP2} and M_{CN2}-M_{CN3}, and a resistor R_{bias}. The voltage V_{ref}, which is generated by a bandgap reference circuit, and the feedback loop composed of the OPAMP EA_2 and the transistor M_{CN1}, will determine the current I_{bias}, where I_{bias} is equal to V_{ref}/R_{bias}. The sinking current I_a is designed to be m times that of the I_{bias}, so that the slew-rate (SR) for the rising edge of V_{PWM} will be

$$SR = \frac{dV_{PWM}}{dt} = \frac{I_a}{C_{FP}} = \frac{m \cdot I_{bias}}{C_{FP}} = \frac{m \cdot V_{ref}}{C_{FP} \cdot R_{bias}} \quad (3)$$

where C_{FP} is the on-chip capacitor between the drain and gate of M_P. From (3), the slew-rate of V_{PWM} can be controlled with the external resistance R_{bias} for the rising edge of V_{PWM}.

The DAW output stage is shown in Fig. 7. The DAW power PMOSFETs M_{P1}~M_{P6} is controlled by the DAW gate drive. When M_P is turned off, M_{P1}~M_{P6} are turned on gradually to provide the current flowing through the inductance L of the output LC filter and thus increase the fall time of V_{PWM} during the switching transient time. After the switching transient time,

M_{P1}~M_{P6} can be parallel- combined with M_P to reduce the total turned-on resistance of the output stage for reducing the conduction power loss. Therefore, the circuit overhead of the DAW output stage is only the DAW gate drive, and not M_{P1}~M_{P6}.

The internal supply node V_{DDI} is inside the package and can't be accessed by measurement. Figs. 8(a) and 8(b) show SPICE waveforms of the PWM output node, V_{PWM}, and the internal and external supply nodes, V_{DDI} and V_{DDE}, with the conventional and the new SRC output stage, respectively. With 10nH parasitic inductance, the internal supply bouncing for the conventional and the new SRC output stage are 1.32V and 0.76V for the PWM rising edge, while they are 4.03V and 0.77V for the falling edge, respectively. The SPICE-simulated V_{DDE} in Figs. 8(a) and 8(b) show that the supply bouncing can be reduced from 0.33V to 0.19V for the PWM rising edge, while the supply bouncing can be reduced from 1.00V to 0.20V for the falling edge, respectively. Due to the external bypass capacitor C_{BP} being added to provide the charge for rapid changing of the supply current during PWM switching, the supply bouncing of the internal supply node V_{DDI} will be much larger than that of the external supply node V_{DDE}.

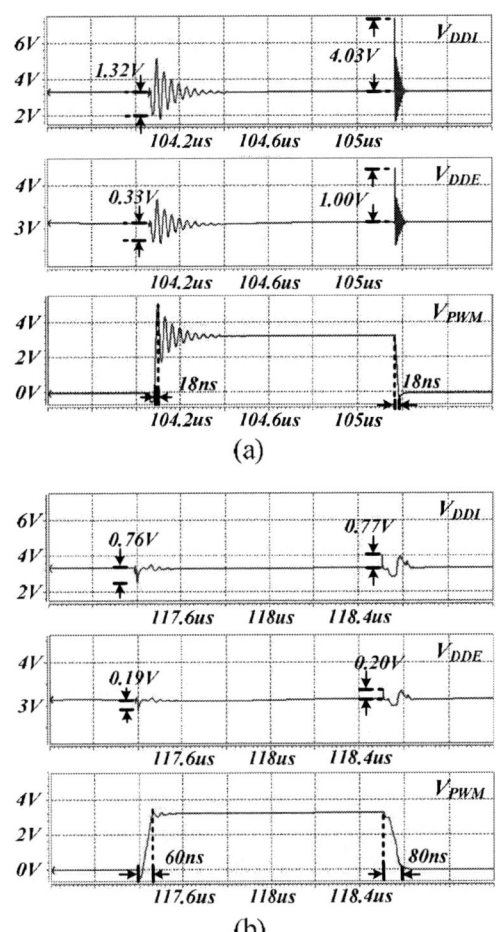

Fig. 8. The SPICE verification waveforms of V_{DDI}, V_{DDE}, and V_{PWM} with (a) the conventional output stage, and (b) the new SRC output stage, respectively.

978-1-4244-9019-6/11 $26.00 © 2011 IEEE

IV. Measurement Results

The converter is designed and fabricated with a 0.35μm 1P4M 3.3V mixed-signal CMOS process and the total active chip area is 2.3mm², as shown in Fig. 9. The circuit overhead of the SRC output stage, including the feedback capacitor C_{FP}, is 0.035mm², which is only 1.52% of the whole chip. The DAW output stage can be paralleled-combined with the original power MOSFET for obtaining even smaller on-resistance of the power MOSFET. Therefore, the new SRC output stage is cost-effective and is of low-complexity design.

The measured input voltage range, the output voltage range, and the load current range of the designed switching DC-DC converter are 2.4~3.6V, 0.6V~3V, and 0~700mA, respectively. With a switching frequency of 500 kHz, the measured line and load regulations are 0.68% and 0.175%, respectively. The measured rise time and fall time by using the conventional output stage and the SRC output stage are illustrated in Figs. 10(a) and 10(b), respectively. With the conventional output stage, the rise and fall times of the PWM output are 10ns and 15ns, respectively, while the rise and fall times for the SRC output stage are 60ns and 80ns, respectively. The rising and falling edge supply bouncing with the conventional output stage are 0.50V and 0.44V, respectively, while that of the SRC output stage are 0.30V and 0.25V, respectively. From the measured supply bouncing reductions outside the IC package, it is believed that the internal supply bouncing is significantly reduced by using the new SRC output stage instead of the conventional design. The performance parameters of the new SRC output stage are summarized in Table I.

V. Conclusions

In this paper, a new SRC output stage is proposed to increase the rise and fall times of PWM output. According to the measurement results, the proposed SRC output stage can reduce the external supply bouncing of the converter by 40% with negligible circuit overhead. This result reveals that the internal supply bouncing can be significantly reduced so that the EMI and reliability problem can be improved. The proposed SRC output stage can also be applied to other PWM related applications, such as DC-DC converters, LED drivers, and class-D amplifiers.

Acknowledgment

The authors would like to acknowledge fabrication support provided by National Chip Implementation Center (CIC), Taiwan.

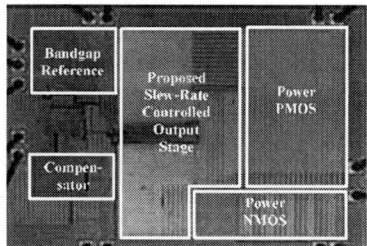

Fig. 9. The chip photo of the designed converter with the SRC output stage

Table I. The performance summary of the new SRC output stage

Items	Conditions	Results
Process		0.35um 1P4M CMOS Process
Chip area	control part	0.035 mm² (only 1.52% overhead)
PWM rise / fall time	conventional	10 ns / 15 ns
	with SRC	60 ns / 80 ns
Internal supply bouncing (SPICE)	conventional	1.32V / 4.03V (rise / fall)
	with SRC	0.76V / 0.77V (rise / fall)
External supply bouncing (measured)	conventional	0.50V / 0.44V (rise / fall)
	with SRC	0.30V / 0.25V (rise / fall)

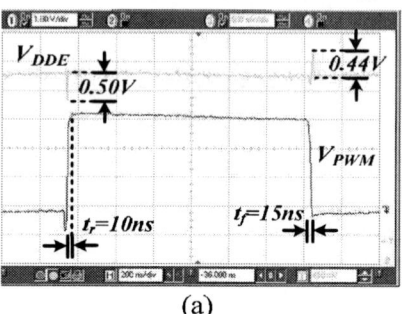

(a)

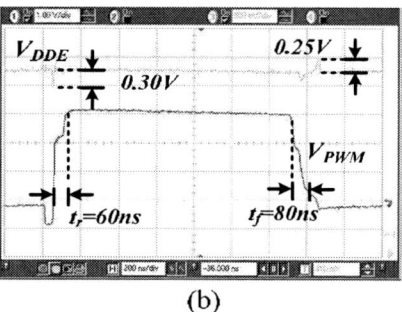

(b)

Fig. 10. The measured waveforms for the rising and falling PWM output and their supply bouncing with (a) the conventional output stage, and (b) the SRC output stage, respectively. The scales of the V_{DDE} and V_{PWM} are 0.5V/div and 1V/div, respectively, while the x-axis scale is 30ns/div.

References

[1] F. Garcia, P. Coll, and D. Auvergne, "Design of a slew rate controlled output buffer," *IEEE Int. ASIC Conf.*, pp. 147-150, Sep. 1998.

[2] B. Deutschmann and T. Ostermann, "CMOS output drivers with reduced ground bounce and electromagnetic emission," *IEEE 29th European Solid-State Circuits Conf. (ESSCIRC)*, pp.537-540, Sep. 2003.

[3] C. S. Choy, C. F. Chan, and M. H. Ku, "A feedback control circuit design technique to suppress power noise in high speed output driver," *IEEE Int. Symp. on Circuits and Systems (ISCAS)*, pp.307-310, May 1995.

[4] R. Senthinathan and J. L. Prince, "Application specific CMOS output driver circuit design techniques to reduce simultaneous switching noise," *IEEE J. Solid-State Circuits,* vol. 28, pp. 1383–1388, Dec. 1993.

[5] C. S. Choy, M. H. Ku, and C. F. Chan, "A low power-noise output driver with an adaptive characteristic applicable to a wide range of loading conditions," *IEEE Journal of Solid-State Circuits*, vol. 32, no. 6, pp. 913-917, June 1997.

[6] S. K. Shin, W. Yu, Y. H. Jun, J. W. Kim, B. S. Kong, and C. G. Lee, "Slew-rate-controlled output driver having constant transition time over process, voltage, temperature, and output load variations," *IEEE Trans. Circuits and Syst. II,* vol. 54, no. 7, pp. 601-605, July 2007.

[7] S. K. Shin, W. Yu, B. S. Kong, C. G. Lee, Y. H. Jun, and J. W. Kim, "A slew rate-controlled output driver having a constant transition time over the variations of process, voltage and temperature," *IEEE Custom Integrated Circuits Conference (CICC)*, pp. 231-234, Sep. 2005.

[8] *EN55022 – "Information Technology Equipment. Radio Disturbances Characteristics. Limits and Methods of Measurement"*, 2007.

Temperature Dependence of Device Mismatch and Harmonic Distortion in Nanoscale Uniaxial-Strained PMOSFETs

Jack J.-Y. Kuo, William P.-N. Chen, and Pin Su

Department of Electronics Engineering & Institute of Electronics, National Chiao Tung University, Hsinchu, Taiwan

Abstract—This paper examines the temperature dependence of mismatching and harmonic distortion properties in nanoscale uniaxial strained pMOSFETs. Our results reveal that the temperature dependence of drain current mismatch as well as harmonics distortion can be modulated by uniaxial strain. In the high gate-voltage overdrive ($|V_{gst}|$) linear region, the compressively-strained device shows smaller increment in drain current mismatch than the unstrained counterpart as temperature decreases. In the high $|V_{gst}|$ saturation region, opposite to the unstrained case, the drain current mismatch of the compressively-strained device decreases with temperature. The underlying mechanism is the larger temperature sensitivity of carrier mobility for the strained device. The larger temperature sensitivity of carrier mobility may also results in larger temperature sensitivity of the harmonic distortion amplitudes. Our study may provide insights for analog circuit design using advanced strained devices.

I. INTRODUCTION

Device mismatch and harmonic distortion are important analog metrics that may limit the achievable accuracy in analog applications and mixed-mode integrated circuits [1]-[4]. Regarding the temperature dependence of MOSFET mismatching properties, Andricciola *et al.* [2] have shown that as temperature decreases, both the threshold voltage mismatch ($\sigma\Delta V_{th}$) and the normalized current factor mismatch ($\sigma(\Delta\beta)/\beta$) increase. In addition, Mennillo *et al.* [3] have suggested that the enhanced current factor mismatch as temperature decreases can be attributed to the increased Coulombic scattering. As strained-silicon is widely used in state-of-the-art CMOS technologies [5]-[6], however, the impacts of strain on the temperature dependence of mismatching and harmonic distortion properties for nanoscale transistors are rarely known and merits investigation. In this work, we examine the drain current mismatch and harmonic distortion properties of uniaxially-strained pMOSFETs under various temperatures. Our results will be explained by the measured carrier mobility and the extracted Hooge parameter from the low frequency noise spectrum.

II. DEVICES AND MEASUREMENT

Co-processed uniaxial strained and unstrained PMOSFETs [7]

are investigated in this study. The strained device features compressive uniaxial Contact Etch Stop Layer (CESL) and SiGe source/drain. For the transistors with gate length $L_{gate} = 65$ nm, the saturation drain current (I_{dsat}) of the strained device is improved more than 100% as compared with its control counterpart. The mismatching properties were measured from identical devices in a matching pair configuration on 60 dies. Low frequency noise measurements were carried out using the BTA9812 measurement system.

III. DEVICE MISMATCH

Fig. 1 shows the Pelgrom plot of $\sigma\Delta V_{th}$ under various temperatures for the strained and unstrained devices. The geometries of the devices are $W/L_{gate} = 1\mu m/54nm$, $0.3\mu m/54nm$, and $0.15\mu m/54nm$. Note that the L_{gate} for strained devices needs to be the same in order to keep similar strain in the channel because the channel strain is gate-length dependent in process-induced strain silicon devices [6]. The linear relationship between $\sigma\Delta V_{th}$ and $(WL_{gate})^{-1/2}$ indicates a random-dopant-fluctuations origin $\sigma\Delta V_{th}$ [10]. In addition, the strained device shows similar temperature dependence of $\sigma\Delta V_{th}$ as compared with its control counterpart. Since the $\sigma(\Delta I_d)/I_d$ inthe low $|V_{gst}|$ regime is mainly determined by the $\sigma\Delta V_{th}$ [8], the similar temperature dependence of $\sigma\Delta V_{th}$ results in similar temperature dependence of $\sigma(\Delta I_d)/I_d$ for the strained and unstrained devices as shown in Fig. 2.

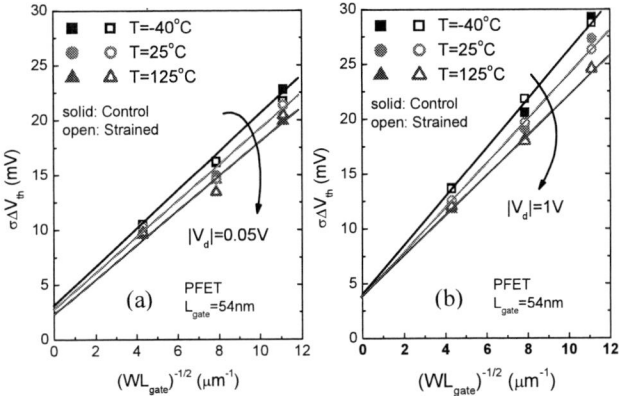

Fig. 1 The strained device shows similar temperature dependence of $\sigma(\Delta V_{th})$ as that of the control device at (a) $|V_d|$=0.05V, and (b) $|V_d|$=1V.

978-1-4244-9019-6/11 $26.00 © 2011 IEEE

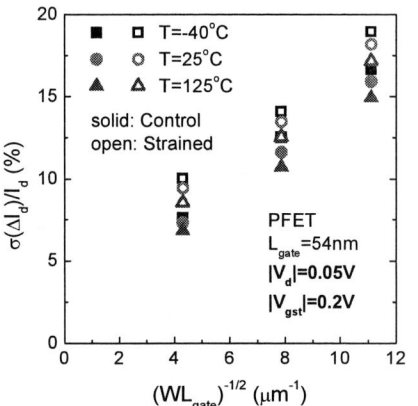

Fig. 2 The strained device shows similar temperature dependence of $\sigma(\Delta I_d)/I_d$ as that of the control device at $|V_{gst}|$=0.2V and $|V_d|$=0.05V.

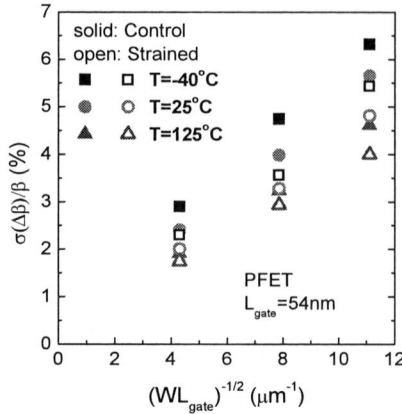

Fig. 4 Pelgrom plot of $\sigma(\Delta\beta)/\beta$ showing smaller temperature dependence of $\sigma(\Delta\beta)/\beta$ for the strained device.

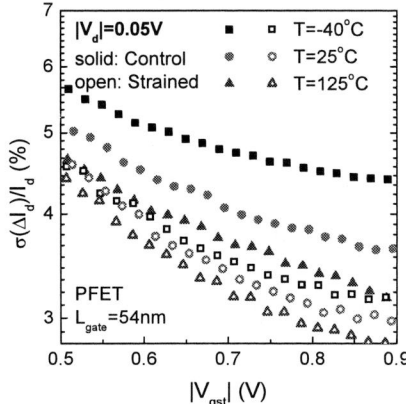

Fig. 3 The strained device shows smaller temperature dependence as compared with the unstrained one in the $|V_{gst}|$ regime with $|V_d|$=0.05V.

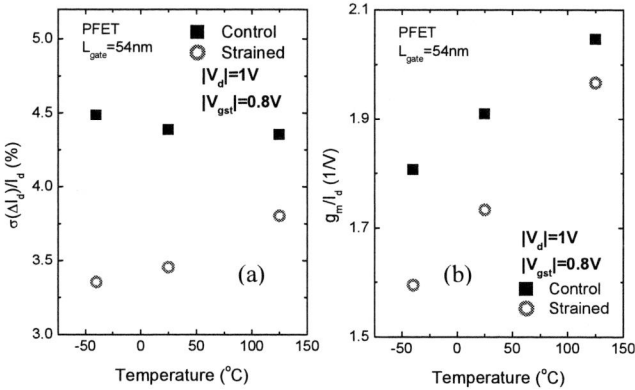

Fig. 5 (a) $\sigma(\Delta I_d)/I_d$ vs. temperature characteristics showing reduced $\sigma(\Delta I_d)/I_d$ for the strained device as temperature decreases. (b) The strained device shows larger reduction in g_m/I_d as temperature decreases.

In the high $|V_{gst}|$ regime, however, the strained device shows different temperature dependence of $\sigma(\Delta I_d)/I_d$ as compared with its control counterpart. It can be seen from Fig. 3 that in the high $|V_{gst}|$ linear regime the $\sigma(\Delta I_d)/I_d$ increases with decreasing temperature for both the strained and unstrained devices. The temperature dependence of $\sigma(\Delta I_d)/I_d$ in Fig. 3 can be explained by the temperature dependence of $\sigma(\Delta\beta)/\beta$, as shown in Fig. 4. The increased $\sigma(\Delta\beta)/\beta$ with decreasing temperature has been attributed to the Coulombic scattering [2]-[3]. However, it should be noted that the $\sigma(\Delta I_d)/I_d$ of the strained device in Fig. 3 exhibits smaller temperature dependence than that of the unstrained one. The smaller temperature dependence of $\sigma(\Delta I_d)/I_d$ for strained devices results from the smaller temperature dependence in $\sigma(\Delta\beta)/\beta$ (Fig. 4). It is the significantly enhanced β for the strained device as temperature decreases that reduces the temperature sensitivity of $\sigma(\Delta\beta)/\beta$ and $\sigma(\Delta I_d)/I_d$ for the strained device. The larger temperature sensitivity of β present in the compressively strained PFET results from the larger temperature sensitivity of

carrier mobility, which will be further discussed in next section.

Fig. 5(a) compares the temperature dependence of $\sigma(\Delta I_d)/I_d$ for the strained and unstrained devices in the high $|V_{gst}|$ dependence of $\sigma(\Delta I_d)/I_d$ for the strained device is opposite to saturation regime. It can be seen that the temperature that of the unstrained one. In other words, the compressive strain has changed the temperature trend in drain current mismatch and the $\sigma(\Delta I_d)/I_d$ decreases with temperature for the strained PFET. In the high $|V_{gst}|$ saturation regime, both the threshold voltage mismatch and current factor mismatch are relevant to the $\sigma(\Delta I_d)/I_d$ [8]. The decreased $\sigma(\Delta I_d)/I_d$ for the strained device results from the larger reduction in g_m/I_d reduction as temperature decreases (Fig. 5(b)), while the $\sigma(\Delta I_d)/I_d$ of the unstrained device is weakly dependent on temperature because of the opposite temperature dependence of $\sigma\Delta V_{th}$ (Fig. 1) and g_m/I_d (Fig. 5 (b)). The larger reduction in g_m/I_d as temperature decreases for the strained device is also a consequence of the larger temperature sensitivity of carrier mobility.

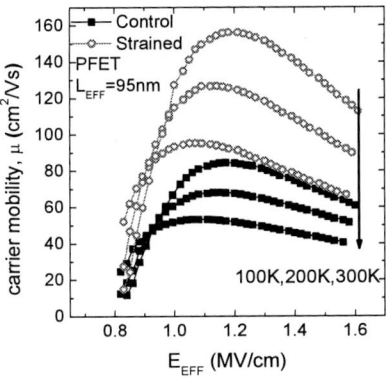

Fig. 6 Extracted carrier mobility showing larger temperature sensitivity for the strained device.

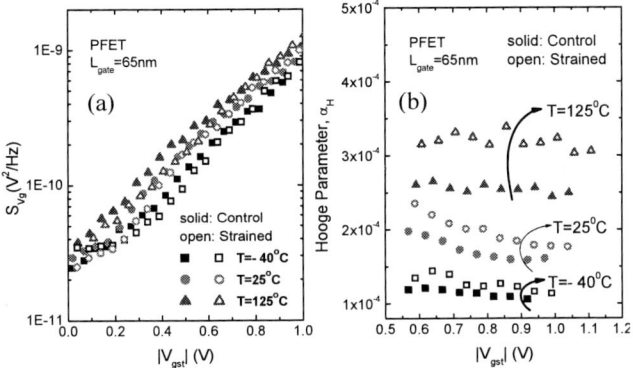

Fig. 7 (a) S_{Vg} versus $|V_{gst}|$ at f=10Hz and $|V_d|$=0.05V. (b) The strained device shows larger Hooge parameter as compared with the unstrained one.

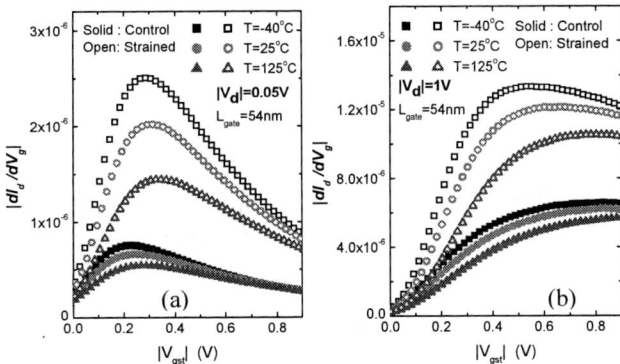

Fig. 8 Comparison of HD1 amplitude in term of dI_d/dV_g in (a) the linear region ($|V_d|$=0.05V) and (b) the saturation region ($|V_d|$=1V) for the strained and unstrained devices.

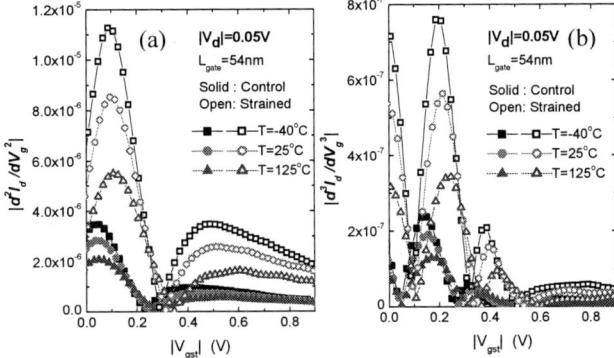

Fig. 9 Temperature dependences of (a) d^2I_d/dV_g^2 and (b) d^3I_d/dV_g^3 in the linear region ($|V_d|$=0.05V) for the strained and unstrained devices.

IV. TEMPERATURE DEPENDENCE OF CARRIER MOBILITY AND HOOGE PAREMETER

Fig. 6 compares the temperature dependence of the extracted carrier mobility for the strained and unstrained devices [11]. It can be seen that the temperature sensitivity of carrier mobility is larger for the strained device as compared with its control counterpart. It is plausible that as temperature increases, the compressively strained PFET has less holes to populate states near the band edge where the conductivity effective mass along the channel direction is smaller. Therefore, the observed carrier mobility enhancement decreases with increasing temperature [12].

In addition, larger temperature sensitivity shown in the strained device also infers a more phonon-scattering dominant carrier mobility. Based on Hooge's carrier-mobility-fluctuations origin low frequency noise model [13], the magnitude of Hooge parameter (α_H) indicates the dominance of phonon scattering. Fig. 7(b) shows the extracted Hooge parameter for the strained and unstrained devices. It can be seen that the strained device has larger Hooge parameter than the

unstrained one. This result is consistent to the temperature sensitivity of the carrier mobility shown in Fig. 6. Note that the Hooge parameter is extracted from the carrier-mobility-fluctuations origin low frequency noise, which behaves linear gate bias dependent input-referred voltage spectral density S_{Vg} (as shown in Fig. 7(a)).

V. HARMONIC DISTORTION

The harmonic distortion characteristics for the strained device are evaluated through the amplitude of the first-, second-, and third-order of harmonic distortions (HD1, HD2, and HD3, respectively). Since the amplitudes of harmonic distortions are related to the drain current by successive orders of differentiation with respect to the gate voltage under low frequency operation [4], dI_d/dV_g, d^2I_d/dV_g^2 and d^3I_d/dV_g^3 are extracted to investigate the impact of strain on HD1, HD2, and HD3 and their temperature dependence.

Fig. 8 (a) and (b) compare the dI_d/dV_g for the strained and unstrained devices in the linear region ($|V_d|$=0.05V) and saturation region ($|V_d|$=1V), respectively. It can be seen that the dI_d/dV_g for the strained device is significantly larger than that of

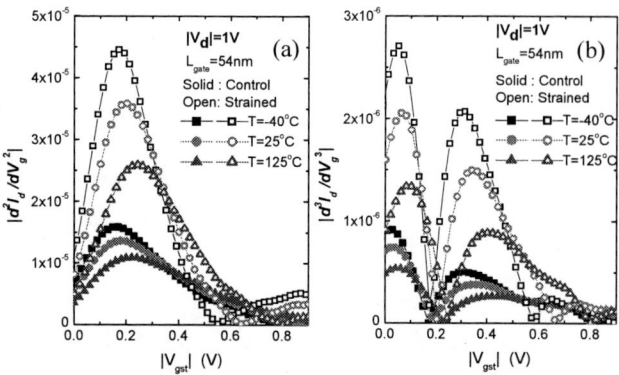

Fig. 10 Temperature dependences of (a) d^2I_d/dV_g^2 and (b) d^3I_d/dV_g^3 in the saturation region ($|V_d|$=1V) for the strained and unstrained devices.

the unstrained one, which implies a larger HD1 for the strained device. Besides, the difference of dI_d/dV_g between the strained and unstrained device is larger in the linear region than in the saturation. This is because the dI_d/dV_g in the linear region and saturation region are determined by the carrier mobility and saturation velocity, respectively. With the adoption of strained silicon, the enhancement in carrier mobility is larger than in saturation velocity [14]. In addition, Fig. 8 (a) and (b) also indicate larger temperature sensitivity of the HD1 amplitude for the strained device, which can be attributed to the phonon scattering limited carrier transport mechanism (Fig. 6) as explained in pervious section.

Fig. 9 (a) and (b) compare the HD2 and HD3 amplitudes in the linear region in terms of d^2I_d/dV_g^2 and d^3I_d/dV_g^3, respectively. It can be seen that the strained device has larger harmonic amplitudes and higher temperature sensitively as compared with the unstrained one. Regarding the HD2 and HD3 amplitudes in the saturation region, Fig. 10 (a) and (b) show similar d^2I_d/dV_g^2 and d^3I_d/dV_g^3 characteristic as in the linear region. Nevertheless, the increase in harmonic amplitudes as well as the temperature sensitively for the strained device in the saturation region is smaller than in the linear region due to velocity saturation.

VI. CONCLUSION

We have investigated and analyzed the device mismatching properties and harmonic distortion characteristics of nanoscale uniaxial strained pMOSFETs under various temperatures. Our result indicates that the drain current mismatch versus temperature trend for the strained device is different from the unstrained one. In the high $|V_{gst}|$ linear regime, the compressively-strained device shows smaller increment in drain current mismatch than the unstrained counterpart as temperature decreases. In the high $|V_{gst}|$ saturation region, opposite to the unstrained case, the drain current mismatch of the compressively-strained device decreases with temperature. The underlying mechanism is the larger temperature sensitivity of carrier mobility for the strained device, which leads to larger

Hooge parameters in the low frequency noise spectrum of the strained device. The larger temperature sensitivity of carrier mobility also results in larger temperature sensitivity of the harmonic distortion amplitudes. Our study may provide insights for analog circuit design using advanced strained devices.

ACKNOWLEDGMENT

This work was supported in part by the National Science Council of Taiwan under Contract NSC99-2221-E-009-174, and in part by the Ministry of Education in Taiwan under ATU Program.

REFERENCES

[1] A. Bhavnagarwala, S. Kosonocky, C. Radens, K. Stawiasz, R. Mann, Q. Ye, and K. Chin, "Fluctuation limits & scaling opportunities for CMOS SRAM cells," *IEDM Tech. Dig.*, Dec. 2005, pp. 659–662.

[2] P. Andricciola and H. P. Tuinhout, "The Temperature Dependence of Mismatch in Deep-Submicrometer Bulk MOSFETs," *IEEE Electron Device Lett.*, vol. 30, no. 6, pp. 690–692, Jun. 2009.

[3] S. Mennillo, A. Spessot, L. Vendrame, L. Bortesi, "An Analysis of Temperature Impact on MOSFET Mismatch," *Proc. ICMTS*, 2009, pp. 56–61.

[4] D. Navarro, Y. Takeda, M. Miyake, N. Nakayama, K. Machida, T. Ezaki, H.J. Mattausch, and M. Miura-Mattausch, "A Carrier-Transit-Delay-Based Nonquasi-Static MOSFET Model for Circuit Simulation and Its Application to Harmonic Distortion Analysis," *IEEE Trans. Electron Devices*, vol. 53, no. 9, p. 2025, Sep. 2006.

[5] C. Hu, "Device challenges and opportunities," *VLSI Symp. Tech. Dig.* 2004, pp.4-5.

[6] S.E. Thompson, M.Armstrong, C.Auth, M.Alavi, M.Buehler, R.Chau, S.Cea, T.Ghani, G.Glass, T.Hoffman, C.-H.Jan, C.Kenyon, J.Klaus, K.Kuhn, Zhiyong Ma, B.Mcintyre, K.Mistry, A.Murthy, B.Obradovic, R.Nagisetty, N.Phi, S.Sivakumar, R.Shaheed, L.Shifren, B.Tufts, S.Tyagi, M.Bohr, Y.El-Mansy, "A 90-nm logic technology featuring strained-Silicon," *IEEE Trans. Electron Devices*, vol. 53, no. 5, p. 1010, May 2006.

[7] J. J.-Y. Kuo, W. P.-N. Chen, and P. Su, "A Comprehensive Investigation of Analog Performance for Uniaxial Strained PMOSFETs," *IEEE Trans. Electron Devices*, vol. 56, no. 2, pp. 284-290, Feb. 2009.

[8] J. J.-Y. Kuo, W. P.-N. Chen, and P. Su, "Investigation and Analysis of Mismatching Properties for Nanoscale Strained MOSFETs," *IEEE Trans. Nanotechnology*, vol. 9, no. 2, pp. 248-253, Mar. 2010.

[9] J.A. Croon, M. Rosmeulen, S. Decoutere, W. Sansen, H.E. Maes, "An easy-to-use mismatch model for the MOS transistor," *IEEE J. Solid-State Circuits*, vol. 37, no. 8, pp.1056, Aug. 2002.

[10] T. Mizuno, J. Okamura, and A. Toriumi," Experimental Study of Threshold Voltage Fluctuation Due to Statistical Variation of Channel Dopant Number in MOSFET's " *IEEE Trans. Electron Devices*, pp.2216, Nov. 1994.

[11] W. P.-N. Chen, J. J.-Y. Kuo, and P. Su, "Impact of Process-Induced Uniaxial Strain on the Temperature Dependence of Carrier Mobility in Nanoscale pMOSFETs," *IEEE Electron Device Lett.*, vol. 31, no. 5, pp 414-416, May 2010.

[12] X. Yang, S. Parthasarathy, Y. Sun, A. Koehler, T. Nishida, and S. E. Thompson, "Temperature dependence of enhanced hole mobility in uniaxial strained p-channel metal–oxide–semiconductor field-effect transistors and insight into the physical mechanisms," *Appl. Phys. Lett.*, vol. 93, no. 24, p. 243 503, Dec. 2008.

[13] L.K.J. Vandamme and F.N. Hooge, *IEEE Trans. Electron Devices*, vol. 55, no. 11, p. 3070, Nov. 2008.

[14] J. Kuo, W. Chen, and P. Su, "Investigation of analog performance for process-induced-strained pMOSFETs," *Semicond. Sci. Technol.*, vol. 22, no. 4, pp. 404–407, Apr. 2007.

A 8-bit 50-Msamples/s Switched-Current Pipelined ADC with Residue Generator and Interlaced stage

Guo-Ming Sung, *Member, IEEE* and Ying-Tzu Lai
Department of Electrical Engineering, National Taipei University of Technology,
1, Sec. 3, Chung-Hsiao E. Rd., Taipei, Taiwan, R.O.C.
Email: gmsung@ntut.edu.tw

Abstract—**This paper presents a 8-bit 50-MHz sampling rate switched-current pipelined analog-to-digital converter (ADC) in a standard 0.35-μm 2P4M CMOS process. Not only a new residue generator is proposed to cancel the sub-DAC circuit, but also an interlaced arrangement is adopted to improve the transmission error in a seven-stage pipelined ADC. That is, the odd stage adopts the traditional structure and the even stage employs the proposed residue generator. The simulated results reveal that power dissipation is 160mW and sampling rate is 50 MHz at a supply voltage of 3.3 V. As a sinusoidal waveform with 1 MHz sampling rate is adopted, a signal to noise distortion ratio (SNDR) of 48 dB and an effective number of bits (ENOB) of 7.7 bits are demonstrated. Additionally, the differential nonlinearity (DNL) of -0.4 LSB ~ +0.3 LSB and the integral nonlinearity (INL) of -0.7 LSB ~ +0.8 LSB are presented with a chip area of roughly 1.59 × 1.63 mm².**

Index Terms—**Switched-current, pipelined analog-to-digital converter, residue generator.**

I. INTRODUCTION

The cost and power consumption are always the important parameters to estimate the performance of electronic products. For ADC, the pipelined ADC belongs to high-speed conversion rate and high-resolution categories, which is used to increase the resolution and operating speed. In this work, we use the 1.5 bits per stage technique to accomplish each stage of pipelined ADC. This circuit architecture not only improves the quantization error, but also decreases the delay time of DAC, the error and offset of amplifier. Additionally, a new residue generator is proposed to replace the traditional pipelined ADC. The purpose is to reduce the error caused by DAC and to increase the operating speed. Notably, a current mirror is adopted to supply all currents for current comparator, DAC current signal, and current source of S/H.

Figure 1 shows the traditional internal structure of conventional pipelined ADC [1]. According to benefit of 1.5 bits per stage, the magnification of output reside is multiplied by 2. The sub-ADC produces two binary codes (bin1, bin2) in one stage and bin2 will be added to the next stage to get right digital code, which complete with digital error correction circuit.

Figure 2 depicts the internal circuit of the proposed residue generator. A conventional pipelined ADC will separate the input signal into two paths, one is connected to the internal part of sub-Flash ADC to generate a digital output code, and the other is connected to multiply-by-two amplifier. The disadvantage is that a sub-DAC is in need. However, there is no sub-DAC in the proposed residue generator. That is, the input signals will be split into two paths, one is connected to Flash ADC to generate a set

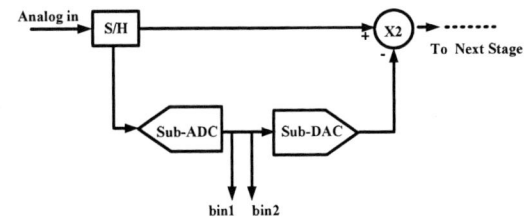

Fig. 1. Architecture of the traditional 1.5 bits/stage of pipelined ADC

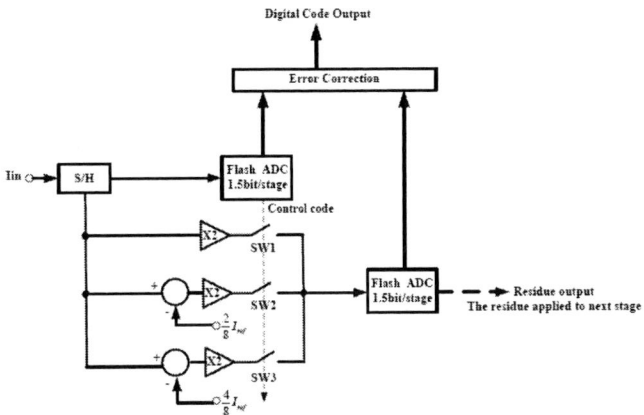

Fig. 2. Internal circuit of the proposed residue generator.

of control code, and the other is connected to three multiply-by-two amplifiers and subtractors. Notably, the control code is used to guarantee that only one switch will be turned on at a cycle. The proposed residue generator will provide a constant level according to the turn-on cycle of the switch.

II. RESPONSE TIME OF PROPOSED PIPELINED ADC

Figure 3 presents the response time of a stage in a conventional pipelined ADC [2]; and that Fig. 4 displays the response time of a stage in the proposed pipelined ADC. As shown in Fig. 3, we can find that one of the two paths certainly reaches the subtractor in advance after sampling and it needs to wait for the conjoint signal on the other path passing through comparator, Sub-ADC and sub-DAC. There is a delay error occurs at subtractor. The delay error will be a serious problem to limit the operating speed as the cascaded stages were added considerably. Comparing with conditional pipelined ADC, the sampled input signal of the proposed pipelined ADC, as shown in Fig.4, works without waiting step and shortens the response time by removing the sub-DAC. In a word, the input signal can be simultaneously digitalized to generate digital code, and be subtracted and amplified in sub-ADC to avoid the DAC delay.

978-1-4244-9019-6/11 $26.00 © 2011 IEEE

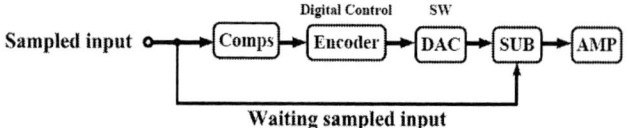

Fig. 3. Response time per stage for a conventional pipelined ADC

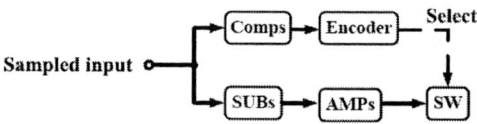

Fig. 4. Response time per stage for a proposed pipelined ADC

III. SWITCHED CURRENT S/H CIRCUIT

The data sampling techniques can be divided into two main categories, switched-capacitor (SC) [3] and switched- current (SI) [4-5]. For SC, the sampling work can be completed with the parasitical capacitor between the spread lines, whereas SI is with the parasitical capacitor between gate and source of MOSFET (C_{gs}). This implies that the chip area of SI is smaller than that of SC.

Figure 5 shows the current transmission circuit and its equivalent error model. In this equivalent model, the impedances between drain and source of M_1 and M_2 are equal to $1/g_{ds1}$ and $1/g_{m2}$, respectively. Then the current gain, i_{d2}/i_{d1}, can be given by

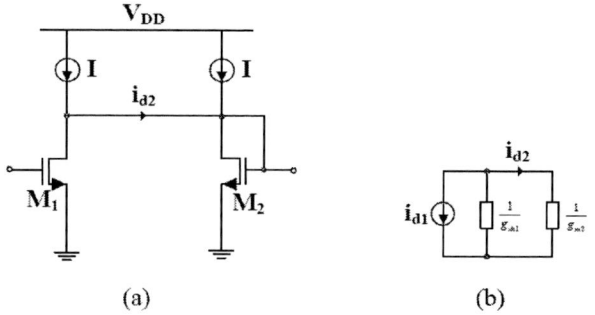

(a)   (b)

Fig. 5. (a) The current transmission circuit; (b) equivalent error model.

$$\frac{i_{d2}}{i_{d1}} = -\frac{g_{m2}}{g_{m2}+g_{ds1}} = -\frac{1}{1+\dfrac{g_{ds1}}{g_{m2}}} = -\frac{1}{1+\varepsilon} \quad (1)$$

where the transmit error ratio ε is defined as g_{ds1}/g_{m2}. A traditional first generation (FG) SI memory cell inherits with high input impedance of $1/g_{m1}$ [7]. Hence, a current feedback technique can be used to decrease the input impedance of SI memory cell [8]. Figure 6 shows the circuit schematic of the proposed switched-current feedback memory cell. Notably, those constants γ, β and k are width-to-length ratios (W/L) of M5, M6 (or M7) and M2, respectively. Figure 7 shows the simplified small-signal equivalent circuit of the proposed memory cell. As the sampling switch is on, the current transfer function is given by

$$\frac{i_{out}}{i_{in}} = \frac{g_{m5}}{g_{m1}+\left(\dfrac{g_{m6}g_{m4}g_{m2}}{g_{m7}g_{m3}}\right)} = \frac{\gamma}{\beta}\cdot\frac{1}{1+k} \quad (2)$$

Thus, the input impedance can be given by

$$r_{in} = \frac{1}{(1+k)g_{m1}} \quad (3)$$

If $|i_{in}| = |i_{out}|$ is on demand, we need to properly adjust the width-length ratio γ, β and k, as

$$\gamma : \beta : k = 1 : 0.5 : 1 \quad (4)$$

Then the input impedance can be simplified to

$$r_{in} = \frac{1}{2\bullet g_{m1}} \quad (5)$$

Notify that the input impedance of the proposed memory cell is reduced to half of traditional cell. Then the operating speed will be increased due to the small time constant.

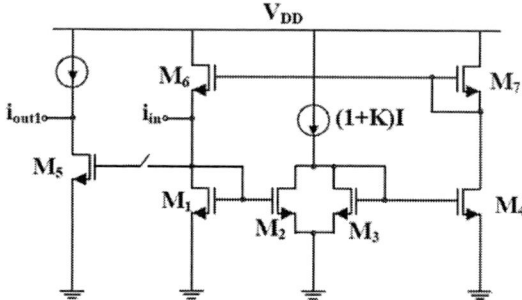

Fig. 6. Circuit schematic of proposed switched-current feedback memory cell.

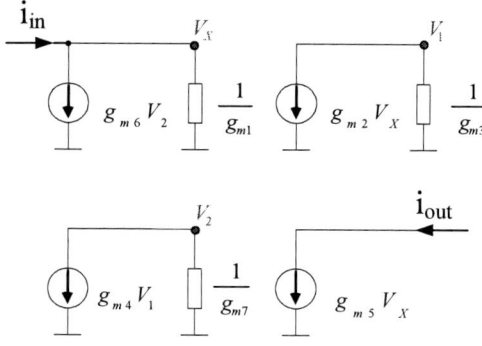

Fig. 7. Equivalent circuit of proposed switched-current feedback memory cell.

IV. ARCHITECTURE OF PIPELINED ADC

Nowadays, the best performance of the switched-current pipelined ADC is about 2.5 GSPS [9] and the best resolution is approximately 13 bits [10]. Unfortunately, the power dissipation is too high to be acceptable. Thus, a current mirror is used to provide all currents with low power dissipation.

As shown in Fig. 8, a seven-stage pipelined ADC is presented. In this ADC, each stage is adopted with 1.5 bits, except for 2 bits at the last stage. Furthermore, we need to point out that the odd stage adopts the traditional structure and the even stage employs the proposed residue generator. Finally, the output digital code will be presented with delay string in each stage properly and have a correct output after passing through

978-1-4244-9019-6/11 $26.00 © 2011 IEEE   164

the error-correction circuit.

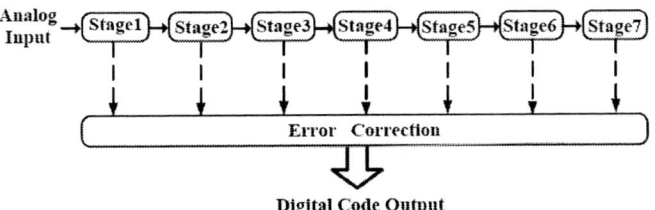

Fig. 8. Block diagram of the proposed seven-stage pipelined ADC.

Figure 9 shows the circuit diagram of odd stage. In this stage, the sampled current signal will be copied into three paths; one is connected to the current delay circuit and then feeds into the subtractor, and the others are connected to the comparator and then deeds into the sub-DAC. Thus, a set of thermometer codes are presented before decoder circuit; and that a two-bit output code is performed after decoder circuit. The DAC circuit will convert the thermometer code into analog signal and subtract with the input signal to produce the residue output. By passing through the multiply-by-two amplifier, the residue output will be amplified by two times and be transmitted to the next stage.

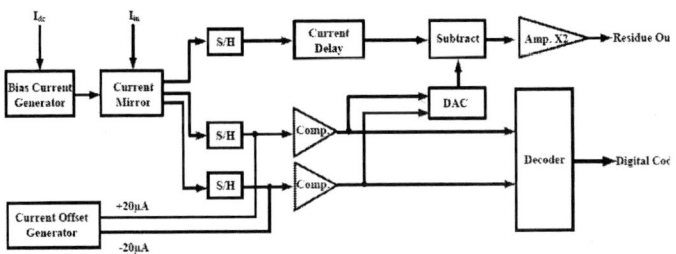

Fig. 9. Circuit diagram of odd stage for conventional pipelined ADC.

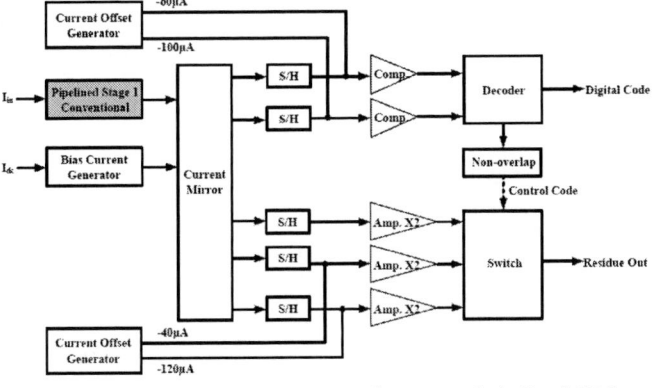

Fig. 10. Circuit diagram of even stage for proposed pipelined ADC

Figure 10 indicates the circuit diagram of even stage whose input signal is the residue output of the previous odd stage. As shown in Fig. 10, five paths are generated. Two paths are connected to comparators to produce thermometer codes. Those thermometer codes will not only be converted into digital output codes by decoder circuit, but also be used to control the switch with non-overlap circuit. The others are connected to the residue generator, which is shown in Fig. 2. Notify that the current offset generator must be added to generate an equal current offset.

Passing through the multiply-by-two amplifier and conveying to the control codes, which are generated with previous two paths, the residue output of even stage is presented.

Figure 11 demonstrates the current amplifier which is designed with current mirror technique. The main function is to let the residue signal be amplified two times and be transferred to the next stage. Unfortunately, the output current will be affected with channel length modulation and transmission error. To resolve those impacts, the gain boost technique is adopted to increase the output impedance, R_{out}. This technique can improve the error of current amplifier effectively.

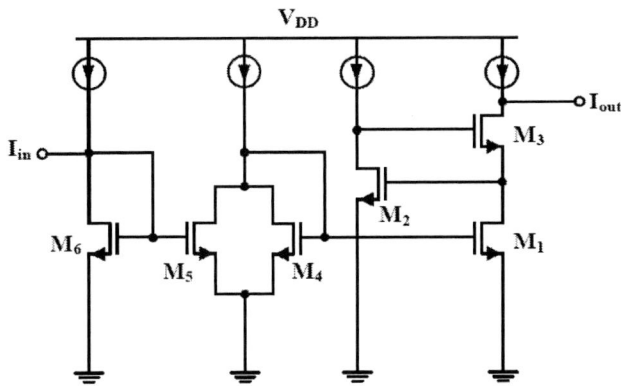

Fig. 11. Proposed circuit of current amplifier

Figure 12 shows the sub-DAC circuit for conventional ADC. It controls the switch using the thermometer code. The advantage of this circuit is that high output-swing is achieved easily and equal output current can be controlled with thermometer code. Notably, each switch controls the same current very well and the W and L of MOS will be chose a small chip area with low poer dissipation.

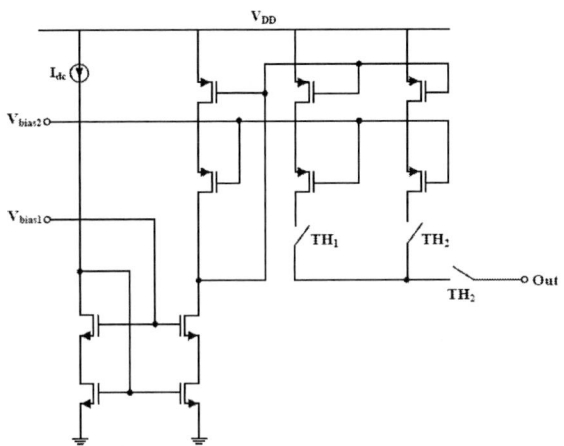

Fig. 12. Sub-DAC circuit for conventional ADC

V. SIMULATION RESULTS AND DISCUSSIONS

After the circuit integration has been completed in the proposed switched-current pipelined ADC, the simulated results are summarized in Table I. The simulation results present that the signal to noise distortion ratio (SNDR) is around 48 dB and the effective number of bit (ENOB) is about 7.7 bits with the

TABLE I
SIMULATED RESULTS OF PROPOSED SWITCHED-CURRENT PIPELINED ADC.

Technology	0.35μm 2P4M
Resolution	8 bits
Full scale current	-80 μA ~ +80 μA
Unit LSB current	0.625 μA
$f_{s,max}$	50 MHz
INL range	(-0.7, 0.8) LSB
DNL range	(-0.4, 0.3) LSB
SNDR (f_s = 50 MHz f_{in} = 1 MHz)	48 dB
SFDR (f_s = 50 MHz f_{in} = 1 MHz)	50 dB
Power consumption	161 mW
V_{DD}	3.3 V
Chip core area	0.66×0.68 mm^2

input frequency of 1MHz. The power spectrum density of the proposed switched-current pipelined ADC is shown in Fig. 13.

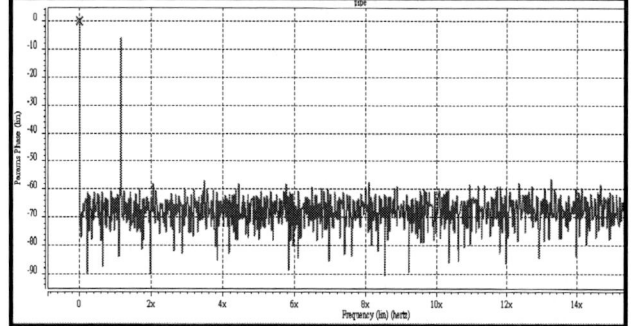

Fig. 13. Power spectrum density of proposed switched-current pipelined ADC.

The DNL varies from -0.4 to +0.3 LSB and the INL changes from -0.7 to +0.8 LSB. Those results reveal that the converted procedure of the ADC system is without missing code. Figures 14 and 15 depict the simulated results of DNL and INL, respectively.

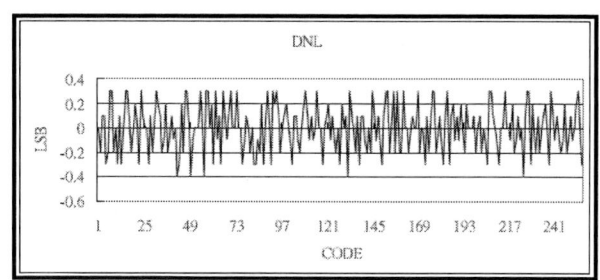

Fig. 14. Simulated result of the DNL

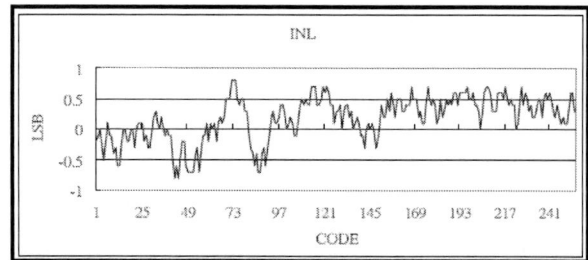

Fig. 15. Simulated result of the INL

VI. CONCLUSIONS

In this paper, a new residue generator is proposed to replace the even stages of the pipelined ADC, except for conventional type in odd stages. The proposed architecture can improve the transmission error. Furthermore, all the current sources will be completed with current mirror technique to accomplish an 8-bits high-speed pipelined ADC with low power dissipation. The simulation results show that the SNDR, maximum INL, maximum DNL and maximum sampling rate are 48 dB, 0.8 LSB, 0.4 LSB and 50 MHz. respectively, at the power supply of 3.3 V and the input frequency of 1 MHz; and that the power dissipation is about 161mW and the core area is about 0.66 mm × 0.68 mm.

ACKNOWLEDGMENT

The author would like to thank the National Science Council of the Republic of China, Taiwan, for financially supporting this research under Contract No. NSC 98-2221-E-027-110. The Chip Implementation Center (CIC), Taiwan, is appreciated for fabricating the test chip.

REFERENCES

[1] T. Ito, T. Ueno, D. Kurose, T. Yamaji and T. Itakura, "A 10-bit, 200-MSPS, 105-mW pipeline A-to-D converter," *IEICE Electronics Express*, vol. 2, No. 15, pp. 429-433, July 2005.

[2] S. Park, "*Design of a fast pipelined A/D converter in CMOS technology,*" Ph.D. dissertation, Univ. Arizona, Arizona State, May 1997.

[3] C. Wang, Z. Li, X. Liu, B. Li and Y. Li, "A Capacitor-Mismatch -Insensitive Switch-Capacitor Amplifier for Pipeline ADC," *7th International Conference on ASIC*, Oct. 2007, pp. 473–476.

[4] G. M. Sung, J. H. Tzeng, C. S. Liao and S. C. Shu, "A Low-power 7-b 33-Msamples/s Switched-current Pipelined ADC for Motor Control," *IEEE APCCAS*, Dec. 2006, pp. 171–174.

[5] B. Sedighi and M. S. Bakhtiar, "An 8-bit 300MS/s Switched-Current Pipeline ADC in 0.18μm CMOS," *IEEE ISCAS*, May 2007, pp. 1481–1484.

[6] Y. Sun and F. Lai, "Low Power High Speed Switched Current Comparators for Current Mode ADC," *International Symposium on Communications and Information Technologies*, Oct. 2007, pp. 222–225.

[7] G. Nikandish, B. Sedighi and M. S. Bakhtiar, "Performance Comparison of Switched-Capacitor and Switched-Current Pipeline ADCs," *IEEE ISCAS*, vol. 2, Oct. 2007, pp. 2252–2255.

[8] N. Tan, *Microelectronics Research Center Ericsson Components AB Sweden, Switched-Current Design and Implementation of Oversampling A/D Converters*, Kluwer Academic Publishers, 1997.

[9] M. Wang, C. I. H. Chen and S. Radhakrishnan, "Low-Power 4-b 2.5-GSPS Pipelined Flash Analog-to-Digital Converter in 130-nm CMOS," *IEEE Transactions on Instrumentation and Measurement*, vol. 56, no. 3, 2007, pp. 1064 – 1073, 2007.

[10] J. S. Wang and C. L. Wey, "A low-voltage low-power 13-b pipelined switched-current cyclic A/D converter," *Proceedings of the IEEE 2nd Dallas CAS Workshop on Low Power/Low Voltage Mixed-Signal Circuits and Systems*, 2001, pp. P15-18.

Continuously Auto-Tuned and Self-Ranged Dual-Path PLL Design with Hybrid AFC

Min Wang, Bo Zhou, Woogeun Rhee, and Zhihua Wang

Institute of Microelectronics, Tsinghua University, Beijing, China

Abstract-This paper describes a dual-path phase locked loop (PLL) design which automatically tunes to capture the desired frequency as well as the input control voltage range of the dual-path LC voltage-controlled oscillator (VCO). A hybrid automatic frequency calibration (AFC) circuit provides digital frequency calibration and mixed-mode continuous frequency tuning. Since the hybrid AFC circuit independently controls the coarse-tuning control voltage of VCO in this structure, the VCO input control voltage range can be preset before the PLL is locked. Therefore, an optimum loop filter voltage range can be chosen to have good charge pump matching or VCO gain linearity. Simulation results verify that the self-ranged control can be achieved by adjusting the output range of the limiting amplitude amplifier (LAA) during the PLL transient mode.

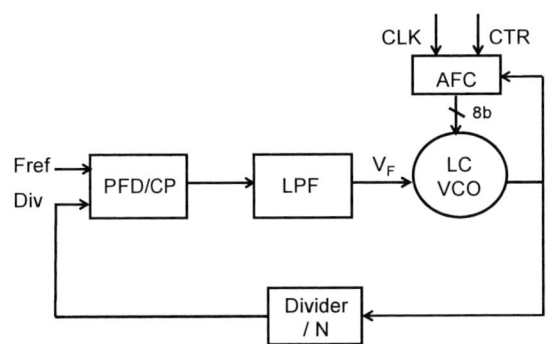

Fig. 1. Conventional PLL with LC VCO and AFC.

I. INTRODUCTION

Low noise and wide tuning range are fundamental trade-off in the voltage-controlled oscillator (VCO) design. Although having much better noise performance than the ring VCO, the LC VCO suffers from a narrow tuning range. The trade-off between the noise performance and the tuning range in the LC VCO design can be overcome by having programmable capacitor array [1]-[2] or by adopting a dual-path control [3]-[5]. When the programmable capacitor array is used, an automatic frequency calibration (AFC) circuit is used to have the center frequency of the VCO close to the desired output frequency [2]. This digital calibration is done before the normal operation of the phase-locked loop (PLL) and considered a coarse tuning procedure. The number of bits for the digital calibration is typically limited to six bits since large number of capacitors increases parasitic capacitance substantially. On the other hand, the dual-path PLL shown in Fig. 2 offers continuous tuning over entire frequency range but suffers from noise coupling due to high VCO gain in the coarse-tuning path [4]. Also, the dual-path PLL creates additional pole and zero frequencies within the bandwidth, resulting in more complicated PLL design than the conventional one.

In this paper, we proposed a hybrid AFC method which offers both digital coarse calibration and mixed-mode continuous frequency tuning with high resolution. Besides, the independent frequency control by the proposed method makes it possible to set the loop filter control voltage range to a certain value regardless of the output frequency, which is useful to obtain optimum voltage range to maximize charge pump matching or VCO gain linearity.

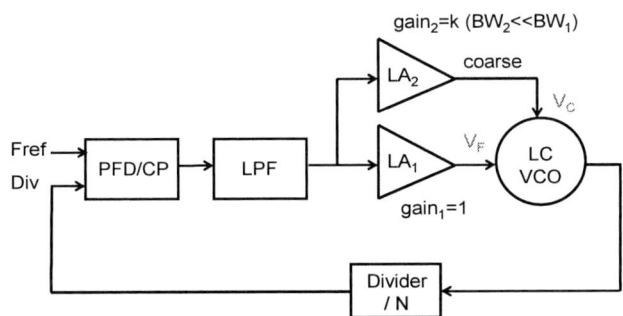

Fig. 2. PLL with continuously tunable dual-path LC VCO.

The paper is organized as follows: Section II explains the basic principle of the proposed Dual-path PLL structure with hybrid AFC method. Section III presents the detailed circuit implementation. In Section IV, simulation results are shown, followed by conclusions in Section V.

II. HYBRID AFC

Fig. 3 shows a conceptual block diagram of the proposed Dual-path PLL method. The LC VCO has three control inputs; digital input for the capacitor array, analog coarse-tuning input V_C, and analog fine-tuning input V_F. The digital control for the capacitor array is the same as the conventional AFC method, while two analog inputs are similar to those of the dual-path LC VCO [3]-[5]. Different from the dual-path PLL, the analog coarse-tuning input is not a part of the PLL control path and controlled by the mixed-mode frequency-locked loop (FLL) circuit. Hence, depending on the initial settling condition of the coarse-tuning voltage V_C, different fine-tuning voltage V_F

978-1-4244-9019-6/11 $26.00 © 2011 IEEE

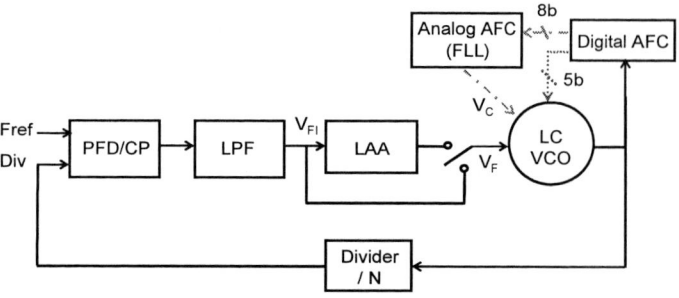

Fig. 3. Basic concept of the proposed Dual-path PLL.

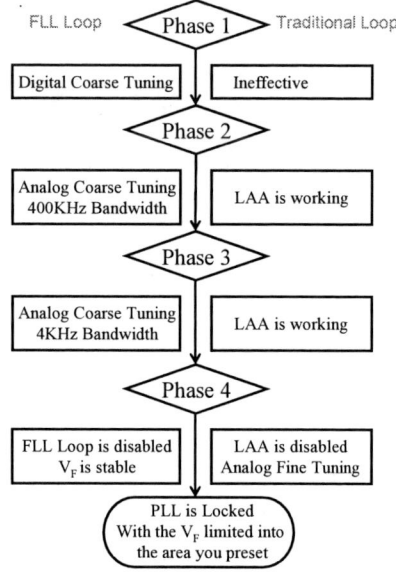

Fig. 4.   Dual-path PLL operation flow chart

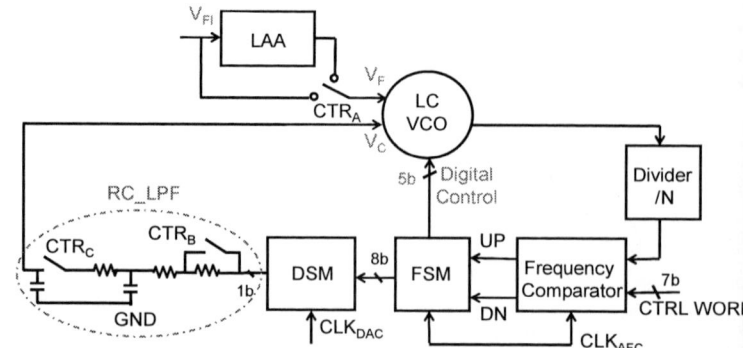

Fig. 5. Block diagram of VCO and hybrid AFC.

single control voltage as the V_F range is forced to a certain value. When the FLL is fast locked, the dual-path PLL enters the third phase operation mode. The FLL bandwidth is set to the normal value 4 KHz which is much lower than the PLL bandwidth at the time 200us so that the PLL stability is not affected. Then the dual-path PLL enters the fourth phase operation mode. At the same time 500us, the LAA and the FLL Loop are disabled, so the VCO is controlled by the loop filter as it is in the normal PLL operation. As a result, the PLL control voltage is close to the value which is preset by the LAA during the third phase. Since the bi-level output of the LAA can be easily made digitally programmable, self-ranged VCO control is realized by the proposed PLL method.

III.   IMPLEMENTATION

Fig. 5 shows a detailed block diagram of the hybrid AFC circuit and the 2GHz VCO in 0.18μm CMOS. Behavior mode is used for the digital blocks to verify the function of the proposed AFC method. The AFC circuit consists of the frequency comparator, the 13-bit FSM, the 2nd-order Δ–Σ modulator (DSM), and the RC low-pass filter (RC_LPF). In the first phase, only 5-bit output of the FSM is used for digital calibration of the VCO capacitor array. The 32 frequency bands controlled by the 5-bit digital input cover the frequency range of 1GHz. That is, each band has about 35MHz tuning range. After digital calibration is done, the mixed-mode FLL is enabled. At the same time, the PLL control voltage V_F is connected to the LAA with preset output range, which prevents the PLL from performing frequency acquisition. Since the fine-tuning path cannot provide sufficient frequency tuning, the FLL tracks the frequency change and settle the coarse-tuning voltage V_C. The 3dB corner frequency of the 2nd-order RC low-pass filter is first set to 400 KHz, which is higher than the PLL bandwidth of 200 KHz. When the FLL is locked, the 3dB corner frequency of the RC low-pass filter is set to 4 KHz. During the whole working process, the clock of the frequency comparator and FSM is 1MHz and the 2nd-order Δ–Σ modulator uses the 40MHz clock frequency. This change is done by switching the resistor value as depicted in Fig. 5.

can be obtained. A limiting amplitude amplifier (LAA) is employed to preset the V_F range and make the V_C stabilized first. Then, the LAA and the FLL Loop is disabled making the PLL stabilized with the corresponding V_F based on the previous V_C.

Fig. 4 shows a detailed operation flow. The first phase is the same as the conventional AFC procedure. The finite-state machine (FSM) outputs 5 bits from the MSB followed by the 5-bit capacitor array to select the desired frequency band out of 32 available frequency bands. At this moment, the mixed-mode FLL is not activated. When the right digital input of the capacitor array is determined, the FLL is enabled and the dual-path PLL scheme enters the second phase operation mode. During the second phase, the range of the fine-tuning voltage V_F is limited by the LAA. The LAA output is programmable so that the V_F range can be set to the preferred value. During this mode, the bandwidth of the FLL needs to be 400 KHz higher than the PLL. Hence, the dynamic bandwidth control is used for the FLL by controlling the 3dB corner frequency of the loop filter. Even though the PLL has two independent control voltages in the feedback loop, the FLL can find a

978-1-4244-9019-6/11 $26.00 © 2011 IEEE   168

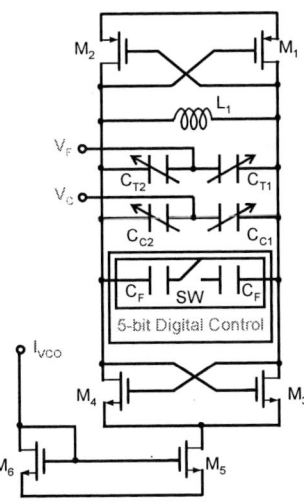

Fig. 6. LC VCO schematic.

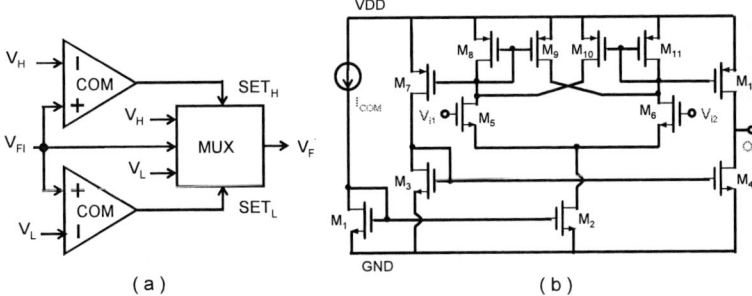

Fig. 7. (a) LAA schematic. (b) COM schematic

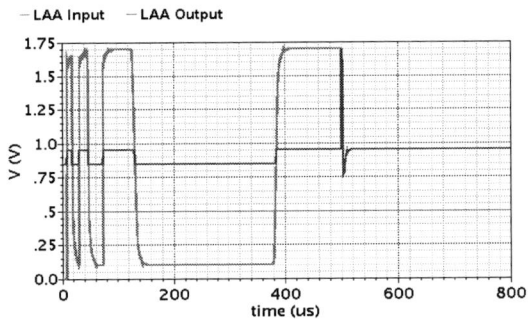

Fig. 8. LAA input and output waveforms.

In the FLL loop, the frequency divider can be shared with the main frequency divider in the PLL. The frequency comparator compares the divider output with the system clock frequency. Then, the output signals of the frequency comparator updates the FSM with the right control word. The 8-bit output from the LSB is fed to the $\Delta-\Sigma$ modulator which controls the coarse-tuning varactor after the RC filter for fine resolution and pseudo randomization. Hence, the latter part of the FLL is considered $\Delta-\Sigma$ digital-to-analog converter (DAC). When the coarse-tuning voltage V_C is stabilized, the FLL bandwidth is set to the normal bandwidth of 4 KHz. Since the FLL bandwidth is much lower than the PLL bandwidth, the FLL takes care of large-signal operation from now while the PLL tracks the small-signal information and determines the overall transfer function. As the PLL operates with the stabilized coarse-tuning voltage V_C, it mostly tracks the phase variation with minimal frequency acquisition. Therefore, the preset control voltage V_F does not change much and can be maintained even after the PLL bandwidth is higher than the FLL bandwidth. In this way, arbitrary control voltage V_F can be set for any output frequency.

Fig. 6 shows the schematic of the hybrid VCO, which is based on a complementary NMOS and PMOS cross-coupled negative-g_m topology. Two asymmetric spiral inductors are put in series to form symmetrical configuration. An accumulation-mode varactor is used for the analog fine-tuning path and an 8-bit weighted MIM capacitor array is designed for the coarse-tuning path. The MIM capacitor instead of the accumulation-mode varactor is used to reduce substrate noise coupling. The analog fine-tuning path is designed to have the nominal tuning gain of about 10MHz/V, and the digital coarse-tuning path is designed to have the nominal tuning gain of about 100MHz/V.

Fig. 7 (a) shows the schematic of the LAA which consists of two analog comparators and a digital MUX. Two threshold voltages set the upper and bottom range of the control voltage or V_H and V_L. Those threshold voltages can be made digitally programmable in the actual design. If the input voltage is

higher than the upper threshold voltage V_H, the first comparator output SET_H steers the MUX output V_F to the highest boundary value; if the input voltage is lower than the bottom threshold voltage V_L, the comparator output SET_L is to hold the MUX output V_F for the minimum boundary value. Fig. 8 shows the simulated input and output waveforms of the LAA with the preset range of 0.85-0.95V.

IV. PLL BEHAVIORAL SIMULATION

To verify the usefulness of the proposed architecture, closed-loop PLL behavioral simulations are done. Table I shows loop parameters used in the simulation.

TABLE I   SIMULATION PARAMETERS

Parameter	Value	Parameter	Value
Fref	8MHz	CLK_AFC	1MHz
I_CHP	320uA	CLK_DSM	40MHz
PLL BW	200kHz	CTRL	01111110 (62)
Fo	2GHz	RC	400KHz/4KHz
K_VCO,F	10MHz/V	K_VCO,C	100MHz/V

Fig. 9 shows transient settling behavior of the PLL with the V_F range of 0.85-0.95V with 1.8V supply voltage. It is assumed that digital calibration by the AFC is already done

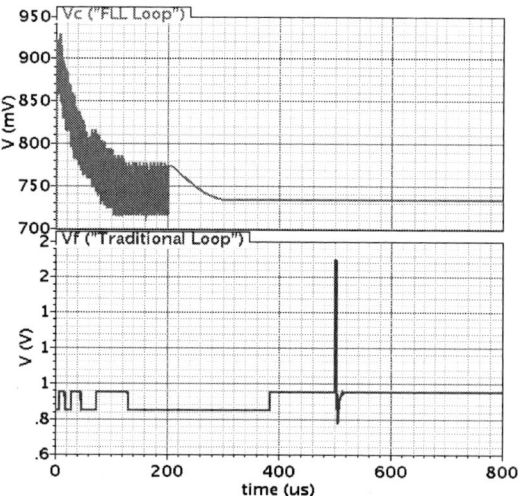

Fig. 9. Transient settling behavior for locking at 1.984GHz with 0.85-0.95V input range (V_F=0.95V and V_C=0.735V).

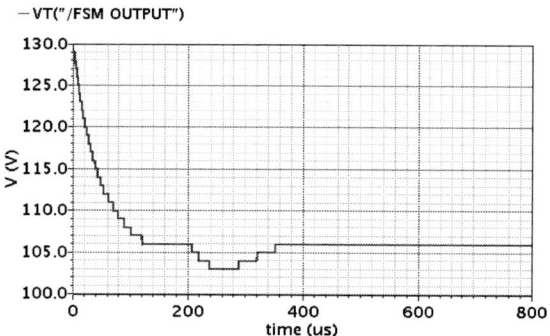

Fig. 10. FSM Output.

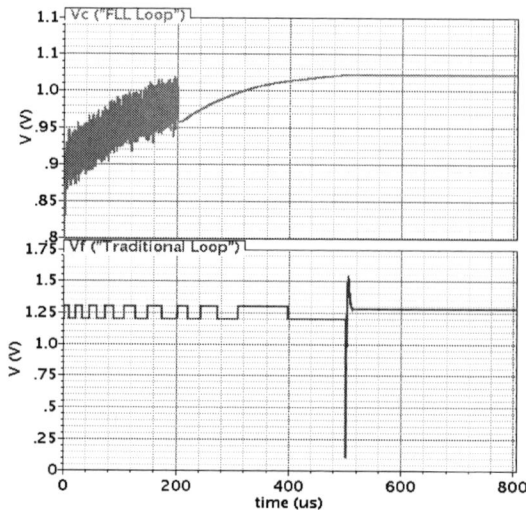

Fig. 11. Transient settling behavior for locking at 2.016GHz with 1.20-1.30V input range (V_F=1.28V and V_C=1.022V).

10MHz/V) and the V_C (0.122V x 100MHz/V). The simulation results show that the V_F range of 0.45-1.4V can be preset regardless of the output frequency.

V. CONCLUSION

The dual-path PLL method for continuously tunable and self-ranged LC VCO design is proposed. By combining the traditional digital AFC with the mixed-mode Δ–Σ FLL circuit, seamless frequency calibration is achieved. Since the proposed AFC method operates independently within the VCO, it enables separate control voltage range setting in the PLL. The benefit of controlling the VCO input voltage range is that we can choose the analog LPF voltage range to get the optimized performance for the varactor linear range or the charge pump matching performance. Behavioral simulation results show that the control voltage range can be preset from 0.45V to 1.4V for the same output frequency without affecting the normal PLL operation.

and that the phase 2 and 3 modes are verified in the simulation. At the beginning, the LAA sets the V_F range of 0.85-0.95V. The bottom plot shows V_C settling by the mixed-mode AFC circuit or FLL. The FLL with high clock frequencies and high 3dB corner frequency of the RC filter, the V_C quickly settles to around 0.75V. At 200μs, the FLL bandwidth is set to the normal bandwidth and the V_C quickly settles to 0.735V. At 500μs, the LAA and the FLL are disabled remaining the V_C of 0.735V. With reduced 3dB corner frequency of the RC filter from 400 KHz to 4 KHz, the V_C has much less voltage ripple. As shown in Fig. 9, the V_F has a sudden voltage jump during transition but reaches to the final value of 0.95V which is between 0.85V and 0.95V as expected. Fig. 10 shows the digital output of the FSM. The FSM output is stabilized to the value 106 showing that the digital part of the FLL works.

Fig. 11 shows transient settling behavior of the PLL for the same output frequency with the V_F preset range of 1.2-1.3V. The V_F settles to 1.28V while the V_C settles to 1.022V. The desired locking frequency is 2.016GHz by the V_F (0.38V x

REFERENCES

[1] T.-H. Lin and W. Kaiser, "A 900-MHz 2.5-mA CMOS frequency synthesizer with an automatic SC tuning loop," *IEEE J. of Solid-State Circuits*, vol. 36, pp. 424-431, Mar. 2001.

[2] H. Lee, *et al.*, "A Δ-Σ fractional-N frequency synthesizer using a wideband integrated VCO and a fast AFC technique for GSM/GPRS/WCDMA applications," *IEEE J. of Solid-State Circuits*, vol. 39, pp. 1164-1169, July 2004.

[3] R. Nonis, N. Da Dalt, P. Palestri, and L. Selmi, "Modeling, design, and characterization of a new low jitter analog dual-tuning LC-VCO PLL architecture," *IEEE JSSC*, vol. 40, pp. 1303-1309, June 2005.

[4] W. Rhee, *et al.*, "A uniform bandwidth PLL using a continuously tunable single-input dual-path LC VCO for 5 Gb/s PCI Express Gen2 application," in *Proc. IEEE A-SSCC*, Nov. 2007, pp. 63-66.

[5] Y. Sun, *et al.*, "Low-noise fractional-N PLL design with mixed-mode triple-input LC VCO in 65nm CMOS," in *Proc. IEEE Radio Frequency Integrated Circuits Symp.*, May 2010, pp. 61-64.

An Integrated HDTV Predictive Pixel Compensator for H.264/AVC Decoder

Ting-Chi Tong and Yun-Nan Chang
Department of Computer Science and Engineering
National Sun Yat-sen University
Kaohsiung, 80424, Taiwan.

Abstract—In this paper, a highly efficient pixel compensator architecture for the H.264/AVC standard is proposed which can provide both inter and intra prediction functions for luma and chroma components of pixels. By decomposing the algorithms used for both prediction methods into small micro-operation steps, a suitable common arithmetic unit architecture capable for performing these operations has been determined. Next, taking into account the possible reference sample transfer scenario, the overall compensator architecture consists of some buffers and multiple common arithmetic units considering is proposed. Since both arithmetic units and the intermediate data buffer for both inter and intra prediction processes have been shared, our integrated design can achieve more than 31% reduction of gate count compared with the sum of the separate designs. Our design can also lead to more than 37% saving of gate count compared with the previous designs. Our compensator can decode the videos up to HDTV resolution, and be applied for the dedicated H.264 hardware codec for various consumer devices.

I. INTRODUCTION

With the enormous growing number of multimedia applications, how to achieve better video compression has always been an important issue. In recent years, the proposal of H.264/AVC standard [1] has drawn a lot of attention as it can significantly outperform the previous video compression standards. It supports more flexibility in the selection of motion compensation block sizes as small as 4x4 and quarter-sample-accuracy which can lead to smaller residual energy after exploring the temporal redundancy. In addition, for the prediction of intra frame samples, it adopts more extrapolation modes to better explore the spatial correlation of samples based on various directions. Therefore, it has been shown [2] that H.264/AVC allows bit savings of about 50% compared to previous video coding standards like MPEG-1 and MPEG-2. With the improvement on the compression ratio of the new standard, the computational complexity also increases significantly. Therefore, how to design the efficient hardware accelerator especially for the core pixel compensation process has become an important issue. In the past, many designs have been proposed which focus on optimizing the implementation of the individual inter-prediction or the intra-prediction function. However, since during the video decoding process, the application of inter and intra prediction function is exclusive such that if both designs can be efficiently integrated together, the overall hardware cost can be reduced. Although both prediction modes have been realized in some literature [3], [4], [5], [6] the actual hardware sharing and integration of both

functions can only be found in [7]. In this paper, we propose another different architecture called *integrated predictive pixel compensator* (IPPC) by exploring the common sharing of functional units as well as data storages used for both inter and intra prediction modules.

The rest of the paper is organized as follows. Section II first discuss the algorithms of inter and intra prediction function. By decomposing the algorithms into a series of sub-steps, the common data-path architecture can be determined and illustrated in Section III. The implementation result of the proposed IPPC and its comparison with other designs will be given in Section V. Finally, some conclusion of this paper is given in Section VI.

II. PIXEL PREDICTION ALGORITHM

This section will review the algorithms used in the pixel prediction for inter and intra frames in order to find out the possible common data-path for the integrated predictor design.

A. Inter prediction process

Inter prediction is used to predict a block of pixels from one of more previously encoded video frames or fields according to the given location offset denoted as the motion vector. The motion vector can be non-integer, and the minimum resolution allowed is a quarter pixel. For non-integer vector, the samples have to be produced by interpolating the neighboring reference samples. For example, the luma component of the pixels with horizontally half-pixel offset $P_{i+\frac{1}{2},j}$ can be computed according to the following equation:

$$P_{i+\frac{1}{2},j} = \begin{aligned}(r_{i-2,j} - 5 \times r_{i-1,j} + 20 \times r_{i,j} + 20 \\ \times r_{i+1,j} - 5 \times r_{i+2,j} + r_{i+3,j} + 16) >> 5\end{aligned} \quad (1)$$

where $r_{i,j}$ represents the reference sample at the integer coordinate (i, j). As for the pixels with quarter-pixel offset, they can be computed by simply taking the average of the neighboring reference and predicted half-pixel samples obtained by (1).

To generate one half-pixel sample based on (1) requires six multiply-accumulation (MAC) operations. Consider the general video systems with 32-bit memory and bus-width, the required six input reference samples are located in three different but successive memory words, it takes at least three cycles to obtain these samples. Therefore, considering the hardware utilization due to the data transfer issue, the original

978-1-4244-9019-6/11 $26.00 © 2011 IEEE   171

six-tap filtering operation in (1) now will be decomposed into the following three micro-steps:

$$
\begin{aligned}
s1: \quad t &= r_{i-2,j} - 5 \times r_{i-1,j} + 16 \\
s2: \quad t &= t + 20 \times r_{i,j} + 20 \times r_{i+1,j} \\
s3: \quad P_{out} &= (t - 5 \times r_{i+2,j} + r_{i+3,j}) >> 5
\end{aligned}
\tag{2}
$$

Each micro-steps defined in (2) can be performed at different clock cycles as illustrated in Fig. 1(a). Similarly, the column-wise filtering operation can also be realized by grouping two required input samples in the same column together to be executed by the same 2-tap MAC unit every other two cycles. It seems that the throughput of the column-wise filtering will be only a half compared with the row-wise; however, it can be easily doubled by interleaving the computation of another output sample into those idle cycles. Some data-alignment circuits have to be used to buffer and align the received reference samples from the bus according to the required reference sample order shown in Fig. 1.

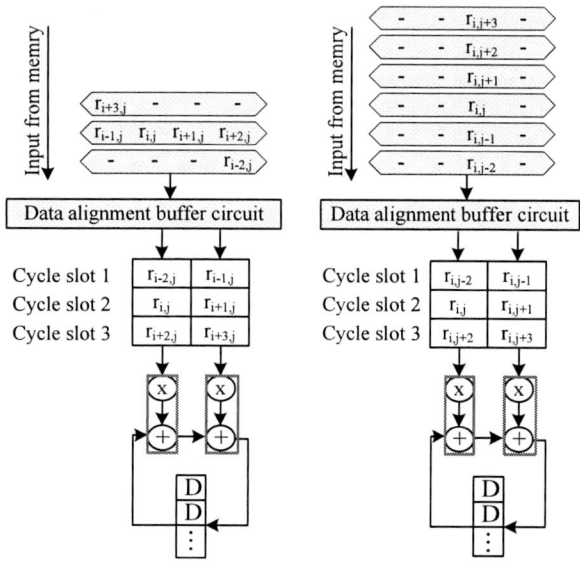

Fig. 1. An example of proposed filter architecture: (a) the processing schedule pattern for horizontal filtering, (b) the processing schedule pattern for vertical filtering.

To predict the chroma component of the non-integer pixels $P_{i+dx,j+dy}$ applies different distance weighting equation based on the four corner neighboring integer pixels as follows:

$$
\begin{aligned}
P_{i+dx,j+dy} = \quad & [d_x d_y r_{i+1,j+1} + (8 - d_x) d_y r_{i,j+1} \\
& + (8 - d_x)(8 - d_y) r_{i,j} \\
& + d_x (8 - d_y) r_{i+1,j}] >> 6
\end{aligned}
\tag{3}
$$

where dx and dy represent the fractional part of pixel coordinate which have the minimum resolution of 1/8. Direct implementation of chroma interpolation may require eight multiplications per sample. However, (3) can also be decomposed into a series of computations as follows:

$$
\begin{aligned}
s1: \quad t1 &= 8 \times r_{i,j} + (r_{i+1,j} - r_{i,j}) \times d_x \\
s2: \quad t2 &= 8 \times r_{i,j+1} + (r_{i+1,j+1} - r_{i,j+1}) \times d_y \\
s3: \quad P_{out} &= [8 \times t1 + (t2 - t1) \times d_y + 32] >> 6
\end{aligned}
\tag{4}
$$

The number of multiplication can then be reduced from 8 to 6. The possible value of dx and dy is from zero to seven. Each step in (4) can also be realized by a 2-tap MAC unit in one cycle.

B. Intra prediction process

Instead of using the samples of the other frames to predict the pixels, the intra interpolation is used to extrapolate the pixels from the neighboring samples of the same frame. As mentioned before, H.264 standard provides flexible size of the basic processing block. Therefore, the algorithm used to predict the intra samples varies with the size of block which can be luma 4x4, luma 16x16, or chroma 8x8. Fig. 2 illustrates all the possible extrapolation approaches used in intra prediction. For the intra prediction of the luma 4x4 block,

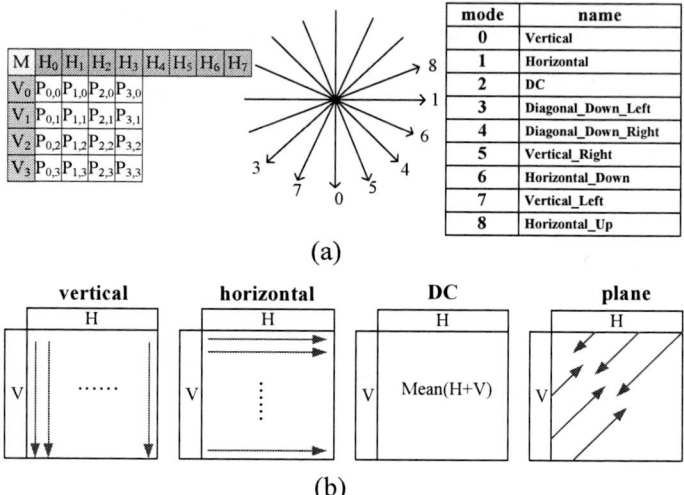

Fig. 2. Various intra prediction modes provided for (a) 4x4, (b) 16x16 and 8x8 blocks.

there are total 13 reference samples being used which are labelled as $V_0 \sim V_3$, M and $H_0 \sim H_7$. There are nine modes provided which can be used to predict pixels in different directions. By categorizing the extrapolation equations [1] of these modes, it can be found that any pixels $P_{i,j}$ in the predicted 4x4 block can be computed based on either one or the combination of the following equations:

$$
\begin{aligned}
P_{out} &= (r_1 + 2 \times r_2 + r_3 + 2) >> 2 \\
P_{out} &= (r_1 + r_2 + 1) >> 1 \\
P_{out} &= r_1
\end{aligned}
\tag{5}
$$

where r_i represents one of the corresponding reference samples. For example, the luma pixels under the DC mode can be computed as:

$$
P_{i,j} = \quad (H_0 + H_1 + H_2 + H_3 + V_0 + V_1 + V_2 + V_3 + 4) >> 3
\tag{6}
$$

which can be realized by executing the second equation listed in (6) four times. Each equation can be realized by a simple arithmetic unit which contains three adders.

The methods used for the intra prediction of the other two block sizes as shown in Fig. 2(b) are similar to luma 4x4

978-1-4244-9019-6/11 $26.00 © 2011 IEEE

except for one special extrapolation approach called *plane mode* which is the most complex equation to implement so far. The extrapolation equation applied in plane mode for luma 16x16 is shown as follows:

$$P_{i,j} = \quad clip1((a + b \times (i - 7) + c \times (j - 7) \\ +16) >> 5) \quad 0 \le i, j \le 16 \quad , \tag{7}$$

where the coefficients a, b, and c will remain constant for the entire 16x16 block. However, they have to be first derived from the neighboring reference samples. The computation of these coefficients can also be decomposed into several micro-steps. After obtaining these coefficients, the computation procedure according to (7) becomes relative simple. It can be computed incrementally by one of the following equations:

$$\begin{aligned} P_{i+1,j} &= \quad P_{i,j} + b \\ P_{i,j+1} &= \quad P_{i,j} + c \end{aligned} \tag{8}$$

Therefore, after computing the value for the corner pixel $P_{0,0}$ which equals to $clip1((a - 7b - 7c + 16) >> 5)$, the remaining pixels can be computed iteratively by adding an offset of either b or c to its previously predicted neighboring pixels. In summary, we can realize (7) in multiple steps by decomposing it into the combination of the following equations:

$$\begin{aligned} t &= coef \times r_1 + t \\ t &= r_1 \pm r_2 \\ t &= (5 \times r_2 + 32) >> 6 \\ t &= (r_1 - r_2 + 16) >> 5 \end{aligned} \tag{9}$$

where $coef$ can be 16 or any number from one to eight.

III. PROPOSED IPPC DESIGN

In previous section, the algorithms used to predict the inter or intra pixels for both luma and chroma component have been reviewed. The original equations used to predict the pixels have been decomposed into the combination of some basic functions based on the concern of either the input data transfer bandwidth or the computational complexity. Based on the decomposition results shown in (2) (4) (5) (9), it can be found that each basic function involves no more than three reference samples, and it will either add a constant or the previous temporary result followed by some shift function. Therefore, these basic functions can be reorganized as the following equation form:

$$\begin{aligned} temp = \quad &[c_1 \times r_1 + c_2 \times r_2 + c_3 \times r_3 \\ &+const \,(or\; temp)] >> SV \end{aligned} \tag{10}$$

where r_i represents the reference sample while c_i represents the coefficients. The variable $temp$ and $const$ represent the temporary computational result, and one of the predefined constants. Finally, SV denotes the scaling factor used for various prediction algorithms. Based on (10), a unified accumulation-based arithmetic unit capable for providing all the basic function is proposed and shown in Fig. 3. This unified unit consists of only four adders without the multipliers. It should be noted that although in (10), it seems a general three-tap MAC unit will be required. However, the coefficients c_i are not general variables. For most of time, they are some simple

constants; only on occasion either c_1 or c_2 can be a variable input. However, the range of the variable value is from zero to eight, so all these coefficients can be realized by simple shift-and-add design and controlled by *ctrl* signals in Fig. 3.

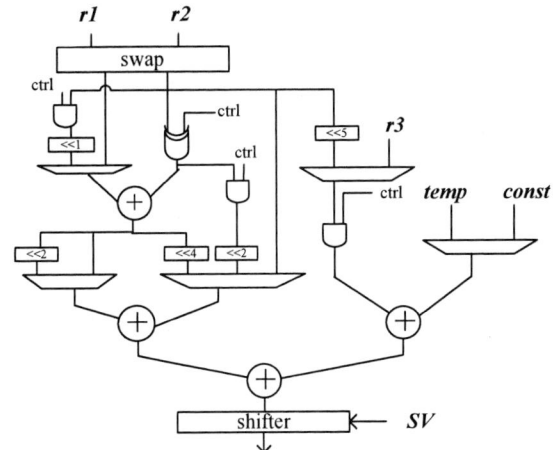

Fig. 3. The architecture of the proposed unified arithmetic unit.

From the discussion given in Section II, it can be found that the inter prediction for the luma component of the pixel requires the most computational demand. In addition, it also requires the most average number of the reference samples. Therefore, the overall design of IPPC should consider the performance of the inter-luma prediction processing first. According to the simple diagram shown in Fig. 1 or the practical MAC arithmetic unit shown in Fig. 3, the interpolation throughput can be only up to one sample per three cycles which is too slow for most of the general video applications. The similar design idea to Fig. 1 can be extended in parallel to support faster throughput. Assuming the luma interpolation for a 4x4 block of pixels, the nine reference samples used for the row-wise filtering processing are distributed into three different 32-bit memory words. It takes at least three cycles to obtain the nine samples from the memory. Therefore, under the bandwidth restriction of the reference samples, the ideal processing throughput of the interpolator should be four samples per three cycles. Therefore, the filtering architecture shown in Fig. 1 or Fig. 3 can be unfolded by a factor of four to obtain the block-level interpolator module. Furthermore, since inter prediction may require both horizontal and vertical interpolations, the unfolded architecture can be further duplicated as shown in Fig. 4 to allow both directions of interpolation being executed in parallel.

The overall IPPC shown in Fig. 4 consists of two processing cores $G1$ and $G2$, and each core contains four arithmetic units like the one shown in Fig. 3. The scheduling of $G1$ and $G2$ cores used for inter-prediction is similar to the one shown in [8]. The *row-wise buffer* is used to buffer the reference samples per row received from the system bus while the *intra buffer* stores neighboring samples of the extrapolated pixels. The *data alignment and selection units* are responsible for selecting data for further processing. The processing core $G1$

978-1-4244-9019-6/11 $26.00 © 2011 IEEE

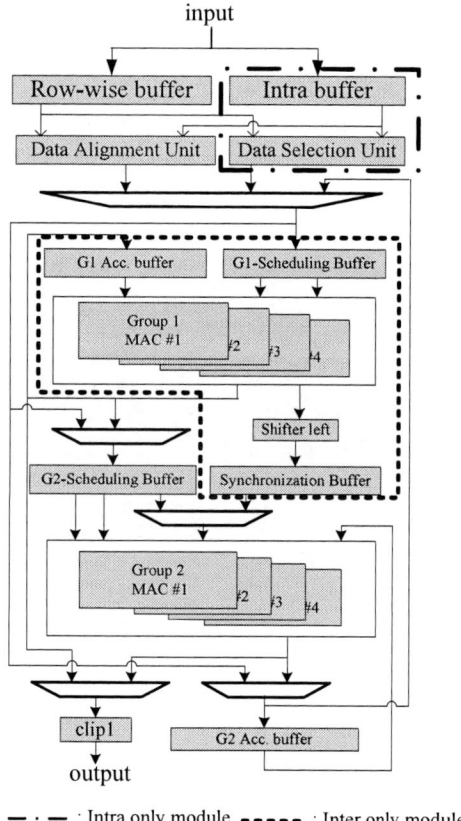

Fig. 4.   The overall block diagram of the proposed IPPC architecture

— · — : Intra only module   ▪▪▪▪ : Inter only module

is responsible for the computation of horizontal interpolation of inter prediction. The other core $G2$ is responsible for the vertical interpolation of inter prediction, and the intra prediction. The *accumulation buffer* is used to store the temporary computation results while *scheduling buffer* is used to store the input reference samples which will be used again for $G2$ core.

IV. EXPERIMENTAL AND COMPARISON RESULT

The proposed IPPC design has been implemented in Verilog HDL and synthesized using Synopsys Design Compiler with TSMC $0.18\mu m$ cell library. It passes the gate level simulation with 100MHz. The required number of processing cycles are from 160 to 480 per macro-block for inter prediction, and from 67 to 192 per macro-block for intra prediction. For the general test videos, the processing throughput of our IPPC can support the decoding of video resolution up to HDTV (1920x1088). The total number of gate count is 33.5K. If we further consider the individual function of inter or intra prediction only, the required gate count can be found in the last column of Table I. The marginal cost for extending the circuit for inter prediction only to provide both prediction functions is only 5.9K. In overall, the integrated design can lead to about 31% saving of gate count compared with the sum of both separate designs. Table I also compares the proposed integrated architecture with other designs reported in the literature which can also support both inter and intra prediction functions. As it can be found

that our design non only can achieve good performance for HDTV resolution, but it is also the most compact design. The only integrated design in the past [7] does not provide the actual implementation result. However, the number of cycles it takes for inter prediction is 608 cycles per macro block which is longer than our proposed.

TABLE I
COMPARISON OF VARIOUS DESIGNS WHICH PROVIDE BOTH INTER AND INTRA PREDICTION FUNCTIONS.

	[3]	[4]	[9]	[10]	[11]	Ours
Tech.	.18um	.13um	.25um	N/A	0.18um	.18um
Library	TSMC	UMC	N/A	N/A	TSMC	TSMC
Fre.(MHz)	87	250	54	37	120	100
Max res.	1920x 1080	1920x 1080	352x 288	640x 480	2048x 1024	1920x 1080
Inter#	41.1K	50.8K	80.5K	47.1K	69.6K	27.6K
Intra#	12.7K	26K	32.1K	24.6K	28.7K	21K
Total#	53.8K	72.8K	112.6K	71.7K	98.3K	33.5K
Integrated	N	N	N	N	N	Y

V. CONCLUSION

In this paper, an integrated VLSI design for both inter and intra pixel prediction has been proposed. Both prediction circuits have shared the fundamental arithmetic units and some internal buffer as well. The integration of both circuits can achieve more than 31% saving in the gate count compared with the separate designs. Our IPPC design can support the decoding of H.264 video up to HDTV resolution. Compared with the previous designs, our design approach can lead to over 37% saving of the gate count.

REFERENCES

[1] Joint Video Team (JVT) of ISO/IEC MPEG and ITU-T VCEG, *Draft ITU-T recommendation and final draft international standard of joint video specification*, JVTG050, May 2003.

[2] J. Ostermann, J. Bormans, P. List, D. Marpe, M. Narroschke, F. Pereira, T. Stockhammer, and T. Wedi, "Video coding with H.264/AVC: tools, performance, and complexity," *IEEE CAS Magazine*, vol. 4, no. 1, pp. 7–28, First Quarter 2004.

[3] J. W. Chen, C. C. Lin, J. I. Guo, and J. S. Wang, "Low complexity architecture design of H.264 predictive pixel compensator for HDTV application," in *Int. Conf. on Acoustics, Speech and Signal Processing (ICASSP06)*, Toulouse, France, May 2006, pp. 932–935.

[4] M. Alle, J. Biswas, and S. K. Nandy, "High performance VLSI implementation for H.264 Inter/Intra prediction," in *Int. Conf. on Consumer Electronics*, Las Vegas, USA, Jan. 2007, pp. 1–2.

[5] K. Xu and O. C. sing Choy, "A power-efficient and self-adaptive prediction engine for H.264/AVC decoding," *IEEE Trans. VLSI Syst.*, vol. 16, no. 3, pp. 302–313, 2008.

[6] X. Wang, X. Cui, and D. Yu, "A power-efficient prediction hardware architecture for H.264 decoding," in *ICIEA*, June 2010, pp. 2121–2126.

[7] C. H. L. et al, "An unified systolic architecture for combined inter and intra predictions in H.264/AVC decoder," in *IWCMC*, Vancouver, July 2006.

[8] Y.-N. Chang and T.-C. Tong, "An efficient design of H.264 inter interpolator with bandwidth optimization," *Journal of Signal Processing Systems*, vol. 53, no. 3, pp. 435–448, 2008.

[9] S. Park, H. Cho, H. Jung, and D. Lee, "An implemented of H.264 video decoder using hardware and software," in *IEEE Custom Integrated Circuits Conference*, California, USA, Sept. 2005, pp. 271–275.

[10] S. Yoon, S. Park, and S. Chae, "Implementation of a H.264 decoder with template-based communication refinement," in *2006 IEEE Asia Pacific Conference on Circuits and System*, Singapore, Dec. 2006, pp. 570–573.

[11] T. C. Chen, C. J. Lian, and L. G. Chen, "Hardware architecture design of an H.264/AVC video codec," in *ASP-DAC*, Jan. 2006, pp. 750–757.

978-1-4244-9019-6/11 $26.00 © 2011 IEEE

IBM zEnterprise™ Energy Efficient 5.2Ghz Processor Chip

Authors: H. Wen[1], J. Warnock[2], Y. Chan[3], G. Mayer[4], B. Truong[1], T. Strach[4], T. Slegel[3], S. Carey[3], G. Salem[3], F. Malgioglio[3], D. Malone[3], D. Plass[3], B. Curran[3], Y.-H. Chan[3], M. Mayo[3], W. Huott[3], P. Mak[3]

1. IBM Systems and Technology Group, Austin, TX
2. IBM Systems and Technology Group, Yorktown Heights, NY
3. IBM Systems and Technology Group, Poughkeepsie, NY
4. IBM Systems and Technology Group, Boeblingen, Germany

Abstract— **The IBM zEnterprise z196 processor chip is an energy efficient high-frequency, high-performance design that implements 4 processor cores optimized for maximum single-thread performance. Chip energy efficiency is improved by 25% compared to the previous 65nm design, which enables the processor chip to run at product frequency of 5.2 GHz, providing a significant performance boost for the z196 system.**

This paper discusses the enablement of a high frequency and high performance design with a focus on energy consumption challenges and solutions. Design tradeoffs for processor speed, performance and energy consumption were optimized during project concept phase and practiced through detailed implementation stages on all aspects of the processor design. Various high speed circuit techniques were deployed to achieve the high frequency goal and improve overall energy efficiency. A comprehensive power methodology was developed to calculate leakage and dynamic power dissipation at various workloads. Sustained thermal power as well as instantaneous peak power was analyzed and worked on throughout the entire design process. The final design stays within the system power constraints and achieves a 5.2Ghz product frequency.

I. INTRODUCTION

THIS paper is focused on the work that was dedicated to improve energy efficiency of the z196 processor chip. The energy efficiency improvement was a key contributor for the delivery of the 5.2Ghz, high performance and high reliability z196 processor chip [1-3]. The z196 high-end system node is hosted on a glass ceramic multichip module (MCM) which comprises six four-core processor (CP) chips and a pair of system controller (SC) chips. Compared to the MCM for the previous system, the z10 [4-5], the z196 MCM has one more CP chip. Although the z196 CP chip has the same number of cores as the z10 chip, each core contains 50% more logic and needs to support a frequency goal that is 20% higher than that of the z10 core. In addition, new design features were added to ensure and improve z196 system reliability [1-3].

However, the z196 MCM power consumption limit, set by the system cooling constraints remains the same as its predecessor. The design team accepted the challenge and took unprecedented actions to reduce the processor power consumption. As a result, energy efficiency of the processor chip for the z196 system is improved by 25% compared to its predecessor.

II. PROCESSOR CHIP OVERVIEW

Fig. 1 is the floorplan overview of the z196 core processor chip. The chip has 4 processor cores. Each core has a 64KB Icache, a 128KB Dcache and is accompanied by a private 1.5MB SRAM L2 cache, and 24MB of DRAM L3 cache is shared by all 4 cores. The CP chip also contains 2 co-processors (CoP), each shared by a pair of cores for data encryption and compression.

In addition, as shown in fig. 1, there is a DD3 (double-data-rate three) memory controller (MCU), an IO bus controller (GX) and 2 sets of SC IO busses (SC0/1 IOs) that communicate between each CP and the pair of SC chips. The memory port (MCIO) on the left edge communicates through high-speed differential buses to DIMMs [3] operating at 4.8 GT/s (billion transfers per second). The CP-SC SMP fabric buses and GXIO buses on the right edge are unidirectional and single-ended, operating at 2.6GT/s.

The CP chip is manufactured in a 45nm SOI CMOS technology similar to that used by other IBM servers [6-8]. The resulting CP chip has a die size of $512mm^2$ and contains 1.4B transistors.

Power work started with chip voltage domain planning. The processor chip as shown in fig. 1 has 5 supply voltage rails. A primary rail, referred to as Vdd (not shown in fig. 1), is distributed across the entire chip and supplies all logic devices. All SRAM and DRAM cells receive a 2nd rail, referred to as Vcs. Differential MCIOs are supplied with their own 2nd rail Vmem, while single-ended SCIOs and GXIOs also receive a 2nd rail, Vio. The 5th rail is used to supply pervasive logic that is needed during chip test and diagnostics. It should be noted Vmem, Vio and the pervasive supply are shared between all 6 CP chips and 2 ES chips on a MCM, while each processor chip has its own logic supply Vdd and SRAM supply Vcs. The availability of chip unique Vdd and Vcs rails minimizes the chip power for the required frequency goal.

978-1-4244-9019-6/11 $26.00 © 2011 IEEE

III. POWER ANALYSIS METHODOLOGY

A comprehensive power methodology was developed to calculate leakage and dynamic power dissipation at various workloads and track power reduction progress. Analysis of chip thermal power running various workloads as well as instantaneous peak power was critical to ensure a robust design of chip power grids as well as the system power delivery structure.

Fig. 2 shows the flow of the power analysis methodology used for z196 chip design as well as for other IBM server designs [6-8]. The analysis started with the contents of the previous design, scaling the data according to the new design requirements as well as the new process technology parameters including device leakage and capacitance.

The power was partitioned into 2 major components: switching (AC) power and leakage (DC) power. The AC and DC power was calculated for each transistor according to its switching and biasing conditions. Furthermore, the effective switching capacitance (Ceff), as an AC power parameter, and the effective leaking width, as a DC power parameter, were characterized and stored for each macro, where the macro, containing a collection of circuits, is the smallest placeable object on the chip floorplan. This way, chip AC and DC power could be easily calculated at any given voltage, frequency and temperature condition.

The initial data is labeled as the contract stage data, shown on the left side of fig. 2. The contract stage data represents a power target for each logical or physical partition. The sum of all partition targets is the target for the chip which needs to fit within the system power constraints with the chip at its frequency goal.

As the design progressed, actual design data was collected, as illustrated on the right side of fig. 2. The actual power data included a full physical design (PD) netlist, signal activity data resulting from logical simulation and the macro level power models. The PD netlist included all manufacturable devices. Activity data from logic simulations varied with the workload.

The power calculation engine shown in the middle of Fig. 2 takes the input either from the contract stage or actual design data and calculates AC power using

$$AC\_Power = 1/2 \times Ceff \times VDD^2 \times F \times SF \qquad (1)$$

Where VDD is the voltage value a macro receives, F is the clock frequency. SF is the average switching factor of all the signals driving the macro input ports. Note that SF is characterized relative to the macro clock cycle and therefore ranges from 0 to 1 for all static signals which change their states only once every clock cycle. By this definition, the SF of the clock net is 2.

DC power was calculated using a leakage model provided by the process technology team. The resulting power data at thermal, peak and test conditions were used for chip and package power grid design.

IV. POWER REDUCTION

In order to meet MCM power constraints, chip power was analyzed and budgeted during project concept planning. At the micro-architecture level, the length of a single pipeline stage, as measured in technology-independent fanout-of-4 inverter delay units, was kept the same as the previous design [4-5], in order to leverage the full technology improvement for frequency improvements.

Chip clock grid design and distribution has a significant impact on the chip power. In order to reduce clock grid power, the z196 processor chip has multiple clock grids as shown in fig. 3. Highest frequency grids only cover the 4 core areas, while the majority of chip area is covered with a grid running at ½ the core grid frequency. This resulted in about a 30% clock grid power reduction compared to the previous design with one single frequency grid. It should be noted that MCIO as well as GXIO regions have separated grids that support asynchronous operations as required by the system architecture.

The use of on-chip DRAM for the L3 cache drastically improved cache power efficiency while simultaneously increasing bandwidth and performance [9]. For the equivalent amount of chip area, embedded DRAM reduces leakage power by 60% at triple the cache capacity and at a higher performance level compared to a conventional SRAM cache implementation. This led to about a 4% overall reduction of the total chip power.

From the logic design perspective, fine grained clock gating was implemented throughout core, nest and cache designs, which enables clocks of any latch group to be turned on or off on a cycle-by-cycle granularity. Average clock gating for the processor core exceeds 60% for typical performance workloads. Special focus was given to array structures to reduce power consumption. Array clocks only run while arrays are being accessed for read or write. For typical workloads, dynamic clock gating reduces total chip switching power by 25%.

On the physical design side, the pervasive use of pulsed clocks for clocked storage elements (CSE) [10] as well as fine tuning of the local clock networks led to an overall CSE power reduction of about 15% compared to the previous design which had already started pulsed-clock CSE usage.

Leakage power was worked down by employing MOSFETS with multiple threshold voltages (Vt). In general, low threshold voltage MOSFETs are higher speed, but dissipate more leakage power. In order to achieve an optimal tradeoff between speed and leakage, 4 devices with threshold voltage characterized as super-high (SVT), high (HVT), regular (RVT) and low (LVT) were implemented on the z196 CP chip. Less than 1% of devices were low threshold voltage devices, which were only used on the most timing critical paths. The selective usage of multiple threshold voltage devices reduced total chip leakage by about 30%, proving to be a power efficient way to improve processor speed.

A semicustom design flow [11] enabled rapid tuning for

timing and power optimization at the near end of the design closure. The flow includes Vt swapping of low level logic gates and adjustment of individual device widths for timing and power tuning. Automated design tuning resulted in about 10% reduction of the total physical design power.

On the array design side, innovative ideas reduced power consumption without sacrificing the high frequency goal. Fig. 4 describes the array structure change from the prior-art with bit interleaving fig. 4 a) to the current design with set-interleaving fig 4 b). For an array read, the arrangement in fig. 4 a) activates a total number of 112 bits in two rows from the upper and lower subarrays, while the new arrangement in b) only activates 56 bits from one row of the two subarrays. This structure change reduced the number of column bits being activated as well as the number of active wordlines by 50%, thus reducing switching power by 50% [12]. Note the new structure in fig. 4 b) also eliminates bit decoding logic, as the bit selection information is embedded in the expanded wordline addresses. The Dcache, Icache, Directory, and Translation Look-aside-Buffer (TLB) arrays on the z196 processor chip were implemented with this new array structure.

Fig. 5 a) shows the resulting leakage power breakdown by device type. The processor is mainly designed with HVT devices. LVT leakage is only 0.4% of the total chip leakage. The total leakage of a chip at nominal process running typical workloads is approximately 30% of the total chip power. Fig. 5 b) shows the total chip power (AC+DC) breakdown by component. Note that, due to the high operating frequency, about 70% of the total power is consumed by the 4 cores and core clock grids.

V. POWER GRID NOISE REDUCTION

The z196 processor chip contains more than 10 microfarads decoupling capacitance using deep trench (DT) capacitors. The DT capacitors were fabricated with the same process technology as the on-chip DRAM and provided about 25 times more capacitance than the traditional oxide capacitors. As a result, switching noise on voltage supply rails is reduced significantly. Fig. 6 a) shows the reduced noise on the z196 voltage rail compared to the noise observed on z10 processor chip b). It should be noted that leakier chips have less switching noise due to the noise damping effect created by leakage. Comparisons at equivalent leakage levels show that the reduction from z10 to z196 can be as large as 50% due to the 20x increase in decoupling capacitance on z196 chip. The reduction in the supply noise led to about 15% chip power reduction at the targeted frequency.

VI. CHIP POWER DATA

Fig. 7 shows the relative processor power measured in an actual system setting running at the 5.2 Ghz product frequency and a maximum activity workload. For each z196 processor chip, the logic and array supply voltages were adjusted according to the process speed to achieve the constant

frequency goal and minimize the chip power. As a result, chip power is almost constant across the process speed range as shown. The fluctuation in the data represents leakage variation at a given process speed

Figure 8 shows z196 processor power variation running various workloads. Max power was characterized with a workload that produced the maximum core activity, while typical power is what will be observed by customers running actual workloads. Typical power is about 85% of the max power.

It should be noted that the z196 processor maintains the same power saving mechanism, via millicode control, that was first implemented in z990 system. This power saving feature is invoked when millicode detects that a core is in the architected wait state, and power saving bits are turned on which halt instruction execution, thereby automatically turning off most running clocks and arrays. The power saving due to this mechanism is illustrated by the power difference between Quasi-idle and final Idle state. An architected wait state occurs when an Operating System has no useful work for a core, or for all spare cores, or after the initial microcode load (IML) of the system.

VII. CONCLUSION

We improved the processor energy efficiency by 25% despite high frequency and high RAS requirements. The power analysis methodology, first introduced here to system z processor development has proven its validity and accuracy. At the system level, z196 system power stayed within the same power envelop as z10, while the system capacity per watt is improved by more than 70% [13]

VIII. ACKNOWLEDGMENT

We thank the entire dedicated system z design team around the globe. We also thank the IBM technology development, manufacturing, as well as the IBM electronic design automation team to make this project a success.

REFERENCES

[1]   J. Warnock et al, "A 5.2 GHz Microprocessor Chip for the IBM zEnterprise System", ISSCC 2011, paper 4.1

[2]   B. Curran, "The Next-generation System z Micro-Processor, Hot Chips 2010,

[3]   B. Curran, "The zEnterprise 196 System and Microprocessor", to be published in EEE Micro, 2011

[4]   C. K. Shum et al. , "Design and Microarchitecture of the IBM System Z10 Microprocessor" IBM J. Res. & Dev. Vol. 53 Issue 1 (August 2008)

[5]   C.F. Webb, "IBM z10: The Next-Generation Mainframe Microprocessor," IEEE Micro, vol. 28, no. 2, 2008, pp. 19-29

[6]   D. Wendel et al, "Implementation of POWER7™: A Highly Parallel and Scalable Multi-Core High-End Server Processor", ISSCC 2010, p. 102

[7]   D. Wendel et al, "POWER7™, a Highly Parallel, Scalable Multi-Core High End Server Processor", JSSC, Jan. 2011, p. 145

[8]   D. Wendel et al, The Power7TM Processor SoC, ICICDT, 2010, p. 71

[9]   J. Barth et al, "A 45nm SOI Embedded Dram macro for the POWER7™ Processor 32MByte On-Chip L3 Cache", ISSCC 2010, 19.1

[10]   J. Warnock et al, "POWER7™ Local Clocking and Clocked Storage Elements", ISSCC 2010, p. 178

[11] G. Northrop et al., ''A Semi-custom Design Flow in High-Performance Microprocessor Design,'' Proc. 38th Design Automation Conf. (DAC 01), ACM Press, 2001, pp. 426-431.

[12] Y. Chan, US Patent Pending.
[13] [13] M. Andres et al, "zEnterprise energy efficiency and energy management improvements" to be published in IBM J. Res. & Dev. 2011

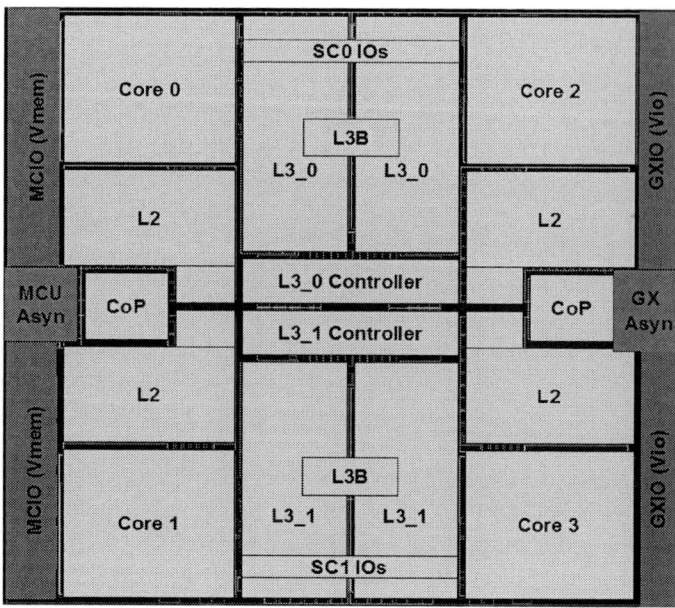

Figure 1: Floorplan overview of z196 processor chip

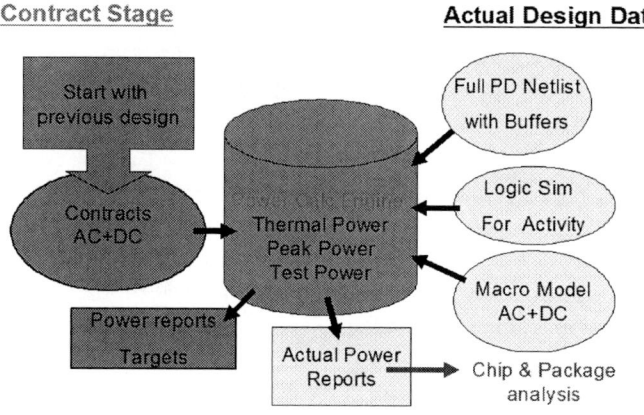

Figure 2: Power Analysis Methodology Flow

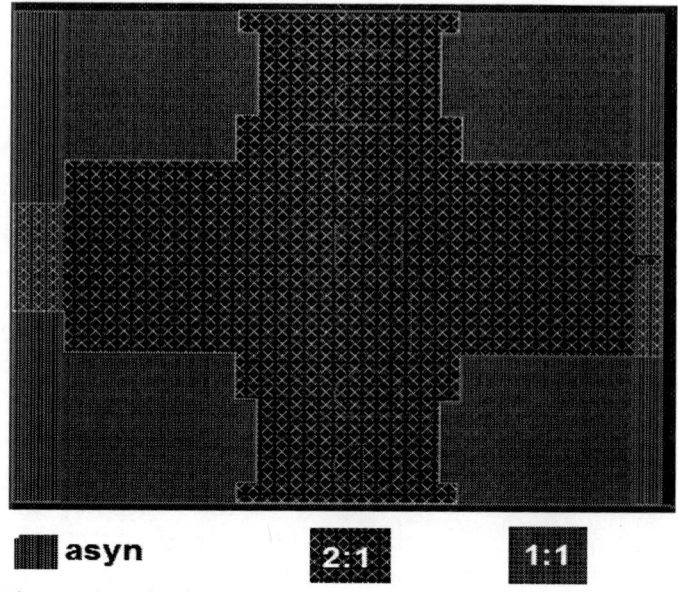

Figure 3: Clock grids of z196 processor chip

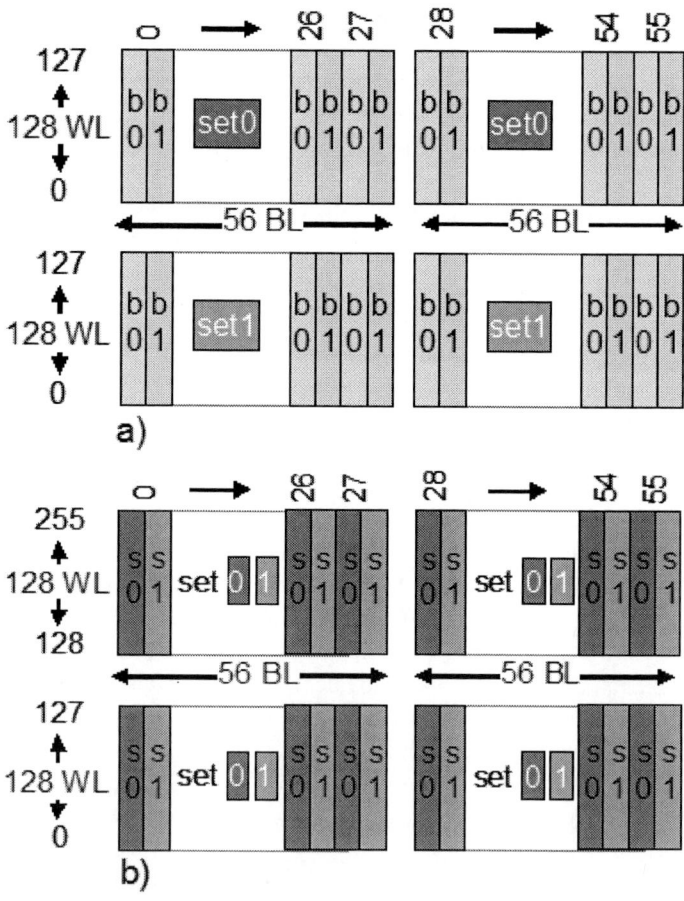

Figure 4: a) prior art with bit interleave, b) new array structure with set interleave, reducing switching power by 50%.

978-1-4244-9019-6/11 $26.00 © 2011 IEEE   178

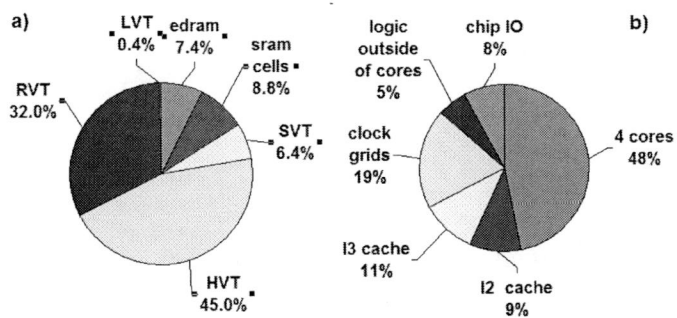

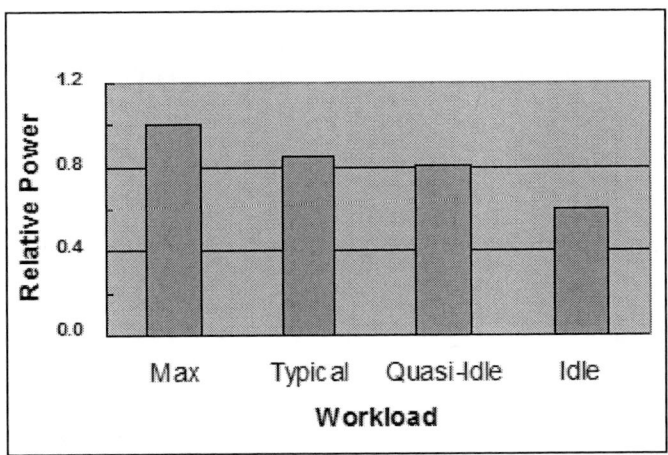

Figure 5: a) Leakage breakdown by device type, b) Total power break down by component

Figure 8: z196 processor power data running various workloads

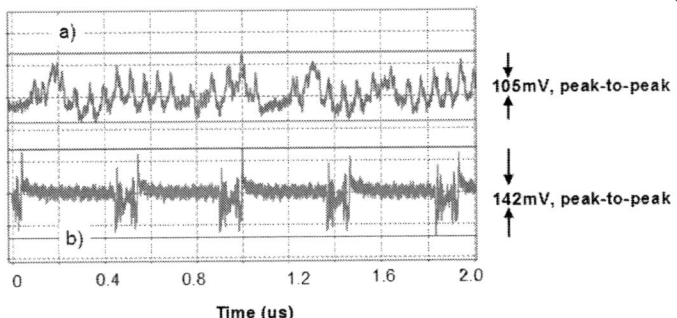

Figure 6: a) vdd supply noise on z196 processor chip, b) vdd supply noise on z10 processor chip

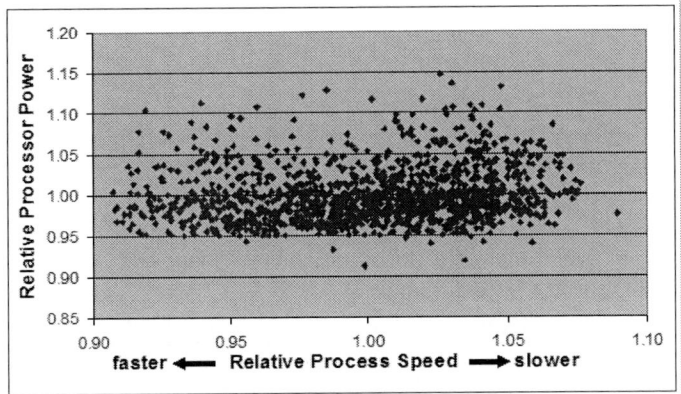

Figure 7: z196 processor power running max Workload

Ultra-Low power FIR filter using STSC-CVL logic

Sajib Roy, Md. Murad Kabir Nipun, and J Jacob Wikner, *Linköping University, Sweden*

Abstract— The paper shows the implementation of digital FIR filter using ultra-low power logic components. Source coupled logic is used and operated at sub-threshold region to achieve low power consumption while keeping a satisfactory output swing. The STSCL (sub-threshold source coupled logic) circuit is added with controllable voltage-level feature to minimize overall leakage current flow, including both gate leakage and sub-threshold. Seven-stage ring oscillators are implemented in CMOS, STSCL and our proposed logic at similar supply voltage to observe the differences with power consumption for the proposed technique came at nW range. Later on the FIR was design in both CMOS and proposed with measurement results shown in the paper. All measurements for are shown using 65 nm process technology, at a supply voltage of 0.5 V.

Keywords; sub-threshold source couple logic, controllable voltage level, gate leakage, PDP, FIR filter.

I. INTRODUCTION

Ultra-low power requirements have become a necessity in aspects of both digital and mixed-signal circuits and systems. The demand is extremely high in sectors of bio-engineering and sensors. Achieving such low power requires minimization of leakage current flow and implementation of logic gates with new topology that would reduce the dominating leakages that are currently the bother of power consumption in CMOS logic. This paper focuses on the motivation of using differential logic such as SCL, at a sub-threshold region with controllable voltage level feature as advantage in reducing total leakage of a system. Operating at sub-threshold region gives the flexibility to run gates at low supply voltage as current density for sub-threshold MOSFET operation is small. Basic CMOS gates are also used to utilize this advantage, however, it has a drawback. It degrades the output voltage logic swing. In such case even though power consumption decreases but delay of the logic gates rises thus worsening the PDP (power delay product). Thereby low supply voltage operation without too much swing degradation requires to deviate from using conventional CMOS circuits to better and reliable logic such as the proposed one in the paper.

In this paper we investigate the gate leakage and it's variation with MOS process, perform measurements and simulation on ring oscillator circuit with 3 different topologies including the proposed technique and shown the results. The new topology is elaborately explained and designed at 65 nm process to implement a FIR filter.

I. STSCL – SOURCE COUPLED LOGIC AT SUB-THRESHOLD REGION

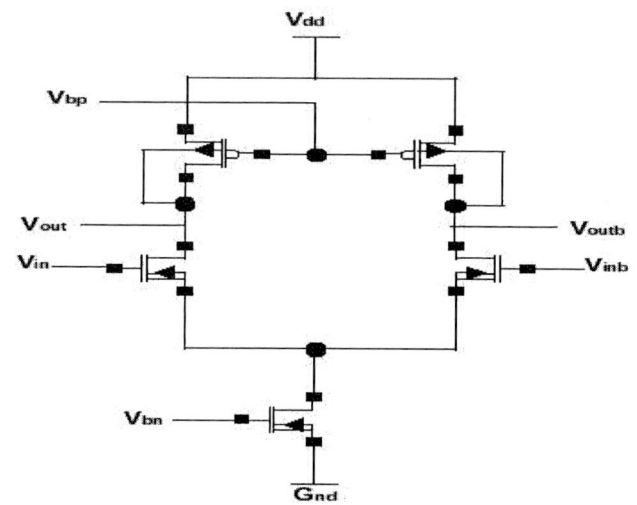

Figure 1 STSCL Inverter [1]

Figure 1 shows the representation of basic STSCL inverter [1] with differential input/output gates. This cause complexity in wiring of gates in designing a whole system, but topology like STSCL allows sub-threshold leakage current (important for CMOS logic) consideration to be an afterthought due to the fact, in this topology only junction leakage and gate-to-channel leakage are dominating.

In CMOS logic lowering supply voltage allows power consumption to be low, as power consumption being quadratically dependent on the supply. But this has drawback and cause the voltage swing at output to degrade. In case of STSCL that operates at weak inversion region of MOS devices it requires a minimum of 0.15 V to completely switch the output of a next logic stage. Thus supply voltage can be reduced to a very low value without endangering the swing. One drawback that may arise is the speed of operation which is directly proportional to the bias current, I_{bias}, (equation 1). It is generated from the bias circuit as shown in figure 2 whereas the overall power consumption is reversely dependent on I_{bias}. This forces a trade-off at choosing the correct bias current for achieving an optimum PDP value.

978-1-4244-9019-6/11 $26.00 © 2011 IEEE   180

$$t = V_{swing} \square C_L / I_{bias} \qquad (1)$$

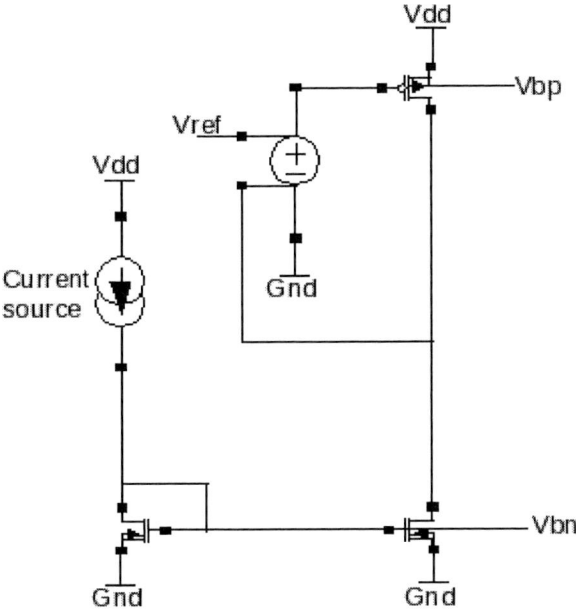

Figure 2 STSCL Bias circuit

The PMOS devices in this topology acts as a high load resistance, which is necessary in maintaining the desired output voltage for extremely low supply. The body terminal for the PMOS is connected to its drain that gives a high resistive value (high enough for a reasonable swing at an I_{bias} of nA range) in M Ohm range [1]. This value can be varied and larger resistance can be achieved by lowering the size of the PMOS device compare to the NMOS.

II. PROPOSED DESIGN

The circuit in figure 3 shows the proposed STSCL logic with a controllable voltage level feature. This provides the minimization of any sort of leakage components between supply and ground for operation of logic gates in a system [2]. During stand-by operating mode, with the logic still functioning in sub-threshold region, the NMOS with larger channel length creates a virtual ground which lowers the overall supply to ground voltage. The larger channel length MOSFET with high gate oxide is used to avoid gate leakage flow through the gates. This NMOS is fed with a V_{bn} signal from the bias circuit which navigates the device to operate. Alongside this NMOS, a PMOS diode is connected in parallel which allow the drain voltage node for the source coupled NMOS to be adjusted. This benefit during stand-by operation where logic circuits have zero input and the connected PMOS diode will allow the additional gate leakage to be adjusted by reducing the gate-to-source voltage. This causes the off-state current flow to be very low.

The gate leakage is mainly composed of three components, gate to body, gate to source/drain and gate to channel, I_{gc}, all of which depends on the operation mode. Considering gate leakage for the proposed logic within the two main modes of operation described above, the I_{gc} is more significant. This current shows dependency on the gate-to-drain voltage [3]. Here this gate-to-drain voltage equals voltage drop across the PMOS diode which is comparatively lower than supply voltage. Thus reduction in gate leakage is achieved.

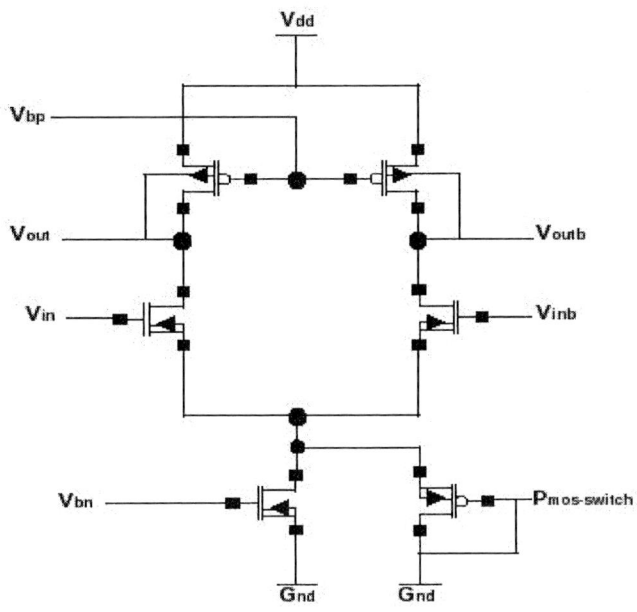

Figure 3 STSC-CVL Inverter (sub-threshold source coupled controlled voltage level logic)

A. Seven stage ring oscillators

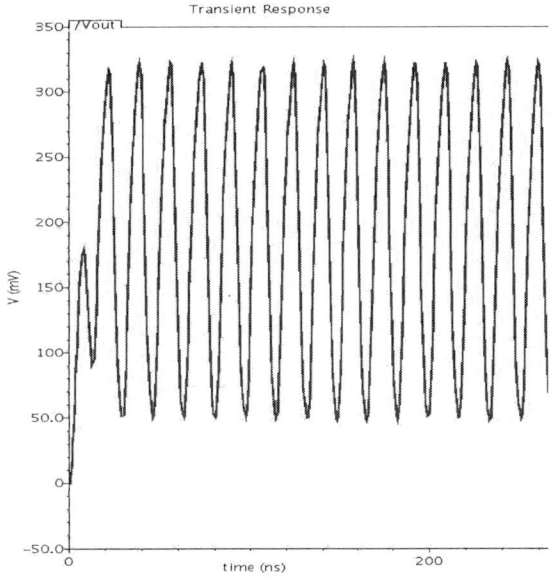

Figure 4 Output response for 7 stage STSC-CVL

To understand the proposed logic and compare its advantage over CMOS and STSCL logic, seven stage ring oscillators are designed using the three different logic circuits. A similar output frequency was maintained with supply voltage of 0.5 V. Table 1 shows the results for the power consumption and output frequency, with figure 4 showing the output response for the seven stage ring oscillator implemented with the proposed STSC-CVL logic.

Logic	CMOS	STSCL	STSC-CVL
Frequency (MHz)	65	60	60
Power Consumption (uW)	4.98	1.83	0.17

TABLE I.   COMPARISON OF POWER CONSUMPTION FOR THE SEVEN STAGE OSCILLATOR USING THREE DIFFERENT LOGIC

B. Implemented logic blocks in STSC-CVL

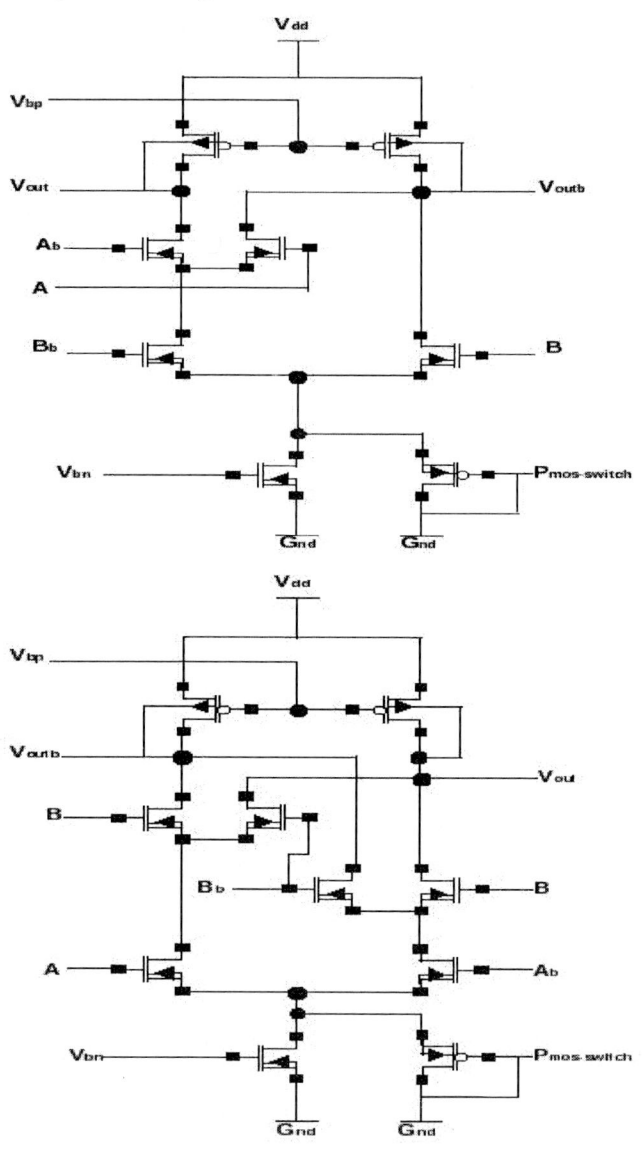

Figure 5 STSC-CVL OR gate (top), XOR gate (bottom)

The previous figures (in figure 5) shows few of the STSC-CVL logic gates , used to form logic blocks including a Full

adder (FA), d-flipflop (DFF), 5bit Serial-Parallel multiplier and 5th order FIR filter.

C. FIR filter using STSC-CVL

Further justifying the proposed technique a 5th order FIR filter was constructed with a sampling frequency of 20 MHz. The filter response along with its block diagram is shown in figure 6 and figure 7 respectively, with designing done in CMOS, STSCL and the proposed technique. A 5bit serial-parallel multiplier [4] is designed as a requirement for the filter design, with fixed coefficient values denoted in CSD (Canonical signed digit) [5]. Simulations are performed at $V_{dd} = 0.7$ V and 0.5 V, with temperature at 70 deg Celsius. I_{bias}, is maintained above 10nA for obtaining the desired filter output.

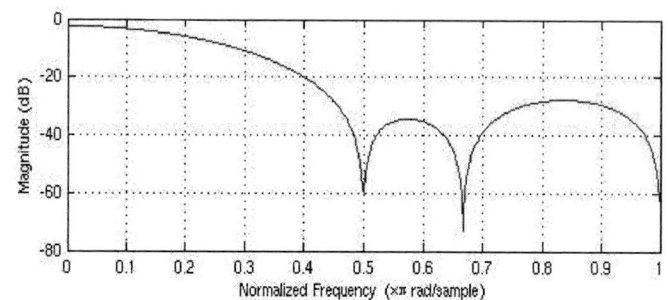

Figure 6 Filter Magnitude response

The PDP results for the simulation for the all three logic are shown in Table 2 and 3 respectively for 0.5 V and 0.7 V power supply

Logic $V_{dd} = 0.7$ V	CMOS	STSCL	STSCL-CVL
Power Consumption (uW)	19.32	9.2	7.82
Delay(ns)	2.86	3.13	2.98
PDP(fJ)	55.25	28.8	23.3

TABLE II.   SIMULATION RESULTS FOR 0.7 V SUPPLY

Logic $V_{dd} = 0.5$ V	CMOS	STSCL	STSCL-CVL
Power Consumption (uW)	8.52	1.23	0.95
Delay(ns)	10.13	15.03	13.93
PDP(fJ)	85.05	18.48	13.23

TABLE III.   SIMULATION RESULTS FOR 0.5 V SUPPLY

The obtained results suggest the effectiveness of the proposed topology in achieving ultra-low power levels in a digital system, in this case a FIR filter.

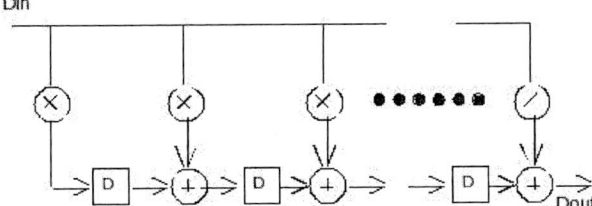

Figure 7. Block diagram 5th order FIR Filter

III. CONCLUSION

The new approach for the implementation of the FIR filter certainly fulfills the ultra-low power criteria as the results of the simulation clearly show it. This technology utilizes the advantages of the SCL logic running at sub-threshold region with additional features of voltage level controlling. Few things to be noted is that further optimization, reduction of power and an improved PDP can be achieved if other low power techniques like implementation of compound logic gates, pipelining, etc are utilized.

ACKNOWLEDGMENT

The authors would like to thank Lars Wanhammar, Oscar Gustafsson and Niklas Andersson for their contribution to the completion of this paper.

REFERENCES

[1] Tajalli. A, Brauer. E.J, Leblebici. Y, Vittoz. E, "Subthreshold Source-Coupled Logic Circuits for Ultra-Low-Power Applications," *Solid-State Circuits, IEEE Journal of*, vol.43, no.7, pp.1699-1710, July 2008

[2] Tawfik. SA, Kursun. V, "Low Power and High Speed Multi Threshold Voltage Interface Circuits," *Very Large Scale Integration (VLSI) Systems, IEEE Transactions on*, vol.17, no.5, pp.638-645, May 2009.

[3] Cao, K.M, Lee, W.-C, Liu. W, Jin. X, Su. P, Fung, S.K.H, An. J.X, Yu. B, Hu. C, "BSIM4 gate leakage model including source-drain partition," *Electron Devices Meeting, 2000. IEDM Technical Digest. International*, vol., no., pp.815-818, 2000

[4] M. Vesterbacka, K. Palmkvist, and L. Wanhammar, "Realization of Serial/Parallel Multipliers with Fixed Coefficients," *Proc. National Conf. on Radio Science*, RVK'93, Lund Institute of Technology, Lund, Sweden, pp. 209-212, 5-7 April 1993.

[5] He, S, Torkelson, M, "FPGA implementation of FIR filters using pipelined bit-serial canonical signed digit multipliers," *Custom Integrated Circuits Conference, 1994, Proceedings of the IEEE 1994*, vol., no., pp.81-84, 1-4 May 1994.

Design of a Low-Cost Floating-Point Programmable Vertex Processor for Mobile Graphics Applications Based on Hybrid Number System

Shen-Fu Hsiao, Chan-Feng Chiu, and Chia-Sheng Wen

Department of Computer Science and Engineering
National Sun Yat-Sen University
Kaohsiung, Taiwan

Abstract

Recent OpenGL ES 2.0 API Specification for embedded systems graphics operations requires programmable vertex shader and fragment shader to process vertex and pixel data. Calculation of dot-product for two vectors and transcendental functions for a scalar are two fundamental arithmetic operations in the vertex processing. Since some complicated arithmetic operations in binary number system (BNS) turn into simple operations of addition and/or multiplication in the logarithmic number system (LNS), we present a low-cost design of a floating-point programmable vertex processor based on hybrid BNS and LNS. The proposed design achieves at least the same (or even higher) precision with much lower cost compared with recent similar implementations.

Keywords: graphics processors, OpenGL ES 2.0, vertex shader, computer graphics, 3D graphics.

I. Introduction

Computer graphics is known to be computation-intensive, and thus graphics processing unit (GPU) has been widely used in PC and other server machines to speed-up 3D graphics operations. For example, Nvidia has come out with a series of high-end GPU designs with performance comparable to supercomputers [1]. However, these GPUs, with hundreds of streaming processing cores, are not suitable to handheld devices due to the huge power consumption and area cost. Recently, there is increasing demands for real-time 3-D graphics in mobile devices [2-6]. These mobile GPU designs usually contain vectored processing unit (VPU) for dot-product (DP) calculation, a key operation in graphics transformation. Another computation-intensive operation in 3-D graphics is the lighting that usually requires operations of logarithm, exponentials, square-root and reciprocals, which are usually realized in special function unit (SFU).

To meet the requirement of graphics operations in resource-limited embedded systems, open source graphics application programming interfaces (API) standard OpenGL ES (Embedded Systems) has been announced recently [7]. OpenGL ES 1.0 defines fixed function graphics pipeline while OpenGL ES 2.0 provides more programmability for more realistic graphics effects. Vertex shader and fragment shader are the two major programmable parts in OpenGL ES 2.0 that perform per-vetex and per-fragment operations specified by OpenGL shading language.

In this paper, we focus on the design of a low-cost programmable vertex shader graphics processor that supports OpenGL ES 2.0 graphics standard. Sec. II describes some recent designs of vertex processors. The proposed design is explained in Sec. III. The experimental results and comparison are given in Sec. IV, followed by a conclusion.

II. Related Work

As shown in Tab. 1, some complicated arithmetic operations in traditional binary number system (BNS) turn into simple operations in logarithmic number system (LNS) [8]. On the other hand, computation of simple arithmetic functions (such as addition/subtraction) is not suitable in LNS, but can be easily implemented in BNS. Thus, hybrid LNS and BNS are recently adopted to realize the graphics operations for handheld graphics processors [1-3]. However, the conversions between BNS and LNS need to be done fast enough so that they will not degrade the overall performance of the operations. In [1-3], the LNS-BNS converters are implemented using small look-up-tables (LUTs) followed by some shifts and additions/subtractions. However, accuracy is sacrificed to make these converters realized in only one pipelined stage with simple LUTs, hardwired shifts, and adders/subtractors. For example, the log converter (LOGC) and anti-log converter (ALOGC) in [1-3] are accurate to only around 8 fractional bits.

Tab. 1: Arithmetic operations in BNS and LNS domains.

Operations	BNS	LNS
multiply	$X \times Y$	$log_2(X \times Y) = log_2 X + log_2 Y$
divide	X/Y	$log_2(X/Y) = log_2 X - log_2 Y$
Square-root	$\sqrt{X}$	$log_2(\sqrt{X}) = \frac{1}{2} log_2 X$
power	X^Y	$log_2(X^Y) = Y \times log_2 X$
add	$X + Y$	$log_2(X + Y) = log_2 X + log_2(1 + 2^{log_2 Y - log_2 X})$
subtract	$X - Y$	$log_2(X - Y) = log_2 X - log_2(1 + 2^{log_2 Y - log_2 X})$

Another category of mobile graphics processor designs focus on the exploration of various processor architectures including multi-thread, single instruction multiple data (SIMD) or very long instruction word (VLIW) [4-5]. In these processors, vectored hardware multipliers-accumulator and special function units (SFU) are two major hardware units to realize the dot-product and other complicated arithmetic functions in BNS. The SFU adopts LUT-based piecewise polynomial function evaluation methods to achieve the required precision.

In this paper, we will present a low-cost vertex shader processor designs based on hybrid number system (HNS) by exploiting the individual features of BNS and LNS in computing arithmetic operations required in graphics processing. In particular, the accuracy of the datapath components in the LNS domain is carefully selected to support the precision limited by the data conversion units between BNS and LNS.

III. Proposed HNS Vertex Processor

Fig. 1 shows the overall architecture and the instruction set of the proposed vertex shader processor and its interfaces with ARM micro-controller and other memory and buffer units including vertex attribute buffer (VAB) and vertex output buffer (VOB). The datapath core consists of a Special Function Unit (SFU) implemented in hybrid LNS/BNS (HNS) and a general function unit (GFU) implemented in BNS. SFU executes instructions related to dot-product, multiplication, division, and other complicated arithmetic operations. GFU is in charge of instructions of basic addition/subtraction, logic operations, magnitude-comparison, branching, and data access.

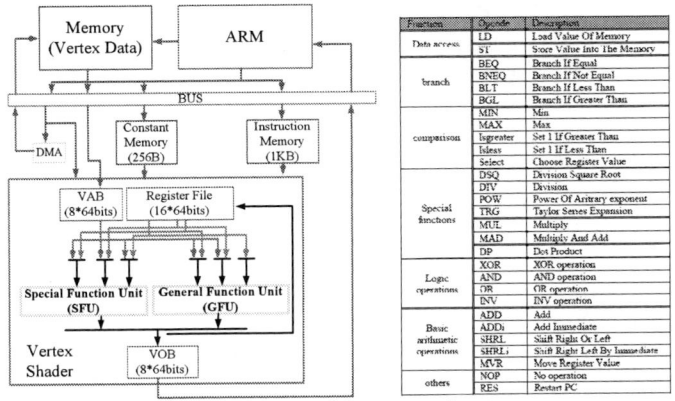

Fig. 1: Overall architecture of the proposed vertex processor.

In this paper, the input format is in IEEE 754 half-precision floating-point (H-FLP) with 1 sign bit, 5 exponent bits (with bias of 15) and 10 significant bits (with hidden 1), as shown in Fig. 2. The 16-bit FLP input in BNS is converted to LNS with 1 zero bit, 1 sign bit, 5 integer bits, and 10 fractional bits (Q5.10 format). In this paper we design conversion hardware that satisfies the precision of up to 10 fractional bits. The subsequent hardware needs to maintain the accuracy limited by the conversion hardware.

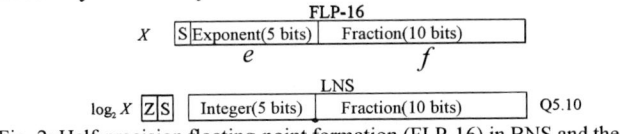

Fig. 2: Half-precision floating-point formation (FLP-16) in BNS and the Q5.11 format in LNS (with zero bit and sign bit).

Fig. 3 shows the architecture of the SFU that contains four operation channels with each channel executed in four pipelined execution stages. In each channel, the operands in BNS are converted to those in LNS via log converters (LOGC). Afterward, the multiplications, divisions, dot-product calculations, and other complicated arithmetic operations in the BNS domain are performed i via adders and multipliers of proper precision in the LNS domain. Since the floating-point addition is done in BNS, we need to perform alignment by computing the difference of exponents of two floating-point numbers. This is done in the block marked by *compute shift*, in the second pipeline stage and the FLP-AU (exponent) after anti-logic converter (ALOGC) in the third stage. The fourth pipeline stage of FLP-AU (mantissa) performs the basic arithmetic operations in BNS along with the post-normalization and rounding required to generate output in IEEE 754

half-precision FLP format. In the following, we describe the design of the major components in the SFU.

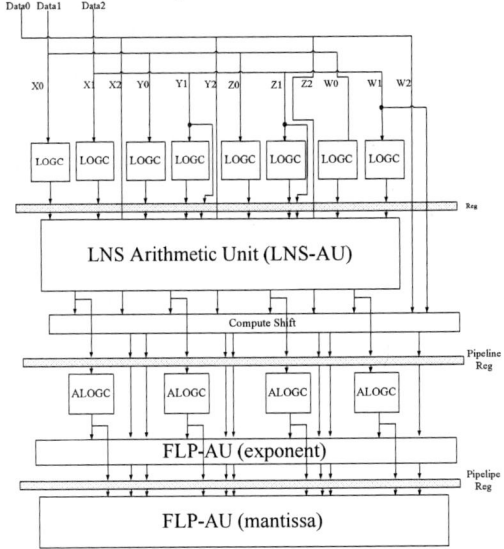

Fig. 3: Overall pipelined architecture of the SFU.

(A) LOGC

Fig. 4 is the design of the log converter (LOGC) that change the magnitude of a H-FLP number $X = 1.f \times 2^{e-15}$ in BNS to $\log_2 X = \log_2(1.f) + (e-15)$ in LNS. The fractional part of $\log_2 X$ is $\log_2(1.f)$ which can be computed using degree-1 piecewise polynomial approximation for function evaluation, i.e., $\log_2(1.f) \approx b + a \times (1.f)$. However, since our target accuracy is only 10 fractional bits, we can represent the piecewise polynomial coefficients in binary signed digit (BSD) with minimum number of non-zero digits [8]. Then the multiplication of the BSD coefficient can be realized using shift and add/subtract operations. We found that at most three non-zero digits are required to represent the polynomial coefficients to meet the precision requirement, i.e., $\log_2(1.f) \approx b \pm (1.f \gg p) \pm (1.f \gg q) \pm (1.f \gg r)$.

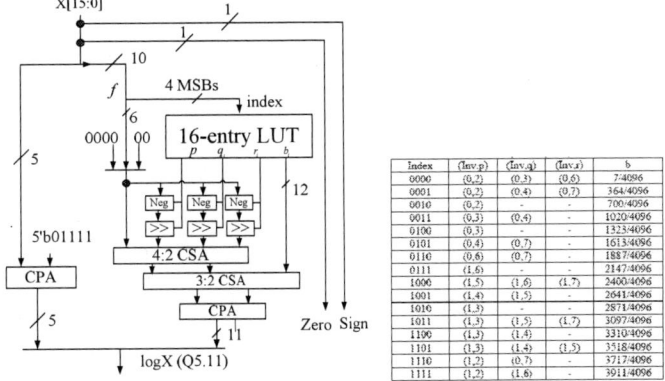

Index	(Inv.p)	(Inv.q)	(Inv.r)	b
0000	(0,2)	(0,3)	(0,6)	7/4096
0001	(0,2)	(0,4)	(0,7)	364/4096
0010	(0,2)	-	-	700/4096
0011	(0,3)	(0,4)	-	1020/4096
0100	(0,3)	-	-	1323/4096
0101	(0,4)	(0,7)	-	1613/4096
0110	(0,6)	(0,7)	-	1887/4096
0111	(1,6)	-	-	2147/4096
1000	(1,5)	(1,6)	(1,7)	2400/4096
1001	(1,5)	(1,5)	-	2641/4096
1010	(1,3)	-	-	2871/4096
1011	(1,3)	(1,5)	(1,7)	3097/4096
1100	(1,3)	(1,4)	-	3310/4096
1101	(1,3)	(1,4)	(1,5)	3518/4096
1110	(1,2)	(0,7)	-	3717/4096
1111	(1,2)	(1,6)	-	3911/4096

Fig. 4: Design of log converter (LOGC) from BNS domain to LNS domain.

(B) LNS Arithmetic Unit (LNS-AU)

Fig. 5 shows the architecture of the LNS-AU where the 4 multiplications $x_i y_i$ (in dot product $\sum_{i=1}^{4} x_i y_i$), powering x^y, or 4 multiplications $a_i x^{k_i}$ (in Taylor series expansion $\sum_{i=0}^{4} a_i x^{k_i}$ for function evaluation) are performed in LNS respectively by four additions $\log x_i + \log y_i$, one multiplication $x * \log y$, or four multiplications-additions $\log a_i + k_i * \log x$ (k_i are small integers expressed by at most 6 bits). The single- or four-way multiplication selector is to select proper data from the radix-4 Booth encoder for the one multiplication of 16×16 for powering or four multiplications of 16×4 for Taylor series expansion. The carry save adder (CSA) trees and the subsequent carry-propagate adders (CPA) are used to generate the single output of $x * \log y$ at the second channel, or the four outputs $\log a_i + k_i * \log x, i = 1, 2, 3, 4$, at the four channels.

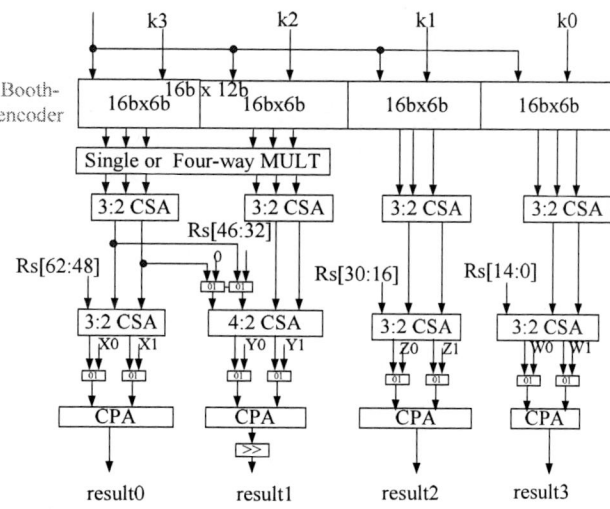

Fig. 5: Architecture of the LNS Arithmetic Unit (LNS-AU)

(C) ALOGC

Fig. 6 shows the architecture of the anti-log converter (ALOGC) that transforms the data from LNS domain to BNS domain. The input $\log_2 X$ of format Q5.10 is divided into the 5-bit integer part I and the 10-bit fractional part F (F<1), i.e., $\log_2 X = I + F$. Thus, after ALOGC, we have $2^{\log_2 X} = 2^I + 2^F$ where 2^I contributes to the exponent part of floating-point representation while 2^F can be calculated by degree-1 piecewise polynomial approximation with LUT in a similar way as in the design of LOGC.

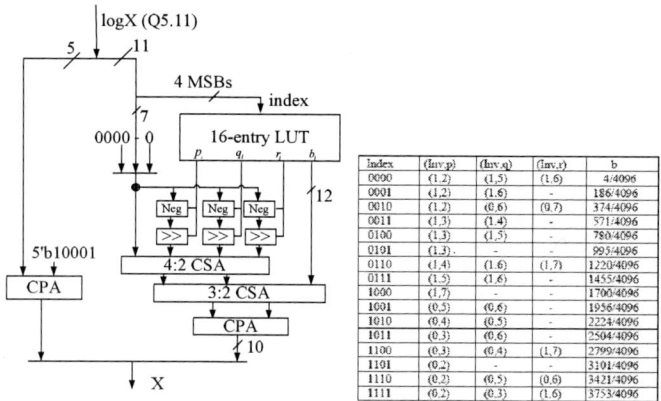

Fig. 6: Design of anti-log converter (ALOGC).

(D) FLP Arithmetic Unit (FLP-AU)

Fig. 7 is the architecture of the FLP domain that is divided into two parts: exponent part and significant parts, executed in two pipelined stages. The exponent part performs alignment shifts and the significant part performs fixed-point addition followed by normalization and rounding. Afterward, the final results are packed to IEEE 754 half-precision floating-point representation in the block named *combine*.

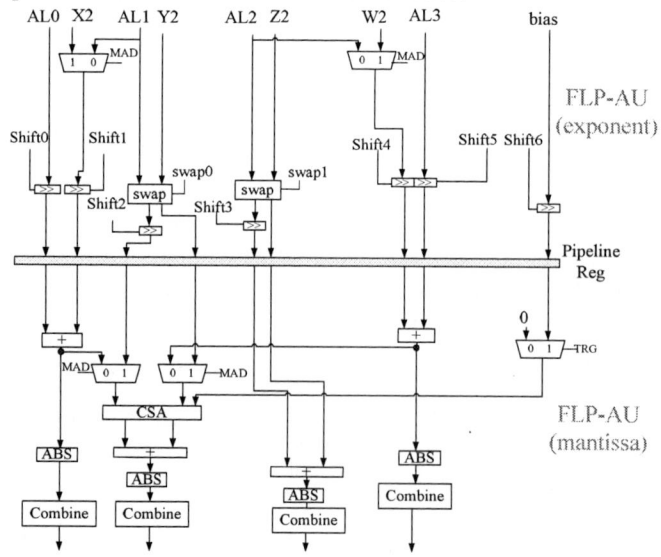

Fig. 7: Architecture of the FLP Arithmetic Unit (FLP-AU)

(E) Pipelining

The proposed vertex processor is divided into seven pipeline stages including fetch, decode, execution, and write-back as shown in Fig. 8. The execution of SFU is split into four pipelined stages (EX0-EX3) and the execution of GFU takes only two stages (EX0 and EX3). The write-back stage includes the write-mask to selectively write the results of four channels to the individual vector components. Swizzle of vector components are also supported in EX0 stage.

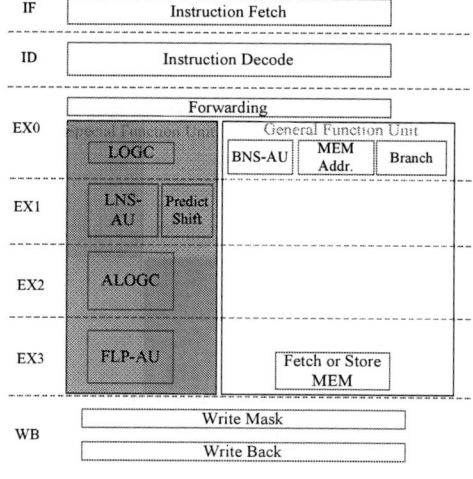

Fig. 8: Pipelining of the vertex processor.

IV. Experimental Results

Tab. 2 shows the synthesis results of the proposed vertex processor using TSMC 0.18um cell library. Among the total gate count of 71.45K, The core datapath described in Sec. III including vectored SFU and related registers takes only 33 K gates. The processor can work at 200MHz with power consumption of around 3 mW.

Tab. 3 compares the maximum absolute errors of the proposed vertex processor with recent similar designs [3-4] by running some fundamental operations such as vectored

978-1-4244-9019-6/11 $26.00 © 2011 IEEE

multiplication, matrix-vector multiplication, and dot product. We observe that the proposed vertex processor has smaller error. This is because we use more entries in the LUTs of LOGC and ALOGC in order to achieve H-FLP precision of at least 10 fractional bits and carefully select the bit-widths (Q5.10) of the arithmetic units in the LNS domain to support the precision set by the converters. On the other hand, the BNS-LNS converters in [3][4] has only about the same precision as ours, but the internal data format in the LNS domain is Q.5.26, which is not necessary.

Tab. 4 compares with [3] the area cost of the key components of the vertex processor in unit of fall adder (FA) based on the estimation given in [3]. The SFU in our design contains 8 LOGC, a LNS-AU, four ALOGC, and one FLP-AU. The fixed-point design in [3] has similar architecture with ours except for the internal data precision of Q5.26 in the LNS domain. But as shown in Tab. 3, our design has slightly better precision, but required only about 50% area cost compared to the design in [3].

Tab. 5 compares, under the same TSMC 0.18um technology and the same area/delay constraints, the area-delay product for the key components in the SFU, including LOGC, LNS Domain, and ALOGC, as shown in the first three execution stages. The area-delay product is more than 60% smaller than the designs in [4] because of the efficient design of the SFU.

Tab. 6 makes comparison of various LNS-based designs of the special function unit for vertex shader processors. The area information in [2-4] is derived assuming 1 gate (4 transistors) has area of 10 um^2 based on Artisan TSMC 0.18um standard cell library. In [2][3], the fixed-point SFU based on hybrid LNS and BNS (HNS) can compute vector multiply, multiply-and-add, divide, divide-square-root, dot product, and some complicated arithmetic functions (such as log, power, and trigonometric). The HNS SFU is extended to floating-point format with integer general function unit (GFU) to realize a streaming processor for vertex processor design in [4]. Since all these HNS designs require LOG and ALOG for conversion between LNS and BNS, the precision is determined by the accuracy of these converters. In [2-4], the LUT-based LOG and ALOG is executed in one pipelined stage with limited precision. So, we only need to perform the subsequent LNS arithmetic operations satisfying the precision. This is the major reason of area saving in our design.

Tab. 2: Area profiling of the proposed vertex processor.

Hardware Units	Components	Area (gates)	Area (Percentage)
Special Function Unit (SFU)	LOGC*8	7.83K	10.9%
	LNS Domain	5.17K	7.3%
	ALOGC*4	3.88K	5.4%
	Compute Shift	1.58K	2.3%
	FLP-AU(exponent)	1.39K	1.9%
	FLP-AU(mantissa)	2.3K	3.2%
General Function Unit (GFU)	BNS-AU	8.5K	11.9%
	Branch	0.62K	0.9%
	Pre_mem	0.28K	0.4%
Register File		10.1K	14.1%
VAB+VOB		10.1K	14.1%
Pipeline registers		13.49K	18.9%
others		6.21K	8.7%
Total		71.45K	100%

Tab. 3: Comparison of precision of three major instructions.

Instr.	operation	Max. absolute errors	
		[3][4]	Ours
MUL	Multiply	0.00096	0.0008
MAT	Matrix-vector multiply	0.00386	0.0025
DP	Dot Product	0.00386	0.0025

Tab. 4: Comparison of the major components in SFU.

Area (in unit of FA)	[3]	Ours
LOGC	292	102
LNS-AU	1339	505
ALOGC	222	103
FXP/FLP-AU	132	481
SPU Total	4695	2214

Tab. 5: Comparison of area-delay product.

	Area (gates) * Delay (ns)		Saving rate
	[4]	Ours	
LOGC	2957*2.3 = 6801.1	1001*2.3 = 2302.3	66%
LNS	13405*3.38 = 45308.9	4261*3.38 = 14402.2	68%
ALOGC	1810*1.55 = 2805.5	1004*2.3 = 2231	20%

Tab. 6: Comparison of various HNS-based SFU designs.

	Speed (MHz)	Area (K gate count)	Function	FXP/ FLP
[2]	210	93	SFU only	FXP
[3]	232	79	SFU only	FXP
[4]	200	242	SFU+others	FLP
Ours	200	33	SFU only	FLP
Ours	200	71	SFU+others	FLP

V. Conclusion

We propose a vertex processor design based on hybrid binary number system (BNS) and logarithmic number system (LNS). The vertex processor contains two major parts: special function unit (SFU) and general function unit (GFU).The special function unit (SFU) computes complicated arithmetic operations in the LNS domain where the arithmetic units are designed carefully so that they can meet the precision constrained by the log converter (LOGC) and anti-log converter (ALOGC) for BNS and LNS data conversion.

Acknowledgement

This work is supported by Taiwan's National Science Council under grant NSC 98-2220-E-110-005 and NSC 99-2221-E-110-055. EDA tools are supported by Chip Implementation Center (CIC).

References

[1] J. Nickolls and W. J. Dally, "The GPU Computing Era," *IEEE Mirco*, pp. 56-69, Mar./Apr., 2010.

[2] B.-G. Nam, H. Kim, and H.-J. Yoo, "A Low-Power Unified Arithmetic Unit for Programmable Handheld 3-D Graphics Systems," *IEEE Journal of Solid-State Circuits*, vol. 42, no. 8, pp. 1767-1778, Aug. 2007.

[3] B.-G. Nam, H. Kim, and H.-J. Yoo, "Power and Area-Efficient Unified Computation of Vector and Elementary Functions for Handheld 3D Graphics Systems," *IEEE Trans. Computers*, vol. 57, no. 4, pp. 490-504, Apr. 2008.

[4] B.-G. Nam and H.-J.Yoo "An Embedded Stream Processor Core Based on Logarithmic Arithmetic for a Low-Power 3-D Graphics SoC," *IEEE Journal of Solid-State Circuits*, vol. 44, no. 5, pp. 1554-1570, May 2009.

[5] C.-H. Yu, et al., "A 186-Mvertices/s 161-mW Floating-Point Vertex Processor with Optimized Datapath and Vertex Cache," *IEEE Trans. VLSI Systems*, vol. 17, no. 10, pp. 1369-1382, Oct. 2009.

[6] S.-H. Kim, et al., "A 116fps/74mW Heterogeneous 3D-Media Processor for 3-D Display Applications," *IEEE Journal of Solid-State Circuits*, vol. 45, no. 3, pp. 652-667, Mar. 2010.

[7] OpenGL ES Spec. available at http://www.khronos.org/opengles/

[8] B. Parhami, *Computer Arithmetic: Algorithms and Hardware Designs*, 2nd ed., Oxford University Press, New York, 2010.

A Low Jitter Active Body-Biasing Control-based Output Buffer in 65nm PD-SOI

Dimitri Soussan[1,2], Sylvain Majcherczak[1], Alexandre Valentian[2], Marc Belleville[2]

[1] STMicroelectronics Crolles, 850 rue Jean Monnet, 38926 Crolles, France
[2] CEA LETI, Minatec campus, 17 rue des Martyrs, 38054 Grenoble, France
e-mail : dimitri.soussan@st.com

Abstract—**This paper proposes a specific low jitter and high speed Ouput interface which takes advantage of the Partially Depleted Silicon-on-Insulator technology while avoiding its drawbacks related to floating body effects. Thanks to an active body-biasing control technique, the additional jitter related to PD-SOI history effect, as well as the higher static leakage current compared to bulk technology, are more than compensated. In depth analyses are presented to highlight the robustness of this technique with respect to the other solutions considering various capacitive loads and temperatures.**

Keywords—**Active Body-biasing Control, Output Buffer, Jitter, PD-SOI, History effect.**

I. INTRODUCTION

Partially Depleted Silicon-on-Insulator (PD-SOI) technology is attractive for its performance improvement at same supply voltage when using Floating Body (FB) devices [1][2]. Such performance gain in core digital design is also effective in Input/Output (I/O) circuits. Furthermore, another main motivation for using SOI technology here, is the immunity to latch-up effect which brings the opportunities to get transistors closer to one another and to remove guard rings.

On the other hand, due to the floating body effects, drawbacks emerge such as history effect and higher static power consumption, which are respectively non-existent and less significant in bulk technology. The history effect is due to threshold voltage variation caused by floating body voltage fluctuation [3]. Within high speed I/O interface design, timing uncertainty due to history effect could be non-negligible as jitter is a key parameter in I/O design.

One simple solution is to use Body Contacted (BC) devices, with body tied to either Vdde (pMOS) or Gnde (nMOS), in order to suppress those unwanted effects. BC devices require no additional process step, but only a specific layout; the cost, compared to FB devices is a loss of performances and a larger silicon area.

A more advanced solution, called Active Body-biasing Control (ABC)[4], consists in dynamically modulating the threshold voltage (V_T) of transistors. However, the existing solution offers a limited control of the transistor to be body biased since during one half-period, the body remains floating which induces additional static leakage current and jitter. In addition, this solution was not specifically designed for I/O interface.

On the contrary, the ABC circuit proposed in this paper takes into account I/O concerns by providing a better body control in order to diminish jitter and to reduce static leakage current.

This paper is organized as follow: firstly, section II describes a standard output buffer topology that this work is based on. Next, section III deals with the jitter, a key parameter of I/O design, and with the impact of floating body devices. Then, section IV presents the proposed ABC-based output buffer, followed by section V where simulation results are shown and discussed. Finally, conclusions are drawn in the last section.

II. OUTPUT MODE CIRCUIT

I/O circuits are usually bidirectional, including an output mode and an input mode. This paper is focused on the first one. The Output mode circuit is divided into two parts, as shown in Fig. 1. The upper one is dedicated to the output buffer pMOS drive whereas the lower one is dedicated to the nMOS drive.

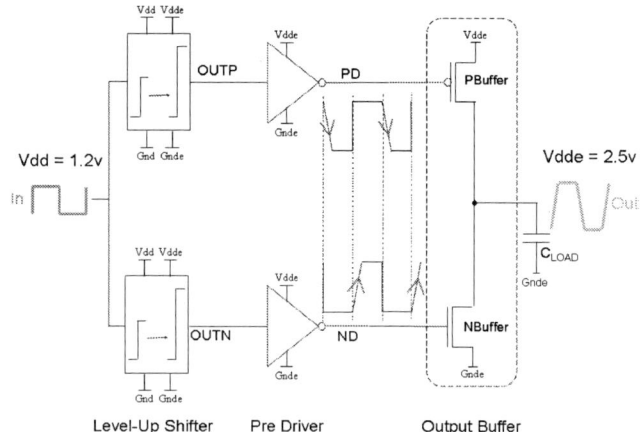

Figure 1. Output mode circuit topology.

It aims at transmitting the non-inverting logic data coming from the core toward an external load. First, the input signal is split into two paths and level-up shifted with respect to the external supply voltage (Vdde). The resulting signals are rise time and fall time controlled, respectively for NBuffer and PBuffer, so as to make sure that both transistors are not turned on simultaneously. Hence, a waste of power consumption is avoided during switching time of the output buffer where a large current has to be provided to the large load. The other reason to achieve fall-time and rise-time controlled signals is to control the di/dt of the buffer. Higher di/dt leads to a higher voltage drop of the supply voltage due to parasitic inductance. Finally, a buffer provides output impedance control.

In this work, the level shifters and pre-drivers transistors are body contacted with bodies tied to Vdde or Gnde accordingly. Only the output buffer will be subject to modification, as further detailed in section V.A.

III. KEY CHARACTERISTICS

A. Data-dependant jitter

Within I/O design, output jitter is one of the main parameters to be considered. As bit time is continuously lowered with the speed increase of data communications, total jitter, including random and deterministic jitter, has to be taken into account in order to keep high signal integrity. Deterministic jitter can be divided into three categories: periodic jitter, duty-cycle-dependant jitter and Data-Dependant (DD) jitter[5]. In this paper, the jitter analysis will only be focused on DD jitter since it's the one reflecting the timing variation induced by data-type signal.

Fig. 2 (a) shows the output signal of a Body Contacted (BC) output buffer and Fig. 2 (b) illustrates its resulting eye diagram from which the DD jitter can be calculated. To highlight the data dependency of output buffer, long-bit/short-bit alternation, acting as a data-type signal, is provided on the input. For large capacitive load and during short-bit sequence, there is not enough time (due to a lack of drive current) for the output signal to reach supply voltages before the next transition, as opposed to long-bit sequence. Hence, the eye diagram exhibits various initial conditions which lead to timing uncertainty. DD jitter is calculated at Vdde/2. The larger the capacitive load, the more difficult for the output signal to reach supply voltages during short-bit sequence and the more the DD jitter is impacted. Let us notice that this assertion is not only true for PD-SOI technology but it is for BULK technology as well.

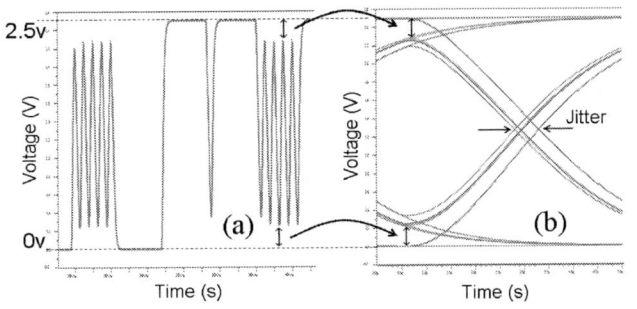

Figure 2. (a) Example of an output signal of BC Output Buffer @ C=90pF, T_{BIT}=5ns & 25°C (b) Resulting eye diagram.

B. Floating Body Impact on I/O Design

Due to floating body effect, history effect arises which leads to timing uncertainty depending on the body potential (V_B) fluctuation that occurred during the past history. As shown in body effect equation depicted in (1), V_T depends on V_{SB}, the source to body voltage. If V_B rises, then V_T falls which speeds up the transition time. Conversely, if V_B falls, then V_T rises which increases the transition time.

$$V_T = V_{T0} + \gamma(\sqrt{|2\Phi_F + V_{SB}|} - \sqrt{|2\Phi_F|}) \qquad (1)$$

As a consequence of timing uncertainty, history effect is expected to impact the output DD jitter of FB-based output buffer. To analyze the history effect impact, FB transistors are used in the output buffer depicted in Fig. 1.

Although more accurate approaches exist [6], 1st/2nd switch method [7] yields in good enough timing characterization to take into account this effect. It consists in carrying out two simulations of the same circuit: a first one with logic '0' on the input as initial condition (DC0) and the second one with logic '1' (DC1). However, to achieve proper DD jitter calculation, all edges have to be taken into account within one simulation. Hence, the input signal considered for DD jitter calculation is depicted in Fig. 3. A first data-type signal is transmitted with DC0 condition. Then, logic '1' is held for 10s until DC1 equilibrium state is reached, before starting the second data-type transmission based on DC1 condition.

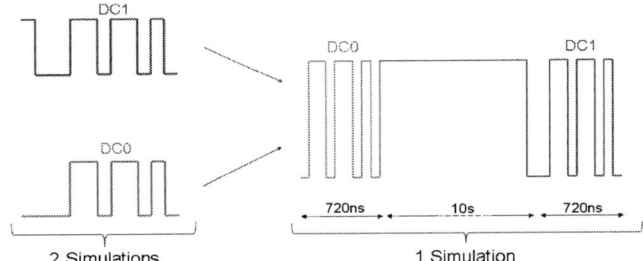

Figure 3. Input data-type signal allowing jitter calculation and considering history effect impact.

At this stage, two sources of DD jitter have been introduced: a first one induced by long-bit/short-bit effect and a second one resulting from history effect (only for FB devices). Long-bit/short-bit effect increases with larger capacitive loads. As a consequence, smaller capacitive loads have been used to minimize this effect in order to only investigate the impact of history effect on DD jitter of FB output buffer. The capacitive load used is 10pF. Fig. 4 (a) shows the floating body voltages of pMOS and nMOS of FB output buffer at DC1 and DC0 conditions, using 1st/2nd switch method with two simulations. It can be observed that the body biasing are not the same between DC1 and DC0 conditions for both the pMOS (ΔV_{BP}) and the nMOS (ΔV_{BN}). Thus, we can expect transient timing differences, resulting in a rising time variation (15ps) caused by ΔV_{BP} (249mV) (1) and a falling time variation (21.6ps) caused by ΔV_{BN} (349mV) (2), as shown in Fig. 4 (b) where the history effect impact on DD jitter of the output signal is depicted.

978-1-4244-9019-6/11 $26.00 © 2011 IEEE

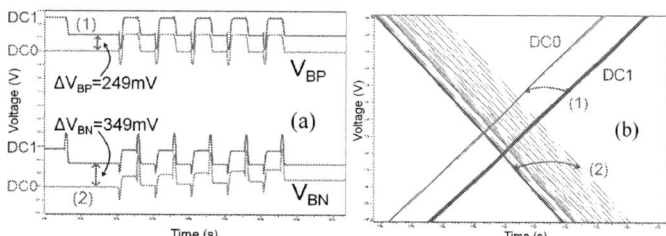

Figure 4. (a) ΔV_B of pMOS and nMOS FB output buffer (b) History effect impact on DD jitter of the output signal @ C=10pF, T_{BIT}=5ns & 25°C.

Along with history effect, FB devices induce another main drawback. On one hand, FB PD-SOI transistors are known to exhibit a better drive current (I_{ON}) than the one in BULK or in BC PD-SOI. On the other hand, the self-body biasing of FB transistors also induces higher static current mainly due to lowered V_T.

In order to cope with both the history effect and the high static current impact, the simplest solution is the use of BC transistors, whose behavior is equivalent to BULK transistors, by biasing the body to Gnde (nMOS) or to Vdde (pMOS) accordingly. This makes V_T rise at the expense of a loss of dynamic performance.

IV. PROPOSED ABC CIRCUIT

The aim of the proposed ABC circuit is to take benefit of both BC and FB advantages. To this end, V_T is dynamically modulated over the time through body biasing so as to improve transition time by lowering V_T during switching time, and to reduce static current by increasing V_T and pulling the body to source voltage (V_{BS}) down to 0V during steady state.

In Fig. 5, only the nMOS part of the output buffer is shown since the pMOS part is symmetrically the same. The proposed ABC circuit allows a better control of V_T as opposed to FB transistors and as compared to the previous ABC solution where the body is floating for one half-period[4].

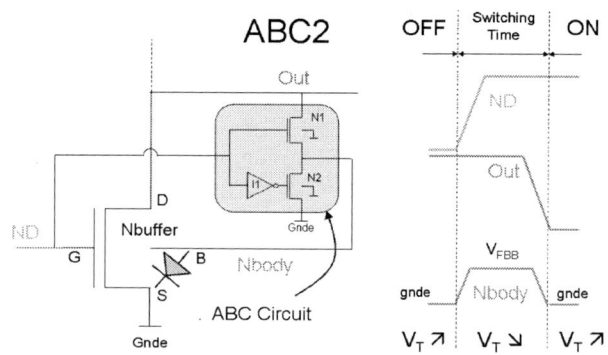

Figure 5. nMOS part of the output buffer and its ABC circuit.

The ABC circuit controlling Nbuffer consists of two nMOS N1 and N2, both body contacted to Gnde, and one BC inverter I1. At OFF-state, when ND signal is low, N1 is turned off and N2 is turned on. Thus, Nbody is low. During switching time, while Nbuffer is being turned on, N1 is turned on which makes current flow through N1. Hence, Nbody tends to be high, as Out is initially high. Then, Out goes low which consequently brings back Nbody to low. In this phase, during the switching time, V_T has been lowered and I_{ON} current has

increased. Consequently, Nbuffer switches faster from OFF-state to ON-state. At ON-state, when ND signal is high, N1 is on and Nbody is pulled down to the Out node potential. Thus, Nbody is grounded.

At I/O cell level (not shown), one additional improvement consists in connecting the signals triggering both ABC circuits of Pbuffer and Nbuffer to the previous stages in order to anticipate the switching of the corresponding buffer transistor. This technique is called Look-Ahead ABC [8]. Let us notice that the area of the two ABC Circuits represents only ~1.5% of the output buffer area. As for the ESD issue, the already available protection diode was sized to offer a lower resistive discharge path compared to the output buffer one. This will also be effective in protecting the additional ABC circuit, provided that care is taken in the layout phase to place it away from the major discharge paths.

V. SIMULATION RESULTS

A. Comparative results

Simulations at schematic level were performed to compare four different output buffer structures based on: BC devices, FB devices, previous ABC circuit and proposed ABC circuit. Level-Up Shifters and Pre Drivers are body contacted whatever the output buffer structure. A comparative result is shown in table 1. DD jitter is calculated with a capacitive load of 50pF and T_{BIT}=5ns.

TABLE I. DATA-DEPENDANT JITTER, STATIC CURRENT AND NORMALIZED RISE TIME @ C=50PF, T_{BIT}=5NS & 25°C.

	BC	FB	PREVIOUS ABC [2]	PROPOSED ABC
Rise time wrt. BC	------	-7%	-10%	-10%
DD Jitter	5.9 ps	22 ps	16 ps	3.7 ps
Static Current	4.74 nA	375 nA	375nA	4.78 nA

BC-based output buffer is appointed as reference structure for performance evaluation. One can say that FB-based output buffer is quite faster than BC-based output buffer. Nevertheless, this structure is consuming extra static power and has a greater DD jitter, respectively due to lowered V_T and history effect impact. Previous ABC-based output buffer offers only mitigated improvement as compared to FB-based output buffer, essentially due to still floating body effect during one half-period. Finally, proposed ABC-based output buffer offers 10% dynamic performance gain while improving static leakage current and reducing DD jitter in the same order than BC-based output buffer, at the expense of an additional ~4% dynamic current consumption.

B. Data-dependant jitter considering capacitive load

As discussed in section III, we can denote two contributions to DD jitter: history effect (only for FB devices) and long-bit/short-bit effect. Fig. 6 illustrates the DD jitter of BC, FB and proposed ABC-based output buffer (ABC2) at 25°C as a function of the external capacitive load, from 40pF to 90pF. It aims at highlighting both contributions. For small load, the major contributor to DD jitter is the history effect. As

978-1-4244-9019-6/11 $26.00 © 2011 IEEE

a result, FB output buffer is the most impacted. On the other hand, for increasing values of capacitive load, long bit/short-bit effect is taking over the history effect. The structure providing the less drive current exhibits the highest DD jitter. As BC-based output buffer is the slowest structure, this solution is the most impacted.

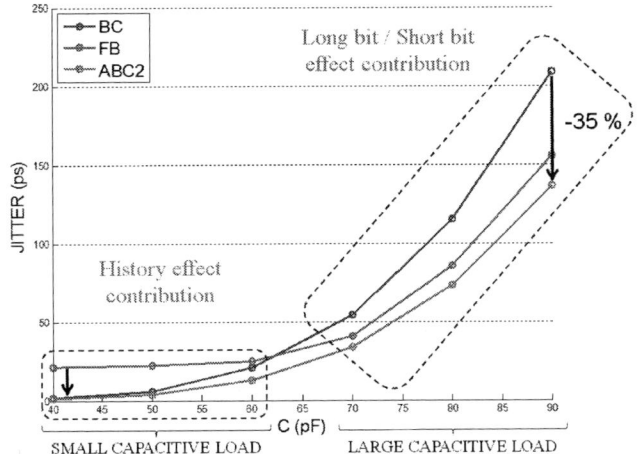

Figure 6.   Jitter versus the external capacitive load (C) of BC, FB and ABC2 based output buffer @ T_{BIT}=5ns & 25°C.

As a conclusion, thanks to controlled history effect technique, proposed ABC-based output buffer exhibits a reduced DD jitter with respect to FB-based output buffer. Furthermore, thanks to speed improvement, proposed ABC technique exhibits 35% DD jitter improvement with respect to BC-based output buffer.

C. Data-dependant jitter considering temperature

A deeper analysis involving temperature is carried out in this part. Jitter is depicted in Fig. 7 as a function of the capacitive load and the temperature for BC, FB and proposed ABC-based output buffer (ABC2). The latter remains the best solution for extreme temperatures with respect to other solutions.

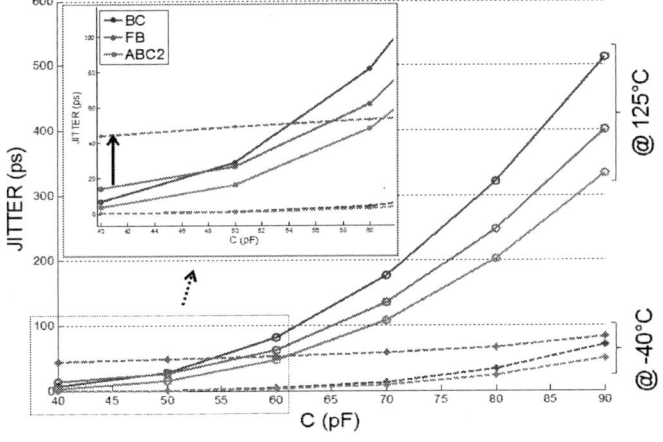

Figure 7.   Jitter versus the external capacitive load (C) of BC,FB and ABC2 based output buffer @ T_{BIT}=5ns, -40°C & 125°C

At 125°C, carriers are slower which leads to a degradation of dynamic performance, and thus to a poorer DD jitter

(regular line) as compared to DD jitter at -40°C (dashed line). Moreover, history effect tends to be mitigated with the increasing operating temperature. Hence, long-bit/short-bit is the main contributor to DD jitter.

At -40°C, carriers are much faster which improves dynamic performance. Hence, DD jitter values are generally lower than the ones at 125°C, except for FB-based output buffer. Indeed, at 40pF and 50pF, this structure exhibits the worst DD jitter due to higher history effect impact.

History effect depends on the operating temperature. Indeed, the floating body biasing depends on the charge fluctuation due to, among other, recombination/generation currents which, in turn, are highly dependent to temperature variation. The higher the temperature, the higher the recombination current is, leading to less charge in the body when DC1 equilibrium state is reached [9]. V_{BS} after DC1 is lower as well as ΔV_B, since V_{BS} after DC0 is always at 0V. Hence, timing variation is narrowed causing a reduction of history effect impact. Conversely, the lower the temperature, the higher the history effect is. DD jitter of FB-based output buffer is very sensitive to history effect with smaller capacitive load, as discussed in section V.B. For very low temperature, history effect is even becoming the main cause of DD jitter over long bit/short bit effect.

VI. CONCLUSION

An active body-bias control circuit has been proposed within an I/O interface circuit. This solution offers the advantages of both FB and BC transistors. It brings higher performance and low static power consumption along with controlled history effect, only by adding a small circuitry as compared to the large transistors of the last I/O stage. In depth analyses have shown that the proposed ABC circuit remains the best solution whatever the capacitive load and the temperature. In one case, FB output buffer exhibits the worst DD jitter at lower temperature and lower capacitive load due to increasing impact of history effect. On the other hand, BC output buffer is the one showing the worst DD jitter at higher temperature and higher capacitive load due to increasing long-bit/short-bit effect caused by lower dynamic performance.

REFERENCES

[1]   Aipperspach A.G. et al, "A 0.2-μm, 1.8-V, SOI, 550-MHZ, 64-b PowerPC microprocessor with copper interconnects," JSSC 1999.

[2]   J. Le Coz et al,. "Comparison of 65nm LP Bulk and LP PD-SOI with Adaptive Power Gate Body Bias for an LDPC Codec", ISSCC 2011.

[3]   Casu, M.R.; Flatresse, P; "History effect characterization in PD-SOI CMOS gates," SOI CONF 2002.

[4]   In-Young Chung; Young-June Park; Hong-Shick Min; , "A new SOI inverter for low power applications," SOI CONF 1996.

[5]   http://electronicdesign.com/article/test-and-measurement/ characterize_jitter_measurements_on_10g_signals.aspx

[6]   V. Liot., P. Flatresse, J.M. Fournier, M. Belleville, "Advanced CAD methodology for history effect characterization in partially depleted SOI libraries," Solid-State Electronics 49, 2005, pp. 1466-1476

[7]   K. Bernstein, N J. Rohrer, "SOI Circuit Design Concepts" Kluwer Academic Publishers, Dordrecht, 2000, pp.69-74

[8]   Iijima M.et al;. "A technique for high-speed circuits on SOI using look-ahead type active body bias control," ISCAS 2004.

[9]   Duckhyun Chang et al.; "Temperature dependent hysteretic propagation delay in FB SOI inverter chain," SOI CONF 1999.

Adaptable Stimulus Driver for Epileptic Seizure Suppression

Ming-Dou Ker[1,2], Wei-Ling Chen[1], and Chun-Yu Lin[1]

[1] Institute of Electronics, National Chiao-Tung University, Hsinchu, Taiwan
[2] Department of Electronic Engineering, I-Shou University, Kaohsiung, Taiwan

Abstract – The novel implantable stimulus driver for epileptic seizure suppression with low power design and adaptive loading consideration was proposed in this work. The stimulus driver consisted of the output stage, charge pump system, and adaptor can constantly provide 40-µA output stimulus currents, as the electrode impedance varies within 10~300 kΩ. The performances of this design have been successfully verified in a silicon chip fabricated by a 0.35-µm 3.3-V/24-V CMOS process. The power consumption of this work was only 1.1~1.4 mW. The proposed stimulus driver has been integrated into closed-loop epileptic seizure monitoring and controlling system for animal test.

I. INTRODUCTION

Epilepsy, one of the common neurological diseases, is caused by transient abnormal discharge in brain [1]. There are more than 40 types of epilepsies, and these epilepsies classify by location of seizures, syndromes, and causes. Each of epilepsies has unique seizure type, age group, diagnosis, and treatment. Epilepsy is usually treated by pharmacologic treatments. However, every kind of medicines may lead to side effects, such as blurry vision, dizziness, headaches, and fatigue [2]. Besides, some patients do not respond to medicines. For these patients, they may take surgical treatment. However, surgical treatment is not suitable for every patients, because it is risky that may cause functional loss.

Except for pharmacologic and surgical treatments, the electrical stimulations have been investigated recently [3]. It has been demonstrated that the abnormal discharge signal to cause epilepsy can be suppressed by electrical stimulation before epileptic seizures happen. Advantages of electrical stimulation are flexible, recoverable, and non-destructive. Fig. 1 shows the block diagram of an implantable stimulus driver for epileptic seizure suppression. The considerations of stimulus driver include safety, reliability, charge balance, voltage compliance, and density of stimulus site. Some studies revealed that unbalanced stimulus current causes net charge stores in body and leads to problems of pH shift, ionic charges near the implanted electrodes, and erosion of the electrode material [4]. To prevent from these problems, several implantable stimulus drivers have been presented. The aim of these works concentrated on balance of anodic pulse and

cathodic pulse. The methodology, dynamic current balancing [5], and feedback DAC calibration, have been used for minimizing mismatch between anodic and cathodic currents. Besides, in order to achieve large voltage compliance to close fixed power supply and maintain high output impedance to hold the constant stimulus current irrespective of highly variety of stimulus site and tissue/electrode impedances, improved current sources have been presented. The fully cascade and wide swing cascade current sources are used in output stage of stimulus driver to increase output resistance [6]. Another voltage controlled resisters (VCR) current source gains large voltage compliance close to the fixed power supply by utilizing MOS transistors in deep triode region [6]. However, if variance of output voltage caused by variety of electrode impedance exceeds fixed power supply, stimulus current will decreases dramatically.

In this work, the implantable stimulus driver for epileptic seizures with low power design and adaptive loading consideration is proposed. The new stimulus driver with adaptor can detect the variation of electrode impedance to minimize power consumption. This proposed stimulus driver can deliver charge-balanced bi-phase stimulus current by two leads electrode per stimulus site with single supply voltage (V_{DD} = 3.3V). A 0.35-µm 3.3-V/24-V CMOS process is used in this work for the chip implementation.

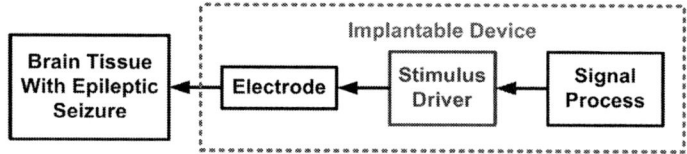

Fig. 1. Block diagram of implantable stimulus driver.

II. NEW PROPOSED DESIGN

The new implantable stimulus driver for seizure control with consideration of power consumption and adaptability on electrode impedance is proposed. As shown in Fig. 2, the proposed stimulus driver consists of the output stage, charge pump system, adaptor, and two leads electrode set. The output stage is realized with 24-V devices, while the adapter is realized with 3.3-V devices. The targeted stimulus current is 40 µA, which is able to suppress epileptic seizures of Long-Evans rats [7]. The effective electrode impedance is fully resistive and it will vary from 100 kΩ to 250 kΩ [7]. That

---

This work was partially supported by National Science Council (NSC), Taiwan, under Contract of NSC 99-2220-E-009-021, and by the "Aim for the Top University Plan" of National Chiao-Tung University and Ministry of Education, Taiwan.

978-1-4244-9019-6/11 $26.00 © 2011 IEEE

is to say, stimulus voltage and required power supply are much higher than the operating voltage (V_{DD} = 3.3V). Depending on electrode impedance, the proposed stimulus driver with the adaptor regulates the high operating voltage for output stage of 24-V devices by feedback of the adaptor. The charge pump system is designed to control the charge pump circuit to generate the required high operating voltage (V_{CC}).

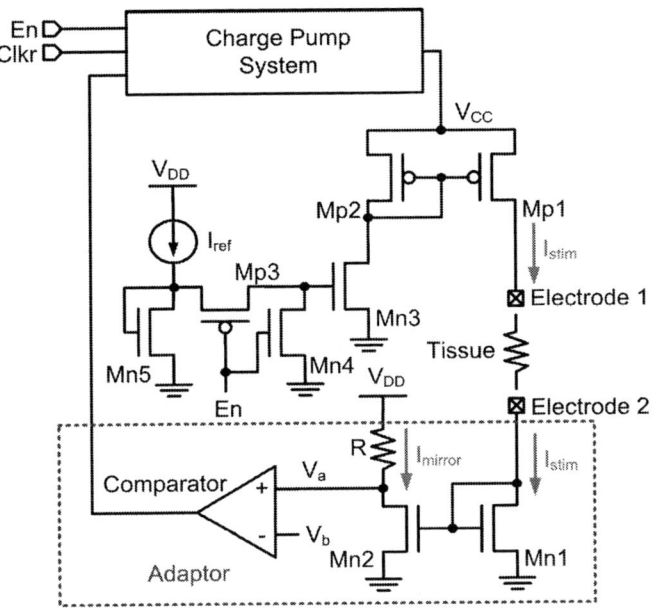

Fig. 2. Circuit schematic of the proposed stimulus driver.

The stimulus current source consists of a current mirror (Mp1 and Mp2) and bias circuit. During the stimulus driver "turn-on" interval, enable (En) is low (0V), Mn4 is switched off, Mp3 is switched on, and Mn3 is biased through Mp3. The stimulus current (I_{stim}) is delivered by Mp1. During the stimulus driver "turn-off", enable (En) is high (3.3V), Mn4 is switched on, and Mn3 is connected to ground through Mn4, no stimulus current is delivered.

The proposed stimulus driver adopts two leads electrode set per stimulus site (electrode 1 and electrode 2) to generate bi-phase stimulus current. Whenever the stimulus current is delivered by Mp1 and flows into electrode 1, the direction of stimulus current is controlled by the on-chip switches of two leads electrode set. As illustrated in Fig. 3, two leads electrode set consists of four switches and two electrodes. While anodic current is required, the switch_anodic is switched on and switch_cathodic is switched off. Anodic stimulus current flows from the top site of tissue to the down site of tissue. While cathodic current is required, the switch_anodic is switched off and the switch_cathodic is switched on. The cathodic stimulus current flows from the down site of tissue to the top site of tissue. After stimulus current passes through tissue, electrode 2 is used to collect the stimulus current back to the adaptor.

The adaptor of stimulus driver consists of a current mirror (Mn1 and Mn2), resistor (R), and comparator. While stimulus

current (I_{stim}) is collected by electrode 2 to pass through Mn1 of current mirror, the gate of Mn2 is biased to induce the proportional current (I_{mirror}). The adaptor utilizes I_{mirror} flowing through poly resistor (R = 41.25kΩ) to generate a voltage signal (V_a), which can be expressed as

$$V_a = V_{DD} - I_{mirror} * R \qquad (1)$$

Another voltage signal (V_b) which is used to compare with V_a is $0.5 \times V_{DD}$. Whenever the stimulus driver is turned on to stimulate the tissue, the comparator in the adaptor compares these two voltage signals (V_a and V_b) and distinguishes the amplitude of stimulus current. The operating voltage (V_{CC}) for stimulus driver is controlled by the output of comparator.

When the electrode impedance is increasing, the stimulus current is decreased. The insufficient stimulus current leads to V_a higher than V_b, the output of comparator will become high (3.3V), and charge pump system is activated to provide stimulus current source with higher voltage until stimulus current reaches 40 µA. When electrode impedance is decreasing, the V_a is also decreasing. The larger stimulus current leads to V_a lower than V_b, the output of comparator will become low (0V), and charge pump system is inactivated to provide stimulus current source with lower voltage until stimulus current returns 40 µA.

The charge pump system of stimulus driver consists of the 4-stage charge pump circuit, buffer, clock control, and on-chip output loading capacitance (CL), as shown in Fig. 4 [8]. The 4-stage charge pump circuit uses 24-V devices with deep N-well, and the rests of the circuit use 3.3-V devices. The clock control is utilized to generate interweaved clock signals which depends on the reference clock (Clkr), enable signal (En), and the feedback signal from the comparator. There is a compromise between CL size and clock frequency, so the used clock frequency is 25 MHz.

At the beginning of stimulation, there is no charge stored at the output loading capacitance (CL) of charge pump system, and output voltage of charge pump system (V_{CC}) is initially 0V. The stimulus current is approaching 0 A; therefore, as Eq. (1) shown, V_a is higher than V_b and charge pump system is activated. Therefore, the output voltage of charge pump system increases and stimulus current is delivered by the current source. The charge pump system keeps activated until the stimulus current is slightly higher than 40 µA to cause V_a lower than V_b, that leads to the output of comparator becomes low (0V). Meanwhile, the stimulus current reaches the designed amplitude, and the output voltage of charge pump system equals to the required operating voltage (V_{CC}). The inactivated charge pump system causes that stimulus current and output voltage of charge pump decrease again. Therefore, by changing state of charge pump system constantly, the output voltage of charge pump system, as well as the operating voltage for stimulus driver, keeps at the required operating voltage.

The stimulus driver with the adaptor can match the least required voltage. The proposed stimulus driver with adaptor

978-1-4244-9019-6/11 $26.00 © 2011 IEEE

provides any amplitude of electrode impedance with least operating voltage for 24-V device, which is sufficient to deliver the designed stimulus current, but not provides a fixed operating voltage for 24-V devices.

The proposed bi-phase stimulus driver to suppress epileptic seizure with adaptive loading design and low-power consideration has been fabricated in a 0.35-μm 3.3-V/24-V CMOS process. Fig. 5 shows the chip photograph of the fabricated stimulus driver. The layout size of the proposed stimulus driver is about 1000×700 μm².

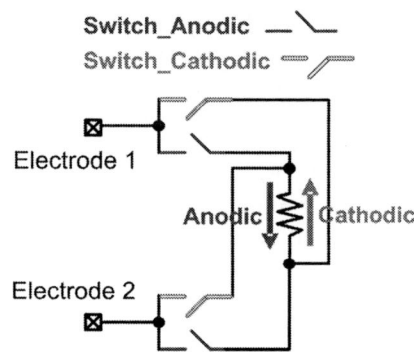

Fig. 3. Two leads electrode set consists of four switches and two electrodes.

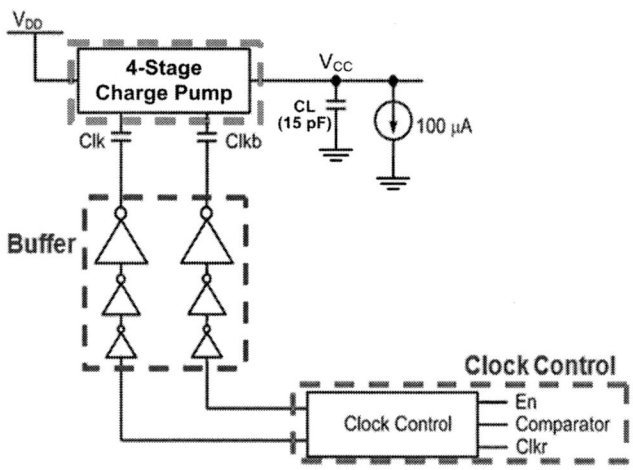

Fig. 4. Circuit schematic of charge pump system.

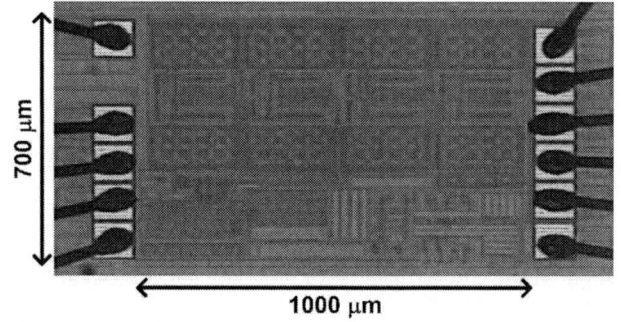

Fig. 5. Chip photograph of the fabricated stimulus driver in a 0.35-μm CMOS process.

III. EXPERIMENTAL RESULTS

Under measurement, Agilent E3631A is used to provide the fixed 3.3-V V_{DD}. Hp 33120A is used to supply the 25-MHz clock signal and to enable the implanted stimulus driver. Tektronix 3054B is used to observe stimulus voltage/current of the stimulus driver.

While enable signal is given, the proposed design starts to deliver the stimulus current. Fig. 6 summarizes the measurement results of stimulus currents when the electrode impedance varies from 10 kΩ to 300 kΩ. The anodic and cathodic stimulus currents match the specification of 40 μA under the various range of electrode impedance. Fig. 7 summarizes the operating voltage verses electrode impedance. The average power consumption of the stimulus driver is 1.1~1.4 mW. All of measurement results are summarized in Table I.

The proposed stimulus driver is further integrated into closed-loop epileptic seizure monitoring and controlling system for animal test [7]. The sketch diagram of measurement setup for animal test is shown in Fig. 8. In this experiment, the stimulus current is conducted by a 4-microwire bundle, each made of Teflon-insulted stainless steel. Whenever the system detects an epileptic seizure, the proposed stimulus driver is activated by a trigger signal to stimulate the Long-Evans rat. Fig. 9 shows the test result, where the epileptic seizure with abnormal discharge was detected during 1~6 seconds, and the system triggered the proposed stimulus driver to stimulate the Long-Evans rat. After stimulation, the intensive and rapidly brain activity was suppressed.

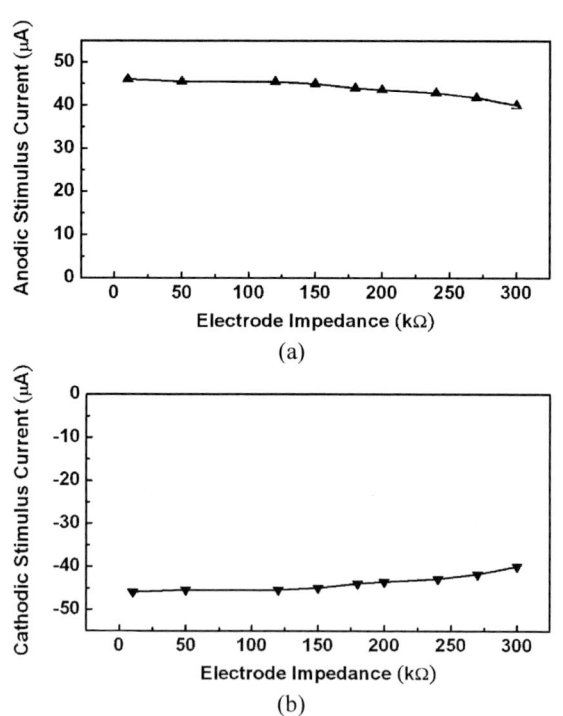

Fig. 6. (a) Anodic and (b) cathodic stimulus currents under different electrode impedances.

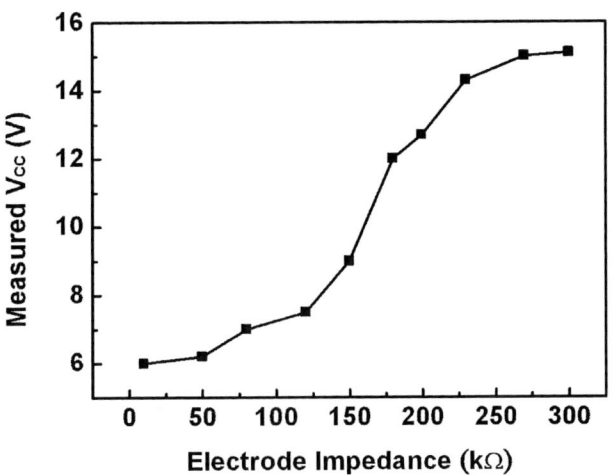

Fig. 7. Output voltage of charge pump system under different electrode impedances.

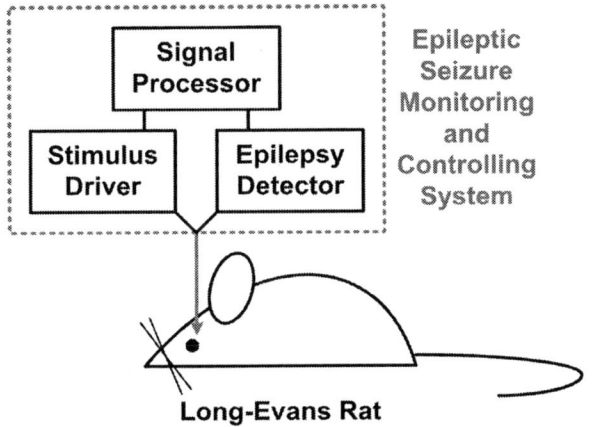

Fig. 8. Measurement setup for animal test.

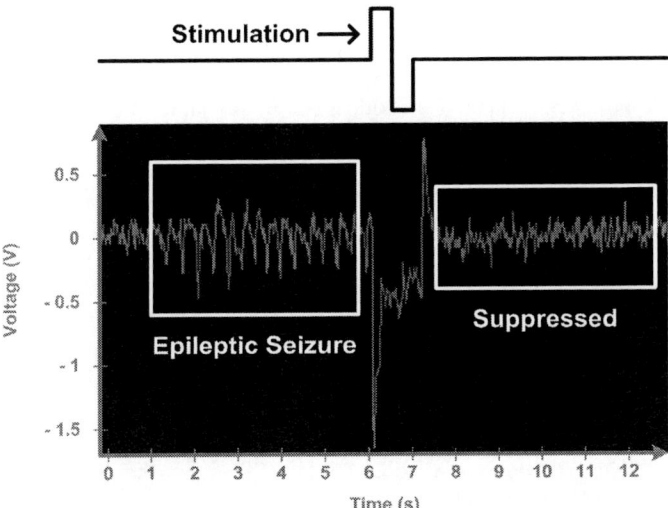

Fig. 9. Experimental result on EEG signal of Long-Evans rat.

Table I. Summary on the Proposed Stimulus Driver

Technology	0.35-µm 3.3-V/24-V CMOS Process
Layout Area	1000 µm x 700 µm
Available Stimulus Current	~ 40 µA
Electrode Configuration	Two Interface Leads Per Site
Electrode Impedance	10~300 kΩ
Supply Voltage	3.3 V
Power Consumption	1.1~1.4 mW

IV. CONCLUSION

Design of bi-phase stimulus driver to suppress epileptic seizure with adaptive loading consideration is proposed. The stimulus driver consists of the output stage, charge pump system, and the adaptor. While the electrode impedance varies from 10 kΩ to 300 kΩ, the proposed stimulus driver can constantly deliver 40-µA stimulus current. The power consumption of this work is only 1.1~1.4 mW. The stimulus driver is successfully integrated into closed-loop epileptic seizure monitoring and controlling system to verify its performance.

REFERENCES

[1] P. Hese *et. al.*, "Automatic detection of spike and wave discharges in the EEG of genetic absence epilepsy rats from Strasbourg," *IEEE Trans. Biomedical Engineering*, vol. 56, no. 3, pp. 706-717, Mar. 2009.

[2] G. Cascino, "Epilepsy: contemporary perspectives on evaluation and treatment," in *Proc. Mayo Clinic*, 1994, pp. 1199-1211.

[3] W. Stacey and B. Litt, "Technology insight: neuroengineering and epilepsy - designing devices for seizure control," *Nature Clinical Practice Neurology*, vol. 4, pp. 190-201, Feb. 2008.

[4] J. Gwilliam and K. Horch, "A charge-balanced pulse generator for nerve stimulation applications," *J. Neuroscience Methods*, vol. 168, no. 1, pp. 146-150, Feb. 2008.

[5] S. Guo and H. Lee, "Biphasic-current-pulse self-calibration techniques for monopolar current stimulation," in *Proc. IEEE Biomedical Circuits and Systems Conference*, 2009, pp. 61-64.

[6] M. Ghovanloo and K. Najafi, "A compact large voltage-compliance high output-impedance programmable current source for implantable microstimulators" *IEEE Trans. Biomedical Engineering*, vol. 52, no. 1, pp. 97-105, Jan. 2005.

[7] C. Young *et. al.*, "A portable wireless online closed-loop seizure controller in freely moving rats," *IEEE Trans. Instrumentation and Measurement*, vol. 60, no. 2, pp. 513-521, Feb. 2011.

[8] M.-D. Ker *et. al.*, "Design of charge pump circuit with consideration of gate-oxide reliability in low-voltage CMOS processes," *IEEE J. Solid-State Circuits*, vol. 41, no. 5, pp. 1100-1107, May 2006.

Gate-Driven 3.3V ESD Clamp Using 1.8V Transistors

Guang-Cheng Wang, Chia-Hui Chen, Wen-Hsin Huang, Kuo-Ji Chen, Ming-Hsiang Song and Ta-Pen Guo

Taiwan Semiconductor Manufacturing Corp.

Hsinchu, Taiwan, R.O.C.

Abstract—**A new gate driven 3.3V ESD clamp circuit using 1.8V transistor is proposed. This new clamp circuit is suitable for ESD protection of legacy 3.3V I/O interface circuit in SOC chips which use only 1.8V I/O transistors. This clamp along with 3.3V I/O have been demonstrated in 40nm 1.8V process. Life-time test can pass 1000-hours prolonged operation. ESD/Latch-up can pass HBM 3KV, MM 300V, and +/-200mA current triggering and 4.95V (1.5 x VDD) over-voltage test.**

Index Terms— **gate-driven, gate oxide over-stress, human-body model (HBM), machine model (MM)**

I. INTRODUCTION

Logic SOC chips in advanced technology integrate I/O signal interfaces of different operation speed and voltages ranging from 1.5V, 2.5V to 3.3V or even higher. To support high speed interface operation requirements, below 40nm, 1.8V I/O device is usually adopted. For the legacy I/O signal interfaces higher than 2V, it is usually implemented by stacking 1.8V I/O transistors [1], [2] to prevent from introducing 2nd thicker I/O devices to save the cost. To design an effective ESD power clamp for these higher voltage I/O interface using lower voltage transistors become an important issue in chip design. Ideal ESD power clamp should have low trigger on voltage, high immunity for latch up and enough margin for reliability under high voltage operation. Several proposals based on substrate triggered snapback scheme were proposed [3], [4], but these clamps could be false triggered in latch up event and didn't address on the performance of life time tests.

In this paper, a new 3.3V ESD clamp using pure 1.8V I/O transistors is proposed. This ESD clamp is fabricated in 40nm 1.8V process and serves as 3.3V power rail to VSS ESD protection circuits in 3.3V I/Os. This design has been silicon verified and proven to pass human-body-model 3KV, machine-model 300V in whole chip level ESD zapping. In latch up characterization, this 3.3V ESD clamp along with 3.3V I/Os pass +/-200mA current triggering and 4.95V (1.5xVDD) over-voltage test. This ESD clamp along with 3.3V I/Os also pass 1000-hrs prolonged operation test and shows no increase in stand-by leakage measurement.

II. ESD PROTECTION CIRCUITS

A. Operation Under Normal Circuit Operation Condition

Fig.1 shows the proposed 3.3V ESD clamp. This clamp is composed of pure 1.8V I/O transistors and should be kept off in normal operation to avoid unnecessary current leakage path. ESD clamp should be designed so that it wouldn't suffer any reliability issue in normal operation. The bigFET MOS (MN1,MN2) of this ESD clamp are designed in cascode structure and are the main ESD discharge path. In normal operation, both 1.8V and 3.3V power are on. 1.8V power may come from external supply or generate from on-chip regulator. In the former case, power up sequence should be set in such a way that 1.8V power should be ready before 3.3V power is on. In the latter case, 1.8V power should track 3.3V to ensure gate oxide won't be over-stressed. The gate voltage of MN2 (Vd) is pulled down to 0V by ESD detection circuit R3, MN4 and MN3 so that bigFET MOS is turned off in normal operation. The gate of MN1(Vb) is tied to 1.8V through resistor R1 to ensure gate oxide won't be over-stressed. Vc is 1.8V-Vtn. Vtn denotes the threshold voltage of MN1. In this way, Vgs and Vgd of MN1 and MN2 are kept within 1.8V. MP4 is designed to be as a capacitor. Its source, drain and bulk are all tied to 3.3V. To avoid gate oxide over-stress, Va is tied to 1.8V through resistor R2. MP2 and MP3 are implemented to isolate net a from pulling down to ground. The Vgs and Vgd of all transistors in this clamp are all kept within 1.8V. The bulk of all PMOS transistors is biased by control circuit MP5, MP6 to track the maximum voltage of 1.8V and 3.3V to avoid leakage current through parasitic diode of PMOS.

B. Operation Under ESD Transient

In ESD zapping of 3.3V-VDD to ground, ESD charge would couple up the nodal voltage Va through MP4 and turn on MP1, MP2. This high Va passing through MP1, MP2 can turn on MN1, MN2 and achieve ESD discharge purpose. MP4 which acts as a capacitor together with R4 are the main trigger circuit for MN1, MN2. Both devices dominate the RC-time constant of this clamp. In addition to the "channel" discharge path, the parasitic bipolar of MN1, MN2 can be triggered on and enter "snap-back" mode. Resistors R1, R2, R3 and R4 are of the order of Mega-Ω. The gate voltage of MN3 is close to 0V due to the vicinity of R3 and MN4 in ESD zapping so that MN3 is kept off and MN2 can be turned on by ESD charge. MP1 is implemented so that ESD charge can have similar impedance on the way to turn on both MN1 and MN2. If the bulk of MP1 and MP2 is tied to fixed power (1.8V or 3.3V) instead of being biased by control circuit (MP5, MP6), the ESD charge would leak through parasitic diodes of MP1,MP2 and fail to turn on MN1,MN2.

C. Design and Simulation of ESD Clamp

The ESD design window of 3.3V ESD clamp should be that both trigger on voltage and holding voltage are higher than maximal operation voltage (3.63V) and lower than breakdown

978-1-4244-9019-6/11 $26.00 © 2011 IEEE

voltage of any junctions and the trigger on voltage of internal 1.8V transistors (Vt1~7.2V). However, the cascode structure of bigFET MOS to prevent from gate oxide over-stress shows higher trigger on voltage (9.5V) if not equipped any trigger on circuits and can't protect internal circuit effectively. Capacitor-Resistor trigger circuit (MP4-R4) is implemented to lower down the trigger on voltage of bigFET MOS. The relative ratio of capacitance of MP4 to that of bigFET MOS (MN1,MN2) dominates the nodal voltage Va in ESD zapping and therefore the trigger on voltage (Vt1) of bigFET MOS. The lowest trigger on voltage can be exactly the holding voltage. Hspice-simulation can give some ideas of the relative ratio of capacitance and the clamp's response in ESD event. Fig.2 shows the Hspice-simulated result when human-body-model 3KV ESD is zapped on the ESD bus line of this clamp. Capacitor-Resistor (MP4-R4) trigger circuit indeed works in simulation so that the voltage waveform of 3.3V bus line shows a "channel-discharging" behavior. Internal nodal voltage Vd is almost the same as Vb, which indicates that MN1 and MN2 can be triggered on at the same time.

Fig.3 shows the Hspice-simulated result of trend of channel length of MN1, MN2 and off current with the same ESD finger width. The channel length and width should be designed so that holding voltage, trigger on voltage can stay within ESD design window and leakage current in normal operation is acceptable. In advanced technology below 40nm, the off current of 1.8V I/O transistor increases tremendously with temperature. As shown in Fig.3, off current at 25°C is of the order of nano Ampere, while off current at 125°C is of the order of micron Ampere. The longer the channel length is, the smaller the off current become.

III. EXPERIMENTAL RESULTS

A. Leakage current

The leakage current from silicon measurement shows good correlation with that from simulation as shown in Fig.3.

B. ESD Robustness

Fig.4 shows the TLPG-measured I-V curve of this ESD clamp. This ESD clamp can be turned on as imposed pulse voltage is higher than 1V. As pulse voltage is lower than 4.2V, this ESD clamp discharges by means of channel of MN1, MN2 only. The turn on resistance is about 2.5Ω. As pulse voltage is higher than 4.2V, the parasitic bipolar of MN1, MN2 is turned on and enter "snap-back" mode. The holding voltage (Vh) of this ESD clamp is 4.3V and the turn on resistance is 0.14Ω. The secondary breakdown current I_{t2} of this Clamp is 4.4A. On chip ESD test shows that the 3.3V I/Os can pass HBM 3KV and MM 300V with this 3.3V ESD clamp. A second ESD clamp with gate resistor as shown in Fig.1(b) is served as comparison. The gate of MN1 is tied to 1.8V and gate of MN2 is tied to ground. As shown in Fig.4, this second ESD clamp shows high trigger on voltage ~ 9.2V which is only slightly lower than the trigger on voltage (~9.5V) of cascode bigFET without any trigger on circuit. The 3.3V I/Os with this second ESD clamp fails HBM 2KV and MM 200V.

C. Turn-on Verification

The turn on behavior of the proposed ESD clamp can be verified by imposing a sharply rising voltage pulse with a rise time of 3ns on 3.3V power rail. Fig.5(a) shows that a 3.3V voltage pulse is imposed on this ESD clamp. The 3.3V power is pulled down to 1.5V first and then recover back to 3.3V after 50us. The sharply rising voltage pulse acts like ESD and will trigger on bigFET MN1,MN2 and results in a short circuit path to ground. That's why the 3.3V power is pulled down to 1.5V at first. The parasitic bipolar of MN1, MN2 won't turn on and sustain because the 3.3V amplitude of imposed pulse is lower than the holding voltage 4.3V. Fig.5(b) shows that a 6V voltage pulse is imposed on the same ESD clamp. The 3.3V power is pulled down to 1.5V first and then get higher to 4.3V and stay at that voltage level till the end of pulse. The parasitic bipolar of MN1, MN2 get turned on and sustain because the 6V amplitude of imposed pulse is higher than the holding voltage 4.3V. Fig.5(c) shows that a 6V voltage pulse imposed on the gate-resistor grounded cascode NMOS ESD clamp in Fig.1(b). Since its trigger on voltage is as high as 9.2V, 6V voltage pulse won't trigger it on.

IV. CONCLUSION

Fig.6 shows the layout of the proposed 3.3V ESD clamp using 1.8V I/O transistors with active area 54μm x 158μm. This ESD clamp has been silicon verified and can pass human-body-model 3KV and machine-model 300V in whole chip level zapping. This ESD clamp together with 3.3V I/Os also pass 1000-hrs prolonged operation tests and show no increase in stand-by leakage.

ACKNOWLEDGEMENTS

The authors gratefully acknowledge the technical support of the members in Electrical Lab in TSMC design and technology platform, Pi-Chia Shi, Chester Kuo and Bruce Huang.

REFERENCES

[1] Ming-Dou Ker and Yan-Liang Lin,"2xVDD-Tolerant I/O Buffer with 1xVDD CMOS devices," IEEE CICC, p.539-542, 2009

[2] Hui-Wen Tsai and Ming-Dou Ker ,"Design of 2xVDD-Tolerant I/O Buffer with Consideration of gate oxide reliability and hot carrier degradation," Electronics, Circuits and Systems 2007, pp.1240-1243

[3] Ming-Dou Ker, Chang-Tzu Wang, "Design of 3.3V ESD Protection Circuit in Low-Voltage CMOS Process", IEEE Trans. Device Mater .Rel., vol.9,no.2, pp. 49-58, Mar.2009

[4] Ming-Dou Ker, Wei-Jan Chang, "ESD Protection Design with On-Chip ESD Bus and 3.3V ESD Clamp Circuit for Mixed-Voltage I/O Buffers", IEEE Trans. Electron Devices, vol.55 ,no.6, pp.1409-1416, June.2008

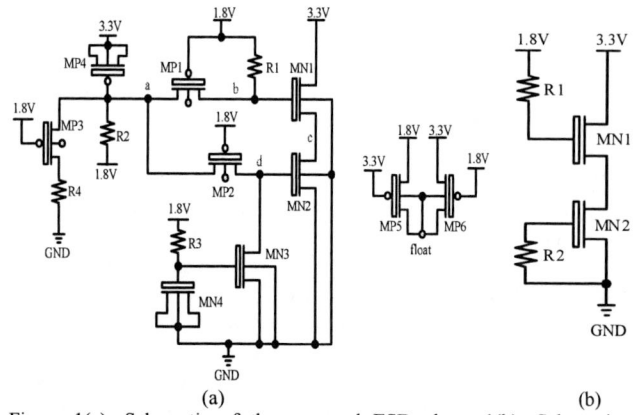

(a)

Figure 1(a): Schematic of the proposed ESD clamp 1(b): Schematic of gate-grounded cascode NMOS

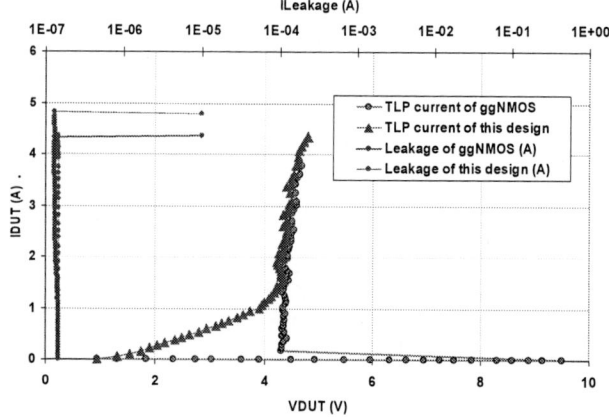

Figure 4: TLPG-measured I-V curve of gate-resistor grounded cascode NMOS and this proposed ESD clamp

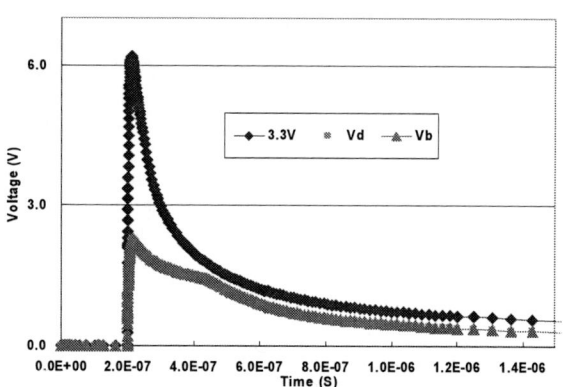

Figure 2: 3.3V bus voltage waveform as human-body-model 3KV is zapped. Vd almost overlap with Vb, which indicates that both MN1,MN2 can be turned on at the same time.

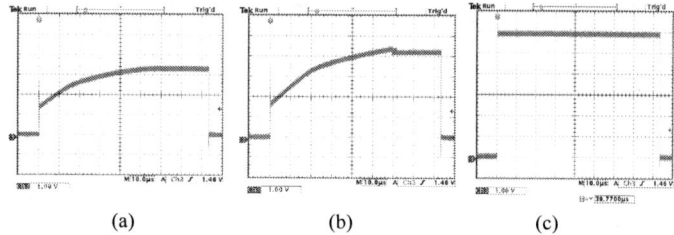

(a)   (b)   (c)

Figure 5(a): Waveform of 3.3V power rail when 3.3V pulse voltage is imposed on this proposed ESD clamp. 5(b) Waveform of 3.3V power rail when 6V pulse voltage is imposed on this proposed ESD clamp. 5(c) Waveform of 3.3V power rail when 6V pulse voltage is imposed on the gate-grounded cascode NMOS ESD clamp.

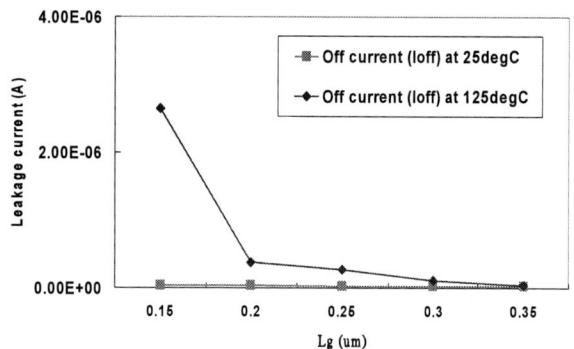

Figure 3: Off current at 25°C, 125°C at 3.3V

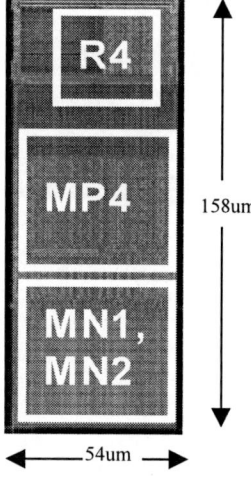

Figure 6: Layout of the proposed 3.3V ESD clamp

978-1-4244-9019-6/11 $26.00 © 2011 IEEE   198

Beta-Matrix ESD Network: throughout End of placement rules?

J. Bourgeat, P. Galy, B. Jacquier

STMicroelectronics , 850 rue Jean Monnet , 38920 Crolles , France

Abstract—**Electrostatic Discharge (ESD) protection for advanced CMOS technologies is based on efficient device Network. But these protection strategies imply some constraint on IO and particularly on the frame and the placement in IO ring. In this context we develop and propose an ESD network with Beta-Matrix power device and its own trigger circuit which are integrated in each IO. We obtain a new local strategy which allows removing all IO placement constraint.**

Index Terms— **ESD, ESD network, SCR, Beta-Matrix**

I. INTRODUCTION

THE scaling down of CMOS process leads to degrade the intrinsic ESD (ElectroStatic Discharge) robustness. It is important to develop a robust and efficient ESD network with optimized ESD devices. For example Silicon Controlled Rectifier structures (SCRs) have superior ESD performances with low on-resistance, high second breakdown current, low parasitic capacitance and reduced footprint area in comparison of classical MOS Switch (MOSSWI) [1-2]. But SCRs or bi-SCRs (Bidirectional SCRs) are mainly used as local protection to derivate an ESD current between IO and ground pins in advanced CMOS digital chips [3-6]. Some works present bidirectional local protection between IO/VDD and IO/GND [7-8].

This work present a new concept where the classical ESD network is replaced by a Beta-Matrix structure providing three bidirectional local protections and allowing three ESD paths between VDD/IO, IO/GND and VDD/GND [9].

Second section presents the global and distributed ESD network and the problematic. Then section three introduces the concept of Beta-Matrix with its trigger circuit. Section four presents ESD/LU qualifications and section V the implementation of the protection in digital IO frame.

II. PREVIOUS ESD NETWORK

A. Global ESD Network

This network [10-11] is composed by one bidirectional protection between VDD and ground (Clamp). Furthermore, to derivate the ESD current to VDD or Ground, two dual-diodes are implemented in each IO. Usually, the clamp includes a big MOSFET of few millimeters, its trigger circuit with an RC filter and buffer stages for positive pulses and a reverse diode. This diode is added to provide a bidirectional protection.

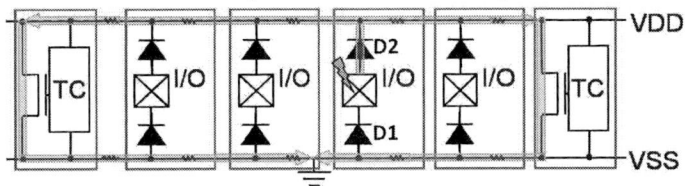

Fig. 1 : Global ESD Network

Fig. 1 illustrates the case where positive discharge is applied between one IO pin and ground. D1 diode is in reverse mode, current flows through D2 diode on the VDD rail to the central protection (clamp). Thus the protection is efficient if the maximal voltage during an ESD event is lower than the minimum breakdown voltage (i.e: gate oxide). The equation is presented below:

$$V_{t1D2} + R_{VDD} \cdot I_{ESD} + V_{t1PC} < BV_{min} \qquad (1)$$

V_{t1D2} is the diode turn on voltage ($\approx 0.6V$), R_{VDD} is the rail resistor between D2 and clamp, V_{t1PC} is the clamp turn on voltage and BV_{min} is the minimum breakdown voltage.

One central protection is not enough to protect all the IC. Indeed, an overvoltage is associated to the rail resistance and a critical distance is defined between two consecutive central protections. One way to limit this overvoltage is the distributed network.

B. Distributed ESD Network

This network consists in distributing a part of the big MOSTFET in each IO [12-13]. A "Trigger" rail connects all NMOS gates. The trigger circuit is implemented in the power IO (clamp) and it is connected between "Boost", "Trigger" rails and ground.

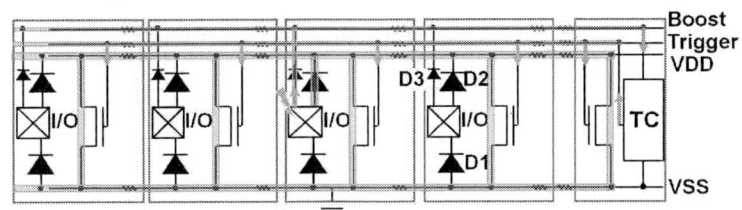

Fig. 2 : Distributed ESD Network

Fig. 2 shows the distributed network during an ESD event between one IO and ground. During this event a little part of

978-1-4244-9019-6/11 $26.00 © 2011 IEEE   199

current flows through D3 diode and "Boost" rail, then the trigger circuit detects the stress and turns all the NMOS on via "Trigger rail". So the ESD current is derivate by D2 diode to VDD and is distributed to NMOS. This solution allows to reduce the rail resistance impact and so the associated overvoltage. This strategy increases the distance between two power IOs (clamps).

C. Problematic

ESD objective is given by the technology node. For the global ESD network, it defines the maximal distance between two clamps (maximal resistance R_{VDD}) and number of IOs. For the distributed ESD network, this placement constraint is reduced since the ESD current flows through NMOS of the nearest IOs and not through VDD rail. However, the trigger circuit is still in IO power, with regular placement, to turn NMOS on homogeneously. This placement constraint depends on signal propagation time through "Boost" rail, activation time of the trigger circuit and propagation time through the "Trigger" rail to activate NMOS. Moreover, the trigger circuit should be able to turn enough NMOS on and should have a commutation speed enough. A trade of is necessary and trigger circuit has to be optimized.

Even if the placement constraint is relaxed, distributed network does not solve the problem completely. On the contrary, it generates new IO frames to integrate "Boost" and "Trigger" rails. The trigger circuit study becomes complex for a negligible gain in comparison of global network.

That is why it is necessary to design a new ESD protection strategy which is easier to integrate in IO frame, without power rail constraint and so without placement constraint. For that reason, all unidirectional elements (like diodes) are removed to bidirectional structures in order to reduce the power rail utilization and limit the associated overvoltage. This strategy (**Fig. 3**) is local with bidirectional protection between VDD/IO and IO/GND and implemented in each IO frame and can be distributed between VDD/GND. On the other hand, the trigger circuit is integrated in each IO to avoid placement constraint.

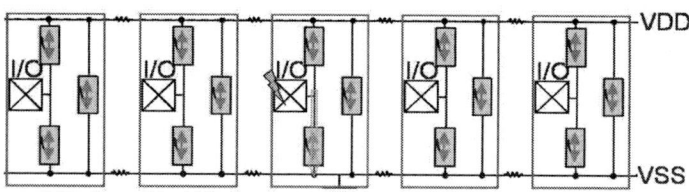

Fig. 3 : New local ESD network

To integrate this strategy in each IO, the protection is based on network of six optimized SCRs called Beta-Matrix and its own trigger circuit [9-10].

III. BETA-MATRIX AND TRIGGER CIRCUIT

A. Beta-Matrix

The Beta-Matrix behavior has been explained previously in [9]. In the same NWELL, each element is a P+/N+ diffusion into PWELL. Due to the topology, each nearest element generate a P+/PWELL/NWELL/DNWELL/PWELL/N+

sequence in forward and a N+/PWELL/NWELL/DNWELL/ PWELL/P+ sequence in reverse path and generate a full symmetrical SCR. Each PWELL is connected to VDD, IO or Ground by the P+ diffusion as it is presented in **Fig. 4**. This configuration gives two identical bi-SCRs between VDD/IO and IO/GND and a secondary diagonal bi-SCR between VDD/GND. Even if the performances and the number of diagonal bi-SCR in comparison of VDD/IO and IO/GND bi-SCR, it is distributed and repeated in each IO, and its robustness increases with the number of IO in IO ring.

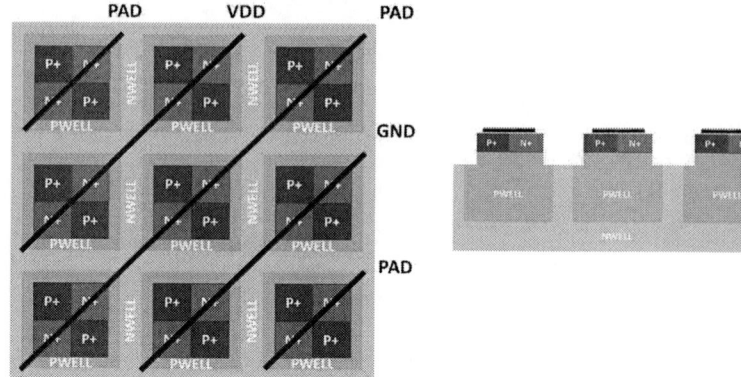

Fig. 4 : TCAD top view of Beta-Matrix with its metallic connection

Other metallic connections are possible to address different protection strategy.

Fig. 5 presents SCRs connection to obtain Beta-Matrix network in schematic.

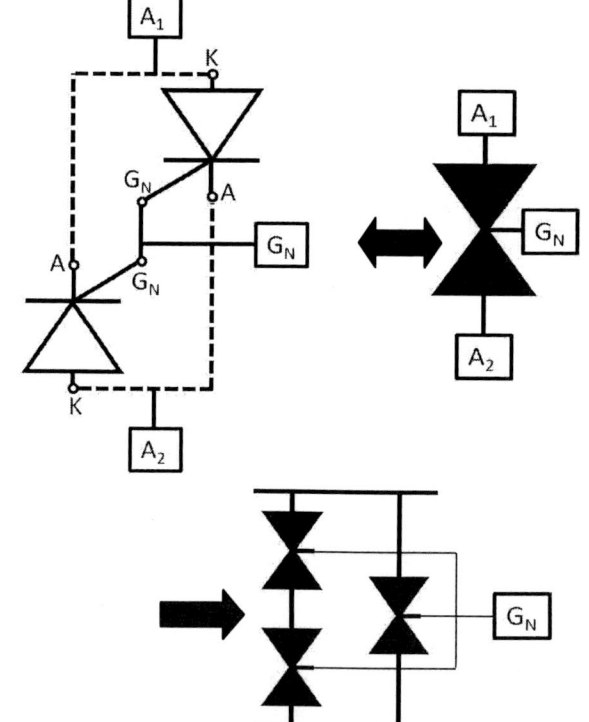

Fig. 5 : Schematic of SCR connection to generate a Beta-Matrix

B. Trigger circuit

To turn Beta-Matrix on, a specific trigger circuit has to be able to provide a trigger current path whatever the two pins stressed when an ESD event occurs. It has been develop thank [3] and [5-6]. The complete protection is presented in Fig. 6.

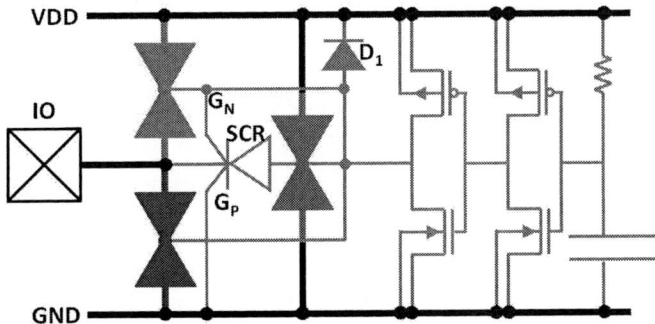

Fig. 6 : Electrical schematic of trigger circuit and Beta-Matrix

Thus, for a positive event between VDD/GND, the trigger circuit is composed by RC filter and an amplifier connected to the Gate N of Beta-Matrix. On the other hand, for a negative polarity a simple reverse diode D_1 activates the structure. Moreover, these elements are used for positive event between IO/GND and negative event between VDD/IO as well. Finally, a current path has to be provided to IO, without generating any leakage current in normal operation mode. An SCR is implemented with its Anode connected to the Gate N+ of Beta-Matrix and with its Cathode shorted to IO. SCR's gate N+ (G_N) is shorted to its Anode. SCR is turned on by its gate P+ which is connected to GND. For ESD event from IO to another IO all the presented current paths are activated.

IV. ESD/LU QUALIFICATION

A. S1/S2: Between IO/GND

The first HBM qualification is realized between IO and GND. The activated Beta-Matrix part is the blue one in **Fig. 7** and the current path to turn on, is shown by yellow arrows.

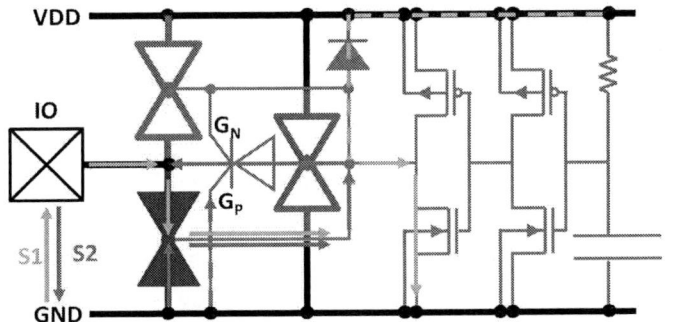

Fig. 7: trigger current paths between IO/GND

The second negative HBM qualification is between IO and GND and is noted S2. The activated Beta-Matrix part is still in blue in Fig. 7 but the current path is now in red. Current flows firstly through the SCR gate P+ (G_P), and then activates the SCR. SCR pulls the current up from the gate N+ of Beta-Matrix and turns Beta-Matrix on.

B. S3/S4: Between VDD/IO

The S3 qualification is between VDD and IO. The activated Beta-Matrix part is the red one in Fig. 8 and the current path to turn on is represented with the purple arrows.

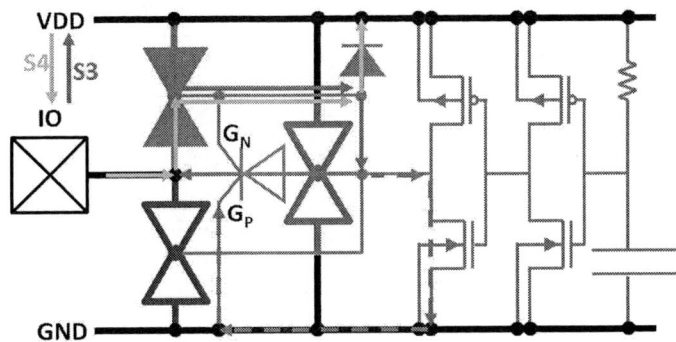

Fig. 8 : trigger current paths between VDD/IO

S4 is a negative HBM between VDD and IO. The activated Beta-Matrix part is still in red in Fig. 8 but the current path is now in green.

C. S5/S6: Between VDD/GND

This HBM qualification is realized between VDD and GND. The activated Beta-Matrix during the event is green in **Fig. 9**. The current path to turn on is the blue arrows. This protection is in parallel in each IO, so the robustness increases with the number of IO.

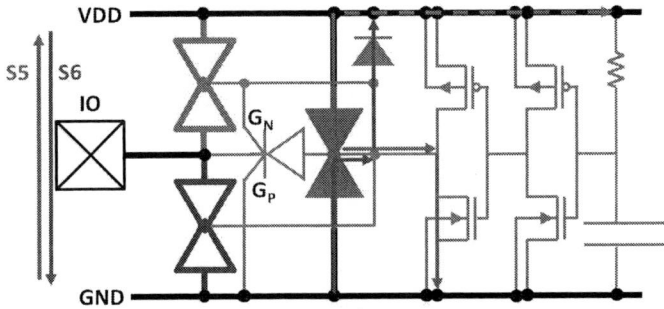

Fig. 9 : trigger current paths between VDD/IO

The negative HBM pulse between VDD and GND is presented with brown arrows.

D. S7: Between IO1/IO2

The last qualification is between two IOs. For this HBM event, the activated Beta-Matrix part is the red one and the blue one in both of IOs. The current paths are the same than IO1/VDD and IO1/GND in one hand, and in other hand are VDD/IO2 and GND/IO2 as seen in Fig. 10.

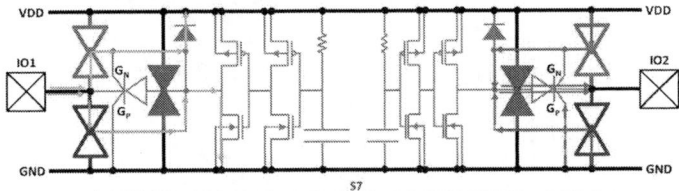

Fig. 10 : trigger current paths between IO1/IO2

For all presented stress combination, the ESD level reached is given by the table in Fig. 11 for two different power domains (GO1=1V and GO2=1.8V). The load to be protected is an output buffer with GO1 or GO2 MOSFET.

		HBM
S1: IO/GND	GO1	> 2kV
	GO2	> 2kV
S2: GND/IO	GO1	> 2kV
	GO2	> 2kV
S3: VDD/IO	GO1	> 2kV
	GO2	> 2kV
S4: IO/VDD	GO1	> 2kV
	GO2	> 2kV
S5: VDD/GND	GO1	> 2kV for 4 IOs
	GO2	> 2kV for 4 IOs
S6: GND/VDD	GO1	> 2kV
	GO2	> 2kV
S7: IO/IO	GO1	> 2kV
	GO2	> 2kV

Fig. 11 : HBM qualifications resume·

This ESD network protects an output buffer with minimum GO1 and GO2 gate length against an event of 2kV HBM whatever the IO pins stressed. This protection is also immune to external Latch Up at 300K and 400K whatever the IO ring configurations (up or down) at 100mA positive or negative injection and power sequence.

V. IMPLEMENTATION

Fig. 12 a) shows a 40um pitch digital IO with a distributed network and in b) the operation to implement Beta-Matrix network. Only two diodes for the ground continuity are kept between these two strategies.

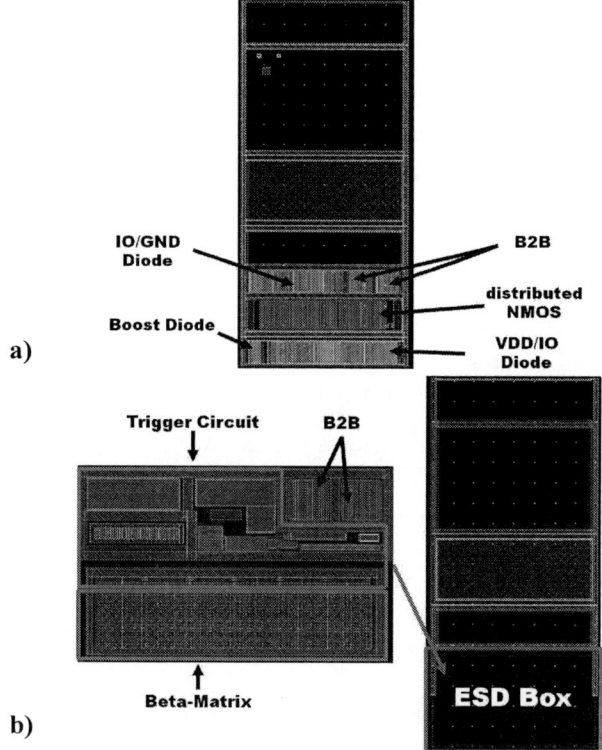

Fig. 12 : Implementation of distributed ESD network in a) and with Beta-Matrix in b) in one 40um pitch digital IO

The trigger circuit is implemented in each IO to turn-on locally Beta-Matrix between IO/VDD and IO/GND. With this solutions all power IOs are removed since there is no clamp or

global trigger circuit for consecutive IO anymore. It is a non negligible area gain in IO ring.

VI. CONCLUSION

This new ESD network is composed by one integrated protection with its own trigger circuit, and are implemented together in each 40um pitch digital IO. The protection area is compatible with the layout footprint of the ESD devices. Thus for CMOS32nm technology, this network protects against 2kV HBM, 100V MM, 500V CDM whatever the pin combinations when the minimum of IOs in the ring is equal to 4 without any Latch Up risk.

This protection strategy is independent of power rail resistance and removes all the placement rules on VDD and GND clamp.

ACKNOWLEDGMENT

The authors would like to thank the C. Richier's team for the ESD qualification.

[1] C. Russ, "GGSCRs: GGNMOS Triggered Silicon Controlled Rectifiers for ESD Protection in Deep Sub-Micron CMOS Process". EOS/ESD symposium, 2001, pp.22.

[2] L. R; Avery, "Using SCRs as Transient Protection Structures in Integrated Circuits". EOS/ESD symposium, 1983, pp.177.

[3] P. Galy, "Development and qualification of ESD protection demonstrators using SCR structure compatible with advanced CMOS technologies". IEW Workshop, 2010.

[4] J. Di Sarro, "Evaluation of SCR-based ESD protection devices in 90nm and 65nm CMOS technologies". IEEE IRPS 348-357, 2007.

[5] J. Bourgeat, "Local ESD Protection Structure based on Silicon Controlled Rectifier achieving very low Overshoot Voltage". EOS/ESD symposium, 2009.

[6] J. Bourgeat, "Evaluation of the ESD performance of local protections based on SCR or bi-SCR with dynamic or static trigger circuit in 32nm", ESREF symposium, 2010.

[7] A.Z.H. Wang, "On a Dual-Polarity On-Chip Electrostatic Discharge Protection Structure", IEEE transactions on electron devices, 2001, pp.978, VOL.48.

[8] P.Galy, "Comparison between isolated SCR & embedded dual isolated SCR power devices for ESD power clamp in C45 50A CMOS technology", ICICDT Symposium, 2010, pp.47.

[9] P.Galy, "β matrix concept for ESD power devices. Demonstrators in C45nm & C32nm CMOS technology", Proposed paper to EOS/ESD symposium, 2011.

[10] R. Merill, "ESD Design Methodology", EOSESD Symposium, 1993, pp. 233.

[11] T. Maloney, "Methods for Designing Low Leakage Power Supply Clamp", EOSESD Symposium, 2003.

[12] M. Stockinger, "Boosted and Distributed Rail Clamp Networks for ESD protection in Advanced CMOS Technologies", EOSESD Symposium, 2003, pp.17.

[13] C. Torres, "Modular, Portable and Easily Simulated ESD Protection Networks for Advanced CMOS Technologies", EOSESD Symposium, 2001.

Design of on-chip Transient Voltage Suppressor in a Silicon-based Transceiver IC to meet IEC System-Level ESD Specification

Ryan Hsin-Chin Jiang, Tang-Kuei Tseng, Chi-Hao Chen and Che-Hao Chuang

Abstract— **The on-chip Transient Voltage Suppressor (TVS) embedded in the silicon based transceiver IC has been proposed in this paper by using 0.8 μm Bipolar-CMOS-DMOS (BCD) process. The structure of the on-chip TVS is a high voltage Dual Silicon-Controlled-Rectifier (DSCR) with ±19V of high holding voltage (Vh) under the evaluation of 100ns pulse width of the Transmission Line Pulsing (TLP) system. The holding current (Ih) of the on-chip TVS is so high that can pass ±200mA latchup testing. Therefore, the on-chip TVS can be safely applied to protect the ±12V of signal level for RS232. The RS232 transceiver IC with on-chip TVS has been evaluated to pass the IEC61000-4-2 contact ±12kV stress without any hard damages and latchup issue. Moreover, the RS232 transceiver IC also has been verified to well protect the system over the IEC61000-4-2 contact ±20kV stress (CLASS B) in the smart scanner and notebook application**

Index Terms— **ESD, RS232, SCR, TVS**

I. INTRODUCTION

THE system level ESD protection of the electronic product is absolutely necessary for reliability and safety concern. The IO interface of the electronics must have the ESD protection solutions for IEC61000-4-2 contact (or air discharge) stress. In general, the TVS arrays are the best choice if the interface IC can not provide the sufficient protection capability. So far, the TVS array is the main stream of the ESD protection solutions due to the efficiency and the cost. Moreover, the system designer can hit the system ESD level by paralleled the off-chip TVS array in the developing stage so that "what kind and how much of the TVS array you choose" is the key factor to achieve the ESD target specification.

Four off-chip TVS devices are necessary for two transmitter outputs and two receiver inputs of the general RS232 transceiver IC, as shown in the Fig.1. The bi-directional TVS array should be taken into consideration for signal level of the

RS232 standard. The RS232 transceiver IC with on-chip TVS have been proposed in the Fig. 2, which can reduce R&D developing time due to the PCB reworking for placement of the off-chip TVS devices. Moreover, the PCB space will be more compact to use RS232 transceiver IC with on-chip TVS.

The structure of the on-chip TVS is a high voltage dual SCR with high Vh for RS232 signal. The layout style is the circle type with small area of 236μm x 236μm. The single chip of the RS232 transceiver IC with on-chip TVS has been evaluated to pass IEC61000-4-2 contact ±12kV discharge by direct pin injection. Moreover, the RS232 transceiver IC with on-chip TVS has been designed in the smart scanner and notebook system, and has been verified to pass ±20kV IEC61000-4-2 (CLASS B) contact discharge without any hard damage and latchup issue[1]~[4].

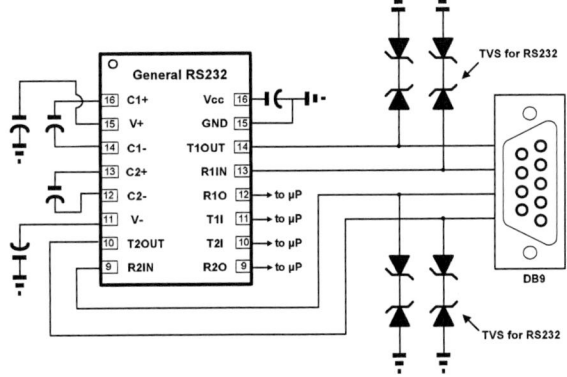

Fig. 1. The traditional application circuit for general RS232 transceiver IC.

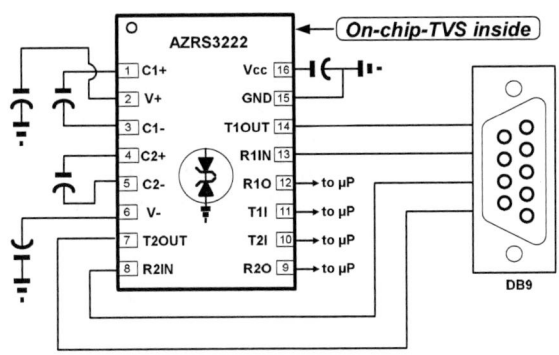

Fig. 2. The simple application circuit for RS232 transceiver IC with on-chip TVS.

Manuscript received February 24, 2011. This work was implemented in the transceiver product of Amazing Microelectronic Corp. and has been issued the patent for the proposed on-chip TVS.

Ryan Hsin-Chin Jiang is the chief of the Amazing Microelectronic Corp., HsinChiu, Taiwan R.O.C. (phone: 886-3-6212071 ext 705; fax: 886-3-5834797; e-mail: ryan@amazingic.com).

Tang-Kuei Tseng, Chi-Hao Chen and Che-Hao Chuang are the member s of the Department of R&D in the Amazing Microelectronic Corp. , HsinChiu, Taiwan R.O.C. (e-mail: tktseng@amazingic.com, chchen@amazingic.com, chehao@amazingic.com).

978-1-4244-9019-6/11 $26.00 © 2011 IEEE   203

II. TVS PROTECTION IN RS232 APPLICATION

A. Off-chip TVS Protection Scheme

The design for system level ESD protection with off-chip TVS has two key points. One is the selection for specification of off-chip TVS, and the other is the PCB layout optimization for placement of the off-chip TVS. For the RS232 application, the maximum signal level of RS232 is ±12V so that the feature of the TVS must be bi-directional TVS with the Vh voltage about 1.2~1.5 times of the signal level (±12V). The layout of PCB must be reworked for location of off-chip TVS. The off-chip TVS should be designed as closed as the DB9 connector for RS232 application to efficiently discharge the current. Moreover, the other layout instructions should follow the off-chip TVS provider.

The turn-on mechanism of the off-chip TVS under the IEC61000-4-2 contact discharge has been shown in the Fig.3. The off-chip TVS should be closed the DB9 connector enough to release energy efficiently through the off-chip TVS. If not, some energy will pass through RS232 transceiver IC by the PCB wire inductor so as to the RS232 chip side may be damaged.

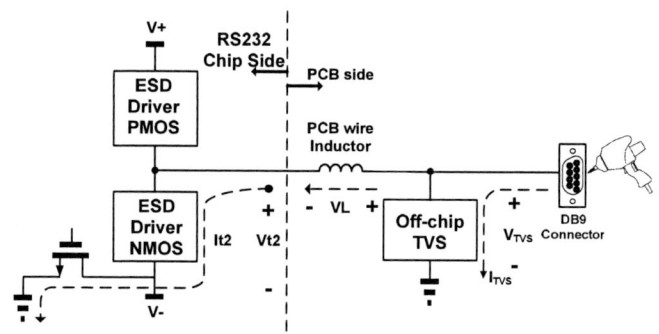

Fig. 3 The protection scheme of the off-chip TVS design under ESD gun contact stress.

For example, the off-chip TVS was stressed through the DB9 connector by IEC61000-4-2 contact Level 4 (±8kV) and three assumptions for devices are listed as follows:

--First, the secondary breakdown current (It2) of RS232 transceiver IC is 2.7A (HBM 4kV) and the secondary breakdown voltage (Vt2) is 20.3V.

--Second, the maximum current flow through the PCB wire inductor is the same with It2 (2.7A).

--Third, the PCB wire inductor is 1nH.

The peak current of IEC61000-4-2 Level4 stress is about 30A within 1ns. So the maximum clamping voltage of the off-chip TVS (V_{TVS}) can be estimated as below (1):

$$V_{TVS} = Vt2 + L* (dIt2/dt) = 23.0V \qquad (1)$$

The V_{TVS} is the important ESD specification for selection of the off-chip TVS. The smaller the V_{TVS}, the more efficiency the off-chip TVS perform under ESD stress. If low V_{TVS} TVS is applied, the component level ESD specification of the general RS232 transceiver IC may be reduced.

B. On-chip TVS Protection Scheme

The silicon implementation for on-chip TVS is more tough than that of the off-chip TVS. Firstly, the process of the on-chip TVS itself must be fully compatible to that of the RS232 transceiver IC. Secondly, the ESD robustness of the output stage should be taken into consideration. Thirdly, any RS232 core circuit connected to output stage (P1) should be blocked by a current limited cell. The overall design architecture of the RS232 transceiver IC with on-chip TVS has been shown in the Fig. 4.

Unlike the off-chip TVS protection scheme, this is not PCB wire inductor as current delay buffer any more but a designed resistor Rp between the on-chip TVS and the ESD driver devices. The resistor Rp can be view as the current limited device under ESD stress. So there is no current delay buffer but a current limit cell refer to the ESD driver MOS.

For the design of the on-chip TVS protection, the layout size of ESD driver NMOS and PMOS should be estimated by the clamping voltage of the on-chip TVS device. If the V_{TVS} of the on-chip TVS is higher, the size of the ESD driver device should be larger for ESD robustness issue. For consideration of the cost issue in this real case, the layout size of the on-chip TVS has been limited by IO area so that the V_{TVS} of the on-chip TVS will be not smaller than that of the off-chip TVS. Therefore, the more robust ESD driver NMOS and PMOS is absolutely necessary.

The design of the RS232 core circuit, which is connected to node of P1, must be connected through the current limited cell. The simple current limited cell is a resistor about 1~10kohm. The selection of the resistor Rp is not preferred in the silicon substrate, the poly resistor for example, due to extra parasitic device.

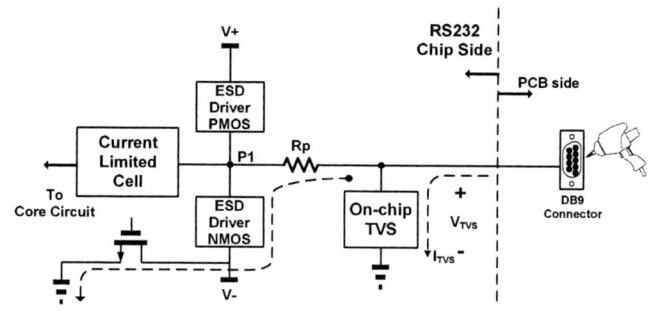

Fig. 4 The protection scheme of the on-chip TVS design under ESD gun stress.

III. EVALUATION FOR THE ON-CHIP TVS

The cross-section view of the on-chip TVS device has been shown in Fig. 5. The on-chip TVS device is high-voltage-triggered DSCR structure [5] ~ [7], which has been silicon proved in the 0.8μm BCD epitaxial process with adjustment of the process recipe. The p-n-p-n structure of the dual SCR device, as the arrow shown in the Fig. 5, is formed by symmetric layers of the P+ in SP, NW, SP and N+ in SP, respectively.

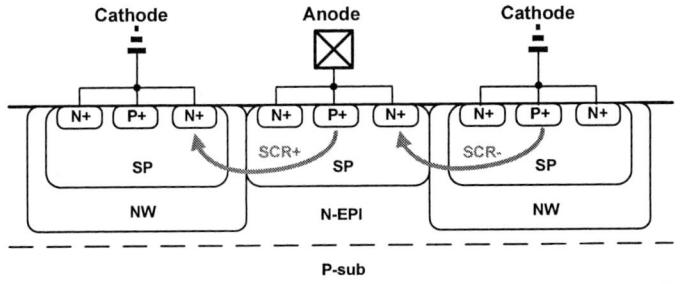

Fig. 5 The cross-section view of high voltage dual SCR device.

The electrical characteristic of the DSCR device has been referred to RS232 standard. The maximum rating voltage of the RS232 input stage of receiver is ±25V so that the breakdown voltage (Vbd) of DSCR device has been designed as 1.2 time of the 25V. The maximum signal level of RS232 transmitter is ±12V, so the Vh of DSCR is designed larger than ±12V to ensure the signal integrity under ESD stress. Moreover, the high Vh voltage (> 12V) of DSCR device can make sure the latchup free for RS232 transceiver IC.

The secondary breakdown I-V curve of the DSCR device has been evaluated by TLP system of 100ns pulse width, as shown in the Fig. 6. The DSCR device under TLP pulsing system has been measured to 17A without any leakage current. According to the TLP I-V curve, the turn-on resistance of the dual SCR device is only 0.62 ohm in secondary breakdown region. The resistance of DSCR is larger than that of general off-chip TVS device because of the limited layout area of the dual SCR device in the IO circuit. The high Vh of the DSCR device is ±19.0V, which is safe for latchup issue. Moreover, the high Ih of the DSCR device is 1.1A, which is larger than maximum current driver of RS232 transmitter. So the RS232 signal will be still keep stable even the DSCR has been triggered.

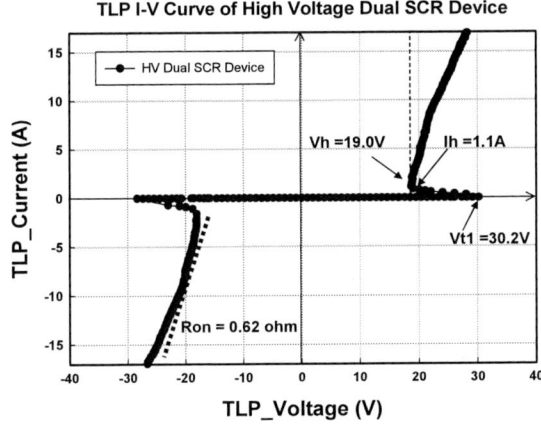

Fig. 6 The TLP I-V curve of high voltage DSCR device under the measurement of the 100ns pulse width.

Once the Vh and the Ron we get, the clamping voltage of the DSCR device can be predicted as below (2) under the energy of the IEC61000-4-2 level4 (±8kV) contact discharge (di/dt = 30A/1ns).

$$Vclamp = Vh + 30A * Ron = 37.6V \qquad (2)$$

Assume both the maximum Rp is 8 ohm and the minimum clamping voltage of the ESD driver device is 16.0V, the maximum current through the driver MOS is 2.7A by 37.6V of Vclamp. Therefore, the ESD specification of the ESD driver NMOS and the switch NMOS is above HBM 4kV to make sure the internal device can still be alive under IEC61000-4-2 contact ±8kV discharge.

The layout style of the on-chip TVS is the circle type for current uniformity consideration, as shown in the center of the Fig. 7. The inner circle is the anode side and the outer circle is the cathode side with the substrate guard ring. To efficiently reduce the layout area, the device under PAD technique has been applied on the DSCR device using two metal layers [8] ~ [10].

The placement of the on-chip TVS is important for efficiently discharge the ESD current. Firstly, the metal width of GND should be designed as wider as possible to reduce the current crowding effect. Secondly, the DSCR device should be placed as closed to the GND pad as possible. Once the GND pad is closer, the wire inductance will be smaller to bypass the current without any current delay. For example, in the IO circuit of Fig. 7, the spacing of both T2 and R2 to GND is more closed than that of both T1 and R1. The T2 and R2 under ESD stress show the better ESD performance than T1 and R1. Therefore, the GND pad should be placed near the T1 and R1 to efficiently discharge the ESD current.

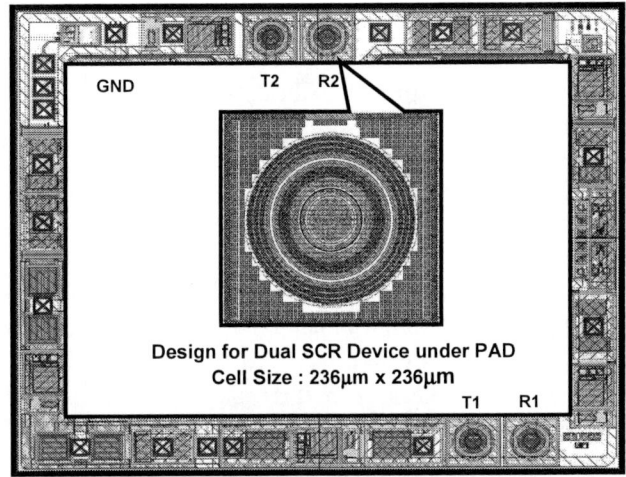

Fig. 7 The whole chip of RS232 transceiver IC with on-chip TVS in the IO circuit. The center cell is the detail layout of dual SCR.

The on-chip TVS of the RS232 transceiver IC has been evaluated to be passed the specification of the IEC61000-4-2 contact ±12kV discharge under the signal-chip verification. After the stress of the IEC61000-4-2 contact ±16kV discharge, the obviously damage spots of the on-chip TVS distribute uniformly in the circle-type layout without internal damages, as shown in the fig. 8. This means that the on-chip TVS device can well protect the internal circuit from ESD stress. The enlarged picture of the fig. 8 shows the "contact spike" located uniformly in the everywhere of the on-chip TVS, that means the turn-on uniformity of on-chip TVS is so good.

978-1-4244-9019-6/11 $26.00 © 2011 IEEE

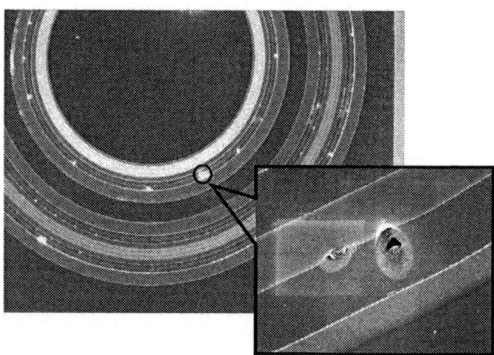

Fig. 8 The SEM failure picture of the on-chip TVS under IEC61000-4-2 contact 16kV discharge.

IV. SYSTEM VERIFICATION FOR RS232 TRANSCEIVER IC

There are two NBs with docking and one RS232 testing board with DB9 connectors in the RS232 verification system, as shown in the Fig. 9. The real-time data transmission from NB-A will be firstly received by RX of the RS232 transceiver IC. By loopback of Rxout to Txin, the real-time data will be transmitted by TX of the RS232 transceiver IC to the NB-B through the 2m RS232 cable. The maximum baud rate is 256kbps with 8 bits data by the NB software control. In normal operation, the real-time data from NB-A to NB-B has been verified to be no error bit in the two thousand bits text file by comparison of the data for each other. In the ESD contact discharge experiment, the cable nears both RX and TX sides of the RS232 transceiver IC will be zapped to evaluate the capability of the on-chip TVS in the data flow condition. In the IEC6100-4-2 ±20kV contact discharge, the real-time data flow will not be stopped but there is about 10% <u>B</u>it <u>E</u>rror <u>R</u>ate (BER) in the two thousand bits. The RS232 transceiver IC has been verified to be no hard damage and latchup issue after ESD stress. If the energy of the contact discharge is lower to ±8kV, the BER will be lower than 1%. To sum up, the proposed RS232 transceiver IC with on-chip TVS protection has been verified to pass IEC61000-4-2 ± 20kV contact discharge (CLASS B) in notebook system. It is reliable and safe to use the DSCR device as the on-chip TVS device in the RS232 application.

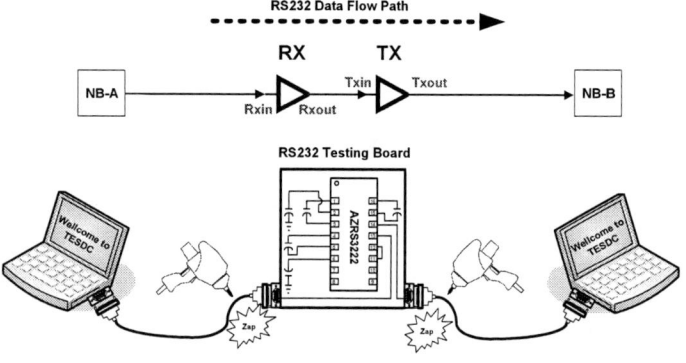

Fig. 9 The system testing setup for RS232 transceiver IC with on-chip TVS in the notebook application.

V. CONCLUSIONS

The high voltage DSCR with high Vh has been proposed in this paper. The RS232 transceiver IC integrated with DSCR device has been verified to be workable solution. Under the off-power condition, the RS232 transceiver IC with on-chip TVS device has been evaluated to pass the IEC61000-4-2 contact ±12kV stress without any hard damages and latchup issue. Moreover, the RS232 transceiver IC with on-chip TVS also has been verified to well protect the notebook system over ± 20kV IEC61000-4-2 contact stress (CLASS B) in the power-on condition.

REFERENCES

[1] M.-D. Ker and W.-Y. Lo, "Methodology on extracting compact layout rules for latchup prevention in deep-submicron bulk CMOS technology," IEEE Trans. Semiconductor Manufacturing, vol. 16, pp. 319-334, May 2003.

[2] I.-C. Lin, C.-Y. Huang, C.-J. Chao, M.-D. Ker, "Anomalous latchup failure induced by on-chip ESD protection circuit in a high-voltage CMOS IC product," Microelectronics Reliability., vol. 43, pp. 1295-1301, 2003.

[3] M.-D. Ker and S.-F. Hsu, "Physical mechanism and device simulation on transient-induced latchup in CMOS ICs under system-level ESD test," IEEE Trans. On Electron Devices, vol. 52, no. 8, pp. 1821-1931, Aug. 2005.

[4] M. Kelly, L. Henry, J. Barth, G. Weiss, M. Chaine, H. Gieser, D. Bonfert, T. Meuse, V. Gross, C. Hatchard, and I. Morgan, " Developing a transient induced latch-up standard fo testing integrated circuits," in Proc. EOS/ESD Symp., 1999, pp. 178-189.

[5] M.-D. Ker and K.-C. Hsu, "SCR device fabricated with dummy-gate structure to improve turn-on speed for effective ESD protection in CMOS technology," IEEE Tran. On Semiconductor Manufacturing, vol. 18, no. 2, pp. 320-327, May 2005.

[6] M.-D. Ker and K.-C. Hsu, "Overview of on-chip electrostatic discharge protection design with SCR-based devices in CMOS integrated circuits," IEEE Trans. On Device and Materials Reliability, vol. 5, no. 2, pp. 235-249, Jun. 2005.

[7] M. P. J. Mergens, C.C. Russ, K. G. Verhage, J. Armer, P. C. Jozwiak, and R. Mohn, " High holding current SCRs (HHI-SCR) for ESD protection and latch-up immune IC operation, " in Proc. EOS/ESD Symp., 2002, pp.10-17.

[8] M.-D. Ker and J.-J. Peng, "Investigation on device characteristics of MOSFET transistor placed under bond pad for high-pin-count SOC applications," IEEE Trans. On Components and Packaging Technologies, vol. 27, no. 3, pp. 452-460, Sep. 2004.

[9] W. R. Anderson, W. M. Gonzalez, S. S. Knecht, and W. Flowler, "ESD protection under wire bonding pads," in Proc. EOS/ESD Symp., 1999, pp.88-94.

[10] G. einen, R. J. Stierman, D. Edwards, and L. Nye, "Wire bonds over active circuits," in Proc. Electronic Components and Technology Conf., 1994, pp. 922-928.

Ryan Hsin-Chin Jiang (M'92) M.S. and Ph.D. degrees from the Institute of Electronics, National Chiao-Tung University, Hsinchu, Taiwan, R.O.C., in 1992, and 1998, respectively.

He served the Computer and Communication Research Laboratories (CCL) and SoC Technology Center (STC), Industrial Technology Research Institute (ITRI), Taiwan, from 1999 to 2003. From 2003 to 2005, he joined in two IC design houses to develop Analog and Mixed-Signal IC products. In 2006, he co-founded the Amazing Microelectronic Corp. in Taiwan and serve with R&D Vice-President. He has proposed many inventions to improve reliability and quality of integrated circuits, which have granted with 31 U.S. patents and 20 ROC (Taiwan) patents. He also has published 4 IEEE journal papers and 20 International conference papers. His current research topics include ESD/EMI/EFT/Surge Protection Design, Analog IC Design, Semiconductor Device Physics.

Dr. Jiang has been elected as the President of Taiwan ESD Association from 2005 to 2009.

Soft Error Modeling, Simulation, and Testing at Advanced Technology Nodes

B. L. Bhuva, W. T. Holman, and L. W. Massengill
Vanderbilt University, Nashville, TN, USA

Abstract— As feature sizes decrease, soft errors are expected to become the dominant failure mechanisms for integrated circuits. This paper discusses the challenges that design and reliability engineers will face with the manufacture and test of ICs at advanced technology nodes.

Keywords-component; Soft errors, testing, neutrons, alpha particles, IC design

I. INTRODUCTION

Future semiconductor manufacturing technologies will result in the fabrication of single-digit nanometer wide structures on integrated circuits. At these device dimensions, the amount of charge required to represent a logic "1" in a digital circuit will be reduced to just a few thousand electrons. Consequently, the loss of less than one femto-coulomb (fC) of charge at a critical node can result in data loss for the circuit, followed by potential circuit malfunction. The ITRS roadmap has identified several failure mechanisms that can cause such a loss of data, with radiation-induced soft errors as a dominant mechanism among them. In fact, FIT (failures in time) rates for soft errors are expected to dominate all other FIT rates in the future. As a result, it is essential to understand the failure mechanisms and mitigation approaches for soft errors. This paper discusses the effects of scaling on soft error modeling, simulation, testing, and mitigation to address concerns for future advanced technologies.

The primary physical causes of single event effects (SEEs) are incident particles such as alpha particles, neutrons, muons, or heavy ions, on silicon (Si) semiconductor devices [1]. When a particle strikes an integrated circuit (IC), it creates electron-hole pairs in the silicon substrate through coulombic interactions with the Si lattice, either directly (alpha particles, heavy ions, etc.) or through secondary reactions (neutrons, high-energy protons, etc.). Most of these electrons and holes recombine and do not affect circuit operation. However, some of the charge may reach a circuit node through drift and/or diffusion processes to create a voltage perturbation that will propagate to other nodes. These perturbations may alter the data stored at critical circuit nodes, resulting in circuit malfunction. The successful mitigation of single-event effects requires an understanding at the physical level (transport of electrons and holes in a Si substrate), at the circuit level (effects of voltage perturbations at a circuit node), and at the system level (incorrect data in a flip-flop or latch affecting the system-level operation) to fully characterize SE errors. With billions of transistors on an IC, any miscalculation at any of the steps in the SE characterization process will result in significant errors in estimated FIT rates. As a result, it is imperative for engineers to have accurate information regarding all aspects of soft error analysis and testing.

Environment

For ICs fabricated using older technology nodes, the amount of charge required to cause an upset was relatively high due to larger nodal capacitances and higher transistor currents. As a result, critical charge to upset a storage cell was of the order of tens of fC. For such critical charge levels, storage cells were vulnerable only to high-energy particles. For advanced technology nodes, the critical charge to upset SRAM cells and DFF circuits is less than one fC. For 28-nm technology, this is further reduced to just a few tenths of one fC. As a result, particle strikes that had little effect on older ICs are now causing upsets in the latest generation of IC technologies. For example, SRAM cells were immune to low-energy proton and muon upsets at the 90-nm technology node. However, SRAM cells at the 28-nm technology node have proven vulnerable to low-energy proton and muon upsets [2].

For any environment, terrestrial or space, the number of low-energy particles is orders of magnitude higher than that for high-energy particles. As the particle energy requirement to cause an upset decreases, the number of particles incident on an IC that is above the critical charge increases. For example, the number of muons in the terrestrial environment is orders of magnitude higher than that for neutrons. As a result, the number of upsets may dramatically increase for future technology generations unless steps are taken to mitigate their effects.

Modeling and Simulation of Soft Errors

The main problem with modeling soft errors and predicting soft error failure rates (SERs) is the number of variables that strongly influence the outcome. SERs are a strong function of fabrication process, incident particle characteristics, circuit design, and system design [1]. Parameters associated with a fabrication process include doping densities, junction depths, parasitic capacitances, parasitic bipolar transistor characteristics, and STI depth. Incident particle parameters include particle type, particle energy, and angle of incidence.

978-1-4244-9019-6/11 $26.00 © 2011 IEEE

Parameters associated with circuit design include individual transistor currents, nodal capacitances, circuit topology, and layout. Parameters associated with system-level designs include error detection and correction capability, critical circuit blocks, and memory block size. All of these parameters strongly influence the final SER for a given system, and all must be taken into consideration for reasonable predictive accuracy.

At the atomic level, an incident particle deposits energy either through direct coulombic interactions (e.g. alpha particles and heavy ions) or through secondary reaction particles (e.g. neutrons and high-energy protons). In all cases, a charge cloud is generated when an ionizing particle traverses through the silicon volume. The manner in which this charge is removed from the Si volume determines the transistor response to the event. Initially, the charge cloud will have a lateral Gaussian profile along the particle track. The initial charge density will far exceed that of the background doping densities of the IC. As a result, a sea of carriers will form without any defined p- or n-doped regions. The charges under the influence of an electric field will be swept away quickly, while the rest will diffuse through the volume. Most of the charge will recombine, but some charge will be collected at individual circuit nodes. Charge within the depletion region of a p-n junction will be swept along by the presence of the electric field, resulting in drift current. Other charge diffusing through the Si volume may reach a depletion region and get collected at a later time, resulting in diffusion current. The amount of charge collected is a strong function of background doping densities (which determines the depletion layer width), the distance from a given p-n junction, and the supply voltage.

The presence of these carriers may introduce additional perturbations in the well (or substrate) potential. Such a perturbation may result in the activation of the parasitic bipolar transistor within an MOS transistor. This parasitic bipolar effect will result in additional charge being collected at a circuit node, resulting in higher voltage perturbations at that node [3]. For advanced technology nodes, the close spacing of drain and source regions in an MOS transistor, combined with distant well/substrate contacts (for the 40-nm technology node, the distance required between a well/substrate contact and a transistor is approximately 30 μm) makes the activation of a parasitic bipolar transistor highly likely for most single-event strikes. The resulting higher charge collections will effectively increase the circuit vulnerability to soft errors.

All of the processes involved for single-event effects (charge generation, charge collection, charge transport, node voltage perturbations, transient propagation, error latching, system-level error propagation, etc.) are very well understood today. However, FIT rate estimation and SE mitigation is still elusive for most engineers because the advances in fabrication technologies add novel failure mechanisms for soft error generation. For example, in advanced technologies, due to close proximity of transistors, multiple circuit nodes will collect charge resulting in multiple voltage perturbations within a circuit. Another mechanism, called pulse quenching, shortens the voltage perturbation due to charge collection by electrically connected nodes [4]. Process parameter variations add another dimension to the problem by making all variables non-deterministic. In addition, the temporal and spatial characteristics of SE transient pulses strongly influence the FIT rates. These SE transients, in turn, are dictated by the fabrication process parameters, layout, and circuit design.

The problem engineers face when trying to estimate SE-related FIT rates is that not all of these factors are known *a priori*. In fact, reliability engineers are typically required to characterize a fabrication process that is not yet commercially available. Additionally, reduced charge requirements to represent a logic state have increased the circuit vulnerability to other particles. Recent research results have reported upsets due to muons for SRAMs fabricated at the 40-nm and 28-nm technology node. With the number of muons in the environment being orders of magnitude higher than any other particle, mitigating muon-induced upsets will prove a formidable challenge for 20-nm and 14-nm technologies.

All of these factors have made the modeling and simulation of basic soft error mechanisms a significant challenge, even without considering predictive analysis at the IC level for FIT rates. Current efforts at addressing these challenges are being developed using TCAD software for 3-D analysis of charge drift, diffusion, and recombination processes after a particle strike. These simulation efforts have been at the forefront in identifying novel mechanisms and explaining transistor-level behavior after a particle strike. Future efforts will require 3-D simulations with multiple transistors (perhaps for an entire sub-circuit, such as a complete flip-flop or multiple SRAM cells) to fully understand the effects of layout and circuit topology on soft errors. New CAD software that can simulate soft errors at the RTL level will also be necessary for the estimation of soft error rates at the system level.

Testing of Soft Errors

Because the mechanisms associated with SE errors are evolving very rapidly with technology, it is very difficult to carry out predictive analysis of modern circuits and systems. TCAD simulations alone may not be sufficient to identify novel failure mechanisms; consequently, experimental results are required for each technology node. Additionally, with the rapid pace of IC process advancement, designers increasingly find themselves designing circuits for technologies that are still in the development phase, and are not completely characterized. For example, designers need to know the FIT rates due to soft errors for the flip-flops being used in the design. But without a test IC, reliability engineers cannot accurately estimate FIT rates due to evolving failure mechanisms. As a result, semiconductor companies must test circuit designs, IC designs, and system designs at great expense. Thus, the greatest challenge faced by individual

978-1-4244-9019-6/11 $26.00 © 2011 IEEE

semiconductor companies with respect to soft errors is the process, and associated cost, of evaluating a circuit design for potential failures that are not fully characterized and/or known.

Another problem faced by reliability engineers is that fabrication houses are reluctant to provide information about a process that is not yet production-ready. This results in the familiar scenario of engineers entering the design phase with little idea of what FIT rates to expect from an upcoming technology node. Consequently, engineers typically resort to the use of data from a previous generation of technology (for example, designers will use data from 65-nm process to estimate FIT rates for 28-nm designs). Since failure mechanisms for soft errors are evolving, data based on 65-nm technologies may not be valid for 28-nm technologies. For example, proprietary data showed that a rad-hard DICE FF design was 100X more resistant to soft errors than a conventional DFF design at 65-nm technologies. However, at the 40-nm technology node, the DICE FF designs were only 5X better than conventional DFF designs [5]. Providing such knowledge to designers during early stages of the design phase is essential to achieve an adequate understanding of SE failure mechanisms. The only way this goal can be accomplished is through collaborative efforts between fabrication houses and their customers to develop failure models, and subsequent mitigation techniques.

Because soft errors are relatively rare in terms of the number of measurable events per day for a given test structure (yet far too common in terms of system-level FIT rates), testing for them requires careful planning and interpretation of data. The number of errors must be sufficiently large to provide a confident estimate of final FIT rates. Increased error requirements increase the cost of testing and data analysis. JEDEC standards must be followed so that results are comparable across different manufacturers and parts.

Recent results have shown that different test facilities for neutron exposures will result in different FIT rates due to differences in neutron spectra. Additionally, smaller target sizes for circuits fabricated at advanced technology nodes means lower cross-sections for each logic cell. Longer exposure times will be required to collect sufficient soft error data, resulting in higher cost for each experiment. Researchers are currently exploring ways to reduce the test cost by developing test ICs with multiple target circuits from multiple companies to reduce test and analysis cost. Such approaches will become the norm in the future for soft error testing.

Mitigation Approaches

Even if all relevant failure modes and mechanisms are understood and able to be simulated, designers must still mitigate the threat of soft errors. If the testing and predictive analysis results in FIT rates that are unacceptable at the system level, designers must make changes in their designs to lower the FIT rates. Once the fabrication process is fixed and the operating environment is chosen, little or nothing can be changed with respect to either of these factors. The only choice designers have is to alter the design to reduce FIT rates, either by modifying critical sub-circuits or altering the architecture of the system.

Simulation algorithms have been developed that will allow designers to identify critical sub-circuits and/or critical storage cells [6]. At the circuit level, many approaches have been proposed, ranging from the hardening of individual flip-flops to the incorporation of EDAC techniques in the design [7]. Unfortunately, all of these efforts impose penalties in terms of increased area, speed, and power requirements for each sub-circuit. At the architecture level, solutions must be application-specific, typically ranging from redundancy of hardware to multiple calculations of data in the presence of an error [8]. As soft errors becoming the most dominant contributor to the overall FIT rates for any system, these approaches must be carefully tailored to each application to obtain optimum performance with minimum design penalty.

Summary

As critical charge requirements decrease for storage cells and logic gates, and operating frequencies of circuits increase, soft errors are expected to dominate over all other failure modes for ICs fabricated at advanced technology nodes. The fundamental failure mechanisms and modes associated with an upset are rapidly evolving with technology, and accurate predictive analysis is becoming very difficult due to the increasingly complex nature of potential failure mechanisms. Similarly, testing requirements are becoming prohibitively costly, and innovative approaches are necessary to efficiently identify future failure mechanisms. Finally, new circuit-level and system-level mitigation approaches must be developed to meet the reliability requirements in the future.

REFERENCES

[1] Short Course at NSRE Conference, July 2001.

[2] B.D. Sierawski, et al., "Muon-induced single event upsets in deep-submicron technology," IEEE Trans. Nucl. Sci., pp. 3273, Dec. 2010.

[3] O. A. Amusan, et al., "Laser verification of charge sharing in a 90 nm bulk CMOS process," *IEEE Trans. Nucl. Sci.*, pp. 3065, Dec. 2009.

[4] J. Ahlbin, et al., "Single-event transient pulse quenching in advanced CMOS logic circuits," IEEE Trans. Nucl. Sci., pp. 3050, Dec. 2009.

[5] T. D. Loveless, et al., "Neutron and proton induced single event upsets for D- and DICE-flip-flop designs at a 40 nm technology node," accepted for publication in *IEEE Trans. Nucl. Sci.*, June 2011.

[6] N. Kaul, et al., "Simulation of SEU transients in CMOS ICs," IEEE Trans. Nucl. Sci., pp. 1514, Dec. 1991.

[7] T. May, "Soft errors in VLSI: Present and future," IEEE Trans. CHMT, pp. 377, 1979

[8] S. Mukherjee, et al., "The soft error problem: an architecture perspective," Proc. of Int. Sym. High-Performance Comp. Architecture, pp. 243, 2005.

Layout Optimization to Maximize Tolerance in SEILA: Soft Error Immune Latch

Taiki Uemura, Tsunehisa Sakoda and Hideya Matsuyama

Abstract— The purpose of this paper is optimization of layout on soft error immune latch (SEILA) for maximizing soft-error mitigation efficiency, and investigating mechanisms of charge collection on multi-node and discussing layout dependence on soft-error. We evaluate soft-error rate (SER) on un-robust latch, conventional robust latch, SEILA with changing well structure, distances from well-contacts, and distance between soft-error critical nodes through neutron acceleration experiments at Osaka Univ. Soft-error mitigation efficiency awfully change with changing layout. In designing robust latches, it is most important for high the mitigation to separated critical nodes with STI and we need to take care on layout especially distance between critical nodes.

***Index Terms*—Soft error, Single Event, SEU, MCU, DICE, SEILA, neutron, alpha, radiation effect.**

I. INTRODUCTION

SOFT-ERROR rates (SERs) increase in modern LSI chips. Main causes of soft-error in terrestrial environments neutrons as secondary cosmic rays and alpha particles from LSI materials [1][2].

An Error Correction Code (ECC) can correct all single bit upset (SBU) and is most effective countermeasure for RAMs [3]. It is difficult to apply ECC to logic circuits, especially sequential elements, because of delay time for error correction and solid data block is necessary for ECC with small overhead.

Many kinds of Radiation-Hardened-By-Design (RHBD) for sequential elements have been proposed [4]-[7]. Dual inter-lock cell (DICE) [4] can mitigate soft-error in sequential element by retaining data with four nodes. The inter-lock type latch is, however, upset when charge is collected on multi-node (CCM).

Mitigation efficiencies of RHBD latches depend on not only circuits but also layouts of the latch cell. Soft error immune latch (SEILA) achieve high soft-error mitigation efficiency by attenuating CCM with designing layout of multi height cell,

such as triple-height-cell (THC) and double-height-cell (DHC), without area overhead. [6] THC may be higher mitigation efficiency than DHC because all PMOS and all NMOS in THC are not included in a well area. However, interconnections on THC are difficult because of narrow width of the cell (Fig. 1) on simple circuit, a few transistors, such as resister. [6][7] As a result, area overheads on THC are higher than on DHC in the simple circuit.

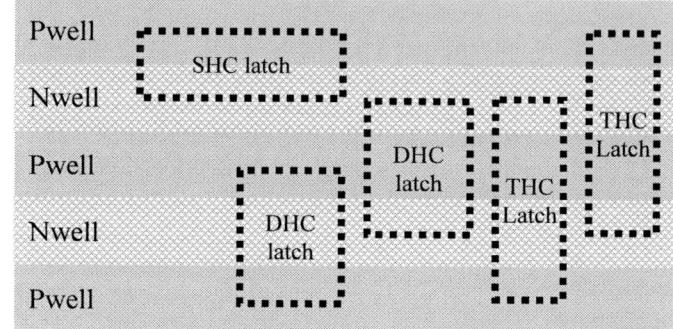

Fig. 1. Layout of SHC (Single-Height-Cell), DHC (Double-Height-Cell) and THC (Triple-Height-Cell) latches.

In spite of same circuit and same layout, soft-error mitigation efficiency is different with well structure (twin-well and triple-well) and distance from well-contacts. [6][8] There are four possible mechanisms of CCM as shown in Table 1. [9] Neutron is easy to induce CCM comparing with alpha ray because induced charge by alpha is not so large. CCM caused by (A), (B) and (C) decreases with voltage increasing although CCM caused by (D) increases with voltage increasing. In any mechanism, rate of CCM increases as technology advances because cell size of sequential elements shrink and distance between critical nodes decreases. The critical nodes on RHBD on sequential element is very close comparing with critical SRAM bit cells on ECC. CCM caused by neutrons on RHBD has become the most serious reliability concern for high reliability electron devices.

TABLE I
POSSIBLE MECHANISM OF CHARGE COLLECTION ON MULTI NODE

Symbol	Quantity
(A)	Successive hits by one ion
(B)	Multi hits by multi ion
(C)	Drift/diffusion charge (charge sharing)
(D)	Parasitic bipolar action

This manuscript received February 7, 2011. This work was supported in part by the reliability engineering department of Fujitsu Semiconductor Ltd.

T. Uemura is with Fujitsu Semiconductor Ltd., Boulder, 50 Fuchigami, Akiruno, Tokyo, 197-0833, Japan (phone: +81-42-532-1379; e-mail: uemura.taiki@jp.fujitsu.com).

T. Sakoda is with Fujitsu Semiconductor Ltd., Boulder, 50 Fuchigami, Akiruno, Tokyo, 197-0833, Japan (phone: +81-42-532-1379; e-mail: t.sakoda@jp.fujitsu.com).

H. Matsuyama is with Fujitsu Semiconductor Ltd., Boulder, 50 Fuchigami, Akiruno, Tokyo, 197-0833, Japan (phone: +81-42-532-1379; e-mail: uemura.taiki@jp.fujitsu.com).

978-1-4244-9019-6/11 $26.00 © 2011 IEEE

The purpose of this paper is optimization of layout on SEILA with DHC, investigating CCM mechanisms on the DHC-SEILA and discussing layout dependence on soft-error. We evaluate SER on un-robust latch, SHC-robust latch, DHC-SEILA latch in twin-well (without deep-N-well) and triple-well (with deep-N-well) with changing distances from well-contacts, and distance of SER critical node through neutron acceleration experiment at Osaka Univ.

II. EXPERIMENT SETUP

We have evaluated soft-error on sequential elements through neutron acceleration experiment. The experiment has been carried out with using spallation neutron beam in Research Center for Nuclear Physics (RCNP) at Osaka Univ. as shown in Fig. 2. The neutron beam irradiates eight DUT boards all together. Two test chips are on the DUT board and sixteen test chips are simultaneously irradiated. The test chip is assembling with plastic package and include latch arrays manufactured with 65nm technologies. Each chip includes more than 200kbit latches. When neutron irradiates to the test chips, clock signal (CK in Fig. 3) of the latches continues to keep high state and the latch retains data.

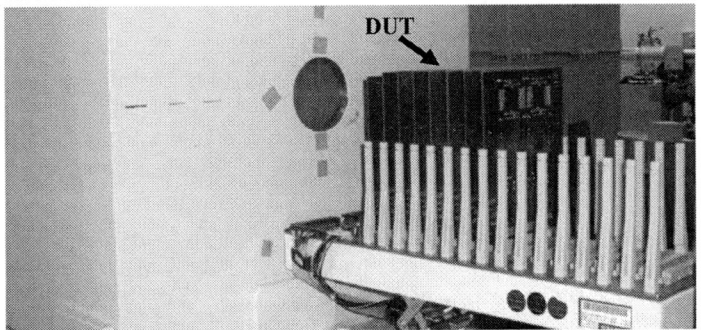

Fig. 2. Setup of neutron acceleration experiment at RCNP at Osaka Univ.

We evaluate SER on four types of latches, (o) un-robust latch, (a) SHC-robust latch, (b1) and (b2) DHC-SEILA as shown in Table 2. The robust latches are data storage structure of inter-lock type. [4] Storage data in latches of this type can not be corrupted by collecting charge on one node. They can be corrupted by collecting charge on two or more nodes.

On (a) SHC-robust latch, critical PMOS drains are not separated by STI as shown Fig. 3(a). On (b1) and (b2) DHC-SEILA, critical PMOS drains are separated with STI as shown in Fig. 3 (b). Dp1 which is distance between the PMOSs on a feed-back gate (P2 and P4 in Fig. 3(b)) is 1um in both (b1) and (b2) DHC- SEILA. Dp2 which are distances between PMOSs on a feed-through gate (P1 and P3 in Fig. 3(b)) are 0.56 um in (b1) and 0.34 um in (b2).

Well-tap(tie)-less cells are mainstream because of achieving large width of transistors with small height in recent design trend. Latches of (o), (b1) and (b2) in the test chip are also well-tap-less cells and need well-taps on out of latch cells. In these test chips, well-tap is put as shown in Fig. 4. There are latches which are near and far position from well-taps in test chips and (a) SHC-robust latch includes well-tap in the cell.

TABLE 2
LATCH LIST ON THE NEUTRON ACCELERATION EXPERIMENT.

Symbol	Latch Type	Cell height	Dp2	Well-tap type
(o)	Un-robust	Single	-	Tap-less
(a)	Robust	Single	-	Tap
(b1)	SEILA	Double	long	Tap-less
(b2)	SEILA	Double	short	Tap-less

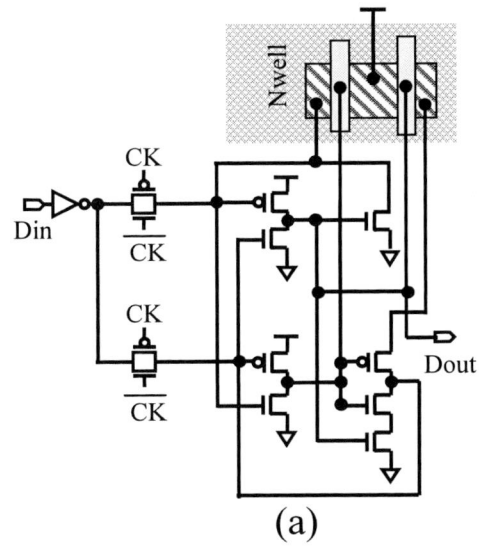

(a)

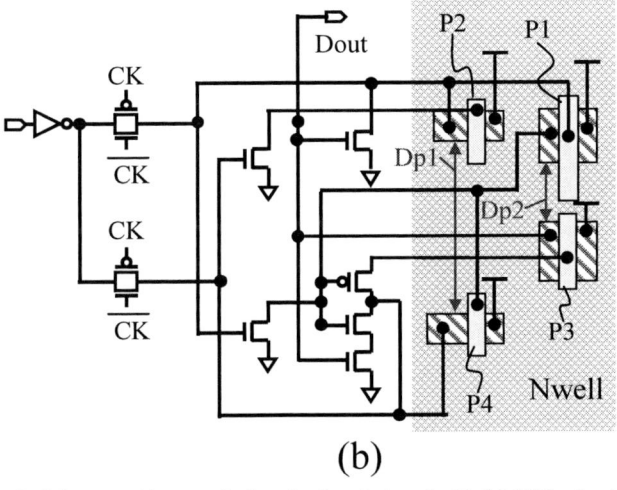

(b)

Fig. 3. Schema and layout of robust latches designed with (a) SHC-robust and (b) DHC-SEILA. The CK in this figure is a terminal for clock signal.

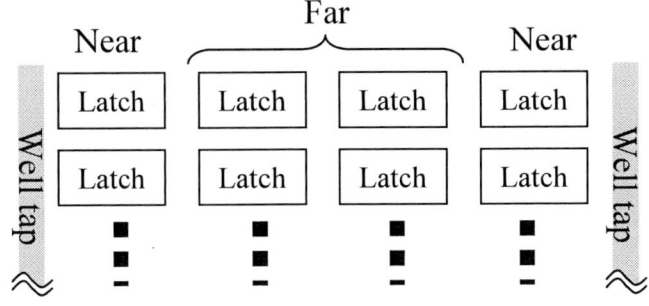

Fig. 4. Latch array layout of the test chip. There are four latches between well-taps. There are latches which are near and far position from well-taps (well-contacts).

978-1-4244-9019-6/11 $26.00 © 2011 IEEE   211

III. EXPERIMENTAL RESULTS

A. Un-robust latches

Soft-error contribution on feed-back gate is higher than on feed-through gate in data retention of latches because critical charge on the feed-back is smaller than on the feed-through. [10] When the un-robust latch retain data of "0" (DATA0), SER contribution on NMOS of feed-back gate is dominant. When the latch retain data of "1" (DATA1), SER contribution on PMOS of feed-back gate is dominant.

On un-robust latch (Table 2(o)), SER on far and near position are not so different in triple-well and SER in triple well is lower than in twin-well as shown in Fig. 5. These results show parasitic bipolar effect (Table 1(D)) in triple-well is lower than in twin-well. In twin-well, SER at far position is higher than at near position on DATA1 as shown in Fig. 5 although dependence of position from well-taps is few on DATA0. Dependence of well-tap position is attributed to parasitic bipolar effect. This result shows that the bipolar effect is observed on PMOS and not on NMOS in twin-well.

Without the bipolar effect, SERs decrease with operation voltage increasing because of critical charge decreasing. Fig. 5 shows that dependence of operation voltage is little in triple-well. There is SER contribution of the bipolar effect on PMOS in the latch although that is smaller than on NMOS.

In SRAMs, generally, SER contribution of the bipolar effect on NMOS is higher than on PMOS and SER on triple-well is higher than SER on twin-well. [11] In the latch, the results show that SER contribution of the bipolar effect on PMOS is higher than on NMOS and SER on twin-well is higher than SER on triple-well. We need to take care of the bipolar effect not only on NMOS but also on PMOS in latches.

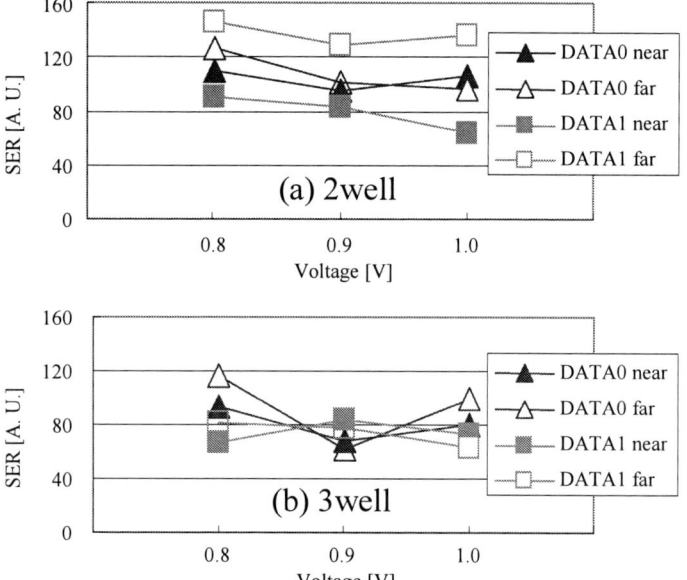

Fig. 5. SER on un-robust latch as shown in Table 2(o) in (a) twin-well and (b) triple-well when retention data are 1 and 0. There are the latches which are near and far position from well-taps. These two figures are same scale on SER (Y-axis).

B. Robust latches

Fig. 6 shows SER mitigation efficiency (SER of a robust latch per SER of an un-robust latch). Mitigation efficiency on (a) latch (Table 2(a) and Fig. 3(a)) is only 10X even thought enough well-contacts because critical nodes of PMOS is not separated by STI and it could affect that CCM caused not only by the bipolar effect (Table 1(D)) but also others (Table 1(B) and (C)). Mitigation efficiency on (b1) is three to four times higher than the efficiency on (b2). There is large SER dependence of Dp2 on same circuit. For high the mitigation in robust latches, separating critical node with STI is most important.

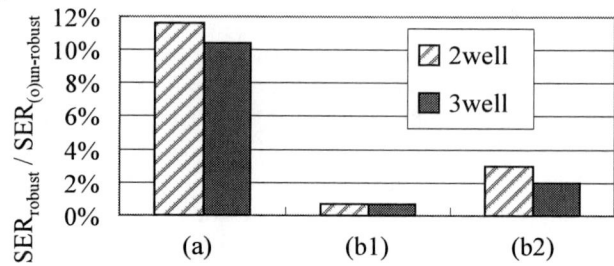

Fig. 6. Mitigation efficiency on (a) SHC-RHBD latch (b1) DHC-SEILA with long Dp2 is and (b2) DHC-SEILA short Dp2 in twin-well (left) and triple-well (right). The details of these latches are shown in Table 2.

Fig. 7 shows SERs on the DHC-SEILAs shown in Table 2 (b1) and (b2). PMOSs on these latches are included in an Nwell area and soft-error critical NMOSs of the latch are not included in a Pwell area and CCM on NMOSs is attenuated by layout technique. [6] Then, CCM on PMOS is dominant for soft-error in these latches. CCM on PMOS on feed-through gate (P1 and P3 in Fig. 3(b)) induces error on DATA1 and CCM on PMOSs on feed-back gate (P2 and P4 in Fig. 3(b)) induces error on DATA0.

In DHC-SEILA with long Dp2 (Table 2(b1)), SER in triple-well is higher than in twin-well, and SER at far position from well-taps is higher than at near position as shown Fig. 7(a) and (b). Parasitic bipolar effects in twin-well are higher than in triple-well and the bipolar effect is on PMOSs. Critical charge (Qc) on PMOSs in DATA1 is smaller the Qc in DATA0. Critical nodes are P2 and P4 in Fig. 3(b) in DATA1, P1 and P3 in DATA0. SER on DATA1 is higher than on DATA0 in triple-well as shown in Fig. 7(a) even distance between critical nodes in DATA1 (P2 and P4 in Fig. 3(b)) is longer than in DATA0 (P1 and P3 in Fig. 3(b)).

In DHC-SEILA with short Dp2 (Table 2(b2)), SER on DATA0 is higher than DATA1 on both twin-well and triple-well as shown in Fig. 7(c) and (d). CCM on P1 and P3 in Fig. 3(b) is dominant on soft-error. Especially in twin-well, SER is eleven to forty three times higher comparing with the (b1) latch.

These results show distances between critical nodes are very sensitive to soft-error. When the distance between the nodes is less than 0.34um, CCM on the nodes awfully contribute to SER and the RHBD latch lose mitigation efficiency of soft-error, especially at far position from well-taps in twin-well because the bipolar effect on PMOS in twine-well is high.

978-1-4244-9019-6/11 $26.00 © 2011 IEEE

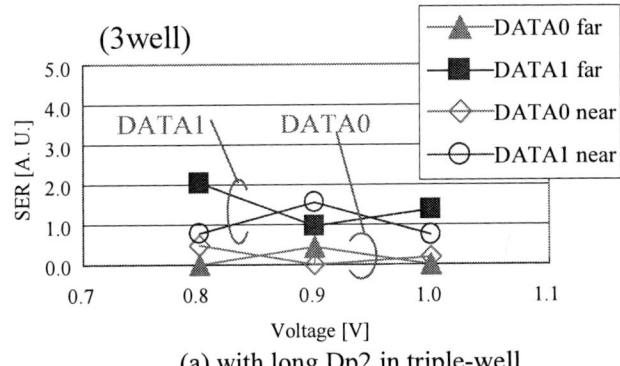

(a) with long Dp2 in triple-well

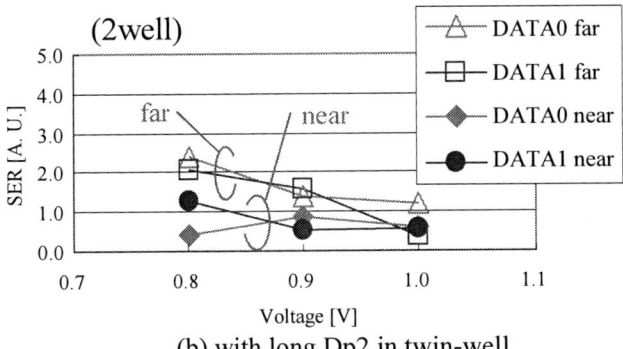

(b) with long Dp2 in twin-well

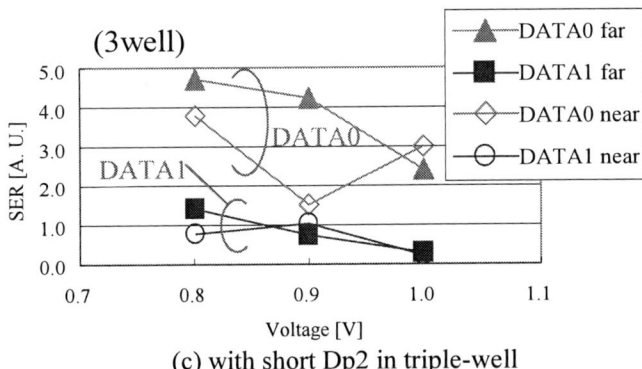

(c) with short Dp2 in triple-well

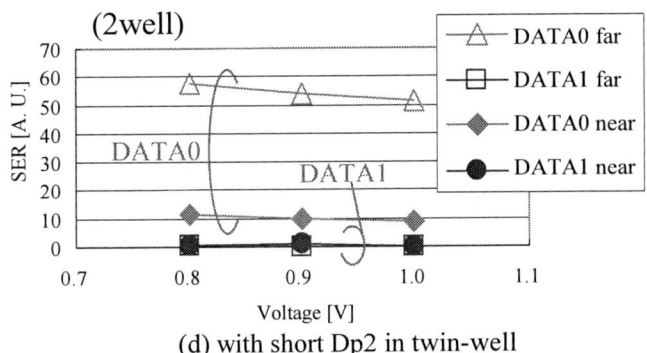

(d) with short Dp2 in twin-well

Fig. 7. SER on DHC-SEILA (a) with long Dp2 (Table. 2(b1)) on triple-well, (b) with long Dp2 (Table. 2(b1)) on twin-well, (c) with short Dp2 (Table. 2(b2)) on triple-well, and (d) with short Dp2 (Table. 2(b2)) on twin-well. Scale of SER (Y-axis) in these four figures are same with Fig. 5.

IV. CONCLUSIONS

We evaluate SER on un-robust latch, conventional robust latch and DHC-SEILA through neutron acceleration experiment at Osaka Univ. In un-robust latch, SER contribution of the bipolar effect on PMOS is higher than on NMOS and SER on twin-well is higher than SER on triple-well. We need to take care of the bipolar effect not only on NMOS but also on PMOS in latches. In robust latches, SER mitigation efficiency awfully changes with layout in robust latches. Separating critical node with STI is most important in designing robust latches. In the DHC-SEILA of which critical nodes are separated with STI, distances between critical nodes are very sensitive to soft-error. When the distance between the nodes is less than 0.34um, CCM on the nodes awfully contribute to SER, especially at far position from well-taps in twin-well. In designing SEILA, we need to care of separation of critical nodes, after that, distance between critical nodes, distance from well contacts and well structures.

ACKNOWLEDGMENT

The authors would like to thank several people. The latches were designed by Mr. A. Nishiwaki, Mr. M. Komaki, and Mr. N. Kakizawa of Fujitsu VLSI. The test chips are designed with advisements of Mr. M. Igeta of Fujitsu Semiconductor. The acceleration test is carried out with Mr. R. Tanabe, Mr. T. Tsuruta and Mr. T. Kato of Fujitsu Semiconductor and with corporation of Dr. K. Takahisa, Dr. M. Fukuda and Dr. K. Hatanaka of Osaka Univ. Fruitful discussions with Mr. H. Shimizu of Fujitsu Semiconductor are greatly appreciated. Finally, we would like to thank Dr. D. Kobayashi of Japan Aerospace Exploration Agency and Dr. S. Onoda of Japan Atomic Energy Agency for grateful discussion.

REFERENCES

[1] T. C. May, and M. H. Woods, "Alpha particle induced soft errors in dynamic memories," *IEEE Trans. Elec. Dev.* vol. 26, pp. 2-9, Jan. 1979.

[2] J. F. Ziegler, "Terrestrial cosmic rays," IBM J. Res. Develop., vol. 40, pp. 19–39, Jan. 1996.

[3] S. Kundu and S. M. Reddy, "On symmetric Error Correcting and All Unidirectional Error Detecting Codes", Trans. on Comput., pp. 752-761, 1990.

[4] T. Calin, et. al., "Upset hardened memory design for submicron CMOS technology", Trans. Nucl. Sci., pp. 2874-2878, 1996.

[5] M. Zhang, et. al., "Sequential Element Design With Built-In Soft Error Resilience," *IEEE VLSI Syst.* vol. 14, pp. 1368-1378, Dec. 2006.

[6] T. Uemura, et. al., "SEILA: Soft Error Immune Latch for Mitigating Multi-node-SEU and Local-clock-SET", *Trans. on IRPS2010*, pp. 218-223.

[7] H. K. Lee, et. al., " LEAP: Layout Design through Error-Aware Transistor Positioning for Soft-Error Resilient Sequential Cell Design", *Trans. on IRPS2010*, pp. 203-212.

[8] J. Furuta, et. al., "Measurement Results of Multiple Cell Upsets on a 65nm Tapless Flip-Flop Array", SELSE 6, 2010.

[9] E. Ibe, et. al., "Spreading Diversity in Multi-cell Neutron-Induced Upsets with Device Scaling", *IEEE CICC* pp. 437-444, 2006.

[10] T. Heijmen, "Soft-Error Vulnerability of Sub-100-nm Flip-flop", IEEE Trans. IOLTS, pp. 247-252. 2008.

[11] G. Gasiot, "Multiple Cell Upsets as the Key Contribution to the Total SER of 65nm CMOS SRAMs and its Dependence on Well Engineering," *NSREC*, No.J-1 2007.

978-1-4244-9019-6/11 $26.00 © 2011 IEEE

Comparative Analysis of Flip-Flop designs for Soft Errors at Advanced Technology Nodes

B. L. Bhuva[1], Senior Member, IEEE, K. Lilja[2], Member, IEEE, J. Holts[3], Member, IEEE, S.-J. Wen[3], Member, IEEE, R. Wong[3], Member, IEEE, S. Jagannathan, Student Member, IEEE, T. D. Loveless, Member, IEEE, M. McCurdy, Member, IEEE, Z. J. Diggins, Student Member, IEEE

Abstract—For advanced fabrication technology nodes, novel single-event related failures are being observed. This paper details efforts to use 3D TCAD simulations to model these failure mechanisms and develop mitigation techniques for flip-flop designs. Simulation, as well as experimental, results are used to show validity of such an approach for future CMOS technologies.

Index Terms—single-event upsets, flip-flop, CMOS, scaling, TCAD simulations.

I. INTRODUCTION

With latest advances in semiconductor manufacturing, novel failure mechanisms may emerge that will dominate (or significantly affect) the overall failure in time (FIT) rates for soft errors (SE) for a given product. The biggest challenge faced by individual semiconductor companies is the process and the associated cost of evaluating a circuit design for such novel failures when they are not fully characterized and/or known. Since the SE FIT rates are dependent on the fabrication process, circuit design and system design, and the fact that the failure mechanisms associated with SE are evolving with technology, it is hard to carry out predictive analysis of circuits and systems. Novel approaches are necessary that can identify SE related failure mechanisms for a technology node and apply these mechanisms to circuit-level designs.

The main hurdle to identification of failure mechanisms and failure modes (and FIT rates) for SE is the complexity of the physical mechanisms associated with it. The primary physical cause of the SE is an incident particle, such as Alpha particles, neutrons, or heavy-ions, on Silicon (Si). When such a particle is incident on an Integrated Circuit (IC), it creates electron-hole pairs

in Si substrate through Coulombic interactions with Si lattice either directly (Alpha, heavy-ions, etc.) or through secondary reactions (neutrons, high-energy protons, etc.). Some of the charges may reach a circuit node through drift and/or diffusion processes to create a voltage perturbation at that node. These perturbations may alter the data stored at the circuit nodes, resulting in circuit malfunction. The problem of understanding the effects of SE arises because one needs a complete understanding at the physical level (transport of electrons and holes in a Si substrate), at the circuit level (effects of voltage perturbation a circuit node), and at the system level (incorrect data in a flip-flop affecting the system-level operation) to fully characterize the error. With billions of transistors on an IC, any miscalculation at any of the steps in the SE characterization will result in large errors in FIT rates. As a result, it is imperative for engineers to have accurate models for predictive simulations.

All of the processes involved for SE (charge generation, charge collection, charge transport, node voltage perturbations, transient propagation, error latching, system-level error propagation, etc.) are very well understood at this point. However, error rate estimation and SE mitigation is still elusive for most engineers because the advances in fabrication technologies add novel failure mechanisms in the soft error generation. For example, for advanced technologies, due to close proximity of transistors, multiple circuit nodes will collect charge resulting in multiple voltage perturbations within a circuit [1]. Another mechanism, called pulse quenching, shortens the voltage perturbation due to charge collection by electrically connected nodes [2]. Process-parameter variations add another dimension to the problem by making all variables non-deterministic [3]. In addition, the temporal and spatial characteristics of SE transient pulses strongly influence the error rates [4]. These SE transients, in turn, are dictated by the fabrication process parameters, layout, and circuit design. The problem engineers face when trying to evaluate SE related error rates is that not all of these factors are known *a priori*. With the rapid pace of

---

Manuscript received February 15, 2011. This work was supported in part by Cisco Systems, Inc.

B. L. Bhuva, S. Jagannathan, T.D. Loveless, M. McCurdy, and Z. Diggins are with the Vanderbilt University, Nashville, TN 37235, USA. (Phone: (615) 3430-3184, e-mail: bharat.bhuva@vanderbilt.edu).

K. Lilja is with Robust Chip, Inc., Pleasanton, CA 94588, USA (e-mail: klas.lilja@robustchip.com).

J. Holts, S.-J. Wen and R. Wong are with Cisco Systems, Inc. San Jose, CA 95134, USA.

978-1-4244-9019-6/11 $26.00 © 2011 IEEE

advancement of technologies, designers need to know the comparative advantages of different flip-flops (FF) being used in the design. Without this knowledge, designers may end up using data from a previous generation of technology (for example, designers are using data from 65 nm process for comparing error rates for different FF cells in their 28 nm designs). Since failure mechanisms for soft errors are evolving, data based on 65 nm technologies may not be valid for 28 nm technologies. For example, a DICE FF design was expected to be 100X better than conventional DFF design at 65 nm technologies. However, at 40 nm technology a DICE FF design was only 5X better than conventional DFF designs [5]. Such knowledge is required by the designers during early stages of the design to achieve desired FIT rates for the final product. One potential solution to such a complex problem is the 3D TCAD simulation of the whole FF cell to estimate the effects of neutrons and alpha particles. This paper discusses such an approach for generating comparative analysis for FF designs at 40 nm technology node.

II. SIMULATION APPROACH

The simulation tools used in this work, Accuro and rExplore from Robust Chip, allowed us to run a very large number of individual single event simulations in an automated fashion, for a complete characterization of the full layout of the FFs. The simulation engine in Accuro uses a full 3D transport model to describe the charge transport and charge collection in the semiconductor (3D TCAD accuracy). The substrate charge simulation is fully and self-consistently coupled to a SPICE/HSPICE compatible circuit simulation. Accuro employs several acceleration methods, including a unique coupling between the circuit and the substrate charge, which makes the simulation up to two orders of magnitude faster than traditional TCAD, while maintaining full TCAD accuracy (the simulation analysis in this work is far beyond the reach of regular traditional 3D TCAD simulation).

As input, the simulation uses information about the layout (GDSII), the netlist (SPICE/HSPICE netlist and compact models), and certain information about substrate and doping. The layout is represented directly in the 3D simulation structure (figure 1), and therefore all layout effects (node placements, spacing, well ties, etc.) are immediately and correctly captured by the simulation. The simulation generally provides very accurate results without any calibration (calibration may be done though, if appropriate measurement data are available before the simulation work starts). In this work no calibration was performed, but a sensitivity study was done providing information on the sensitivity of the single event cross-sections to selected aspects of the doping in the structure.

rExplore is a simulation management tool and GUI which automates the setup, run, and post processing of the simulation project. It was used to setup the initial simulation experiments (see below), manage the simulations, selectively refine the scan grid, calculate and visualize the cross-section regions (maps) as well as cross-sections as a function of LET.

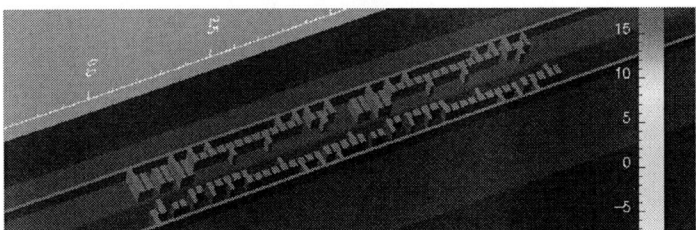

Figure 1. DFF 3D simulation structure (STI oxide removed for visibility)

III. COMPARATIVE ANALYSIS

Two different FF designs were selected for analyzing their soft error performance at 40 nm technology node. Both the designs are conventional DFF designs employing two inverters connected in a loop fashion. Fig. 2 shows the circuit diagram of the basic FF design used in this work. The second design was similar but was specially designed for reduced power requirements and SE vulnerability. The comparative analysis was carried out using the software described above for incident particles with LET ranging from 1 to 30 MeV/cm^2/mg. All particles were assumed to be normally incident on the Si surface. Analysis was carried out for master stage and slave stage separately for all four combinations of data patterns (1-stored-1-input, 1-stored-0-input, 0-stored-1-input, and 0-stored-0-input). For the simulations, the location of the ion hit was scanned across the whole layout in 0.1μm steps in X and Y directions. The scan was then further refined in selective areas (scan steps down to 0.025μm) in order to resolve the sensitive regions. For each ion hit, the full TCAD simulations were performed and the FF state was checked at the end of the simulation to identify an upset state. Such an approach will identify the sensitive area on a layout for which an incident ion with a given LET will cause an upset. The total sensitive area was noted for each ion LET values. Exact failure mechanism responsible for causing the upset was also analyzed whenever an upset was observed.

978-1-4244-9019-6/11 $26.00 © 2011 IEEE   215

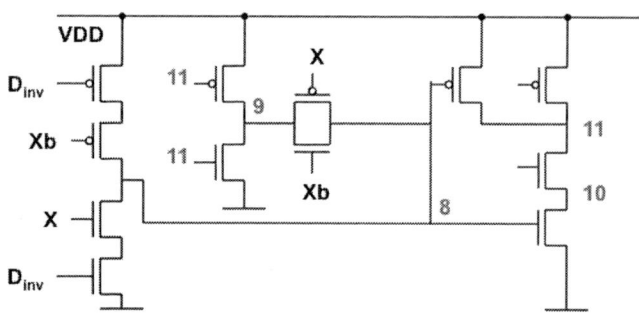

Fig. 2. DFF Schematic for master stage.

Fig. 3 shows the sensitive area information for conventional DFF design on the layout as a function of LET values of the incident particle. The dominant mechanism for this FF for upsets is the collection of charge by individual nodes. As can be seen from Fig. 3, each storage node shows a large sensitive area that is independent of each other. The storage nodes were placed away from each other to avoid charge sharing between the two storage nodes. It has been shown previously that when multiple nodes collect charge, FF vulnerability to SE upsets increases. FFs have a positive feedback path to store data. The single-event

Based on this analysis, which identified the most sensitive and most vulnerable parts of a design along with the physical mechanisms causing the upset, the second design was developed to reduce the overall vulnerability of the design. Another mechanism that is slowly becoming prominent is the pulse-quenching mechanism for advanced technology nodes. Pulse-quenching mechanism deals with two electrically connected circuit nodes when both of them collect charge. This mechanism represents a race condition between the electrical signals propagating from one node to another with the physical charge cloud that is moving between those two nodes in the substrate. The electrical signal is the result of the voltage perturbation created when a sensitive node collects charge after an ion hit. This voltage perturbation will propagate through the circuit at a rate determined by the current drives of the associated transistors and the nodal capacitances. At the same time, the charge cloud generated by the incident ion will diffuse outward to other circuit nodes at a rate determined by the diffusion rate of carriers and the electric field present in the substrate. All OFF transistors will collect charge from this charge cloud generating multiple voltage perturbations in the circuit. These two mechanisms affect each other to either increase or decrease the SE transient pulse width

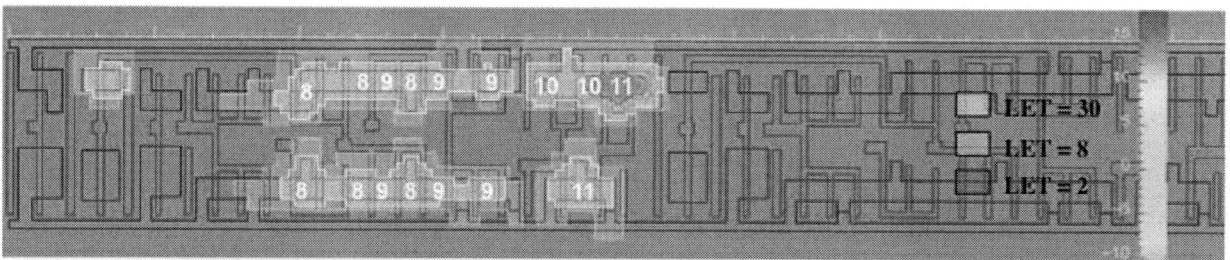

Fig. 3. Sensitive area for the master latch of the DFF on the layout for all LET values

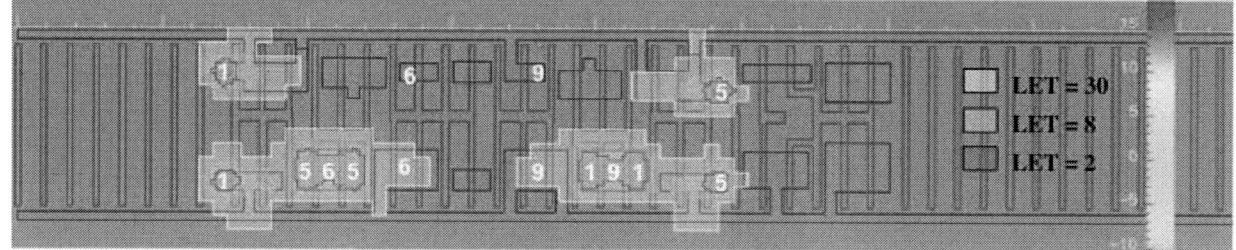

Fig. 4. Sensitive area for the master latch of the low-power FF on the layout for all LET values

voltage perturbation on a storage node must be longer than the feedback loop delay to cause an upset. However, when multiple storage nodes collect charge, the SE transient pulse width required to cause an upset decreases for both the nodes. This results in increased vulnerability for conventional FF designs as well as hardened FF designs such as DICE FF.

depending on the layout and circuit design. For the low-power design, these mechanisms were allowed to interact in such a way as to reduce the resultant SET pulse within the FF design to a minimum thereby reducing the vulnerability of FF. Conventional wisdom is that the low power design will be more susceptible to soft errors because of the lower transistor currents usually associated with low power designs. However,

978-1-4244-9019-6/11 $26.00 © 2011 IEEE

the second flip-flop design was constructed with special considerations for soft errors as noted above and the layout was generated accordingly to reduce the sensitive area. Fig. 4 shows the sensitive area for the low-power flip-fop design as determined by simulations.

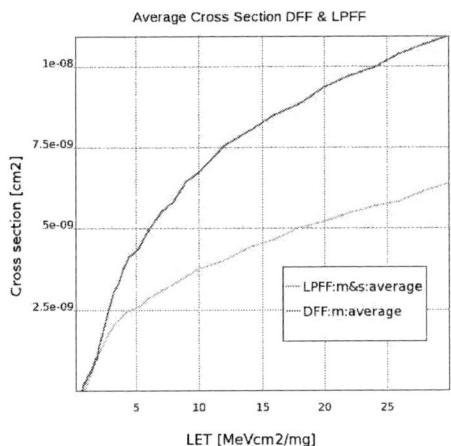

Fig. 5. Cross-section Vs. LET for Master stage of both FF designs.

Upon comparison with the same data from Fig. 3, it is clear that the sensitive area for the second FF at each LET values is lower than that for the conventional DFF design. This will result in lower cross-section area forthe low-power design, yielding lower FIT rates for the FF design. Fig. 5 shows the cross-section for each FF design as a function of LET values for master and slave stages when logic 1 is stored. Such an analysis provide a comparative analysis of flip-flop designs by identifying the most tolerant FF designs during early stages of designs to allow designers to reduce their over all FIT rates.

IV. EXPERIMENTAL VERIFICATION

Both of these designs were fabricated in a 40 nm bulk CMOS technology with exactly the same layout that was used in the comparative analysis. The FFs were designed in a 8K stage shift register fashion. Support circuits were also designed that allowed for operation up to 3 GHz and on-chip error monitoring for the shift registers. The IC was exposed to neutrons at TRIUMF facility and LANSCE facility with total neutron fluence of 1.1×10^{10} /cm^2 and 8.2×10^{10} /cm^2 over multiple days. Alpha particle exposures were carried out with an Americium foil source with 10 µCi strength. Alpha particle exposures were carried out for 12 hours for a total fluence of 1.296×10^9 / cm^2. Table 1 shows the ratio of the error rates observed for all these tests. These experimental results clearly show the superior soft error performance of low power FF as predicted by the

comparative analysis carried out using 3D TCADanalysis.

V. CONCLUSION

3D TCAD analysis is used to compare different designs and to identify the most sensitive areas of flip-flop designs. Based on the results of the simulations, the most dominant mechanisms for causing a failure for DFF design were identified. These results were used to develop a low-power version of the design that was far superior in soft error performance. Experimental results verified the improved radiation tolerance of the low-power version of the design. This project clearly shows that low-power designs do not necessarily result in higher SE vulnerability if proper analysis and design consideration are taken into account.

Table 1: Relative FIT rates due to neutron and alpha particles

Flip Flop design	Relative FIT $= \dfrac{FIT_{FF}}{FIT_{DFF1}}$	
	Neutron	Alpha
DFF	1	1
New DFF	0.68	0.23

REFERENCES

[1] O. A. Amusan, A. F. Witulski, L. W. Massengill, B. L. Bhuva, P. R. Fleming, M. L. Alles, A. L. Sternberg, J. D. Black, and R. D. Scrimpf., "Charge Collection and Charge Sharing in a 130nm CMOS Technology," IEEE TNS, vol. 53 (6), pp. 3253–3258, December 2006.

[2] J. R. Ahlbin, L. W. Massengill, B. L. Bhuva, B. Narasimham, M. J. Gadlage, and P. H. Eaton, "Single-Event Transient Pulse Quenching in Advanced CMOS Logic Circuits," IEEE Transactions on Nuclear Science, vol. 56 (6), pp. 3050-3056, December, 2009.3050-3056, Dec. 2009.jon pulse quenching

[3] A. V. Kauppila, B. L. Bhuva, J. S. Kauppiila, L. W. Massengill, and W. T. Holman, "Impact of Process Variations on SRAM Single Event Upsets," IEEE Transactions on Nuclear Science, Accepted for Publication.

[4] R. Baumann, IEEE Nuclear and Space Radiation Effects Conference Short Course, 2005.

[5] T. D. Loveless, S. Jagannathan, T. Reece, J. Chetia, B. L. Bhuva, M. W. McCurdy, L. W. Massengill, S-J. Wen, R. Wong, and D. Rennie, "Neutron and proton induced single event upsets for D- and DICE-flip-flop designs at a 40 nm technology node," Accepted for publications in IEEE Trans. On Nucl;. Sci., 2011.

Silicon Quantum Well Light-Emitting Diode

S. Saito

Abstract—Monolithic light source is the only missing component to realize all silicon based photonics for high density and low power optical interconnections. In this paper, we will review our attempts to develop light-emitting devices based on silicon quantum wells made by state-of-the-art silicon process.

Index Terms—silicon photonics, quantum confinement, nano-structure, and FinFET.

I. INTRODUCTION

SILICON has been proved to be the most suitable material for electronics in terms of purity, stability, carrier mobility, dopant activation, and so on to achieve high integration density of transistors. But, the application of silicon to photonics is limited because of the lack of a light source due to its in-direct band gap character (Table I). Therefore, compound semiconductor laser diodes are used for global optical network in current data centers (Fig. 1). Both inter- and intra- board optical interconnections are considered to replace the current metal interconnections to reduce the power consumption for high density communication. The demands to increase the transmission density will be accelerated as more core processor units are being integrated within a chip and current hard disk drives are being replaced with solid state disks. The short distance intra-chip optical interconnections will become key technology to realize the on-chip data center [1]. The monolithically integrated light source made of silicon is the last missing component, since the other optical components like the optical waveguide, germanium photo-detector, and modulator are already developed in a past decade.

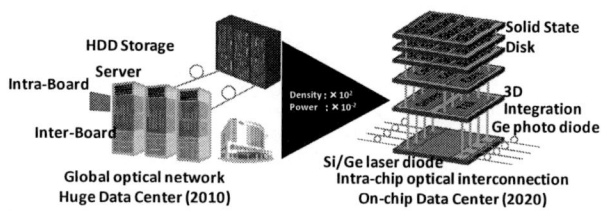

Fig. 1. Future of data center. Optical interconnections are being integrated within a chip to enhance the transmission density and reduce the power consumption for the communication between cores and memory storage.

Manuscript received March 22, 2011. This work was supported by JSPS through the FIRST Program.

S. Saito is with Institute for Photonics-Electronics Convergence System Technology (PECST), Photonics Electronics Technology Research Association (PETRA), and Central Research Laboratory, Hitachi, Ltd., Kokubunji, Tokyo 185-8601, Japan (Phone: +81-42-323-1111; e-mail: shinichi.saito.qt@hitachi.com).

TABLE I
DIMENSIONAL CROSSOVER OF LIGHT-EMITTING SI NANO-STRUCTURE

Structure	Dot	Wire	Well	Bulk
Dimension	0D Direct	1D Direct	2D Direct	3D In-direct
Conduction band structure	Γ	Γ <100>, <110>	Γ (100)	Γ
Quantum confinement	Strong ⟵			Weak
Carrier injection	Tunneling ■■		⟶	Direct contact
Stimulated emission	PL [4]	-	EL [8, 9]	-

One of the ways to overcome the fundamental limitation of the in-direct band gap character of the bulk silicon is to use the low dimensional nano-structures of silicon [2-9] (Table I). With decreasing the dimensionality, the 6-fold degenerate conduction band valleys located away from the Γ point are projected onto the Γ point, and the direct recombination is possible without mediating phonons by quantum mechanical confinements [6]. According to this *valley projection* mechanism, the proper chose of the crystallographic orientation is important; e.g. in 2-dimension, the valleys are projected onto the Γ point within the (100) surface, whereas no valleys is projected onto the Γ point within the (111) surface [6]. These silicon nano-structures are feasible to be made by the current top-down silicon process, as the technology is refined along the scaling.

Another challenge of the electroluminescence (EL) from the silicon was a trade-off between quantum confinement and carrier injections. For the light emission, the lower dimensional structures with stronger quantum confinements are preferable. In fact, the observations of optical gain by photoluminescence (PL) were reported in quantum dots. On the other hand, higher dimensional structures are preferable for efficient carrier injections, since the direct contacts to electrodes are possible, while the surfaces of the low dimensional silicon nano-structures are covered with highly insulating silicon dioxide. It is highly non-trivial whether the intrinsic stimulated emissions can be obtained in 2 dimensional silicon quantum well by carrier injections. The purpose of our work is to examine the possible existence of the stimulated emissions in silicon quantum well (Fig. 2) light-emitting devices [8, 9].

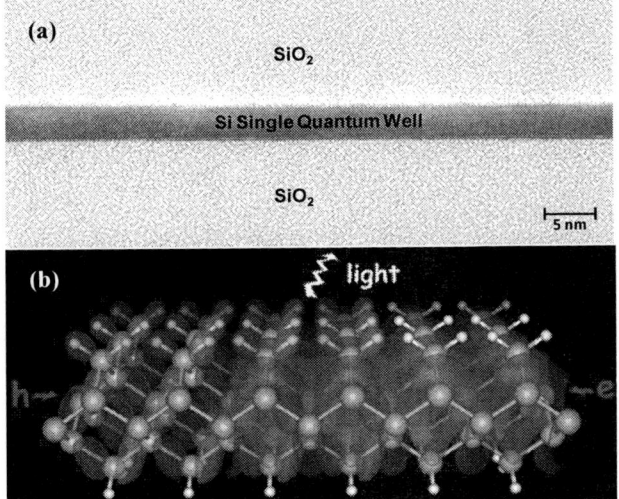

Fig. 2. Silicon quantum well. (a) Transmission electron microscope image of ultra-thin silicon quantum well. (b) Simulated electron and hole wavefunctions within silicon quantum well by first principles.

II. SILICON SINGLE QUANTUM WELL

We proposed the lateral carrier injection scheme, where the Si single quantum well is directly connected to the thick Si electrodes [7, 8] (Fig. 3). The Si single quantum well was made by local oxidation of Si-on-insulator (SOI) substrate. The Si_3N_4 grating was located just above the Si single qunautm well to ensure the evanescent coupling. The high difference of the reflective index between Si_3N_4 and SiO_2 is responsible for the strong optical feedback inside the distributed-feedback (DFB) resonant cavity. The part of the supporting Si substrate is removed by the wet etching to prevent the optical loss to the substrate.

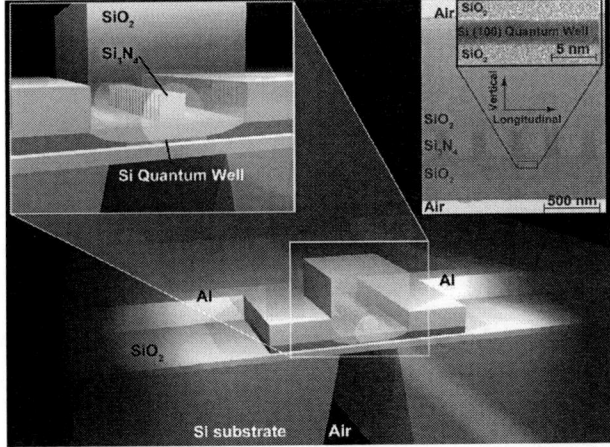

Fig. 3. Si single quantum well light emitting diode. The device consists of the lateral pin diode embedded in the distributed-feedback (DFB) resonant cavity. The grating is made of Si3N4 on the top of the thin SiO_2 layer. The part of the Si substrate is locally removed to enhance the optical confinement. The evanescent coupling to the cavity mode with the Si single quantum well induces the stimulated emissions. Insets show the enlarged view and transmission electron microscope images.

We examined the EL spectra, as shown in Fig. 4. In this experiment, we applied the forward voltage pulse with the pulse width of 100 ns and the period of 1 μs. In the device with the DFB cavity, we observed several peaks from the cavity modes, which were assigned to be the stimulated emissions from the stop band edge. No such modes observed in the reference device without the cavity. The integrated intensity increased super-linearly with currents, which also shows the stimulated emissions.

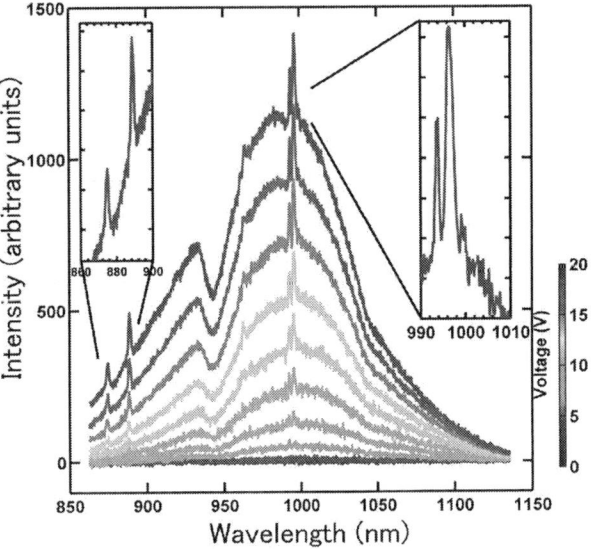

Fig. 4. Stimulated emission spectra from Si single quantum well with resonant cavity operated under application of voltage pulse. EL spectra obtained with integration time of 60 s are shown with applied voltage of 0 to 20 V. Dips around 940 nm originate from optical loss due to fiber and not device.

III. SILICON MULTIPLE QUANTUM WELL: SI FIN

The next requirement towards a practical Si laser diode is to enhance the effective gain due to the evanescent coupling between the guided optical mode and Si quantum well. The obvious approach is to increase the number of quantum wells from the single quantum well to multiple quantum wells. However, it is not easy to make single crystal Si multiple quantum wells whose surface was covered with amorphous SiO_2 by epitaxial growth. Instead, we propose Si multiple quantum well structures suitable to planar Si technologies.

The proposed *Si fin light-emitting diode (FinLED)* [9] is shown in Fig. 5. Instead of a lateral single quantum well, multiple quantum wells, *fins*, are formed along the vertical direction to the substrate. Each side of a fin is connected with a heavily doped Si electrode, forming a lateral pin diode. In this device, thousands of fins can be integrated by simple lithography and etching processes.

The similar multi-fin device structure known as FinFET was proposed as a double gate field-effect-transistor (FET) [10]. The differences between FinLED and FinFET are impurity

978-1-4244-9019-6/11 $26.00 © 2011 IEEE

profiles and absence/presence of gate and waveguide fabrication processes. Therefore, it is possible to integrate both devices solely from the small modification of process steps. The practical challenge for the FinLED is that the width of the fin should be sufficiently small to expect the efficient light emission by quantum confinements, while for the FinFET the width should be larger to avoid the threshold voltage shift.

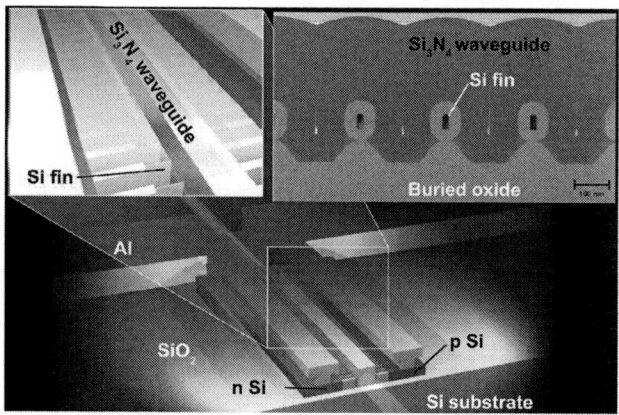

Fig. 5. Si multi-fin light-emitting diode. Insets show the enlarged view and transmission electron microscope image of the device.

The fabricated multi-fin Si by dry etching is shown in the inset of Fig. 5. The fin with the width of less than 5 nm and the height of 50 nm was formed. The fins were made by dry etching and subsequent oxidation processes to reduce the width of the fin. After the formation of multi-fins, the heavily doped electrodes are formed by ion implantations and the activation. Then, the Si_3N_4 film with the thickness of 250 nm was deposited and patterned to become a core of a waveguide located at the center of Si fins. After the metallization of Al/TiN, the H_2 annealing is performed to passivate the interface traps.

The EL spectrum taken from the edge of the waveguide of Si FinLED is shown in Fig. 6. Two peaks at 802.4 nm and 734.4 nm developed significantly with increased currents. These wavelengths correspond to emissions from Si fins with widths of around 1.0 nm according to our theoretical calculations [6]. The integrated intensity increased super-linearly with injected currents indicating stimulated emissions.

In order to understand these spectral peaks, we observed near field images at the edge of the waveguide as shown in Fig. 7 (a). The image shows that the beam spreads vertically to the substrate, although the horizontal width of the waveguide (600 nm) is much larger than the thickness (250 nm). Therefore, the lowest cavity mode with the higher effective index n_{eff} was not dominated (Table II). Instead, the higher modes mainly propagating along the buried-oxide (BOX) layer would be responsible. The calculated mode profiles are similar to the experimental one, as shown in Fig. 7 (b). For these modes, Si Fins work as index and gain coupled DFB structures and stimulated emissions are expected at the stop band edge near the Bragg wavelength of $2 n_{eff} \Lambda$, where Λ (300 nm) is the pitch

of Si Fins. The calculated stimulated wavelengths almost agreed with experimental ones (Fig. 7).

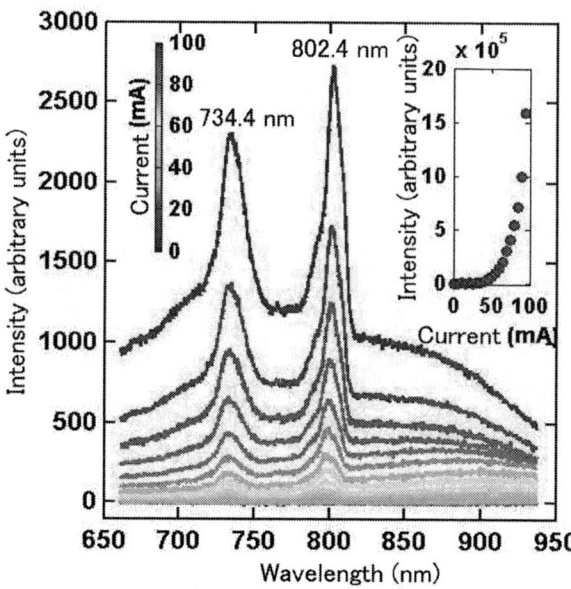

Fig. 6. Electro-luminescence spectra from Si FinLED under constant currents at room temperatures. The inset shows the integrated intensity.

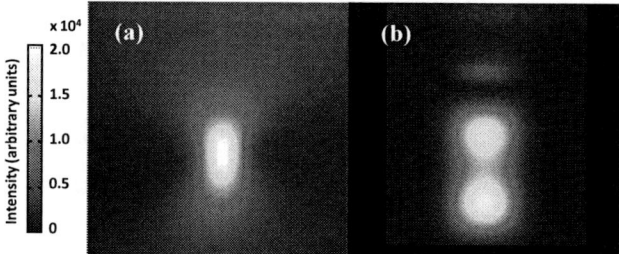

Fig. 7. Near field images of edge emissions from Si FinLED. (a) Experimental images under constant currents of 85 mA. (b) Calculated mode profiles for stimulated emissions.

TABLE II
EXPECTED PROPAGATION MODES IN Si_3N_4 WAVEGUIDE

Mode	0th	1st	2nd	3rd
Profile				
n_{eff} [a]	1.693	1.442	1.333	1.250
λ_{sim} (nm) [a]	1032.1	877.5	800.9	752.5
λ_{exp} (nm) [a]	-	-	802.4	734.4

[a] n_{eff}, λ_{sim} (nm), and λ_{exp} (nm) are the effective refractive index, simulated emission wavelength, and experimental emission wavelength, respectively.

We have proposed a Si quantum well light-emitting diode

fabricated by Si technologies. The experimental results demonstrate the stimulated emissions by current injections in the infrared regime. Si based light emitters should open up a new opportunity for the convergence of photonics and electronics on a Si chip.

ACKNOWLEDGMENT

S. Saito thanks Prof. Y. Arakawa, Prof. S. Iwamoto, and Dr. S. Kako for their enlightening discussions. He also thanks the collaborators in Hitachi for their supports in device fabrication.

REFERENCES

[1] S. J. B. Yoo, "CMOS-Compatible Silicon Photonic Integrated Systems in Future Computing and Communication Systems", 15th OptoElectronics and Communications Conference (OECC), 8D1-1 (2010).

[2] L. T. Canham, "Silicon quantum wire array fabrication by electrochemical and chemical dissolution of wafers", Appl. Phys. Lett. **57**, 1046-1048 (1990).

[3] N. Koshida and H. Koyama, "Visible electrolumninescence from porous silicon", Appl. Phys. Lett. **60**, 347-349 (1991).

[4] L. Pavesi, L. D. Negro, C. Mazzoleni, G. Franzò, and F. Priolo, "Optical gain in silicon nanocrystals", Nature **408**, 440-444 (2000).

[5] D. J. Lockwood and L. Pavesi, in *Silicon Photonics*, p. 1 (Springer, 2004); P. M. Fauchet, *ibid.*, p. 177.

[6] Y. Suwa and S. Saito, "Intrinsic optical gain of ultrathin silicon quantum wells from first-principles calculations", Phys. Rev. B **79**, 233308 (2009).

[7] S. Saito, *et. al.*, "Electro-Luminescence from Ultra-Thin Silicon", Jpn. J. Appl. Phys. **45**, L679-L682 (2006); "Silicon light-emitting transistor for on-chip optical interconnection", Appl. Phys. Lett. **89**, 163504 (2006).

[8] S. Saito, *et. al.*, "Stimulated emission of near-infrared radiation by current injection into silicon (100) quantum well", Appl. Phys. Lett. **95**, 241101 (2009).

[9] S. Saito, *et. al.*, "Stimulated Emission in Silicon Fin Light-Emitting Diode", Solid State Devices and Materials (SSDM), D-5-4, (Tokyo, 2010).

[10] D. Hisamoto, *et. al.*, "FinFET-A Self-Aligned Double-Gate MOSFET Scalable to 20nm", IEEE Trans. Electron Devices **47**, 2320-2325 (2000).

Shin-ichi. Saito received the B. S., M. S., and Ph. D. degrees in the department of physics and applied physics all from Waseda University, in 1995, 1997, and 2000, respectively. From 1998 to 2000, he was a Research Associate at Waseda University, where he studied theoretical condensed matter physics of high temperature superconductivity and strongly correlated electrons systems.

He joined Central Research Laboratory, Hitachi Ltd., in 2000, where he has been engaged in device physics and process technologies of silicon based nanoelectronic and nanophotonic devices. His works include high performance CMOS devices, carrier transport, high-k gate stacks, mobility enhancement, self-organized nanoparticles, flexible thin film transistors, and silicon photonics. He is now a Senior Researcher of Hitachi Ltd. He is the author or co-author of 26 journal papers, 47 international conference proceedings including 9 invited talks, 55 domestic conference presentations including 25 invited talks, and is the inventor or co-inventor of 21 patents, and made 10 domestic press releases. In 2003, he won the SSDM paper award for his analysis on the mobility reduction mechanism in high-k gate dielectric transistors, collaborating with IMEC.

Dr. S. Saito is a member of the Japan Society of Applied Physics (JSAP), the Physical Society of Japan (JPS), and the Institute of Electronics, Information, and Communication Engineers (IEICE).

A Frequency-Shift Readout System for FPW Allergy Biosensor

Chia-Hao Hsu, Yain-Reu Lin, Yue-Da Tsai, Yun-Chi Chen, and Chua-Chin Wang[†], *Senior Member, IEEE*

Department of Electrical Engineering
National Sun Yat-Sen University
Kaohsiung, Taiwan 80424
Email: ccwang@ee.nsysu.edu.tw

Abstract— In this paper, an IgE antigen concentration measurement system using a frequency-shift readout method for a two-port FPW (flexural plate-wave) allergy biosensor is presented. The proposed frequency-shift readout method adopts a peak detecting scheme to detect the resonant frequency. A linear frequency generator, a pair of peak detectors, two registers, and an subtractor are only needed in our system. According to the specification of the FPW allergy biosensor, the frequency sweep range is limited in 2 MHz to 10 MHz. The sensitivity of the peak detector is 0.8 mV. The proposed frequency-shift readout circuit is verified on silicon by using a standard 0.18 μm CMOS technology. The maximal power consumption is 12.94 mW@0.1 MHz clock given by HSPICE simulations.

Keywords—IgE antigen, frequency-shift readout circuit, FPW, resonant frequency, peak detection.

I. Introduction

Due to rapid growth of the biomedical electronics market, *in vitro* bio-analytical applications are quickly developed to help medical staffs to perform pathologic analysis. Many people have been suffered by many allergic diseases, e.g., allergic rhinitis, which may cause uncomfortable feeling. In human serum, concentration of immunoglobulin E (IgE) is an important indicator to show the allergic level therein [1]. Conventionally, many commercial allergy measurement instruments are adopted to measure IgE concentration, e.g., enzyme-linked immunosorbent assay (ELISA) [2], surface plasmon resonance (SPR) [3], and quartz crystal microbalance (QCM) [4] sensing techniques, etc. Unfortunately, these commercial allergy measurement devices require multifarious testing samples, long operation time for sampling analysis procedures, expensive analysis instruments, and lot of analysts. Therefore, a low cost, high speed, and high precision for allergic level estimation is very much needed for those who are suffered.

A two-port allergy biosensor based on an ultrasonic flexural plate-wave (FPW) technique was proposed in [5]. The FPW allergy biosensor adopts the Cr/Au interdigital transducers (IDTs) to be a transmitter and a receiver, which are, respectively, placed on the right and left side on a thin plate. Ac-

cording to the investigation results, the sensitivity of the FPW allergy biosensor is $-8.5 \times 10^7 cm^2 g^{-1}$. Notably, the resonant frequency of the FPW allergy biosensor is variable, which is roughly anti-proportional to the purified human IgE antigen concentration. Thus, the FPW-based allergy biosensor shows another IgE antigen concentration measurement method. In this investigation, a frequency-shift readout system for the two-port FPW allergy biosensor is presented to reduce the operation time and cost. The proposed frequency-shift readout system is realized by a standard 0.18 μm CMOS technology. According to the resonant basics, the output signal amplitude of the FPW allergy biosensor will be maximum when the input frequency is equal to the central resonant frequency. Therefore, a high sensitive peak detector is needed to detect the maximum peak voltage and generate an enable signal to trigger a register to snapshot the frequency value. By calculating the difference between resonant frequencies of sensor1 (with antigen) and sensor2 (without antigen), the frequency-shift value is attained such that the IgE antigen concentration can be estimated. The power consumption of the proposed frequency-shift readout circuit is found to be 12.94 mW at a 0.1 MHz system clock.

II. Frequency-Shift Readout Circuit

The FPW allergy biosensor propagates an acoustic wave via a mechanical thin plate as shown in Fig. 1. Refereing to [5], the resonant frequency of the FPW device can be expressed as follows.

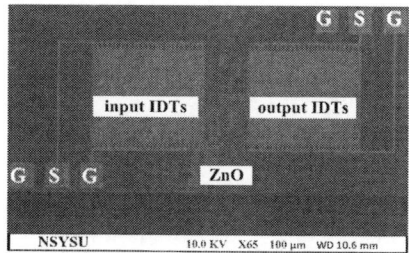

Fig. 1. Photo of the FPW sensor

$$\frac{\Delta f}{f_0} = S_m \Delta m = S_m (MW \times C_S) \qquad (1)$$

[†]Prof. C.-C. Wang is with the Department of Electrical Engineering, National Sun Yat-Sen University, Kaohsiung, Taiwan 80424. (e-mail: ccwang@ee.nsysu.edu.tw). He is the contact author.

978-1-4244-9019-6/11 $26.00 © 2011 IEEE

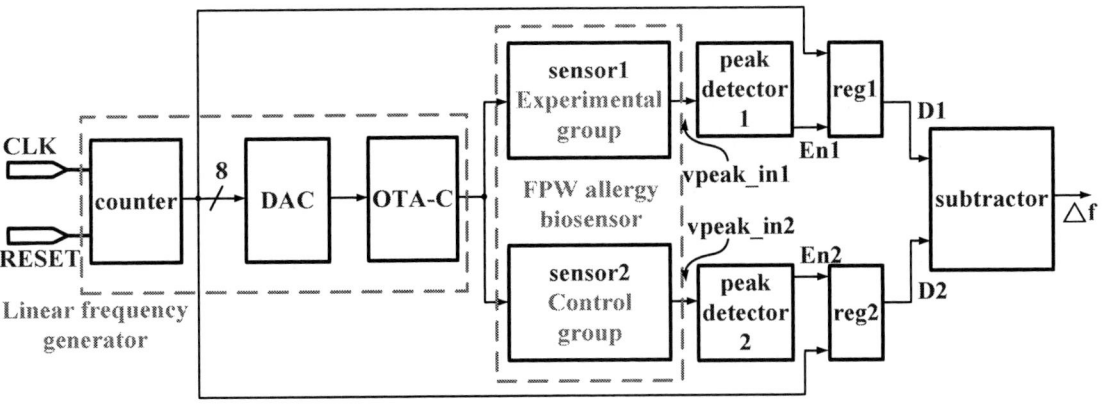

Fig. 2. Schematic of the frequency-shift readout system

where f_0 is the initial resonant frequency, Δf denotes the variation of the resonant frequency, and the mass per unit area and the mass sensitivity are, respectively, denoted as S_m and Δm. MW is the molecular weight and C_S is the surface concentration of the absorptive moleculars. Therefore, Δf can be changed by C_S as well as the IgE antigen concentration . According to the phenomenon of FPW allergy biosensor's frequency shifting, we propose a novel frequency-shift readout circuit to estimate the amount of the frequency shifting. Fig. 2 shows the proposed frequency-shift readout system, which is composed of a counter, a DAC (digital-to-analog converter), an OTA-C oscillator, a pair of peak detectors, two registers, and a subtractor. The detailed description of each subcircuit is explained in following subsections.

A. Linear frequency generator

A sine wave generator is required for the FPW biosensors to generate all of the frequencies in the pre-defined range. As shown in Fig. 2, the counter is a typical digital 8-bit up-counter generating 0 to 256 up counting signal to the DAC. The 8-bit DAC utilizes a current-steering structure with a current complement circuit as shown in Fig. 3. The 8-bit DAC requires only 8 current sources with different sizes instead of $2^8 - 1$ sources. To reduce the error of the 8-bit DAC output voltage, the current complement circuit generates an appropriate complement current to vout. Therefore, the performance of the proposed DAC can be enhanced regarding integral non-linearity (INL) and differential non-linearity (DNL). Given an input binary code, vout can be set to an appropriate potential to provide the OTA-C oscillator a bias voltage.

The schematic of the tunable OTA-C oscillator is shown in Fig. 4 [6]. OTA_VB is given a proper bias voltage to ensure correctness of each OTA's functionality.

Gm1, Gm2, Gm3, and Gm4 are the same operational transconductance amplifier (OTA) as the one shown in Fig. 5. Assuming MP502=MP503 and MN502=MN503, the drain-source currents of MP502 and MP501, i_{dP502} and i_{dP503}, are expressed as follows.

$$-i_{dP502} = i_{dP503} = \frac{g_m}{2}(VP - VN) \qquad (2)$$

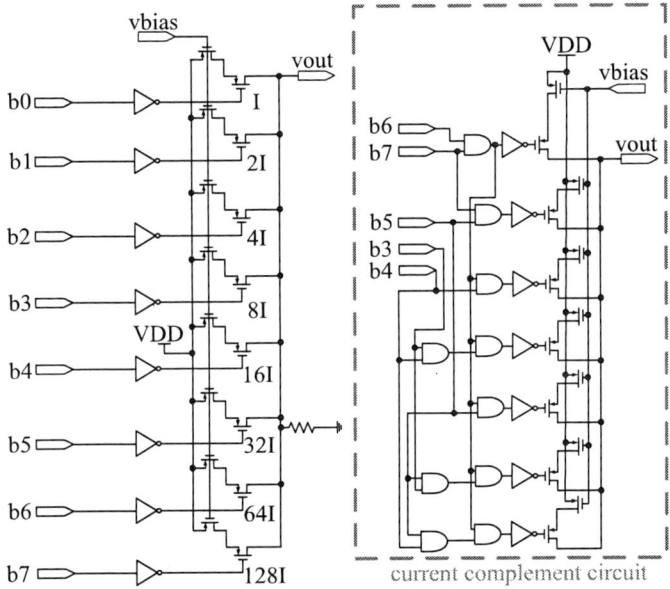

Fig. 3. Schematic of the 8-bit DAC

Furthermore, the gain of the OTA is given by

$$Av = \frac{Vo}{VP - VN} = g_m(r_{oP501} \| r_{oN501}) \qquad (3)$$

If the impedance of output load is smaller than OTA output resistance at a higher frequency, the output current, io, can be written as Eqn. 4.

$$io = i_{dP501} - i_{dN501} \qquad (4)$$

Finally, we can derive the OTA's transconductance as follows.

$$g_{mOTA} = \frac{io}{VP - VN} \qquad (5)$$

However, the transconductance of Gm1 can be adjusted by tuning the bias, Vtune, from output of DAC. Referring to Fig. 4, Gm1-C1 and Gm2-C2 constitutes a 2nd-order RC oscillator with a positive feedback to generate an oscillation signal. On the other hand, Gm3 and Gm4 is used to keep the peak-to-peak

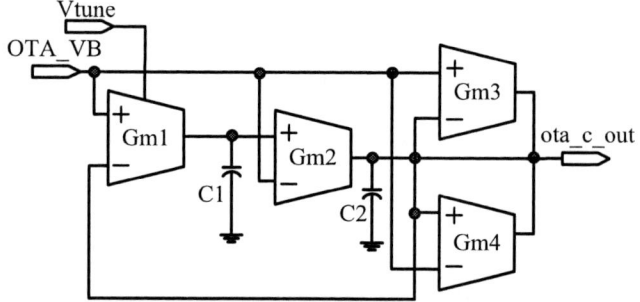

Fig. 4. Schematic of the OTA-C oscillator

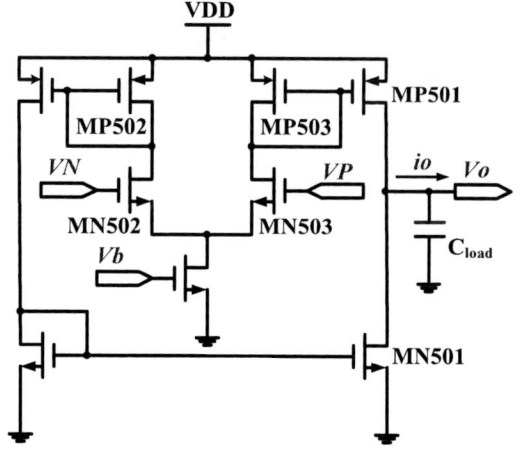

Fig. 5. Schematic of the operational transconductance amplifier

amplitude of the generated sine wave. The frequency tuning range of tunable OTA-C oscillator is limited from 2 MHz to 10 MHz according to the specification of the FPW allergy biosensor.

B. Peak detector

The output signal of the FPW allergy biosensor will reach its peak value when the input frequency equals to the resonant frequency. A peak detector is, then, used to detect the maximum peak from the FPW allergy biosensor output and the determine the frequency value. Fig. 6 shows the proposed peak detector. The detailed operating steps of the peak detector is explained as follows.

Step1: Initially, RESET1, RESET2, and RESET3 are biased at high to discharge C1, C2 and reset the D flip-flop.

Step2: The sine wave from FPW allergy biosensor's output is fed to VIN. When VIN is higher than VPEAK1, OPA1 will turn on MN603. Then, C1 is charged until VPEAK1 ≥ VIN.

Step3: MN603 is off to isolate VPEAK1 from VPEAK2. If VPEAK1 is higher than VPEAK2, OPA2 will trigger the D flip-flop. Then, EN is pulled high to

turn MN603 on. Hence, VPEAK2 is pulled close to VPEAK1 through MN603. If VPEAK1 is not higher than VPEAK2, VPEAK2 keeps the prior high voltage value.

Step4: When VPEAK2 is equal to VPEAK1, RESET3 will be pulled up high to reset the D flip-flop to set EN=0. VPEAK1 and VPEAK2 are isolated again by MN603.

By the above steps, the peak detector can generate the enable signal, EN, to enable the register and store the current counting number in the counter. Therefore, the resonant frequency of the FPW allergy biosensor is detected by the proposed peak detector. The sensitivity of the peak variation is 0.8 mV. By the subtraction reg1 from reg2, the frequency-shift variation, $\triangle$f, can be derived.

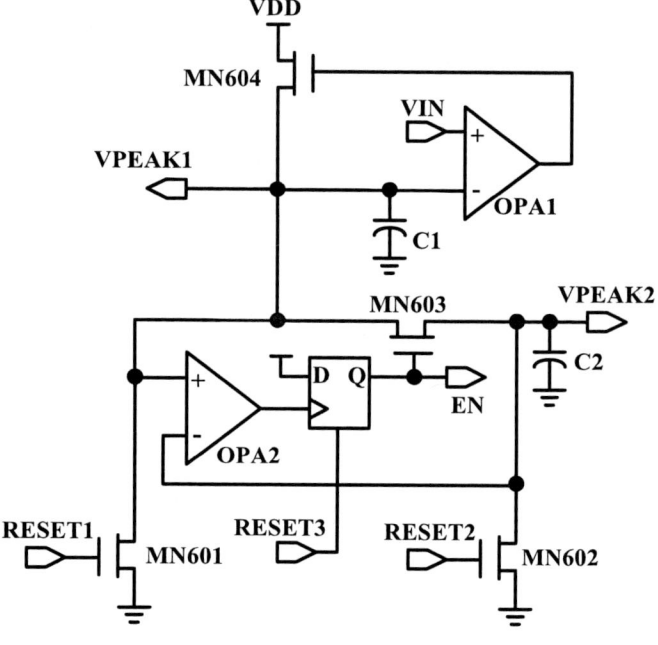

Fig. 6. Simulation of the peak detector

III. IMPLEMENTATION AND SIMULATION

The proposed frequency-shift readout circuit for FPW allergy biosensor is realized on silicon by TSMC (Taiwan Semiconductor Manufacturing Company) standard 0.18 μm CMOS technology. Fig. 7 shows the whole chip layout including I/O PADs of the proposed design. The chip area of the proposed RC5 is 1148 $\times$ 1148 μm^2. The FPW allergy biosensor is converted into an equivalent RLC model, which is added in the proposed frequency-shift readout system to run the HSPICE simulation. The simulation results of the peak detector are illustrated in Fig. 8. When the amplitude of vpeak_in1 is higher than the prior peak, vpeak2 will be pulled to the same position as vpeak1 and $\overline{En1}$ becomes low. Therefore, the register, reg1, is renewed to store the current frequency of the

978-1-4244-9019-6/11 $26.00 © 2011 IEEE   224

Experimental group FPW allergy biosensor. On the other hand, reg2 is renewed to take the current frequency of the Control group FPW allergy biosensor. The frequency-shift variation is calculated by the subtractor. The power consumption is 12.94 mW at a 0.1 MHz clock. The comparison with a similar prior work is tabulated in Table I.

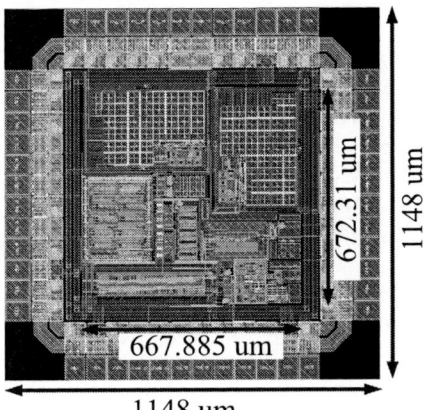

Fig. 7.   Layout view of the proposed frequency-shift readout circuit

TABLE I
COMPARISON WITH PRIOR WORK

	proposed	[7]
Implementation technique	system on chip	on PCB discretes
Detecting method	peak detection	phase detection
Process (μm)	0.18	N/A
Supply voltage (V)	1.8	N/A
Frequency (MHz)	0.1	4.2
Power (mW)	12.94	N/A
Year	2010	2008

IV. CONCLUSION

This paper presents a frequency-shift readout circuit for a two-port FPW allergy biosensor. The linear frequency generator generates a linear frequency sweep fed into the FPW allergy biosensor. The peak detectors are adopted to detect the resonant frequencies of the Experimental group and Control group of the two-port FPW allergy biosensor. The detected resonant frequencies are stored in the registers, reg1 and reg2, respectively. The frequency-shift value is derived by the subtraction of reg1 from reg2. By using the semiconductor technique, the proposed frequency-shift readout circuit is fabricated on a chip by TSMC standard 0.18 μm CMOS technology.

ACKNOWLEDGEMENT

This investigation is partially supported by National Science Council under grant NSC 99-2221-E-110-081-MY3 and EZ-10-09-44-98. It is also partially supported by Ministry of Economic Affairs, Taiwan, under grant 98-EC-17-A-01-S1-104, 98-EC-17-A-19-S1-133, and 98-EC-17-A-07-S2-0010. The authors would like to express their deepest gratefulness

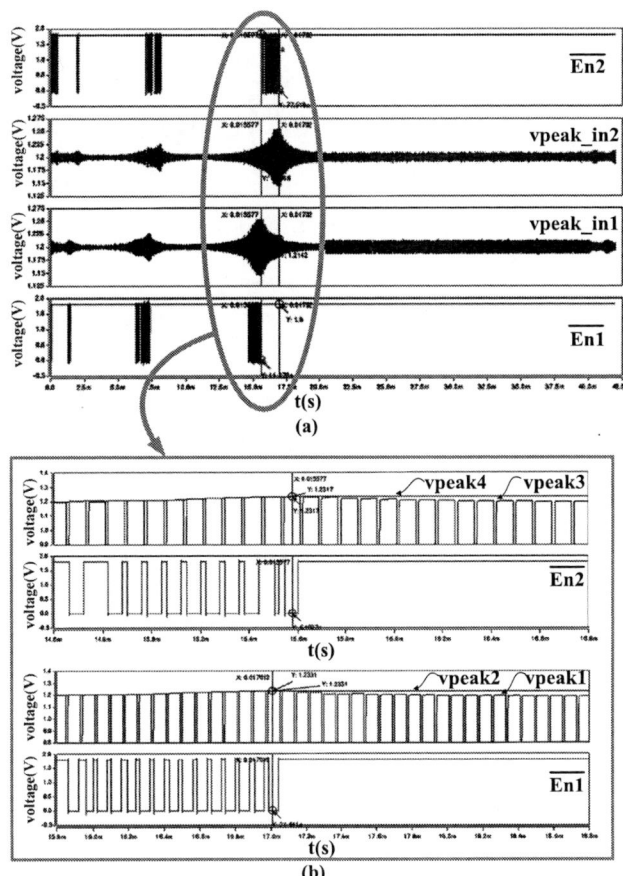

Fig. 8.   Simulation of the frequency-shift measurement

to Chip Implementation Center of National Applied Research Laboratories, Taiwan, for their thoughtful chip fabrication service. Finally, the authors would like to thank NXP Semiconductors (Taiwan) Ltd. for their strong support of the wire bounding equipment.

REFERENCES

[1] H. J. Gould, B. J. Sutton, A. J. Beavil, R. L. Beavil, N. McCloskey, H. A. Coker, D. Fear, and L. Smurthwaite, "The biology of IGE and the basis of allergic disease," *Annual Review of Immunology*, vol. 21, pp. 579-628, Apr. 2003.

[2] R. M. Lequin, "Enzyme Immunoassay (EIA)/Enzyme-Linked Immunosorbent Assay (ELISA)," *Clinical Chemistry*, vol. 51, no. 12, pp. 2415-2418, Sep. 2005.

[3] X. Su and J. Zhang, "Comparison of surface plasmon resonance spectroscopy and quartz crystal microbalance for human IgE quantification," *Sensors and Actuators B: Chemical*, vol. 100, no. 3, pp. 309-314, Sep. 2004.

[4] X. Su, F. T. Chew, and S. F. Li, "Piezoelectric quartz crystal based label-free analysis for allergy disease," *Biosensors & Bioelectronics*, vol. 15, no. 11-12, pp. 629-39, May 2000.

[5] I-Y. Huang and M.-C. Lee, "Development of a FPW allergy biosensor for human IgE detection by MEMS and cystamine-based SAM technologies," *Sensors and Actuators B: Chemical*, vol. 132, no. 1, pp. 340-348, May 2008.

[6] J.A. D. Lima, "A linearly-tunable OTA-C sinusoidal oscillator for low-voltage applications," in Proc. of *2002 IEEE International Symposium on Circuits and Systems*, vol.2, pp. II-408-II-411, 2002.

[7] W.-Y. Chang, P.-H. Sung, C.-H. Chu, C.-J. Shih, and Y.-C. Lin, "Phase Detection of the Two-Port FPW Sensor for Biosensing ," *IEEE Sensors Journal*, vol. 8, no. 5, pp. 501-507, May 2008.

Scaled Nanoelectromechanical (NEM) Hybrid Devices

Hiroshi Mizuta, *Member, IEEE*, Mario A. Garcia-Ramirez, *Member, IEEE*, Zakaria Moktadir,
Yoshishige Tsuchiya, Shunichiro Sawai, Jun Ogi, Shunri Oda, *Member, IEEE*

Abstract— **This paper overviews recent attempts at co-integrating nano-electro-mechanical systems (NEMS) with nanoelectronic devices aiming to add more functionalities to conventional Si devices in 'More-than-Moore' domain and also explore novel physical principles in 'Beyond CMOS' domain.**

Index Terms— **Beyond CMOS, graphene, More than Moore, nanophonon, NEMS, quantum dot, single-electron transistor, suspended-gate**

I. CO-INTEGRATION OF NEMS AND MOSFETs FOR 'MORE-THAN-MOORE' APPLICATIONS

VLSI technology developed and matured over the past decades has been fully exploited to build the vast technology area of micro-electromechanical systems (MEMS). Along with a rapid expansion of the MEMS market, there have also been continuous efforts at making the MEMS smaller (Fig. 1) in order to boost the resonant frequency to GHz and beyond. The appearance of high-frequency nano-electro-mechanical systems (NEMS) is tempting enough for us to consider the co-integration of the NEMS and conventional silicon electronic devices ('*More than Moore*') because we expect such hybrid systems enhance scaling of functional density & performance while simultaneously reducing the power dissipation beyond the conventional CMOS-based systems. In addition, recent emergence of superior graphene-based NEMS (GNEMS) [1] provides more choice of building blocks for the hybrid systems.

A variety of new hybrid NEM-MOS devices have recently been studied for advanced switch, memory and sensing applications (see TABLE I). A pioneering hybrid device is a

Manuscript received March 22, 2011. This work was supported in part by the EU FP7 NEMSIC (Hybrid Nano-Electro-Mechanical / Integrated Circuit Systems for Sensing & Power Management Applications) project and SORST JST (Japan Science and Technology) and MEXT KAKENHI 18310097 and 16206030 Japan.

H. Mizuta is with NANO Group, Electronics and Computer Science, Faculty of Physical and Applied Sciences, University of Southampton, Highfield, Southampton SO17 1BJ, U.K. and School of Materials Science, Japan Advanced Institute of Science and Technology (JAIST), Ishikawa 923-1292, Japan (phone: +44(0)2380 592852; fax: +44(0)2380 593029; e-mails: hm2@ ecs.soton.ac.uk, mizuta@jaist.ac.jp).

M.A.G.-Ramirez, Z. Moktadir, Y. Tsuchiya are with NANO Group, Electronics and Computer Science, Faculty of Physical and Applied Sciences, University of Southampton, Highfield, Southampton SO17 1BJ, U.K. (*phone: +44(0)2380 592852; fax: +44(0)2380 593029; e-mail: hm2@ ecs.soton.ac.uk).

S. Sawai, J. Ogi, S. Oda are with Quantum Nano Electronics Research Center, Tokyo Institute of Technology, Ookayama, Meguro-ku, Tokyo 152-8550, Japan

suspended-gate (SG) FET [2] which features a movable gate electrode located on a conventional oxide / silicon substrate via an air gap. Thanks to unique electro-mechanical pull-in / pull-out operations, the SGFETs exhibit very abrupt electrical switching with a subthreshold swing much smaller than a theoretical limit of 60 mV/dec for MOSFETs as well as extremely low off current. The SGFETs therefore attract much attention in particular for power management applications.

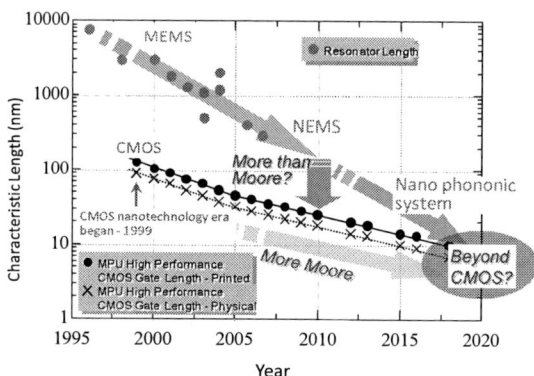

Fig. 1 Recent trend of MEMS/NEMS downscaling along with CMOS miniaturization.

TABLE I. VARIOUS NEM-MOS HYBRID FUNCTIONAL DEVICES

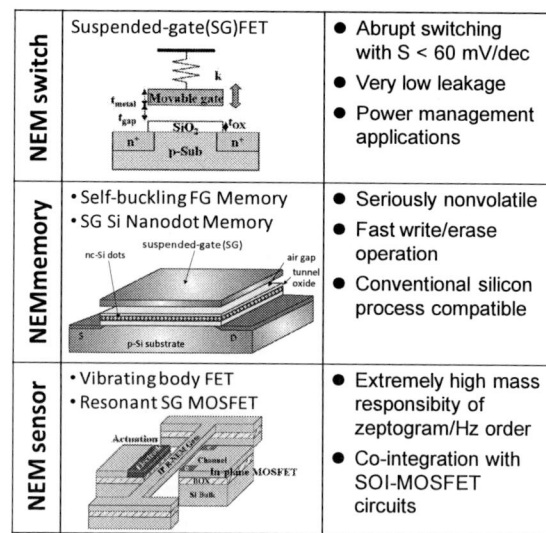

As for the memory applications, two types of novel high-speed and nonvolatile NEM-MOS hybrid memory devices have been proposed in order to realize nonvolatile RAM. The bistable floating gate (FG) NEM memory [3] was first proposed, which features a self-buckling SiO_2 FG with embedded Si nanodots as single-electron storage. The self-buckling FG is flip-flopped electrically with the gate voltage, and the bistable mechanical states are detected with a readout MOSFET located underneath. The other one is a suspended gate (SG) Si nanodot memory (SGSNM, see TABLE I) [4] in which the SG is pulled in to the FG only when we programme and erase the information. The SGSNM consists of a MOSFET as readout, silicon nanodots as a FG, and a clamped-clampled SG (see Fig. 2) which is isolated from the FG by an air gap and a thin tunnel oxide. For the programming (P) process, a negative gate voltage is applied, and the SG is pulled-in on the FG layer, resulting in electron injection from the SG into the FG. For the erasing (E) process, a positive voltage is applied, and the stored electrons are extracted from the FG.

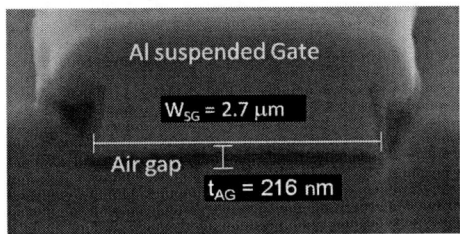

Fig. 2 A cross-sectional SEM image of a fabricated suspended gate .

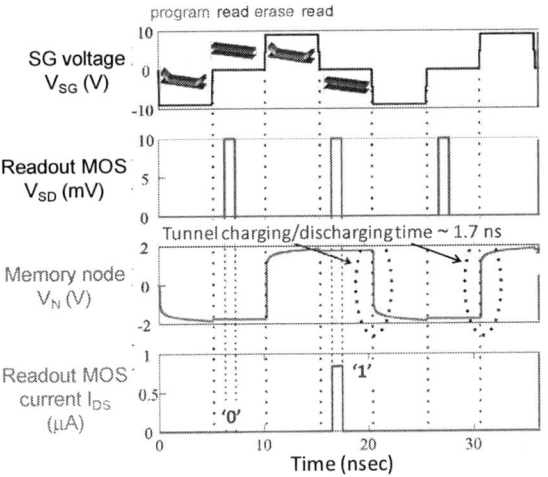

Fig. 3 Program-Read-Erase-Read waveforms simulated for a SGSNM cell with a 1-µm-long suspended gate.

The SGSNM architecture enables to avoid an unfavorable tradeoff between the ON/OFF current ratio and the P/E voltages and therefore facilitates low-power operation compared to the bistable FG NEM memory. Recent theoretical study shows that the SGSNM with a 1-µm-long suspended gate achieves the program/erase times of only few nsecs (Fig. 3), which are three orders of magnitude shorter than those for Flash memory.

As for the sensing applications, a new in-plane resonant

NEM sensor [5] has been proposed based on a mass-detection principle. This nanosensing device features a silicon resonant-suspended-gate and an in-plane MOSFET co-integrated on an ultrathin SOI platform and enables sub-attogram-level mass detection. Despite a number of potential advantages demonstrated for experimental NEM-MOS hybrid devices, there still remain various crucial issues to be clarified, in particular, scalability which takes the influences of nanoscopic Van der Waals force and Casimir effect into consideration [6][7].

II. NEM-SET HYBRID DEVICES AND ATOMICALLY-SCALED NEM SYSTEMS FOR 'BEYOND-CMOS' DOMAIN

A. NEM-SET hybrid devices

By downscaling the NEMS towards a 100-nm-regime and even smaller, we may explore a novel hybrid system of NEMS and single-electron transistors (SETs). There are a number of new phenomena associated with strong coupling of single-electron tunnelling and low-dimensional phonons, such as phonon blockade [8], single-electron quantum shuttle [9] and the quantum ground state of a mechanical resonator [10]. Manipulation of nanophononic states in scaled silicon nanostructures may be exploited to develop a novel approach to thermal management and energy transfer interaction in 'Beyond CMOS' information processing.

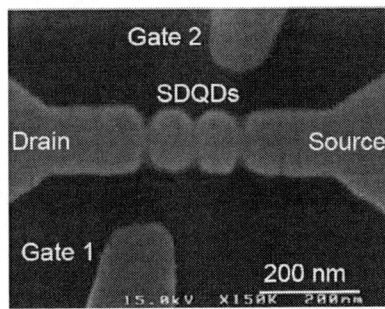

Fig. 4 The SEM image of fabricated SDQDs structure. The lateral dimension of DQDs is about 150 nm.

A suspended quantum dot (QD) SET [11] is particularly attractive to investigate the interaction between single electrons and low-dimensional phonons in the NEM structures. Figure 4 shows suspended double quantum dots (SDQDs) built on a doubly-clamped Si beam with two side gates. The SDQD SET was fabricated on 40-nm-thick heavily-P-doped (2×10^{19} cm^{-3}) silicon-on-insulator (SOI) substrate by using EB lithography [12]. Bias voltages applied to the two side gates, V_{g1} and V_{g2}, control the electro-chemical potentials of the individual DQDs. A single-electron tunnels through DQDs elastically when the chemical potentials align.

Figure 5 shows the drain current measured at 120 mK with V_{ds} = 200 µV as a function of V_{g1} and V_{g2}. The tunnel current is enhanced around the anti-crossing points of two charging lines, resulting in a pair of bias triangles. Fine structures submerged in the bias triangles were recently clarified by decomposing a

978-1-4244-9019-6/11 $26.00 © 2011 IEEE

broad peak across the bias triangle into a few current peaks.

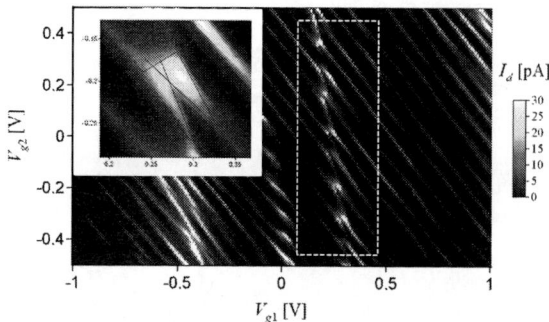

Fig. 5 Drain-to-source current measured at 120 mK with Vds = 200 μV as a function of Vg1 and Vg2. The typical characteristics for serial DQDs were observed in the region shown by a broken-line rectangular. The inset shows a pair of bias triangles observed around one of the anti-crossing regions.

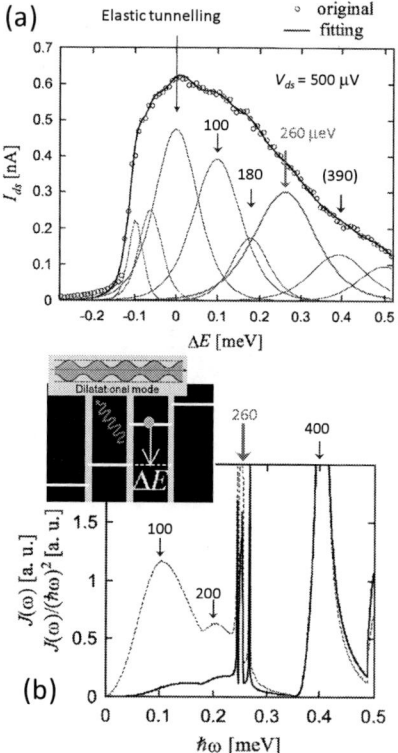

Fig. 6 (a) I_{ds} versus the energy difference ΔE between the double QDs with V_{ds}=500 μV. The original data (circle) are fitted by a sum of peaks. (b) The calculated phonon spectral density (solid line) and its ratio to the square of the phonon energy (dotted line). The arrow shows the van Hove singularities of the dilatational mode that is schematically shown in the inset.

In Fig. 6(a), the largest peak centered at $\Delta E = 0$ corresponds to elastic (coherent) tunneling, and the peaks centered at $\Delta E > 0$ correspond to inelastic tunneling of electrons with phonon emission. By comparing these experimental results with the phonon spectral density calculated for the SDQDs by using the continuum model (Fig. 6(b)), it was found that the observed current peaks were attributable to the enhanced interaction between the electrons confined in the SDQDs and dilatational-

and flexural-mode phonons of the suspended Si beam.

A good agreement between the experimental and theoretical phonon peak energies enables us to understand and even manipulate fundamental energy dissipation processes of tunnelling electrons in Si NEM structures. For example, the peak energy in the spectral density associated with the van Hove singularities of dilatational mode can be controlled via Si slab thickness: the peak energy increases approximately inverse proportional to the Si slab thickness.

B. Atomically-scaled NEM functional systems

Extremely-scaled Si NEMS will eventually exploit electro-mechanical properties of atomically-controlled Si nanostructures. *Ab-initio* simulation of 'nanophonons' has recently been conducted for the H-terminated atomically-thin Si films [13]. Figure 7 shows the dispersion relationship calculated for the 5-atomic-layer-thick film in comparison with those for bulk Si. The phonon bandgap is observed both in the [1 -1 0] and [1 1 0] directions which is defined between the energy maximum of the acoustic branches and the energy minimum of the right above optical-like branch. The formation of the phonon bandgaps are closely related to the Si(001) 2x1 surface dimers which line up in the [1 -1 0] direction on one of the surfaces and in the [1 1 0] direction on the other surface. In contrast to the bulk Si atoms, the surface Si atoms are covalently-bound alternately in the direction and therefore hold the structural and mechanical periodicity of twice as long as that for bulk Si.

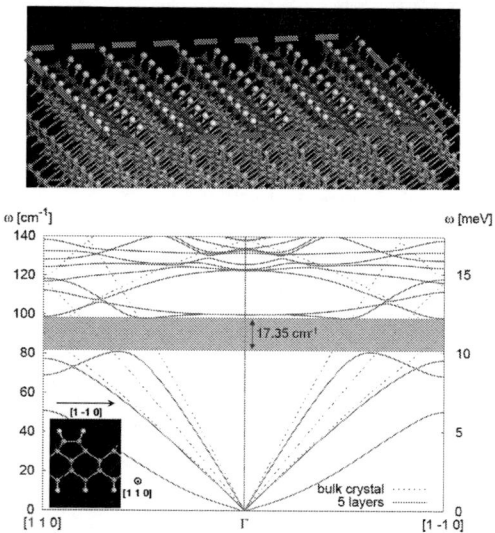

Fig. 7 *Ab-initio* simulation of nanophononic spectra for a hydrogen-terminated ultrathin Si film of 5 atomic layers in thickness. A hatched region shows a phononic bandgap formed both in the (110) and (1-10) directions.

The magnitude of the phononic bandgap decreases with increasing the film thickness and finally vanishes for the thickness above 8 atomic layers. This is because the impact of the surface dimers on the entire dispersion relationships is reduced relative to that of the bulk Si atoms. Those nanophononic properties are quite different from those for bulk silicon and may be exploited as one of approaches to nanoscale thermal management and novel energy transfer interactions. In addition, such periodic nanophononic structures may also work as photonic crystals, referred as optomechanical or phoXonic crystals, and offer a possibility of co-engineering phonons and photons [14], leading to novel nano opto-acoustic device applications. Finally, it is worth mentioning that such atom-scale nanophononic properties can now be explored for an alternative emerging platform, graphene, along with the recent helium ion beam milling technology which enables direct patterning of mono- and multi-layer graphene films (Fig. 8) with a truly nanometer-scale precision [15].

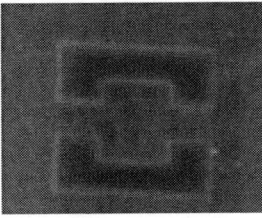

Fig. 8 Graphene NEMS (GNEMS) patterned on a bilayer graphene film by using the helium beam milling technique.

ACKNOWLEDGMENT

The authors wish to thank F. A. Hassani, M. A. Ghiass and N. Kalhor of Univ. of Southampton for valuable discussions and Dr. S. Saito of Central Research Laboratory, Hitachi Ltd. for his valuable contribution to SGSNM fabrication.

REFERENCES

[1] C. Chen, S. Rosenblatt, K. I. Bolotin, W. Kalb, P. Kim, I. Kymissis, H. L. Stormer, T. F. Heinz and J. Hone, 'Performance of monolayer graphene nanomechanical resonators with electrical readout', Nature Nanotech. **4**, 861 (2009)

[2] N. Abelé, R. Fritschi, K. Boucart, F. Casset, P. Ancey and A.M. Ionescu, 'Suspended-gate MOSFET: bringing new MEMS functionality into solid-state MOS transistor', *IEDM Techn. Dig.* 2005, 1075 (2005)

[3] Y. Tsuchiya, K. Takai, N. Momo, T. Nagami, S. Yamaguchi, T. Shimada, H. Mizuta, and S. Oda, "Nanoelectromechanical nonvolatile memory device incorporating nanocrystalline Si dots" J. Appl. Phys. **100**, 094306 (2006)

[4] M. A. G.-Ramirez, Y. Tsuchiya and H. Mizuta, "Hybrid circuit analysis of a suspended-gate silicon nanodot memory (SGSNM) cell", Microelectronic Engineering **87**, 1284-1286 (2010).

[5] F. A. Hassani, C. Cobianu, S. Armini, V. Petrescu, P. Merken, D. Tsamados, A. M. Ionescu, Y. Tsuchiya and H. Mizuta, 'Numerical analysis of zeptogram/Hz-level mass responsivity for in-plane resonant nano-electro-mechanical sensors', in press for Microelectronic Engineering (2011).

[6] F. Michael Serry Dirk Walliser and G. Jordan Maclay., 'The role of the casimir effect in the static deflection and stiction of membrane strips in microelectromechanical systems (MEMS)', J. Appl. Phys. **84**, 2501, (1998)

[7] E. Buks and M. L. Roukes, 'Stiction, adhesion energy, and the Casimir effect in micromechanical systems', Phys. Rev. B **63**, 033402 (2001)

[8] E. M. Weig, R. H. Blick, T. Brandes, J. Kirschbaum, W. Wegscheider, M. Bichler, and J. P. Kotthaus, 'Single-electron-phonon cavity in a suspended quantum dot phonon cavity', PRL **92**, 046804 (2004).

[9] A. Erbe, C. Weiss, W. Zwerger, and R. H. Blick, 'Nanomechanical resonator shuttling single electrons at radio frequencies', PRL **87**, 096106 (2001)

[10] A. D. O'Connel, M. Hofheinz, M. Ansmann, Radoslaw C. Bialczak, M.

Lenander, Erik Lucero, M. Neeley, D. Sank, H. Wang, M. Weides, J. Wenner, John M. Martinis & A. N. Cleland, ' Quantum ground state and single-phonon control of a mechanical resonator', Nature **464,** 697 (2010).

[11] J. Ogi, M. Ghiass, Y. Tsuchiya, K. Uchida, S. Oda and H. Mizuta, "Suspended quantum dot fabrication on a heavily-doped silicon nanowire by suppressing unintentional quantum dot formation ", Jpn. J. Appl. Phys. **49**, 044001 (2010)

[12] J. Ogi, T. Ferrus, T. Kodera, Y. Tsuchiya, K. Uchida, D. A. Williams, S. Oda, and H. Mizuta, "Experimental observation of enhanced electron-phonon interaction in suspended Si double quantum dots", Jpn. J. Appl. Phys. **49**, 045203 (2010)

[13] S. Sawai, S. Uno, M. Okamoto, Y. Tsuchiya, S. Oda and H. Mizuta, "Atomistic study of phonon states in hydrogen-terminated Si ultra-thin films", IEEE Silicon Nanoelectronics Workshop, M0200, Honolulu, June (2008)

[14] Matt Eichenfield, Jasper Chan, Ryan M. Camacho, Kerry J. Vahala, Oskar Painter, 'Optomechanical crystals', Nature **462**, 78 (2009)

[15] Z. Moktadir, S. Boden, H. Mizuta and H. Rutt, 'Graphene for Nano-electro-mechanical Systems', 21st Micromechannics and Micro systems Europe Workshop (MME2010), Enschede, September (2010)

Hiroshi Mizuta (M'89) received the B.S. and M.S. degrees in physics and the Ph.D. degree in electrical engineering from Osaka University, Osaka, Japan, in 1983, 1985, and 1993, respectively. He joined the Central Research Laboratory, Hitachi Ltd., Tokyo, Japan, in 1985, and worked on high-speed heterojunction devices and resonant tunneling devices. From 1989 to 1991, he studied quantum transport simulation, and also from 1997 to 2003 he worked on single-electron devices and associated nanoelectronic devices as the Laboratory Manager and Senior Researcher at the Hitachi Cambridge Laboratory, UK. From 2003 to 2007, he was Associate Professor of Physical Electronics at Tokyo Institute of Technology, Japan. He is currently Professor of Nanoelectronics, NANO Group, University of Southampton, UK, concurrently Professor, School of Materials Science, Japan Advanced Institute of Science and Technology (JAIST). His current research interests include silicon and graphene nanoelectronics, hybrid nanoelectromechanical systems (NEMS), quantum information technology and *ab initio* simulation of atomistic devices. He has published more than 340 scientific papers including books and book chapters and filed over 50 patents. Dr Mizuta is a member of the Physical Society of Japan, the Japan Society of Applied Physics, the Institute of Physics, and the Electron Device Society of the IEEE.

Mario A. García-Ramírez received the B.Eng, from the Instituto Tecnológico de Morelia, "José Maria Morelos y Pavón", México in 2001, his M. S at INAOE, México in 2006 and he is currently working towards his PhD in the Nano Group in the School of Electronics and Computer Science (ECS) at the University of Southampton, UK.

Yoshishige Tsuchiya received the B.S., M.S.,and Ph.D. degrees in multidisciplinary sciences from the University of Tokyo, Japan, in 1996, 1998, and 2001, respectively. From 2001 to 2008, he was an Assistant Professor at the Quantum Nanoelectronics Research Center, Tokyo Institute of Technology, Japan, where he was engaged in research on fabrications and characterizations of silicon nanodevices. Since February 2008 he has been a lecturer at the School of Electronics and Computer Science, University of Southampton, U.K. His current research interests include silicon nanoelectromechanical devices and silicon quantum information devices.

Shunri Oda (M'89) received the B.Sc. degree in physics, and the M.S. and Ph.D. degrees from Tokyo Institute of Technology, Tokyo, Japan, in 1974, 1976, and 1979, respectively. He is currently a Professor in Department of Physical Electronics and Quantum Nanoelectronics Research Center, Tokyo Institute of Technology. He has authored or coauthored more than 200 papers in journals and conference proceedings. His research interests include fabrication of silicon quantum dots by pulsed plasma processes, single-electron tunneling devices based on nanocrystalline silicon, ballistic transport in silicon nanodevices, silicon-based photonic devices, and high-k gate oxide ultrathin films prepared by atomic layer metal–organic CVD. Dr. Oda is a member of the Electrochemical Society, the Materials Research Society, and the Japan Society for Applied Physics. He is a Distinguished Lecturer of the IEEE Electron Devices Society.

Evaluation of DC and AC performance of junctionless MOSFETs in the presence of variability

Xin Qian, Yinglin Yang, Zhiwei Zhu, Shi-Li Zhang, and Dongping Wu

Abstract—In this paper, DC and AC performance of junctionless MOSFETs are extensively examined. A comparison is made between double-gate junctionless MOSFETs and conventional inversion-mode MOSFETs with an emphasis on the variability in performance. Despite clear benefits by eliminating junctions and related junction variabilities, junctionless MOSFETs are found to require double- or multi-gate in order to be fully turned off. They are also significantly more sensitive to variations of channel thickness and channel doping concentration. Though junctionless MOSFETs demonstrate lower driving current and transconductance, they exhibit significantly lower gate capacitances at saturation region and slower degradation of transconductance over gate overdrive.

Index Terms—junctionless, MOSFET, transistor, double-gate, DC, AC, variability

I. INTRODUCTION

According to International Technology Roadmap for Semiconductors (ITRS) [1], the gate length of MOSFETs is forecasted to reach sub-20 nm regime in a couple of years. Three-dimensional transistor structures such as FinFET or Tri-gate MOSFETs are supposed to replace conventional planar MOSFET structure for technology nodes beyond 22 nm owing to their superior performance in control of short channel effects in extremely scaled MOSFETs. For such 3-D device structures, demands on extremely abrupt P-N junctions between the source/drain and the channel regions present an increasingly difficult fabrication challenge. Novel solutions such as metal source/drain [2]-[5] and junctionless MOSFETs [6] have therefore been proposed and extensively investigated. Junctionless MOSFET should in principle be immune to many of the challenges related to the formation of source/drain junctions and hence presents a very promising candidate for future decananometer MOSFET applications. In a junctionless MOSFET, the doping level in the semiconductor source, channel, and drain is identical (n-type: N^+-N^+-N^+ and p-type: P^+-P^+-P^+) and requires no formation of P-N or Schottky junctions between the source/drain and channel. Interesting investigations have been done to demonstrate the functionality and DC performance of nano-scale junctionless MOSFETs [7]-[14]. In this paper, we perform an extensive evaluation of

This work was supported by "National S&T Major Project 02" and the Program for Professor of Special Appointment (Eastern Scholar) at Shanghai Institutions of Higher Learning.

The authors are with State Key Laboratory of ASIC and System, Fudan University, Shanghai 200433, People's Republic of China. (e-mail of the corresponding author Dongping Wu: dongpingwu@fudan.edu.cn).

Shi-Li Zhang is also with Solid-State Electronics, The Ångström Laboratory, Uppsala University, P.O. Box 534, 75121 Uppsala, Sweden.

the DC and AC performance of junctionless MOSFETs with an emphasis on their performance variabilities in comparison to conventional P-N junction source/drain MOSFETs.

II. DEVICE STRUCTURE AND SIMULATION

Silvaco software package is used to construct the device structures and perform the relevant device simulations. Both single-gate and double-gate junctionless nMOSFETs are simulated. For comparison, conventional P-N junction source/drain nMOSFETs with similar geometrical parameters are also studied. The physical gate length (L_G) is set to be 22 nm for simulated MOSFETs unless otherwise mentioned. Detailed geometrical parameters can be found in Fig. 1 for single-gate planar junctionless and conventional nMOSFETs. Similar structures are shown in Fig. 2, but for double-gate junctionless and conventional nMOSFETs.

In order to achieve comparable threshold voltage (V_T) values, the work-functions of the simulated metal gates for junctionless and conventional nMOSFET are set to be 5.15 eV and 4.5 eV, respectively. As shown in Table I, parameters including channel doping concentration, gate length (L_G), effective gate oxide thickness, silicide-to-channel distance and Si channel thickness are varied in order to evaluate their effects on the variability of the simulated devices. The values in the middle represent the default settings for the simulated transistors. The variation of effective gate oxide thickness is performed via alteration of the dielectric constants of the gate oxide. The silicide-to-channel distance is defined as the distance from the silicide-Si interface in the source/drain regions to the gate edge. The silicide at the

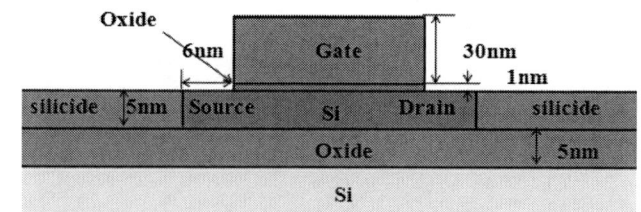

Fig. 1. Schematic cross-section of single-gate junctionless and conventional nMOSFETs on SOI substrate for simulation studies.

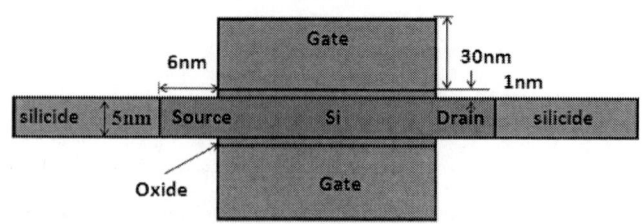

Fig. 2. Schematic cross-section of the simulated double-gate junctionless and conventional nMOSFETs.

978-1-4244-9019-6/11 $26.00 © 2011 IEEE

TABLE I
PARAMETERS FOR THE SIMULATED DOUBLE-GATE NMOSFETS

	Channel doping (cm^{-3})	L_G (nm)	E	Silicide-to-channel distance (nm)	Si thickness (nm)
junction-less	4.40E19	26	4.3	8	6
	4.00E19	22	3.9	6	5
	3.60E19	18	3.5	4	4
conven-tional	3.30E15	26	4.3	8	6
	3.00E15	22	3.9	6	5
	2.70E15	18	3.5	4	4

TABLE II
V_T AND SS OF THE SIMULATED DOUBLE-GATE JUNCTIONLESS AND CONVENTIONAL NMOSFETS WITH VARIED PARAMETERS DEPICTED IN TABLE I

		Channel doping	L_G	ε	Silicide-to-channel distance	Si thickness
junction-less	V_t(V)	0.09	0.17	0.19	0.15	-0.04
		0.15	0.15	0.15	0.15	0.15
		0.21	0.13	0.09	0.15	0.31
	SS(mV/dec.)	62.9	61.5	62.3	62.6	65.2
		62.6	62.6	62.6	62.6	62.6
		62.5	64.7	63.2	62.6	61.9
conven-tional	V_t(V)	0.14	0.16	0.14	0.14	0.14
		0.14	0.14	0.14	0.14	0.14
		0.14	0.13	0.14	0.14	0.16
	SS(mV/dec.)	62.7	61.2	62.3	62.8	64.7
		62.8	62.8	62.8	62.8	62.8
		62.8	65.3	63.4	62.8	61.3

source and drain is simplified as a zero–resistance metal. The electrical contact between the silicides and Si is set to be ohmic for all the cases. For conventional nMOSFETs, the doping profile at source/drain are set to be uniform (n-type dopant: 2×10^{20}/cm^3) from the silicide-Si interface to the vicinity of the gate edge. Thereafter, it follows a Gaussian distribution with a doping concentration gradient of around 2 nm/dec. All the simulated MOSFETs use the same un-calibrated mobility models in order to make reasonable and straightforward comparisons among different devices. As a consequence, the resultant absolute drain current values may differ from other simulation and experimental results.

III. RESULTS AND DISCUSSIONS

A. Single-gate devices

The I_D-V_G characteristics for the single-gate junctionless and conventional nMOSFETs are shown in Fig. 3. The physical gate lengths are set to be 22 nm. Though the Si channel thickness is as thin as 5 nm, the subthreshold slope (SS) of the junctionless nMOSFET is intolerably high and much worse than that of the conventional nMOSFET. This implies that single-gate is unable to properly turn off the channel at the off-state for the junctionless MOSFETs that usually require a uniform channel doping of no less than 2×10^{19}/cm^3. Double- or multi-gate is therefore required to fully deplete the highly-doped channel region in order to keep the off-state source-to-drain leakage current under control.

B. Double-gate devices

Table II summarizes the V_T and SS values of the simulated double-gate junctionless and conventional nMOSFETs with

reference to the parameters listed in Table I. These values assist the subsequent evaluation and comparison between the junctionless and conventional nMOSFETs in a double-gate configuration.

1) DC characteristics

The I_D-V_G characteristics for the junctionless and conventional nMOSFETs are shown in Fig. 4. Interestingly, the two devices exhibit almost identical V_T around 0.15 V and SS around 63 mV/dec. at V_D=0.1V. However, I_D of the junctionless nMOSFET is significantly lower than that of the conventional nMOSFET at the same gate overdrive (V_G-V_T). For example, I_D of the junctionless nMOSFET is less than 50% of that of the conventional one at (V_G-V_T)=0.65 V. A low drive current for the junctionless nMOSFET is mainly caused by the degraded electron mobility due to the unavoidable high channel doping level. Consequently, as shown in Fig. 5, the transconductance of the junctionless nMOSFET is also significantly lower than that of the conventional nMOSFET at the same (V_G-V_T). Notably, a slower degradation of transconductance with (V_G-V_T) is observed for junctionless nMOSFET. This can be well explained by the fact that the decrease in mobility with gate voltage is much less pronounced for the junctionless transistor compared with the conventional transistor as a result of the a lower electric field perpendicular to the current flow [11].

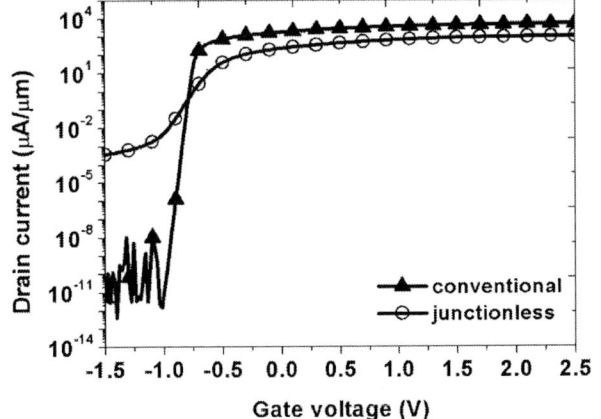

Fig. 3. I_D–V_G characteristics of the single-gate junctionless and conventional nMOSFETs at V_D=0.1V.

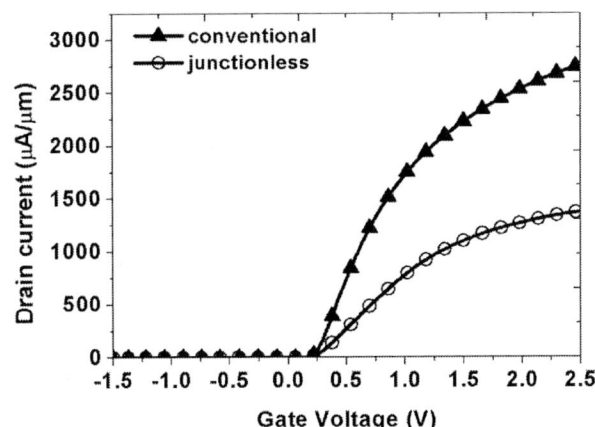

Fig. 4. I_D–V_G characteristics of the double-gate junctionless and conventional nMOSFETs at V_D=0.1V.

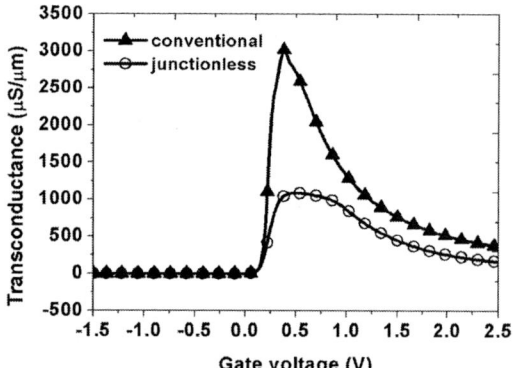

Fig. 5. Transconductances of the simulated double-gate junctionless and conventional nMOSFETs at V_D=0.1V.

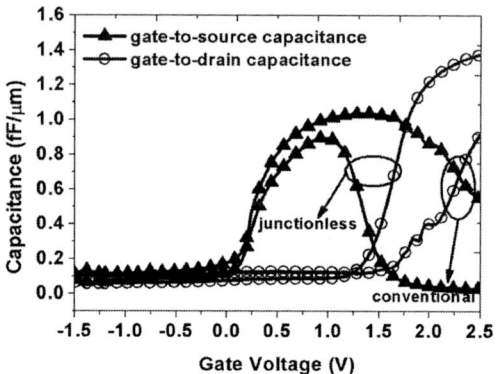

Fig. 6. C_{GS} and C_{GD} vs. V_G for the double-gate junctionless and conventional nMOSFETs at V_D=0.8V.

2) AC characteristics

The gate-to-source (C_{GS}) and gate-to-drain (C_{GD}) capacitances of the junctionless and conventional nMOSFETs are shown in Fig. 6. Interestingly, C_{GS} of the junctionless nMOSFET is clearly smaller than that of conventional nMOSFET at the same V_G. The peak C_{GS} of the junctionless nMOSFET occurs at a significantly lower (V_G-V_T). C_{GD} of the junctionless nMOSFET is also lower than that of the conventional nMOSFET for V_G<1.2 V.

3) Variability

a) Channel doping

The I_D-V_G characteristics for the double-gate junctionless and conventional nMOSFETs are shown in Fig. 7 with the various channel doping levels depicted in Table I. As expected, the conventional nMOSFETs show little change in the I_D-V_G characteristics as a result of a low channel doping level and an ultra-thin Si channel. However, a high sensitivity of V_T to the channel doping level is observed for the junctionless nMOSFETs: a V_T shift by 60 mV is evident for a 10% change in doping level.

b) Gate length

The I_D-V_G characteristics for the junctionless and conventional nMOSFETs are shown in Fig. 8 with different gate lengths depicted in Table I. Both types of the transistors show a relatively small change in V_T and SS, indicating a reasonable control of the short channel effect for both cases.

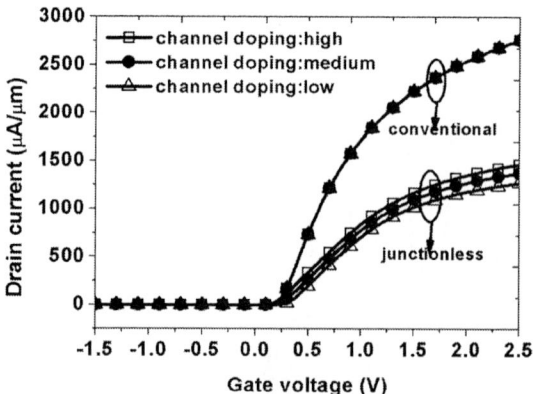

Fig. 7. I_D-V_G characteristics of the double-gate junctionless and conventional nMOSFETs with various channel doping levels at V_D=0.1V.

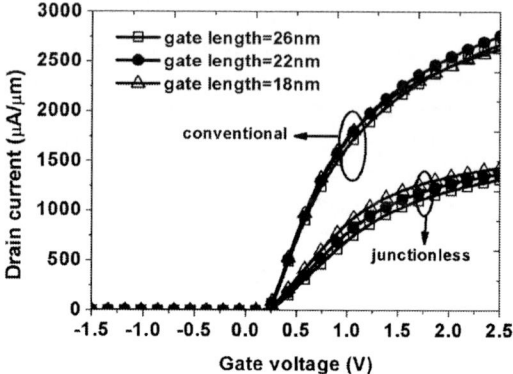

Fig. 8. I_D-V_G characteristics of the double-gate junctionless and conventional nMOSFETs with various gate lengths at V_D=0.1V.

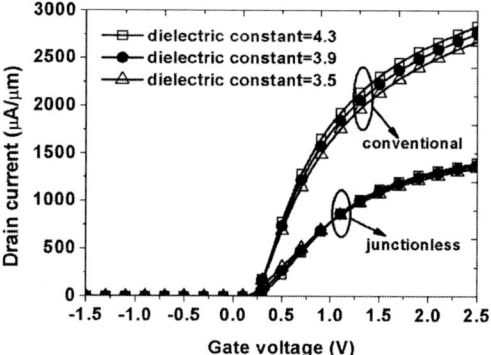

Fig. 9. I_G-V_G characteristics of the double-gate junctionless and conventional nMOSFETs with various dielectric constants at V_D=0.1V.

c) Oxide thickness

The I_D-V_G characteristics of the junctionless and conventional nMOSFETs are shown in Fig. 9 with different dielectric constants depicted in Table I. The effective gate oxide thickness is changed by varying the relative dielectric constant of the gate oxide. As seen in Table II and Fig. 9, compared with the conventional nMOSFETs, the junctionless nMOSFETs clearly show a higher sensitivity of V_T to variations of gate oxide thickness though their I_D at higher (V_G-V_T) seem to be less impacted.

d) Si thickness

The I_D-V_G characteristics for the junctionless and conventional nMOSFETs are shown in Fig. 10 with different Si channel thicknesses depicted in Table I. Here, V_T and SS of the conventional nMOSFETs are found to be insensitive to the variation of the Si channel thickness. However, a large variation of V_T is still observable for the junctionless nMOSFETs as the Si channel thickness varies. A change by around 50 mV in V_T can be found when the Si channel thickness varies by 1 nm.

e) Silicide-to-Channel Distance

The I_D-V_G characteristics for the junctionless and conventional nMOSFETs are shown in Fig. 11 with different silicide-to-channel distances depicted in Table I. As expected, since the source/drain and channel doping profiles have not been changed, the difference in silicide-to-channel distance only changes the S/D series resistance induced by the doped source and drain regions. V_T and SS therefore remain the same when this distance is changed by 2 nm for both junctionless and conventional nMOSFETs. However, I_D at high (V_G-V_T) varies significantly with different silicide-to-channel distances since the source/drain series resistance plays a significant role in I_D for both types of the transistors. For conventional MOSFETs, the variation of the P-N junction between the source/drain and channel is one of the major variation sources for transistor performance [15]. As a comparison, since junctionless MOSFETs have in principle no junctions, they are, as expected, free from variations induced by the P-N junction instabilities.

IV. CONCLUSIONS

Double- or multi-gate structures are necessary for junctionless MOSFETs to suppress the off-state source-to-drain

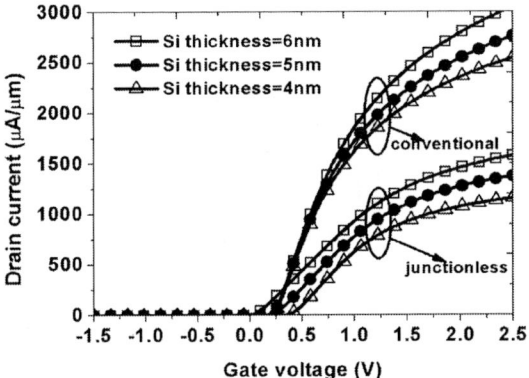

Fig. 10. I_D-V_G characteristics of the double-gate junctionless and conventional nMOSFETs with various Si thicknesses at V_D=0.1V.

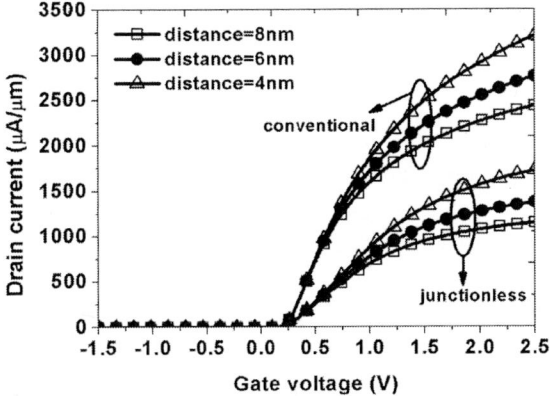

Fig. 11. I_D-V_G characteristics of the double-gate junctionless and conventional nMOSFETs with different silicide-to-channel distances at V_D=0.1V.

leakage current even for an extremely thin channel Si at 5 nm thickness. In comparison with conventional double-gate MOSFETs, double-gate junctionless MOSFETs with a gate length of 22 nm exhibit significantly lower drive current and peak transconductance, but a comparable low subthreshold slope around 63 mV/dec.. For the junctionless MOSFETs, their V_T is significantly more sensitive to variations of channel Si thickness and channel doping concentration. Notably, junctionless MOSFETs demonstrate substantially lower gate capacitances at saturation and a slower degradation of transconductance with gate overdrive. With its unique advantage of elimination of the source/drain junctions, junctionless MOSFETs could be a promising candidate for future sub-10 nm technologies where a 3-D gate structure is most likely required.

REFERENCES

[1] International Technology Roadmap for Semiconductors (ITRS), 2010 update, http://www.itrs.net

[2] J. M. Larson and J. P. Snyder, "Overview and status of metal S/D Schottky-barrier MOSFET technology," IEEE Trans. Electron Devices, vol. 53, no. 5, pp. 1048-1058, May 2006.

[3] J. Kedzierski, P. Xuan, E. H. Anderson, J. Bokor, T.-J. King, and C. M. Hu, "Complementary silicide source/drain thin-body MOSFET for the 20 nm gate length regime," IEDM Tech.Dig., 2000, pp. 57-60.

[4] Z. J. Qiu, Z. Zhang, M. Östling, and S.-L. Zhang, "A comparative study of two different schemes to dopant segregation at NiSi/Si and PtSi/Si interfaces for Schottky barrier height lowering," IEEE Trans. Electron Devices, vol. 55, no. 1, pp. 396-403, Jan. 2008.

[5] J. Luo, Z.-J. Qiu, C. L. Zha, Z. Zhang, D. P. Wu, J. Lu, J. Åkerman, M. Östling, L. Hultman, and S.-L. Zhang, "Surface-energy triggered phase formation and epitaxy in nanometer-thick $Ni_{1-x}Pt_x$ silicide films," Appl. Phys. Lett., vol. 96, no. 3, pp. 031911/1-3, Jan. 2010.

[6] J.-P. Colinge, C.-W. Lee, A. Afzalian, N.D. Akhavan, R. Yan,I. Ferain, P. Razavi, B. O'Neill, A. Blake, M. White, A.-M. Kelleher,B. McCarthy, and R. Murphy, "Nanowire transistors without junctions," Nature Nanotechnology, vol. 5, pp. 225-229, Feb. 2010.

[7] C.-W. Lee, I. Ferain, A. Afzalian, R. Yan, N.D. Akhavan, P. Razavi, J.-P. Colinge, "Performance estimation of junctionless multigate transistors," Solid-State Electronics, vol. 54, no. 2, pp. 97-103, Feb. 2010.

[8] C.-W Lee, A. Afzalian, N.D. Akhavan, R. Yan, I. Ferain, and J.-P. Colinge, "Junctionless multigate field-effect transistor," Appl. Phys. Lett., vol. 94, no. 5, 053511/1-2, Feb. 2009.

[9] C.-W. Lee, S.-R.-N. Yun, C.-G. Yu, J.-T. Park, and J.-P. Colinge, "Device design guidelines for nano-scale MuGFETs," Solid-State Electronics, vol. 51, pp. 505–510, 2007.

[10] A. Afzalian, D. Lederer, C.-W. Lee, R. Yan, W. Xiong, C. Rinn Cleavelin, and J.-P. Colinge, "MultiGate SOI MOSFETs: accumulation-mode vs. enhancement-mode," IEEE 2008 Silicon Nanoelectronics Workshop, 2008, pp.1–2.

[11] J.-P. Colinge, C.-W. Lee, I. Ferain, N.D. Akhavan, R. Yan, P. Razavi, R. Yu, A.N. Nazarov, and R.T. Doriac, "Reduced electric field in junctionless transistors," Appl. Phys. Lett., vol. 96, 073510, 2010.

[12] Y. Cui, Z. Zhong, D.Wang, W. U. Wang, and C. M. Lieber, "High performance silicon nanowire field effect transistors," Nano Lett. 3(2), pp. 149–152, 2003.

[13] Y. Shan, S. Ashok, and S. J. Fonash, "Unipolar accumulation-type transistor configuration implemented using Si nanowires," Appl. Phys. Lett. vol. 91, 093518, 2007.

[14] W. Lu, P. Xie, and C. M. Lieber, "Nanowire transistor performance limits and applications," IEEE Trans. Electron Dev., vol. 55, pp. 2859–2876, 2008.

[15] M.Y. Kwong, R. Kasnavi, P. Griffin, J. D. Plummer, and R. W. Dutton, "Impact of Lateral Source/Drain Abruptness on Device Performance", IEEE Trans. Electron Devices, vol. 49, no. 11, pp. 1882-1890, 2002